TROPICAL PLANT SCIENCE

G.K. Berrie

A. Berrie

J.M.O. Eze

Longman
Scientific &
Technical

Copublished in the United
States with John Wiley &
Sons, Inc., New York

Longman Group UK Limited,
Longman House,
Burnt Mill, Harlow,
Essex CM20 2JE, UK,
and Associated Companies throughout the World.

Copublished in the United States with John Wiley &
Sons, Inc., 605 Third Avenue, New York, NY 10158
0-470-20916-X (Wiley, USA only)

First published 1987

Produced by Longman Group (FE) Ltd
Printed in Hong Kong

British Library Cataloguing in Publication Data

Berrie, G. K.
    Tropical plant science.
    1. Botany
    I. Title   II.Berrie, A.   III. Eze, J. M. O.
    581   QK47

    ISBN 0-582-64705-3

Publishers' acknowledgements
The Publishers are grateful to the following for permission
to use and adapt figures:

Figure 11.10 adapted from *Plant Physiology*, 2nd edn, by
R. G. S. Bidwell and reproduced by permission of
Macmillan Publishing Co. and Collier Macmillan Publishers,
London; Figure 13.3 taken from a 1983 University of Benin
B.Sc. Honours project dissertation *Changes in water status,
and their implications, in a detached and intact first trifoliate
leaf of black-eye bean* by F. Asiegbu.

All other illustrations are original drawings by A. Berrie,
with the exceptions of Figures 5.3, 5.5(a) (b) (c), 5.7,
5.9–5.19, 7.3, 7.4, 10.11, 11.7, 11.9, 11.11–11.15,
12.1–12.9, 13.1, 13.2, 13.4–13.10, 17.6–17.8, 17.22, 18.2,
19.2, 20.3, 22.3, 23.18(a), 23.19(f), 26.1–26.4, 28.1, 28.2
and Map 28.1 which are based on drawings supplied by the
authors.

The Publishers are grateful to the following for their
permission to reproduce photographs:

Professor J. G. Duckett for Figure 5.5; Professor J. L. Hall
for Figure 5.2; C. James Webb for Figure 22.5; and
Professor Fergus B. Wilson for Figure 4.4.

All other photographs, textual and cover, were kindly
provided by G. K. Berrie and A. Berrie.

Authors' acknowledgements

The authors wish to acknowledge the many colleagues, past
and present, who have enriched our knowledge over the
years. Our thanks also go to Mrs Florence Sambo who
typed the manuscript.

Contents

Preface

This book is designed for the use of students in the introductory year or years of degree courses in botany and biology at universities in tropical countries, especially Africa. It is hoped that it will also prove useful to students in teacher training colleges and to teachers of sixth-form biology.

The book assumes a basic knowledge of biology, such as may be obtained by studying for School Certificate or GCE 'O' level biology or general science. No attempt is made here to deal with basic chemical or physical concepts. For instance, although certain terms (e.g. carbohydrate, protein) are explained, we have not tried to deal with chemical bonding or to describe the ionization of electrolytes. It is hoped that most students of botany or biology who use this book will study chemistry at the same time.

Order of treatment of topics

Every college or university department has its own approach to teaching plant science at the introductory level, and its own order of treatment. The order of treatment of topics in this book must represent a compromise. The main sections into which the book is divided correspond to the parts into which a course in this subject is usually divided when a course-credit system is in operation. In order to allow for different sequences of treatment in different departments, we have tried to make it possible to start from the beginning of any section. This has involved some repetition of descriptions and definitions and the inclusion of cross-references between chapters.

The plants

The organisms which are dealt with in this book include all of those which have traditionally been regarded as plants, even though we do not ourselves classify all of them as such. We therefore include viruses, bacteria and fungi as well as green plants.

The plant examples which we cite are almost exclusively tropical plants or crop and garden plants which have become naturalized in the tropics. Specific examples of non-cultivated plants are from the African flora. Related, or similar plants, occur in all tropical countries.

Plant names

The correct use of plant names is very important. For this reason, all plants are referred to by their scientific names first. Common (English) names are added after the scientific names wherever we believe that there is a common name which is widely accepted in anglophone countries, for instance in agriculture, forestry, horticulture or industry. Some writers capitalize the initial letters of common names, others do not. There seems to be no generally accepted procedure. Here, the initial letters of all common names have been capitalized for the sake of uniformity.

Where examples of plants are named or listed in brackets, scientific and common names of the same plant have been separated by a comma and the names of different plants have been separated by a semicolon (e.g. *Zea mays*, Maize; *Arachis hypogaea*, Groundnut).

Terminology

Although some chapters contain many botanical terms which will be new to the elementary student, we have used formal terminology only where it is absolutely necessary, for instance, to enable the student to understand more advanced books. Descriptive, English terms are used whenever possible. However, it should be recognized that the reason for using formal terms is to save words. Every term has a more or less precise meaning and its use makes it unnecessary to describe a structure or a process in full each time it is mentioned. Like the scientific names of plants and animals, many biological terms are based on words from Latin or ancient Greek and are universally understood by scientists.

There are cases in which certain terms are used in slightly different ways by different authors (e.g. prickle, spine, thorn; Chapter 4). We have tried to use such terms in their most widely accepted sense. The meaning of any scientific word used in the book may be found by first looking up the word in the index. The page number printed in heavy type indicates the page on which the word is first used, also printed in heavy type, and where its meaning is described. There is no separate glossary.

Sections

Each section of the book begins with a brief introduction to the group of chapters which it contains. There is also a list of recommended further reading at the end of each section. In each list, we have included only books which should be currently available in bookshops or in college or university libraries.

Chapters

At the end of each chapter there is a list of 'Key words', a brief summary of the contents of the chapter and set of questions on the subject matter of the chapter.

Key words are words which have been used in the chapter and which are of fundamental importance to the understanding of the material in the chapter. In each list of key words, main topics are in heavy type and, after each topic, groups of related words are separated by semicolons and closely related words are separated by commas.

Each chapter summary draws attention to the most important points made in the chapter. The questions which follow are intended to help the student to discover what he has learned after reading the chapter and to find out whether or not his knowledge and understanding are complete enough.

Illustrations

The great majority of the drawings and diagrams have been reproduced from hand-drawn originals by Mrs A. Berrie. Wherever possible, these illustrations have been made from living specimens, or from freshly made preparations. Frequently throughout the book, botanical material is illustrated by both a photograph and a drawing, and this is intended to help the student interpret and illustrate his own slides and specimens.

G. K. Berrie
A. Berrie
J. M. O. Eze

Abbreviations

Less frequently used abbreviations are explained in the text. The following abbreviations are used throughout the book:

Linear measurements
km	kilometre (10^3 m);
m	metre;
cm	centimetre (10^{-2} m);
mm	millimetre (10^{-3} m);
μm	micrometre (10^{-6} m);
nm	nanometre (10^{-9} m);
Å	Ångström (10^{-10} m).

Liquid measurements
l	litre;
ml	millilitre (10^{-3} l).

Temperature
°C	degrees Celsius (degrees Centigrade).

Time (t)
h	hour.

Magnifications of illustrations:
General
× $\frac{1}{2}$	half natural size;
× 1	natural size;
× 2, etc.	two times natural size, etc.

Microscope photographs
VLP	very low power (× 40 to × 60);
LP	low power (× 200 to × 300);
HP	high power (× 500 to × 750);
VHP	very high power (× 1250 to × 1500).

Section A

INTRODUCTION TO PLANT SCIENCE

In this Section, we take a brief look at all those organisms which are traditionally regarded as plants and see how they are named and classified.

1

Looking at plants

Most of the surface of the earth is covered with vegetation. The surface waters of oceans, seas and lakes are full of plant life. The type of vegetation and the types of plants which occur in any particular place are determined by the local environment, but almost everywhere we look there is a great variety of plants. In order to make a scientific study of plant life, we must first reduce the variety of plants to some sort of order.

In the first place, plants, like people, need names. If they have names, we can refer to them easily without having to describe them fully each time. Every language has its own names for plants. We call these 'common names', to distinguish them from the 'scientific names' used by scientists. Some common names, usually those of plants which are especially important as food or medicine, are botanically accurate to the extent that each is applied to only one type of plant. Many common names are less precise and are, therefore, unacceptable for scientific use. For instance, amongst English common names, 'Crinum Lily' is not a lily, 'Asparagus Fern' is not a fern and 'Whistling Pine' is not a pine.

Even when a common name is botanically accurate, it will be different from the names used for the same sort of plant by those who speak different languages.

We may take a common West African plant, known to botanists as *Talinum triangulare*, as an example of confusion that might be caused by the exclusive use of common names. This is a small plant with fleshy, edible leaves. In parts of Nigeria, it is known as 'Ewe', 'Water leaf' or 'Gbologi'. However, the last of these names is also applied to another plant (*Basella alba*) which has fleshy, edible leaves. In Freetown, Sierra Leone, *Talinum triangulare* is known as 'Lagos Gbologi' and the name 'Gbologi' is applied to all plants with fleshy, edible leaves. In the same way, the name 'spinach' is applied to several plants which have edible leaves, both in West Africa and in Europe.

Science is world-wide, so some sort of standard **nomenclature** (system of naming) is necessary, if botanists of all nationalities are to be able to understand each other.

The binomial system of nomenclature

The **binomial** (two-name) system of nomenclature was introduced in 1753 by Carl von Linné, in a book called *Species Plantarum*. The author, better known as Carolus Linnaeus (the latinized form of his name), later went on to apply the same system to animals.

Many of the scientific names of plants and animals which are still in use were first applied by Linnaeus.

Under the binomial system, every plant and animal is known by two names. The first is the name of the **genus** (*pl.* genera). The second is the name of the **species** (*pl.* species).

Species

There are several different ways in which the word 'species' may be defined or described, but a simple one will suit our present purpose. A species is a group of individual plants or animals which differ from each other only in characteristics which depend on their stage of development or on the effects of environmental factors (e.g. light intensity, water supply, availability of nutrients). More precise definitions will be considered in Chapter 27.

Genus

If we examine plants belonging to many different species, we are likely to find that some species resemble each other quite closely and differ to a greater degree from other species. Members of the group of similar species all belong to the same genus.

Writing plant names

When writing plant names, the name of the genus is written first, using a capital initial letter. The species name is written second, with a small initial letter. If they are handwritten or typed, both names are underlined. If printed, they are printed in *italics*, as in this book.

To be absolutely precise, the name, or an abbreviation of the name of the person who named the plant should be written after the species name. For example, the full scientific name of Maize is '*Zea mays* L.', showing that this name was given by Linnaeus, abbreviated to 'L.'.

Plant names themselves can also be abbreviated in several ways. If we wish to refer to a member of a genus without mentioning the actual species, we may leave out the species name and write 'sp.' instead. We may also write 'sp.' if we do not know the name of the species. Thus, '*Talinum triangulare*' or another species of the same genus might be referred to as '*Talinum* sp.'. If we wish to refer to several species of the same genus collectively, the same form is used, but with 'spp.'. For example, '*Talinum* spp.' means 'several or all species of the genus *Talinum*'. Again, if reference has been made to '*Talinum trian-*

gulare', or to other species of the genus *Talinum*, the name may be written as '*T. triangulare*', as long as this does not cause confusion. It might cause confusion if, for instance, reference had also previously been made to *Tridax procumbens*, another common weed.

Systems of classification

Genus and species are the two lowest categories in the classification of plants. A category of classification is called a **taxon** (*pl.* taxa), so we can say they are the two lowest taxa.

The next higher taxon is the **family**. Genera are grouped into families in much the same way that species are grouped into genera. Genera belong to the same family if they resemble each other, in what are thought to be important ways, more than they resemble other genera. Likewise, families are grouped into **orders**, orders are grouped into **classes**, classes are grouped into **divisions** and divisions are grouped into **kingdoms**. Botanists who specialize in the classification of plants (plant taxonomists) also use intermediate taxa (e.g. subkingdoms, subclasses, suborders, etc.).

An International Botanical Congress is held once every five years. Any professional botanist, from any country, may take part. The Bureau of Nomenclature of the International Botanical Congress is recognized as the body which decides on the rules which must be followed when naming plant taxa. It decides upon the endings (**suffixes**) of the names of taxa of different levels (Table 1.1). Sometimes, it also decides to retain (**conserve**) names which do not obey its own rules, usually for historical reasons. For example, the official name for the grass family is 'Gramineae' and not 'Graminaceae', as might be expected.

The first purpose served by any system of classification is convenience for reference. Plants with similar characteristics are grouped together so that reference can be made to any of them. Some older systems of classification were based solely on this need and were, therefore, unnatural. With the general acceptance of the theory of evolution in the second half of the nineteenth century, some botanists and zoologists tried to set out new systems which also took into account what were believed to be evolutionary relationships. In such **natural classifi-**

cations, attempts have been made to reflect evolutionary relationships in the grouping of plants or animals at all taxonomic levels.

Table 1.1 The classification of two important crop plants, using the scheme in Table 1.3.

Taxon	Maize	Groundnut
kingdom	Plantae	Plantae
division	Tracheophyta	Tracheophyta
subdivision	Pteropsida	Pteropsida
class	Angiospermae	Angiospermae
subclass	Monocotyledoneae	Dicotyledoneae
order	Graminales	Leguminosae
family	Gramineae	Papilionaceae
genus	*Zea*	*Arachis*
species	*mays*	*hypogaea*

A great many systems of plant classification have been put forward by different taxonomists at different times. Every taxonomist has his or her own opinion and it is almost true to say that there are as many systems of classification as there are taxonomists. This makes the choice of which classification to use difficult as well as necessary. A student of botany must also understand enough about the principles of classification to recognize the differences between the systems used by the authors of different books.

Two systems of classification, one fairly traditional and one more modern, are shown in Tables 1.2 and 1.3. The first is based on the view that all organisms which are not animals must be plants. The second incorporates some of the knowledge of fundamental differences between organisms which have been brought to light by the application of modern techniques. In the second classification, the range of organisms which were formerly regarded as plants are separated into three kingdoms, Monera, Fungi and Plantae, to stand alongside the kingdom Animalia (the animal kingdom).

Table 1.2 A classification of the principal groups of organisms that are traditionally regarded as plants based on that of Tippo (1942).

Kingdom subkingdom	Division subdivision	Class	Common name
Plantae			plants
Thallophyta			thallophytes
	Cyanophyta		blue-green algae
	Chlorophyta		green algae
	Xanthophyta		yellow–green algae
	Euglenophyta		euglenoids
	Phaeophyta		brown algae
	Rhodophyta		red algae
	Chrysophyta		diatoms, etc.
	Pyrrophyta		dinoflagellates
	Schizomycophyta		bacteria
	Myxomycophyta		slime moulds
	Eumycophyta		true fungi
Embryophyta	Bryophyta		bryophytes
	Tracheophyta		vascular plants
	Psilopsida		psilophytes
	Lycopsida		club-mosses
	Sphenopsida		horsetails
	Pteropsida		—
		Filicinae	ferns
		Gymnospermae	gymnosperms
		Angiospermae	flowering plants

A survey of three kingdoms

In Section B of this book, we are to make a detailed study of flowering plants, which form the greatest part of the present vegetation of the earth. Before this, we shall look briefly at the range of organisms in the kingdoms Monera, Fungi and Plantae, as set out in Table 1.3.

Kingdom Monera

The **Monera** include all **procaryotic** organisms. The cells of procaryotic organisms are simple in structure. There is no nuclear envelope and, therefore, no nucleus in the ordinary sense. Several other structures which are present in the cells of more complex organisms (**eucaryotic** organisms) are also absent.

Bacteria (Eubacteria; Fig. 1.1) are unicellular or have their cells in loose aggregations or filaments. Their cells average about 1 μm in diameter and the maximum cell length, in the largest forms, is 20 μm. Most are **heterotrophic** and obtain their energy and materials from other organisms. These include **saprophytes**, which live on dead organic material, and **parasites**, which live on materials obtained from other living organisms. Some bacteria are **autotrophic** and are able to make their own organic materials. Autotrophic bacteria obtain their energy by photosynthesis (**photoautotrophs**) or by oxidizing inorganic materials (**chemoautotrophs**).

Blue-green algae (Cyanobacteria; Fig. 1.2) are unicellular or have their cells in loose aggregations or filaments. Their cells are usually larger than those of bacteria, up to about 25 μm in diameter. They are all photoautotrophic. Their cells contain the green pigment, chlorophyll a, together with other pigments, but lack chloroplasts.

Kingdom Fungi

Fungi, and all the remaining groups of organisms, are eucaryotic. The cells have nuclei, each limited by a nuclear envelope. Their cytoplasm includes other structures common to eucaryotic organisms but does not include chloroplasts. Fungi are heterotrophic and are either saprophytes or parasites.

Slime moulds (Myxomycophyta; Fig. 1.3) are heterotrophic. They have several stages in their life histories. At one stage, a slime mould consists of many, separate, *Amoeba*-like cells which move around and ingest bacteria. These cells subsequently flow together into a large mass, which continues to

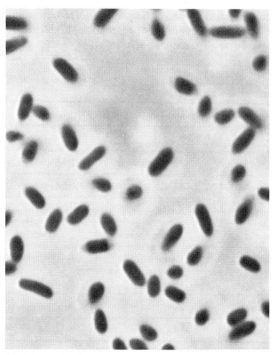

1.1 Eubacteria: the bacterium *Erwinia* sp., which causes stem rot in *Helianthus annuus* (Sunflower) and some other crop plants (VHP)

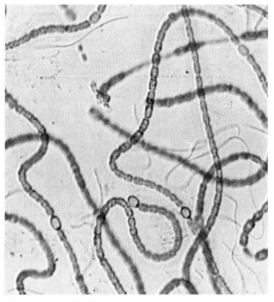

1.2 Cyanobacteria: filaments of *Anabaena* sp. from a soil surface (HP)

Table 1.3 A classification of plants, in which the groups of organisms which are traditionally regarded as plants are divided into three kingdoms. Kingdom Monera includes all procaryotic organisms, in which the cell lacks a nuclear envelope and certain other structures that are found in eucaryotic cells. Fungi are eucaryotic, heterotrophic (parasitic or saprophytic) organisms. Plantae are eucaryotic, autotrophic organisms.

Viruses are omitted from this scheme (see Chapter 22). Lichens, which are composite organisms consisting of a fungus associated with an alga, are usually classified as Fungi, according to the type of fungus which forms the main part of the body.

Kingdom	Division subdivision	Class subclass	Common name or description
Monera			procaryotic organisms
	Eubacteria		bacteria
	Cyanobacteria		blue-green algae (blue-green bacteria)
Fungi			fungi
	Myxomycophyta		slime moulds
	Eumycophyta		true fungi
		Phycomycetes	phycomycetes
		Ascomycetes	sac fungi
		Basidiomycetes	club fungi
		Deuteromycetes	imperfect fungi
Plantae			green plants
	Chlorophyta		green algae
		Chlorophyceae	—
		Charophyceae	stoneworts
	Euglenophyta		euglenoids
	Xanthophyta		yellow-green algae
	Chrysophyta		golden-brown algae
	Bacillariophyta		diatoms
	Pyrrophyta		dinoflagellates
	Cryptophyta		—
	Phaeophyta		brown algae
	Rhodophyta		red algae
	Bryophyta		bryophytes
		Anthocerotae	hornworts
		Hepaticae	liverworts, hepatics
		Musci	mosses
	Tracheophyta		vascular plants
	Psilopsida		psilophytes
	Lycopsida		club mosses
	Sphenopsida		horsetails
	Pteropsida		broad-leaved plants
		Filicinae	ferns
		Gymnospermae	gymnosperms
		Angiospermae	angiosperms, flowering plants
		Dicotyledoneae	dicotyledons, dicots
		Monocotyledoneae	monocotyledons, monocots

1.3 Myxomycophyta: fruiting bodies of a slime mould growing on dead wood in forest (× 2)

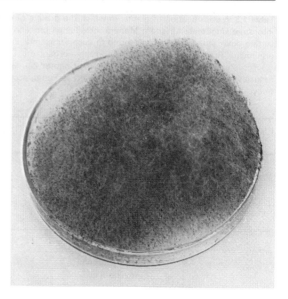

1.4 Eumycophyta, Phycomycets: *Mucor* sp. growing on bread in a petridish

1.5 Eumycophyta, Ascomycetes: fruiting bodies (apothecia) on dead wood in forest (× 3)

move and feed until it forms a solid fruiting body. The fruiting body forms spores, which have cellulose in their walls.

Slime moulds are an interesting and problematic group, since they possess characteristics of fungi and animals. They are commonly seen on rotting wood, or on rotting leaves on the forest floor.

True fungi (Eumycophyta) usually have bodies which are made up of long, branching threads (**hyphae**), but a few are unicellular. The mass of hyphae which makes up the body of a fungus is the **mycelium** of the fungus. Fungi include saprophytes and parasites. They are dispersed by **spores**. Some spores are formed directly from hyphae, without any sexual process. Other spores are formed after a sexual process has taken place.

Phycomycetes (Fig. 1.4) have hyphae with no cross walls (**aseptate** hyphae). They form spores in several ways but do not form macroscopic fruiting bodies (fruiting bodies easily visible to the naked eye).

Sac fungi (Ascomycetes; Fig. 1.5) have hyphae which are composed of many cells arranged end to end. Each cell contains one, two or more nuclei, depending on the species and on the stage in the life cycle. Sexual fusion is followed by the formation of fruiting bodies which are often macroscopic. The fruiting bodies contain sac-like or cylindrical **asci**, each of which releases four, eight, or a multiple of eight **ascospores**.

Club fungi (Basidiomycetes; Fig. 1.6) have hyphae similar to those of sac fungi and many of them form macroscopic fruiting bodies. **Basidiospores** are budded off from the surfaces of club-shaped cells (**basidia**) within or at the surface of the fruiting body, following a sexual process.

Imperfect fungi (Deuteromycetes) are so called because they have no known sexual process and their life cycles are, therefore, apparently incomplete. Their hyphae are like those of sac fungi or club fungi but, without knowledge of the sexual process, it is not possible to place them in either group with certainty. When research workers who study them discover the sexual process in a particular species, that species is then assigned to its proper group.

Lichens (Fig. 1.7) are composite organisms. The **thallus** (plant body of simple structure) of a lichen is formed by an ascomycete or, rarely, a basidiomycete. Within the body of the fungus is a layer of algal cells, which may be either a green alga or a blue-green alga. The fungus protects the alga, the alga carries out photosynthesis and provides food for the fungus. This type of association of different organisms, which results in benefit for both, is called **symbiosis** (Chapter 24). Lichens are dispersed either by small pieces of thallus which include both fungus and alga, or by fungal spores. Fungal spores can establish new thalli only if they germinate close to a place where cells of the necessary alga are already present.

1.6 Eumycophyta, Basidiomycetes: fruiting body of *Lepiota* sp. (× ¾)

1.7 Lichens: *Usnea* sp. (with many narrow branches) and other lichens on a shrub in montane forest (× ¾)

Kingdom Plantae

Plantae are eucaryotic, photoautotrophic organisms.

Algae were once classified together as a class of the division Thallophyta, in which fungi were also included. This was in a system of classification put forward by Eichler in 1883 and still used even in some quite modern books.

The common characteristics of algae, in this sense, are autotrophic nutrition, simple structure and the possession of unicellular (not multicellular) reproductive organs. The application of modern research methods has shown that they should not be included in a single taxon. Blue-green algae are procaryotic and should be separated from other algae on that basis. The remaining algae are separated into several divisions because of fundamental differences in structure and biochemistry.

The name 'alga', spelled with a small initial letter, is still a convenient collective name for these plants, but it must be recognized as a common name, not the name of a taxon in a scientific sense. It is, however, convenient to study algae together, because of their common simplicity and the similarity of their ecology.

Green algae (Chlorophyta; Fig. 1.8) are **unicellular, colonial** (several or many cells grouped together), **filamentous** or **thalloid**. Many unicellular forms, some colonial forms, and the spores and gametes of others, are **motile** and swim by means of two or four **flagella** (hair-like organs) attached at the front end. Like higher plants, their chloroplasts contain chlorophylls *a* and *b*, xanthophylls and carotene. They also have cellulose cell walls and store starch.

Euglenoids (Euglenophyta) are unicellular and swim by means of a single flagellum attached at the front end. They have the same main chloroplast pigments as green algae but have no cell wall and store a different sort of carbohydrate (paramylum).

Although most euglenoids are autotrophic, there is a gullet at the front end which resembles the gullet by which food organisms are ingested in some unicellular animals. Also, some lack chloroplasts and are heterotrophic. Zoologists usually classify euglenoids as flagellates, in the phylum Protozoa. In fact, it is best to regard euglenoids as 'organisms', not necessarily either plants or animals but on the borderline between the two kingdoms.

Yellow-green algae (Xanthophyta) are unicellular, colonial or filamentous. Their chloroplasts contain

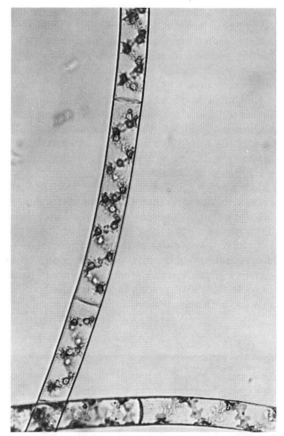

1.8 Chlorophyta: *Spirogyra* sp. from pond water (HP)

chlorophylls *a* and *e*, xanthophylls and carotene. There is a higher proportion of xanthophylls than in green algae and this gives their chloroplasts their yellow-green colour. Some unicellular forms are **amoeboid** (without a cell wall and moving by changing their shape), others have two unequal flagella at the front end. They store oil and their cell walls are of cellulose and pectic substances, sometimes with some silica.

Brown algae (Phaeophyta; Fig. 1.9) include no unicellular forms. They may be filamentous or thalloid or of other form and nearly all live in the sea (are **marine**). Some thalloid forms are very large indeed. Motile spores and gametes have two dissimilar flagella attached at the side. Chloroplast pigments include chlorophylls *a* and *c*, xanthophylls and carotene. There is a high proportion of xanthophylls, sometimes referred to as fucoxanthin, and this gives them their characteristic brown colour.

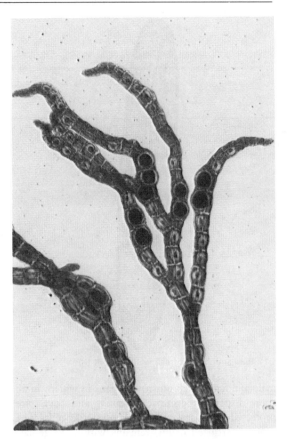

1.9 Phaeophyta: *Sargassum* sp. from below low tide level on the seashore (× 1)

1.10 Rhodophyta: *Polysiphonia* sp., part of the branched thallus, from a rock pool on the seashore (LP)

They store a carbohydrate called laminarin and their cell walls contain cellulose and pectic substances.

Red algae (Rhodophyta; Fig. 1.10) include unicellular and filamentous forms, but in many of the filamentous forms the filaments are bound together to form thalli, which may be large. Nearly all are marine. There are no motile forms and no flagella. Their chloroplasts contain chlorophylls *a* and *d*, xanthophylls and carotene, plus red and blue pigments which combine to give the plants their characteristic colour. They store a starch-like carbohydrate and their cell walls contain cellulose and pectic substances.

Diatoms (Bacillariophyta; Fig. 1.11) are unicellular but their cells sometimes adhere together in loose groups or filaments. Some have motile male gametes, each of which has a single flagellum at the front end. Their vegetative (non-reproductive) cells are non-motile and are either round and usually flat, or elongated. Their chloroplasts contain chlorophylls *a* and *c*, xanthophylls and carotene and have a brownish colour. They store a carbohydrate called leucosine and their cell walls are of silica and formed of two overlapping halves.

Dinoflagellates (Pyrrophyta) are mostly unicellular and motile, but some form short filaments. Motile cells have two flagella attached at one side, of which one passes around the centre of the cell, in a groove. Their chloroplasts contain chlorophylls *a* and *c*, xanthophylls and carotene. They store leucosine. The cell wall, when present, contains cellulose.

Land plants have the same type and arrangement of flagella on their motile cells as green algae. They also have the same chloroplast pigments, store starch and have cellulose cell walls. They have multicellular

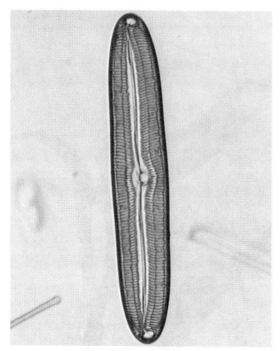

1.11 Bacillariophyta: a fresh water diatom (HP)

male and female reproductive organs in which, in all spore-dispersed groups, sperms and ova are protected by at least one layer of sterile cells. The **zygote** (fertilized ovum) remains within the female structure until it has formed an **embryo**. There are two distinct phases in the life cycle. The sexual phase carries male and female reproductive organs, either together or on separate plants. The other phase grows from the zygote and produces spores, which, in turn, grow into new plants of the sexual phase. Divisions and subdivisions are separated on the basis of life history, structural complexity and dispersal mechanism.

Bryophytes (Bryophyta) have the sexual phase as the main part of the life cycle. This is an independent green plant, attached to the surface on which it grows (the **substrate**) by hair-like **rhizoids**. There are no roots. It bears male and/or female reproductive organs. After fertilization has taken place, the spore-producing plant which grows from the zygote remains attached to and almost entirely dependent upon the sexual plant. The spore-producing plant has a single capsule (**sporogonium**) in which spores are formed. Spores germinate after they have been dispersed and form new sexual

plants. Neither phase has any specialized conducting tissue.

Hornworts (Anthocerotae; Fig. 1.12) are thalloid bryophytes. The sexual plant is strap-shaped, branched and closely attached to the substrate. The spore-producing phase is upright, long and narrow, and grows by the addition of tissue at a point close to its attachment to the sexual plant. The capsule has no stalk.

1.12 Bryophyta: *Anthoceros* sp. growing on soil from a river bank. The flat thallus bears two upright sporophytes, one young, one mature (× 2)

Liverworts (Hepaticae; Fig. 1.13) may be thalloid or may have stems bearing two or three rows of simple leaves. Their capsules are raised above the sexual plant, when ripe, by the rapid elongation of a stalk (seta) which shrivels and dies soon after the spores have been shed.

Mosses (Musci; Fig. 1.14) have stems with simple, spirally arranged leaves. Their capsules are raised on stiff stalks which start to grow upwards while the capsule is still young. The capsule opens by a lid. Both capsule and stalk remain attached to the leafy sexual plant for a long time after the spores have been dispersed.

Vascular plants are so called because they all have

well-developed conducting tissues (**xylem** and **phloem**). They also differ from the bryophytes in that the sexual phase of the life cycle is small, simple and usually short lived, while the spore-producing generation is much more complex and conspicuous.

In older systems of classification, vascular plants were placed in two divisions, the Pteridophyta and the Spermatophyta. In our classification, they are treated as a single division, Tracheophyta, with four subdivisions, Psilopsida, Lycopsida, Sphenopsida and Pteropsida.

Psilophytes (Psilopsida; Fig. 1.15) are spore-dispersed plants with small leaves and no roots. Part of the stem is underground and has rhizoids. From this arise green, usually upright branches which, in living genera, bear small, widely spaced leaves. Fertile branches have **sporangia** (spore containers) above the attachments of some of the leaves. Spores germinate beneath the surface of the ground, where they produce small, cylindrical sexual plants. The sexual plant is able to grow underground, where there is no light, because it becomes associated with a fungus which provides it with food. This type of association of a fungus with an underground part of a plant is called a **mycorrhiza** (Chapter 24). After fertilization, a new spore-producing plant grows up, after only a short period of dependence on the sexual plant.

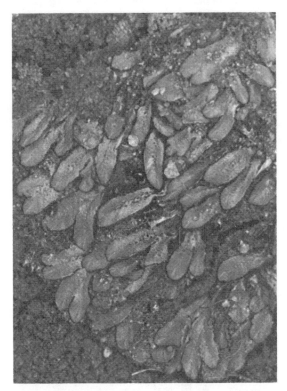

1.13 Bryophyta, Hepaticae: *Riccia* sp. growing on soil in savanna. In this genus, sporophytes have no stalks and remain inside the tissues of the gametophyte thallus (× 1¼)

1.15 Tracheophyta, Psilopsida: *Psilotum* sp. (× ½)

1.14 Bryophyta, Musci: *Octoblepharum albidum*, a common moss growing on a tree branch in rain forest (× 2)

Club mosses (Lycopsida; Fig. 1.16) are so called because they have a superficial resemblance to mosses, to which they are not related. The stem bears roots and many small and usually spirally arranged leaves. Each leaf has a single, unbranched conducting strand (**vein**). Some of the leaves, usually at the tips of branches, have one sporangium each on the upper surface. When spores are released, they grow into small, thalloid sexual plants. After fertilization, new spore-producing plants grow up after only a short period of dependence upon the sexual plant.

Some club mosses (most *Selaginella* spp.) have their leaves arranged in four rows along the stem and produce two sorts of spores. The larger spores grow into sexual plants which have female organs only. The smaller spores produce very small male plants. Plants which produce two types of spores, in this way, are said to be **heterosporous**.

Horsetails (Sphenopsida; Fig. 1.17) occur in tropical Africa only at high altitudes. Each plant has an underground stem, with roots, with green branches above ground. All parts of the stem bear rings of brown, scale-like leaves at intervals. Some of the branches end in cone-like structures in which umbrella-shaped scales, closely packed together, are attached to the stem. The wide end of each scale carries several sporangia on the side facing the stem. Spores germinate to form small, green plants which bear male and female organs. Fertilization is followed by the formation of a new spore-producing plant which is dependent on the sexual plant for only a short time.

Pteropsids (Pteropsida) are vascular plants which have relatively broad leaves with many veins. The subdivision includes ferns (Filicinae) and seed plants (Gymnospermae and Angiospermae).

1.16 Tracheophyta, Lycopsida: *Lycopodium clavatum* growing around the base of a tree (× ¾)

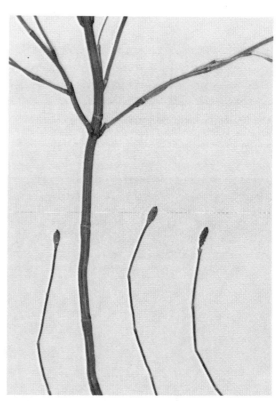

1.17 Tracheophyta, Sphenopsida: *Equisetum ramosissimum*, part of large shoot showing whorled, sheathing leaves and whorled branches, and some detached ultimate branches bearing cones. Collected at an altitude of about 1000 m (× ¾)

Ferns (Filicinae; Fig. 1.18) have stems, usually underground, with roots and large leaves. Some of the leaves bear groups of small sporangia on their lower surfaces. When the spores are released they germinate to form small thalli with male and female organs on their undersurfaces. Fertilization results in the formation of a new spore-producing plant which is dependent on the sexual plant for only a short time.

Seed plants are dispersed by seeds, not by spores. A **seed** is a multicellular structure which contains an embryo plant and food reserves, both enclosed in a protective coat. All of the groups still to be described are seed plants.

Gymnosperms (Gymnospermae) produce their seeds on scale-like structures. The seeds are not enclosed. Living representatives include cycads and conifers.

Cycads (Cycadales; Fig. 1.19) have an upright stem with a strong root system at the base. The leaves are palm-like and are attached in a spiral at the end of the stem. Each plant produces either male or female cones. A male cone consists of a central axis to which many spirally arranged scales are attached. Each scale has many pollen sacs on its lower surface. In African genera (e.g. *Encephalartos* spp.), female cones are similar to male cones, but the scales are larger and each has two, large ovules attached. Each ovule contains a fleshy sexual plant with several female organs. One of these is fertilized by a swimming sperm from the male sexual plant, which is produced by the pollen grain, after the pollen grain has entered the tip of the ovule. Fertilization results in the development of an embryo which remains within the nutritive flesh of the ovule. Once the embryo has developed, the ovule has become a seed. After dispersal, the seed germinates and the embryo grows into a new plant.

Conifers (Coniferales) are woody plants, mostly trees. There are few truly tropical examples, but

1.18 Tracheophyta, Pteropsida, Filicinae: *Arthropteris* sp., a forest fern (× ⅓)

1.19 Tracheophyta, Pteropsida, Gymnospermae: part of a cycad (*Cycas circinalis*) showing a male cone (× ⅓)

many (e.g. *Pinus* spp.) are now planted in the tropics, usually at high altitude, as a source of fast-growing softwood. Most are monoecious and bear both male and female cones, similar to but not identical with the cones of cycads. Pollen and ovule structure, also, are similar to those of cycads, but non-motile sperms are delivered directly into archegonia by pollen tubes similar to those of flowering plants.

Flowering plants (Angiospermae; Fig. 1.20) have their ovules enclosed in an ovary, at the centre of a flower. A flower is a modified shoot (stem with leaves). The female part of the flower (**pistil**) is at the centre and its lowest part, the **ovary**, contains one or more ovules. Next to the pistil are the **stamens**, which produce pollen. Outside the stamens are modified leaf-like parts (**petals, sepals**), which protect the inner parts. These outer parts also, often, serve to attract insects or other animals which carry pollen from flower to flower.

After pollination, each pollen grain delivers two non-motile sperms to the ovule by means of a **pollen tube** and fertilization follows. After fertilization, the ovule develops into a seed and the ovary develops into a fruit, in which one or more seeds are contained. After the seed has been released, it germinates to establish a new plant.

Flowering plants are divided into two subclasses, Dicotyledoneae (**dicotyledons**, often abbreviated to 'dicots') and Monocotyledoneae (**monocotyledons**, often abbreviated to 'monocots'). The two subclasses are distinguished from each other primarily by the number of seed leaves (**cotyledons**) in the embryo. Dicots have two cotyledons and monocots have one cotyledon. Further differences between members of these subclasses are described in Section B of this book.

Key words

Nomenclature: binomial system, genus, species. **Classification**: kingdom, division, class, order, family, genus, species. **Kingdoms**: Monera, Fungi, Plantae. **Plantae**: algae; bryophytes; vascular plants, pteridophytes, gymnosperms, angiosperms; dicotyledons, monocotyledons. **Structure**: procaryotic, eucaryotic. **Nutrition**: autotrophic, photoautotrophic, chemoautotrophic; heterotrophic, saprophytic, parasitic. **Reproduction**: spore; seed, embryo.

1.20 Tracheophyta, Pteropsida, Angiospermae: a flowering shoot of *Gloriosa superba*, a monocot (× ⅔)

Summary

The vegetation of the earth is made up of a great variety of plants. Plants can be named scientifically by using the binomial system. A natural classification brings together related plants into groups (taxa) with similar vegetative and reproductive structures and processes.

Questions

1 What is a taxon? List all the principal taxa, omitting subtaxa.
2 What is the scientific advantage of using scientific names of plants? List some examples of common plant names which are scientifically inaccurate or misleading. Choose common names in any language with which you are familiar and explain in what ways they are inaccurate or misleading.

Further reading for Section A

Bold, H. C. 1977. *The Plant Kingdom*, 4th edn, Prentice-Hall: New Jersey.
McLean, R. C. & Ivimey-Cook, W. R. 1967. *Textbook of Theoretical Botany*, vol. 1, Longmans, Green: London.

Section B

STRUCTURE AND FUNCTION IN FLOWERING PLANTS

In this Section, we study the external form (morphology) and internal structure (anatomy) of flowering plants from the seed and seed germination to the mature plant and the formation of the next generation of seeds. We also cover the relationships between structure and function (physiology) of plant organs and of the plant as a whole. Physiological topics are integrated with structural studies in Chapter 2 and in Chapters 5 to 9. Major physiological processes which affect the plant as a whole are dealt with in Chapters 11 to 13.

2

Seeds and seedlings

The seed is the structure by which flowering plants are dispersed. It develops inside the ovary wall, which itself develops into the fruit (Fig. 2.1).

Seed structure

The most important part of the seed is the embryo. The **embryo** is a miniature new plant, with root and shoot systems, which remains quiescent until the seed germinates. In dicots, the embryo has two seed leaves (**cotyledons**), an embryonic shoot (**plumule**) and an embryonic root (**radicle**). Below the cotyledons, connecting them and the plumule to the radicle, is the **hypocotyl**. This is a transitional zone

between shoot and root and is usually not easy to observe in the embryo. The structure of a monocot embryo is similar, but there is only one cotyledon.

At an early stage of seed development, the embryo is almost always embedded in a nutritive tissue, the **endosperm**. In some seeds, including almost all monocots and many dicots (e.g. *Ricinus communis*, Castor Oil), the endosperm persists until the seed is ripe and acts as a source of food for the embryo after germination. These are **endospermous** seeds (Figs. 2.2 and 2.4).

In other seeds, the endosperm is absorbed by the embryo before the seed is ripe and the food store is transferred to the cotyledons (e.g. *Arachis hypogaea*, Groundnut). These are **non-endospermous** seeds (Fig. 2.3).

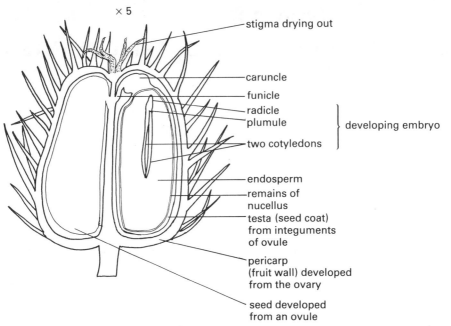

× 5

stigma drying out

caruncle

funicle

radicle

plumule

two cotyledons

} developing embryo

endosperm

remains of
nucellus

testa (seed coat)
from integuments
of ovule

pericarp
(fruit wall) developed
from the ovary

seed developed
from an ovule

2.1 Young fruit of *Ricinus communis*: LS showing
developing seed

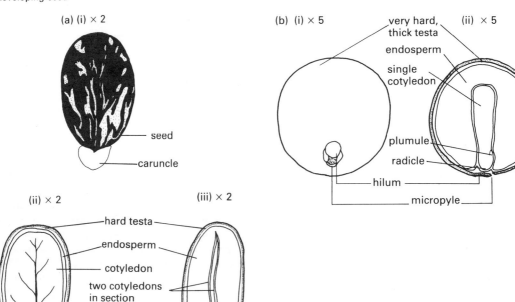

(a) (i) × 2

seed

caruncle

(b) (i) × 5

very hard,
thick testa

endosperm

single
cotyledon

(ii) × 5

plumule

radicle

hilum

micropyle

(ii) × 2

(iii) × 2

hard testa

endosperm

cotyledon

two cotyledons
in section

plumule

hypocotyl

radicle

micropyle

2.2 Endospermous seeds: (a) dicot, *Ricinus communis*,
(i) whole seed, (ii) LS, (iii) LS at right angles to last;
(b) monocot, *Canna indica*, (i) whole seed, (ii) LS

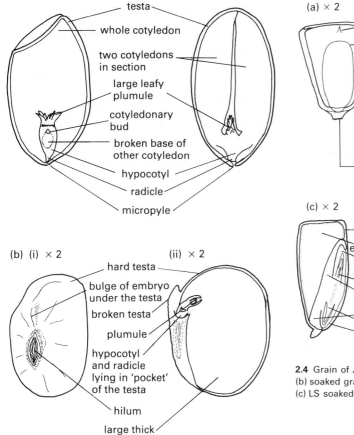

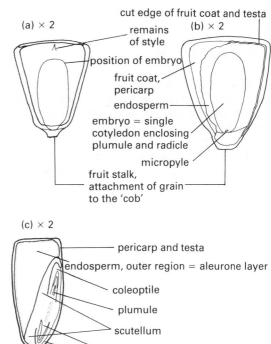

(a) (i) × 2

testa

whole cotyledon

two cotyledons
in section

large leafy
plumule

cotyledonary
bud

broken base of
other cotyledon

hypocotyl

radicle

micropyle

(ii) × 2

(b) (i) × 2

hard testa

bulge of embryo
under the testa

broken testa

plumule

hypocotyl
and radicle
lying in 'pocket'
of the testa

hilum

large thick
cotyledon

(ii) × 2

2.3 Non-endospermous seeds, all dicots: (a) *Arachis hypogaea* (Groundnut), (i) and (ii) LS through different planes; (b) *Voandzeia subterranea* (Bambarra Groundnut), (i) whole seed, (ii) LS seed

(a) × 2

cut edge of fruit coat and testa

remains
of style

(b) × 2

position of embryo

fruit coat,
pericarp

endosperm

embryo = single
cotyledon enclosing
plumule and radicle

micropyle

fruit stalk,
attachment of grain
to the 'cob'

(c) × 2

pericarp and testa

endosperm, outer region = aleurone layer

coleoptile

plumule

scutellum

radicle

coleorhiza

2.4 Grain of *Zea mays* (Maize): (a) whole dry grain;
(b) soaked grain with part of pericarp and testa removed;
(c) LS soaked grain

The main food storage substance in the endosperm of cotyledons is usually starch (e.g. cereals, beans) or oil (e.g. *Ricinus communis; Arachis hypogaea*), plus some . protein. The protein may be dispersed throughout the tissue, or concentrated in a special layer of the endosperm (e.g. the **aleurone** layer of the grain of *Zea mays*, Maize).

The embryo, or embryo plus endosperm, is enclosed in a seed coat (**testa**), which is usually tough. The testa has a scar (**hilum**) on the outside, marking the position at which it was attached to the inside of the fruit. Close to the hilum is the **micro-**

pyle. This is a small hole in the testa and corresponds to the micropyle of the ovule from which the seed was developed (Chapter 14). The function of the testa is to protect the seed contents from desiccation and mechanical injury. In seeds in which the testa is very thin, this function is taken over by a thickening of the inner part of the wall of the fruit. In *Mangifera indica* (Mango), for instance, the hard outer layer of the 'stone' of the fruit is developed from the inner part of the ovary wall (**endocarp**) (Fig. 14.19). The seed, inside the stone, has only a thin, papery testa.

Grains

In cereal grains like those of *Zea mays*, there is only one seed in each fruit and the testa is fused with the pericarp. It is the whole of the fruit which is dispersed. All grains are endospermous (Fig. 2.4).

Germination

Germination of a seed begins with the absorption of water, which causes the seed to swell and bursts the testa. Water first enters through the micropyle, or through the testa if the testa is not impermeable. The first true growth occurs in the radicle, the tip of which emerges through the broken testa in the region of the micropyle and grows downwards into the soil.

In many dicots (e.g. *Lycopersicum esculentum*, Tomato; *Ricinus communis*) the next event is the elongation of the hypocotyl, which raises the cotyledons and endosperm (if any) above the soil, often with the testa still attached. While it is still under the surface of the soil, in darkness, the upper part of the hypocotyl is usually hooked over, so that passage through the soil tends to force the cotyledons together and thus protect the plumule within. Once exposed to light, above the soil, the hypocotyl straightens out. Seeds in which the hypocotyl elongates and carries the cotyledon above the soil are said to have **epigeal** germination.

In the epigeal germination of a non-endospermous seed (Fig. 2.5), the cotyledons spread apart immediately and act as the first green leaves of the new plant. In epigeal germination of endospermous seeds (Fig. 2.6), the cotyledons are still within the endosperm at this stage, where they function to absorb materials from the endosperm. Once the endosperm is exhausted, it falls away and the cotyledons then spread out and function as leaves.

Epigeal germination is not common in monocots. It occurs, for example, in *Allium* spp. (Onion, Leek, Garlic) (Fig. 2.7). Here, the hypocotyl elongates very little but the endosperm and testa are carried up above the soil by the elongation of the lower part of the single cotyledon.

In most monocots (e.g. *Canna indica*, Canna Lily; *Zea mays*) and some dicots (e.g. *Cajanus cajan*, Pigeon Pea; *Mucuna* spp.), the hypocotyl does not elongate and the cotyledons remain below the surface of the ground. This is **hypogeal** germination (Fig. 2.8).

In an endospermous seed with hypogeal germination, the cotyledons act only as absorbing organs. In a non-endospermous seed with hypogeal germi-

nation, they function only as storage organs (Fig. 2.9). In hypogeal germination, it is the plumule which elongates and is first to appear above the soil. The end of the plumule is at first hooked over (**plumular hook**) in the same way as the hypocotyl in epigeal germination.

The seedling

The word 'seedling' does not have a precise scientific definition. It refers to the young plant which appears after germination of the seed, but the point at which a seedling becomes a 'plant' is a matter for subjective decision.

The parts of a seedling which are normally visible above the surface of the ground are different in epigeal and hypogeal germination. In dicots with epigeal germination, the upper part of the hypocotyl, cotyledons and plumule appear (Figs. 2.5, 2.6). In epigeal monocots, the hypocotyl remains very short, the plumule remains at or below the surface of the ground and only the single cotyledon appears (Fig. 2.7). As the plumule starts to grow it forms the **epicotyl**. In the case of hypogeal germination, the cotyledons remain below ground and only the epicotyl appears above the surface.

The cotyledons are the seed leaves of the plant. They are usually simple in shape, but their exact shape is often characteristic of the family or genus to which the plant belongs. For example, the cotyledons of members of the Solanaceae (e.g. *Lycopersicum esculentum*, Tomato, Fig. 2.10) are usually long and narrow, while those of *Ipomoea* spp. (Convolvulaceae) are usually bilobed. Like other leaves, the cotyledons have buds (**axillary buds**) just above their point of attachment, but these often fail to develop into branches, or develop into branches only if the plumule is destroyed.

Leaves formed on the epicotyl are true leaves. The first true leaves are often simpler in shape than leaves which are formed when the plant grows larger. If so, they may be called juvenile leaves (**juvenile foliage**), to distinguish them from the **mature foliage** formed later. Simple, juvenile leaves may very soon be succeeded by mature foliage, a little way up the epicotyl, or there may be considerable delay (e.g. *Eucalyptus* spp., Figs. 3.4, 3.6). The formation of mature foliage sometimes depends on specific environmental factors. For instance, mature foliage is not formed by certain *Ipomoea* spp. if the plants are growing in deep shade.

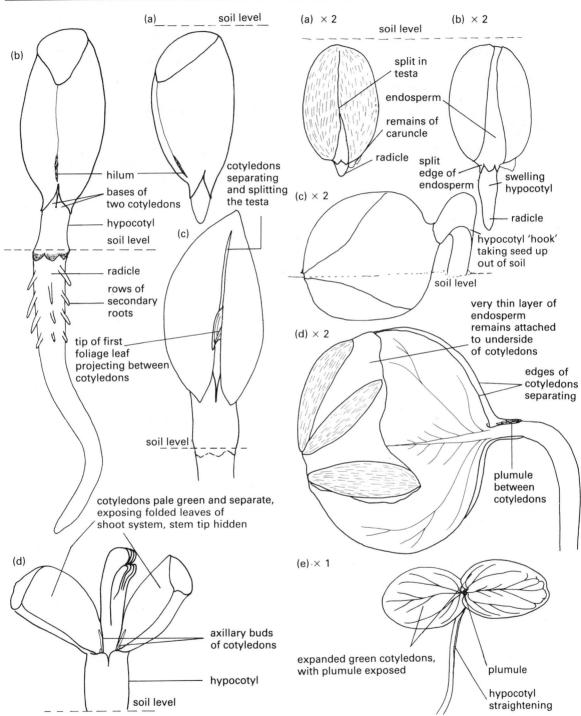

(a)   soil level

(b)

hilum

bases of two cotyledons

hypocotyl

soil level

radicle

rows of secondary roots

tip of first foliage leaf projecting between cotyledons

cotyledons separating and splitting the testa

(c)

soil level

cotyledons pale green and separate, exposing folded leaves of shoot system, stem tip hidden

(d)

axillary buds of cotyledons

hypocotyl

soil level

2.5 Germination: epigeal, non-endospermous dicot, *Arachis hypogaea*, Groundnut

(a) × 2   soil level

(b) × 2

split in testa

endosperm

remains of caruncle

radicle

split edge of endosperm

swelling hypocotyl

radicle

(c) × 2

hypocotyl 'hook' taking seed up out of soil

soil level

very thin layer of endosperm remains attached to underside of cotyledons

(d) × 2

edges of cotyledons separating

plumule between cotyledons

(e) · × 1

expanded green cotyledons, with plumule exposed

plumule

hypocotyl straightening

2.6 Germination: epigeal, endospermous dicot, *Ricinus communis*, Castor Oil

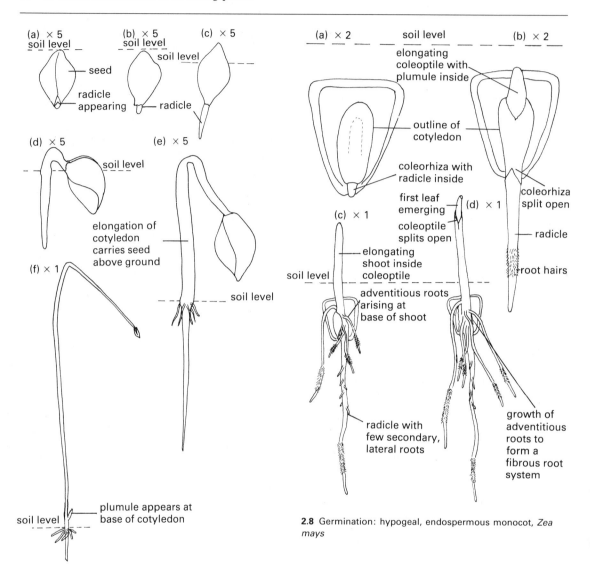

(a) × 5
soil level
seed
radicle appearing

(b) × 5
soil level
radicle

(c) × 5
soil level

(d) × 5
soil level
elongation of cotyledon carries seed above ground

(e) × 5
soil level

(f) × 1
soil level   plumule appears at base of cotyledon

2.7 Germination: epigeal, endospermous monocot, *Allium* sp. (Onion)

(a) × 2   soil level
elongating coleoptile with plumule inside
outline of cotyledon
coleorhiza with radicle inside

(b) × 2
coleorhiza split open
radicle
root hairs

(c) × 1
first leaf emerging
coleoptile splits open
elongating shoot inside coleoptile
soil level
adventitious roots arising at base of shoot
radicle with few secondary, lateral roots

(d) × 1
growth of adventitious roots to form a fibrous root system

2.8 Germination: hypogeal, endospermous monocot, *Zea mays*

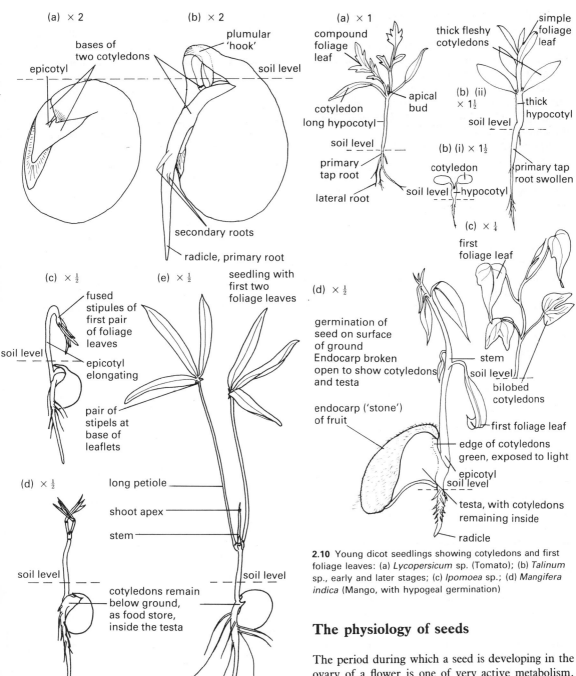

(a) × 2

epicotyl

bases of
two cotyledons

(b) × 2

plumular
'hook'

soil level

secondary roots

radicle, primary root

(c) × ½

fused
stipules of
first pair
of foliage
leaves

soil level

epicotyl
elongating

pair of
stipels at
base of
leaflets

(e) × ½

seedling with
first two
foliage leaves

long petiole

shoot apex

stem

(d) × ½

soil level

soil level

cotyledons remain
below ground,
as food store,
inside the testa

2.9 Germination: hypogeal, non-endospermous dicot,
Voandzeia sp.

(a) × 1

compound
foliage
leaf

cotyledon

long hypocotyl

soil level

primary
tap root

lateral root

apical
bud

thick fleshy
cotyledons

simple
foliage
leaf

(b) (ii)
× 1½

soil level

thick
hypocotyl

(b) (i) × 1½

cotyledon

soil level

hypocotyl

primary tap
root swollen

(c) × ¼

first
foliage leaf

stem

soil level

bilobed
cotyledons

(d) × ½

germination of
seed on surface
of ground
Endocarp broken
open to show cotyledons
and testa

endocarp ('stone')
of fruit

first foliage leaf

edge of cotyledons
green, exposed to light

epicotyl
soil level

testa, with cotyledons
remaining inside

radicle

2.10 Young dicot seedlings showing cotyledons and first
foliage leaves: (a) *Lycopersicum* sp. (Tomato); (b) *Talinum*
sp., early and later stages; (c) *Ipomoea* sp.; (d) *Mangifera
indica* (Mango, with hypogeal germination)

The physiology of seeds

The period during which a seed is developing in the
ovary of a flower is one of very active metabolism.
Sugars and other substances are transported (**translocated**) into the developing seed from the parent
plant. Some of these supply the needs of the developing embryo directly, while the remainder are

converted into starch or oils in the storage tissues of the endosperm or cotyledons.

Like other chemical changes which take place in living tissues, these metabolic activities are controlled by enzymes. Enzymes are proteins which catalyse specific chemical reactions within the cells of organisms. Their action is very sensitive to temperature and acidity (pH) and they are destroyed by temperatures above about 40 °C.

Once the seed has been fully formed, it starts to lose weight as it matures. This loss of weight is mostly due to loss of water as the seed dries out. Depending on the species and prevailing environmental conditions, the water content may be reduced to between 5 and 20 per cent. Water is the medium in which all metabolic processes take place and this loss of water leads to a slowing down of metabolic activity, which continues at a very low rate. The seed thus enters a resting period, the duration of which varies according to species and the prevailing conditions.

Quiescence and dormancy

Some seeds are capable of germinating as soon as certain necessary environmental conditions are satisfied. A seed of this type is said to be **quiescent** until germination begins. Other seeds go into a more prolonged resting period, during which they will not germinate, even if all the usual environmental requirements are satisfied. A seed of this type is said to be **dormant**. The dormant state may be distinguished into primary dormancy and secondary dormancy.

In **primary dormancy**, the seed will not germinate immediately, under any conditions. With **secondary dormancy**, the seed is merely quiescent when it is shed and will germinate immediately under suitable conditions. However, if suitable conditions for germination are not immediately available, the seed goes into a state of secondary dormancy, during which it will not germinate, which persists for some time.

In some plants showing primary dormancy, the embryo is not fully developed when the seed is shed. Such a seed cannot germinate until development of the embryo has been completed. This is known to be the case in many orchids and has also been demonstrated in a number of other plants (e.g. *Fraxinus excelsior*, European Ash).

There are also examples in which the embryo is

prevented from developing further by the presence of substances which inhibit metabolism (**metabolic inhibitors**). Under natural conditions, the rain water that finally washes away these inhibitors also provides ideal soil conditions for the germination of the seeds and for the establishment of seedlings. The juices of many fruits contain compounds which inhibit germination, not only of their own seeds but of other seeds with which they come into contact. For instance, even at 25 times dilution, the juice of the fruit of *Lycopersicum esculentum* (Tomato) has been found to completely inhibit the germination of seeds of *Lepidium sativum* (Cress). A plant **hormone, abscisic acid**, is known to be one such inhibitor and it has been suggested that its primary action is the control of water uptake by the tissues of the embryo.

Sometimes, the internal factors which inhibit germination are not known. No treatment or set of conditions will cause certain otherwise mature seeds to germinate until a certain period of time has elapsed. Such seeds are thought to require **after-ripening** of the embryo. Temperate examples include *Malus pumilla* (Apple) and *Crataegus oxycantha* (Hawthorn). In the latter, the after-ripening period may be reduced from about a year to one month by removing the testa. After-ripening can also be speeded up by low temperature (e.g. 5 °C) and by treatment with acids. In *Lactuca sativa* (Lettuce), dormancy has been broken experimentally by treatment with plant hormones of the type known as **gibberellins**. After-ripening of seeds is assumed to involve some chemical or physical change which is under the control of internal factors.

Primary dormancy may also result from structural constraints. The testa may be impermeable to water or gases, thus limiting metabolic activity in the seed. In such a case, water and oxygen cannot enter the seed and carbon dioxide cannot escape. It is interesting to note that high concentrations of carbon dioxide have been used to induce secondary dormancy, artificially, in some seeds.

The very low metabolic rate within a seed cannot increase until the testa becomes permeable to water. Even when permeability increases, the testa may still be tough enough to resist rupture. This either prevents more water from entering, or stops the radicle and plumule from growing out. Dormancy is broken only when the testa becomes less resistant.

Examples of plants in which most seeds remain dormant for long periods, due to impermeability of the seed coat, are *Caesalpinia pulcherrima* (Pride of Barbados) and *Canna indica* (Canna Lily). If seeds of either plant are sown, a few seedlings appear within a few days or weeks, but most seeds remain dormant in the soil. Further seedlings appear, one or a few at a time, over a period of a year or more. If, however, the testas of the seeds are pierced or partially scraped away (**scarified**) before the seeds are sown, most of the seeds germinate within a few days.

Non-dormancy

Most seeds go into a quiescent state at maturity and many have a short period of dormancy lasting at least until they have been shed from the parent plant. However, the seeds of some species germinate very readily indeed, even before they have been shed. Under moist conditions, the grain of certain varieties of *Zea mays* (Maize) germinate on the cob, within the husk, while still on the parent plant. The non-dormant condition also obtains in some varieties of *Arachis hypogaea* (Groundnut). This characteristic has important economic aspects, for grain and seeds which have germinated before harvest cannot be stored and have little value. Another plant noted for the lack of dormancy, or even quiescence, of its seeds is *Hevea brasiliensis* (Para Rubber). Here, the seeds do not dry out at maturity and start to germinate in the capsule attached to the parent tree, or as soon as they have fallen, regardless of external conditions.

Vivipary

The extreme case of non-dormancy is **vivipary**, when the embryo normally grows into a seedling while it is still attached to the parent plant. To those who live close to a coastal or riverside swampy area the most familiar example of vivipary is *Rhizophora mangle* (Red Mangrove). The fruit of *Rhizophora* sp. contains a single seed in which the embryo continues to grow without any pause at maturity, while still attached to the parent tree. The radicle and hypocotyl grow downward to a length of 20 cm or more. The seedling hangs from the tree for four to six months, while the cotyledons within the testa continue to absorb food from the endosperm. The hypocotyl and radicle eventually detach from the cotyledons and fall from the fruit into the water or mud beneath. The widest and heaviest part of the seedling is close to the lower end, so the seedling always falls vertically, radicle downwards. If it strikes mud, it immediately penetrates and, in effect, plants itself where it lands. If it falls into deeper water, it floats away until it is cast up on to a substrate on which it becomes established, elsewhere.

The physiology of seed germination

The main environmental factors required for the germination of a seed are water, oxygen, a suitable temperature and, sometimes, light or darkness.

Water

The first step in germination is a rapid absorption of water (**imbibition**). The low metabolic level of quiescent or dormant seeds is largely due to their low water content, resulting in the partial dehydration of their proteins, including enzymes. The first water imbibed rehydrates the proteins and activates the enzymes which are present, thus raising the metabolic rate throughout the embryo and in the endosperm, if any.

When water is imbibed to a certain critical level characteristic of the species, water uptake slows and germination continues, to the start of the irreversible processes that lead to growth and development of the embryo. The expansion of the testa is more limited than the swelling of its contents. This causes the testa to rupture and the contents thus obtain free access to water and oxygen.

Oxygen

Respiration of the seed during the water imbibition stage may be **anaerobic**. As soon as the testa bursts and oxygen can diffuse directly in, respiration becomes **aerobic**. The mechanical work involved in the root pushing downwards into the soil and the hypocotyl or plumule pushing upwards through the soil requires energy from the stored food in the seed. This stored food also provides the chemical energy required for the synthesis of new organic materials. This energy is only effectively made available through oxidative processes for which a supply of oxygen is required.

Temperature

For each species, there is a specific temperature range which will permit germination to start and proceed. For most crop plants, the optimum range for germination lies between 20 °C and 30 °C. The lower limit is about 0 °C and the upper limit is about 40 °C. The percentage of germination near lower and upper temperature limits is usually small. The seeds of some temperate plants germinate readily at 10 °C (e.g. *Triticum* spp., Wheat; *Lactuca sativa*, Lettuce). Seeds of *Cucumis sativus* (Cucumber), *Cucurbita pepo* (Vegetable Marrow) and other members of the Cucurbitaceae will not germinate below about 18 °C. Failure of germination at lower temperatures is not, therefore, directly related to the freezing point of water. The upper limit is set by the fact that most enzymes are destroyed (**denatured**) at temperatures above about 40 °C.

The effect of the exposure of seeds to different temperatures before they are sown is modified by other factors. These include duration of exposure, the condition of the seeds (imbibed or dry) and the actual temperature level to which the seeds are exposed. Thus, (a) dry seeds can withstand far higher temperatures than fully imbibed seeds, (b) high temperature applied for a short time has less serious effects than if applied for a longer time and (c) low temperature treatment has as much, if not more, adverse effect on germinability than the high temperature treatment. This last observation, however, ignores the duration factor.

There are cases in which there is a marked improvement in germination rate after various degrees of heating. Seeds of *Hibiscus esculentus* (Okra) and *Elaeis guineensis* (Oil Palm) are familiar examples. In fact, at Oil Palm breeding stations, Oil Palm seeds are routinely given heat treatment before they are sown, to ensure rapid and uniform germination.

Light

The germination of most seeds is not affected by light and darkness. Such seeds are termed **light-indifferent**. Seeds of *Vigna unguiculata* (Cow Pea) and *Zea mays* (Maize) are in this category. On the other hand, the germination of some seeds is slowed or prevented by the presence of light (e.g. some *Allium* spp.). These are **light-hard** seeds. **Light-sensitive** seeds are those for which germination requires light. Seeds of only a few species are known to respond in this way. For example, seeds of *Tridax procumbens*, the common weed, will not germinate unless they are exposed to light when they are moist. The most studied example, because of its strong response, is the variety of *Lactuca sativa* (Lettuce) known as

'Grand Rapids'. Seeds of this variety have been used in experiments on the effects of lights of different wavelengths.

When 'Grand Rapids' Lettuce seeds are moistened and kept in the dark they fail to germinate. If exposed to red light with a wavelength of about 660 nm, they germinate. If, however, exposure to red light is immediately followed by exposure to (invisible) 'far red light', the wave-length of which is about 730 nm, the red light effect is cancelled and germination is decreased or completely inhibited. The reversibility of these light effects is unlimited and the final effect depends simply on which type of light is the last the seeds receive. The pigment which is responsible for absorbing light in this reversible process is **phytochrome**. Phytochrome is important, also, in other processes involving the reactions of plants to light (Chapter 13).

Seed age and germination

After long periods of storage, most seeds show low percentages of germination or fail to germinate at all. For instance, in our laboratories, seeds of *Amaranthus hybridus* (Green) were found to have seventy to ninety per cent germination after storage for eighteen months but only zero to two per cent germination after they had been stored for five years. Seeds remain viable for longer when stored at a temperature below 0 °C or under anaerobic conditions, probably because of the slowing down of metabolic processes under these conditions.

The majority of seeds can survive for only a few years at best, some for only a few days or weeks. Very few seed species remain viable for very long periods. Incredible claims have been made for the germination of grains of Wheat from ancient Egyptian tombs and Maize from ancient American Indian burial mounds, but these are almost certainly not true. There are, however, authentic records for the long-term survival of seeds of certain species, but these survive for no more than a few decades. Seeds of *Amaranthus retroflexus* (Pigweed) and *Portulaca oleracea* (Purslane) have germinated after forty years. Seeds of *Oenothera biennis* and *Rumex crispus* (Yellow Dock) have survived eighty years and remained viable. All of these species are serious weeds of cultivation in the countries in which they grow.

Seeds survive best when they are buried in the ground, probably because the soil provides a very stable environment.

Key words

Seed: embryo; plumule, shoot; radicle, root; cotyledons; testa (seed coat), micropyle, hilum. **Food store**: endosperm, endospermous; cotyledons, non-endospermous. **Germination**: epigeal, hypogeal. **Seedling**: hypocotyl, cotyledons, epicotyl, juvenile leaves. **Seed physiology**: quiescence; dormancy, primary and secondary dormancy, immature embryo, metabolic inhibitor, impermeable testa. **Non-dormancy**: vivipary. **Germination**: environmental factors, water, oxygen, temperature, light.

Summary

A seed contains an embryo, with a food store either in the cotyledons or in endosperm. The seeds of a few plants (viviparous plants) germinate while they are still attached to the parent plant. In most plants, seeds lose water as they ripen and enter a period of quiescence. The quiescent period ends after they are shed, as soon as environmental factors are suitable for germination. The seeds of some plants also have a dormant period, during which they will not germinate under any conditions. Dormancy may result from physical factors such as the possession of a tough seed coat. In other cases, dormancy is due to internal, physiological factors.

Questions

1 Compare the structure of a named endospermous seed with that of a named non-endospermous seed. If possible, obtain examples of two types of seeds, dissect and draw their parts.

2 Compare a dicot seedling with epigeal germination with a dicot seedling with hypogeal germination, at the time when the first green parts appear above ground. Name, (a) the part of the epigeal seedling which is much better developed than the corresponding part of the hypogeal seedling, and (b) the part of the hypogeal seedling which is much better developed than the corresponding part of the epigeal seedling. Apart from the appearance or non-appearance of the cotyledons above ground, what is the fundamental difference between epigeal and hypogeal germination?

3 Distinguish between quiescence and dormancy, and between primary and secondary dormancy.

4 Mangrove trees frequently grow in closely packed clumps, consisting of plants of different ages. How can this be explained by its particular type of vivipary?

5 All living tissues respire, even though respiration in a dry seed may be very slow indeed. What light does this throw on claims that viable Wheat grain is said to have been discovered in Egyptian tombs that have not been entered by man for several thousand years?

6 Suppose you sow some seeds and they do not germinate. What steps would you take to find out why they have not germinated?

3

The morphology of flowering plants

Morphology is the study of the external features of plants and their parts. Every flowering plant consists of a **shoot system** and a **root system**. The shoot system is made up of **stems** and **leaves**. The root system is made up of **roots** only.

The root system

In most dicots, the radicle grows down into the soil to form the **primary root** or **tap root** of the plant. Branches (**secondary roots**) arise from the primary root and these in turn branch again (**tertiary roots**) and so on. The tap root remains as the main axis of the root system. This type of root system, characteristic of dicots, is called a **tap root system** (Fig. 3.1). Dicots which have stems growing along the surface of the ground may also have roots which arise directly from the stem, usually close to a point on the stem where a leaf is attached (e.g. *Oxalis* spp.). These are **adventitious roots**.

In monocots, the growth of the primary root is limited and it branches little. Additional, adventitious, roots grow out from near the base of the stem and the root system is added to in this way. The resulting root system, lacking a prominent tap root, is called a **fibrous root system** (Fig. 3.2).

The shoot system

The shoot system, developed from the plumule of the embryo, consists of stems and leaves. Leaves are attached to stems at **nodes**. A stem consists of alternating nodes and **internodes**. Leaves may be attached at each node singly (**alternate** leaves), in opposite pairs (**opposite** leaves), or more than two at a node (**whorled** leaves). Leaf arrangement may be described in more detail, numerically, by describing the **phyllotaxis** (Fig. 3.3).

The angle formed between the upper side of the attachment of a leaf and the stem above is the **axil** of the leaf. Buds formed in this position are **axillary**

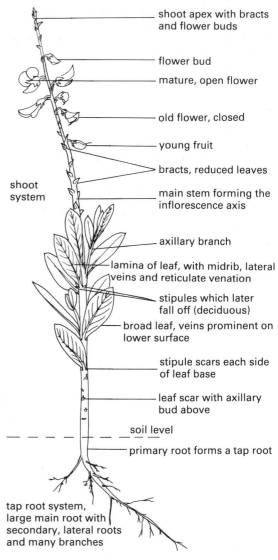

shoot apex with bracts and flower buds

flower bud

mature, open flower

old flower, closed

young fruit

bracts, reduced leaves

main stem forming the inflorescence axis

shoot system

axillary branch

lamina of leaf, with midrib, lateral veins and reticulate venation

stipules which later fall off (deciduous)

broad leaf, veins prominent on lower surface

stipule scars each side of leaf base

leaf scar with axillary bud above

soil level

primary root forms a tap root

tap root system, large main root with secondary, lateral roots and many branches

3.1 Morphology of a dicot: *Crotalaria* sp. (Sunn Hemp) ($\times \frac{1}{4}$)

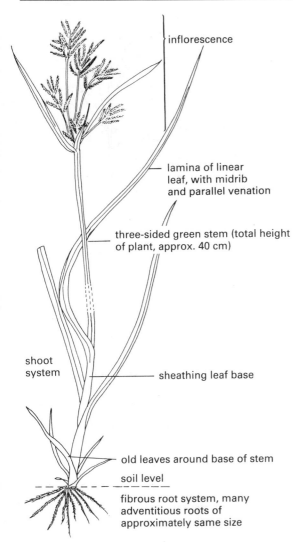

inflorescence

lamina of linear
leaf, with midrib
and parallel venation

three-sided green stem (total height
of plant, approx. 40 cm)

shoot
system

sheathing leaf base

old leaves around base of stem

soil level

fibrous root system, many
adventitious roots of
approximately same size

3.2 Morphology of a monocot: *Cyperus* sp. (× ½)

Each axillary bud, like the **terminal bud** at the tip of a shoot, consists of a growing region (**shoot apex**) protected by developing leaves. Growth of the terminal bud adds to the length of the shoot. Axillary buds are usually dormant for a time but then begin to grow into axillary branches (**lateral branches**). The growth of axillary branches repeats the pattern of nodes and internodes on the main shoot.

Buds, terminal or axillary, may develop into flowers instead of leafy shoots. A flower is a modified shoot. The parts of a flower (sepals, petals, etc.) are modified leaves. The part of the shoot system which bears flowers is the **inflorescence**.

There are two basic patterns of branching of a shoot system, **monopodial** and **sympodial**. In a plant with monopodial branching, each apex remains active and continues to add to the length of the axis of a shoot indefinitely. Axillary branches project sideways and remain shorter than the axis to which they are attached. This usually produces a plant which, in side view, is broad at the base and narrows towards the top (e.g. *Casuarina* spp., Whistling Pines). In a plant with sympodial branching, the apical bud of each branch is regularly used up in the formation of an inflorescence or ceases to grow for some other reason. Further growth is from axillary buds close to the apex. This type of growth usually, but not always, produces a plant which is spreading or bushy (e.g. *Delonix regia*, Flame Tree).

Leaves

The leaf is attached to the node by its **leaf base**. The leaf base is often swollen (**pulvinus**) and may bear stipules. **Stipules** are scale-like or leaf-like outgrowths, one on each side of the leaf base (Fig. 3.4). The leaf base takes a different form in many monocots. In these plants (e.g. *Canna indica*; *Zea mays*) it forms a sheath around the stem. The **sheathing leaf base** of a grass (Fig. 3.5), for instance, is very long and the main part of the leaf is angled away from the stem quite a long way above the node, where the axillary bud is located.

buds. There may be a single bud in each axil or there may be several buds, in a transverse or longitudinal row. Where there are several axillary buds, all but the bud closest to and immediately above the attachment of the leaf are **supernumerary** axillary buds.

(a)

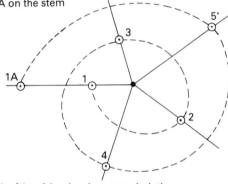

axillary bud

node

internode
= stem between
two nodes

phyllotaxis of (a) = 2/5
Leaf 1 is vertically above
leaf 1A on the stem

from leaf 1 to 2 by the shortest spiral, then
downwards to leaves 3, 4, 5 and 1A equals
twice around the stem, passing through
five internodes = 2/5th phyllotaxis
Leaf 1A is the 5th node going down spirally
from leaf 1.
Angle between leaf insertion is 144°

(b)

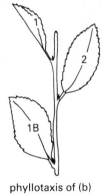

phyllotaxis of (b)
= 1/2
Leaf 1 is vertically
above leaf 1B on
the stem

once around the stem
from leaf 1 to 1B
passes two internodes,
and leaf 1B is at
the 2nd node down
from leaf 1
Angle between leaf
insertion is 180°

(c)

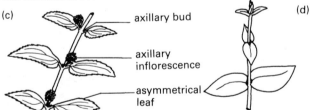

axillary bud

axillary
inflorescence

asymmetrical
leaf

(d)

(e)

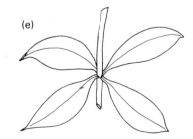

3.3 Leaf arrangement: (a) leaves alternate, phyllotaxis $\frac{2}{5}$:
Talinum triangulare; (b) leaves alternate, phyllotaxis $\frac{1}{2}$: *Sida*
sp. (c) opposite leaves, inserted all in one plane:
Euphorbia hirta· (d) decussate leaves, leaves opposite and
successive pairs at right angles: *Ageratum* sp. (e) whorled
leaves, more than two at each node: *Allamanda* sp.

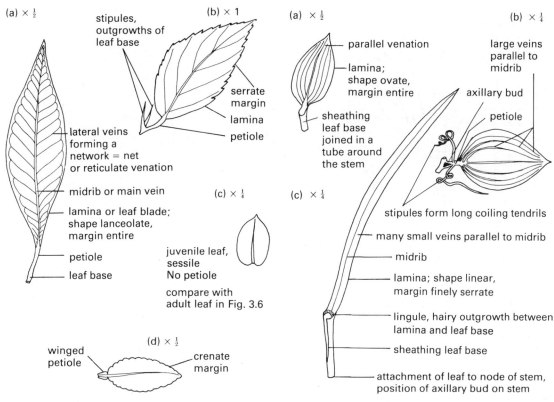

3.4 Leaf morphology, dicots, simple leaves:
(a) *Mangifera indica* (b) *Sida* sp. (c) *Eucalyptus* sp. (d) *Citrus* sp.

3.5 Leaf morphology: monocots: (a) *Commelina* sp. (b) *Smilax* sp. (c) grass

The wide, thin part of the leaf (**leaf blade**) may be attached directly to the leaf base, in which case the leaf is **sessile**. More usually, there is a leaf stalk (**petiole**) which connects the leaf blade to the leaf base and the leaf is said to be **petiolate**.

If the leaf blade is all in one piece, the leaf is a **simple leaf**. If it is divided into several or many separate parts it is a **compound leaf**. The separate parts of a compound leaf are **leaflets**. Different types of simple and compound leaves are illustrated in Figs. 3.6 and 3.7.

The blade of each leaf or leaflet consists of the thin, photosynthetic **lamina**, supported by a system of **veins**. In dicots, there is usually a very conspicuous main vein, the **midrib**, which is continuous with the petiole. A leaf with one main vein is **unicostate**. Some leaves have several large main

veins diverging upwards and are **multicostate** (e.g. family Melastomataceae). The main vein or veins have branches (**lateral veins**) which branch again repeatedly. The finest veins join together and form a network throughout the lamina. Dicots usually have this sort of **net venation** (Fig. 3.8).

Monocots usually have several or many veins which run parallel to each other for the whole length of the lamina. They have **parallel venation** (Fig. 3.9). In some monocots (e.g. *Zea mays*) there is a midrib. In others, the venation of the leaf may resemble that of a dicot (e.g. *Dioscorea* spp., Yams; *Smilax* sp.).

Leaves may remain attached to the node for a long time but it is not unusual for some of the lower leaves, including cotyledons, to fall from the plant at an early stage. Older leaves, or damaged leaves,

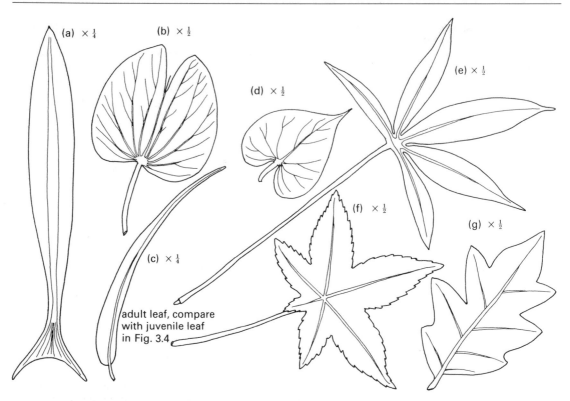

3.6 Leaf morphology, simple leaves, the first a monocot, all others dicots: (a) narrowly lanceolate, parallel venation, wide clasping base: *Dracaena* sp.; (b) bilobed, petiolate: *Bauhinia* sp.; (c) lanceolate, asymmetrical, petiolate: *Eucalyptus* sp.; (d) cordate, petiolate: *Ipomoea* sp.; (e) digitately lobed, the lamina extends around all the lobes: *Manihot* sp.; (f) palmately lobed, peltate (petiole attached to underside of the lamina): *Ricinus communis*; (g) pinnately lobed, base of lamina slightly asymmetrical: *Solanum* sp.

are cut off by a breakdown of tissues, roughly on a level with the surface of the stem, leaving a **leaf scar** on the surface. The dropping of leaves is known as **leaf abscission** and the layer of tissue which breaks down and releases the leaf is the **abscission layer**.

Life cycles and morphology

As we have seen, plants as a whole are classified on the basis of fundamental similarities in structure and physiology and on their probable evolutionary relationships. They may also be put into categories, for descriptive convenience, according to their life cycles and growth habits.

Life cycles

The life cycle of a flowering plant starts with the germination of the seed and ends with the death of the plant, after it has produced seeds once, several or many times.

Ephemeral plants (**ephemerals**) have a very short life cycle, usually only a few weeks. They are characteristic of deserts, where rainfall is infrequent and lasts only for a short time. Some weeds of cultivation grow as ephemerals under dry conditions (e.g. *Ageratum conyzoides*, Goatweed).

Annuals are plants which go through their life cycles in a single growing season. Seeds germinate when the first rain falls after the dry season and seed production is completed by the end of the wet season, after which the plants die.

Biennials are plants which grow for their first season without flowering and produce flowers and seeds during the second growing season, after which they die. Biennials are more common in temperate lands than in the tropics but some temperate plants

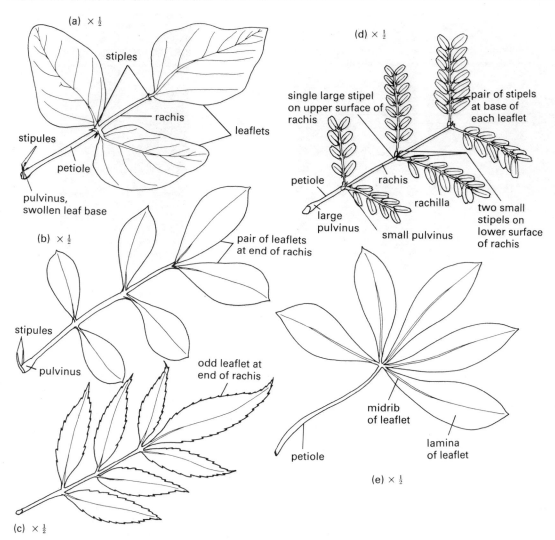

3.7 Leaf morphology, compound leaves, lamina divided into separate leaflets. All dicots: (a) trifoliolate, three leaflets attached to the end of the petiole: *Erythrina* sp.; (b) paripinnate, leaflets all paired, attached along both sides of the central axis (rachis): *Cassia* sp.; (c) imparipinnate, with an unpaired leaflet attached to the tip of the rachis: *Tecoma* sp.; (d) bipinnate, leaflets attached to branches of the rachis (rachillae): *Caesalpinia pulcherrima*; (e) digitate, several leaflets at the end of the petiole: *Ceiba* sp.

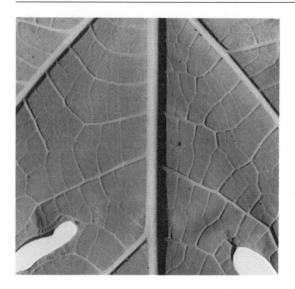

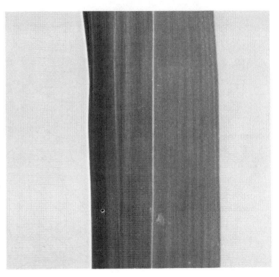

3.8 Dicot leaf, reticulate venation: underside of leaf of *Carica papaya* (Pawpaw)

3.9 Monocot leaf, parallel venation: under surface of leaf of *Bambusa* sp. (Bamboo)

which have been introduced into tropical countries may continue to show the biennial life cycle (e.g. *Daucus carota*, Carrot).

Perennials are plants which live for several or many years. Most produce flowers and seeds many times (e.g. *Tridax procumbens*). A few perennial plants flower only once, just before they die (e.g. Bamboos).

Growth habit

Flowering plants may be broadly separated into herbaceous plants (**herbs**) and woody plants (**shrubs** and **trees**).

Herbs are plants which have only a small amount of woody tissue in their stems and roots. Especially in ecological studies, they may be separated into grasses and grass-like plants (e.g. sedges, rushes) and **forbs** (broad-leaved herbs). Herbs may be upright (e.g. *Euphorbia geniculata*), creeping (e.g. *Tridax procumbens*) or climbing (e.g. *Cardiospermum* sp.). There are ephemeral, annual, biennial and perennial herbs.

Woody plants have woody stems and roots, which become thicker as they grow older. As they become thicker, the original surface tissues are usually replaced by **bark**, which is usually brown. Almost all woody plants are perennials. Woody plants include shrubs, trees and woody climbers.

According to traditional definitions, a **shrub** is a woody plant which branches just above ground level and has no obvious main stem (Fig. 3.10). A **tree** has a conspicuous main stem (trunk) and branches are limited to the upper part (Fig. 3.11) or there are no branches at all (e.g. most palm trees). In practice, however, a very large shrub is regarded as a tree. For instance, the savanna shrub shown in Fig. 3.10 belongs to a species which grows to tree-like dimensions. When so grown, it would be classified as a tree in spite of the fact that it would have more than one 'trunk'.

A **liane** is a woody climber or scrambler with a very long main stem. Lianes attach themselves to other plants in a number of different ways (Chapter 4).

Other classifications of plant form

The categories of plant form given above are satisfactory for most general purposes, though there are plants which it is not easy to place in a category with confidence. For instance, there are plants which are intermediate between the categories of shrub and tree. More detailed classifications exist. Raunkaier's

3.10 Woody plants, shrub: *Acacia* sp., a savanna plant

3.11 Woody plants, trees, *Parkia* sp. (large tree in centre) and *Mangifera indica* (smaller, left)

classification of life forms is probably the most widely used and is to be found in many books on ecology (e.g. Hopkins, 1979: see list of literature at end of Section F).

The morphology of woody plants

Most of the description of plant organs earlier in this chapter is based on herbs, but applies equally to woody plants. There are, however, a few features which are peculiar to woody plants.

As has been stated, the surface of the stem of a woody plant is usually protected by brown bark. The general covering of bark usually has lighter or darker coloured spots scattered over the surface. These spots, **lenticels**, are points at which the cells of the surface tissue are only loosely attached to each other. They allow for the diffusion of gases between the inner tissues and the atmosphere.

Older woody branches have lost their leaves. Leaf scars are usually conspicuous for a time after the leaves have fallen and may be visible throughout the life of the plant (e.g. *Carica papaya*, Pawpaw). In many woody plants, leaf scars disappear as they are covered by new additions of tissue.

Some rainforest shrubs and trees continue to grow throughout the dry season, but most woody plants of forest and savanna discontinue growth until the start of the rainy season. Terminal and axillary buds are then protected by **bud scales**. These are generally not very conspicuous in tropical plants. Those of most forest plants are hardened leaf bases, which replace normal leaves (e.g. *Ouratea flava*, a forest shrub). Some savanna trees form more conspicuous bud scales. Those of *Cussonia* sp., a savanna tree, are modified leaf bases, each with a pair of large stipules (Fig. 3.12).

Many trees and shrubs in the rain forest bear leaves throughout the year. Old leaves die and drop

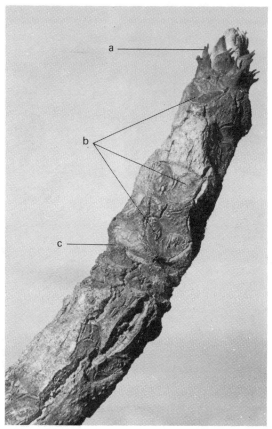

3.12 Deciduous savanna tree: *Cussonia* sp. in dry season condition, with bud scales (a) protecting the apical bud, large leaf scars (b) and bud scale scars from the previous apical bud (c)

Key words

Root system: radicle, primary root, tap root; tap root system, lateral roots, secondary roots, tertiary roots; fibrous root system, adventitious roots. **Shoot system**: plumule, stem, leaves; nodes, internodes; buds, axillary buds, terminal buds; apex; flowers, inflorescence; branching, monopodial, sympodial. **Leaf**: leaf base, stipules; petiole; leaf blade, lamina; simple leaf; compound leaf, leaflets; leaf venation, reticulate venation (net venation), parallel venation; abscission, leaf scar. **Life cycles**: ephemeral, annual, biennial, perennial. **Habit**: herbs; woody plants, shrub, tree, liane; lenticel; bud scales; evergreen, deciduous.

Summary

The external form (morphology) of a flowering plant includes the root system and the shoot system. Monocots usually have a fibrous root system and a non-woody shoot system with few branches bearing elongated leaves with parallel venation. Dicots include herbs and woody plants, usually with a tap root system and much-branched stems. Dicot leaves usually have reticulate venation and may be simple or compound. Life cycles of flowering plants vary from ephemeral herbs to perennial herbs, trees and shrubs. Some are evergreen, others deciduous.

Questions

1 List the morphological features of the root system and shoot system in an herbaceous monocot and a woody dicot.
2 What are the similarities and differences between annuals and perennials in their morphology and life cycles?
3 What are herbaceous ephemeral plants and woody perennial plants?

off, but not at any particular time of year. These are **evergreen** trees and shrubs (e.g. *Entandrophragma* spp., African Mahogany). Other woody plants lose all their leaves at the end of the wet season and remain leafless until the next rains are due to begin. These are **deciduous** trees and shrubs (e.g. *Ceiba pentandra*, Silk Cotton Tree).

4

Modified organs

The primary function of a leaf is to carry out photosynthesis. The primary function of a stem is to support the leaves. The primary function of a root is to attach the plant and to absorb water and dissolved substances from the soil. The structure of each of these three types of organ is usually well adapted to carry out its primary function or functions, as will be shown in Chapters 7, 8 and 9.

This chapter is concerned with the ways in which the morphology of organs may be modified in relation to functions which are not their primary functions in typical dicots and monocots.

We can divide special (i.e. unusual) functions of organs into two main categories. In the first category are those in which an organ is modified so that it carries out what is normally the function of another type of organ, i.e. an interchanged function. For instance, in some plants the stems are the main photosynthetic organs, and therefore carry out what is normally the primary function of the leaf. In the second category, organs are so modified that they are able to carry out additional or substitute functions, such as enabling the plant to climb, or protecting the plant from predators.

Organs may be modified to such an extent that it is impossible to tell whether they are leaves, stems or roots simply from their appearance. There are a few simple rules which usually make it possible to identify the true nature of a modified organ.

A **root** has a root cap and has no nodes and no internodes. If it is not part of the primary root of a plant, it arises from another organ by breaking out from inside the tissues and not as a simple outgrowth of the surface.

A **leaf** is attached to a stem at a node and has one or more buds in its axil. A leaflet is attached to the petiole or rachis of the leaf of which it is part, and there is no axillary bud where it is attached. Stipules are always paired, one at each side of a leaf base.

A **stem** has nodes and internodes and bears leaves, which may be reduced to scales or leaf scars. If it is not part of the main axis of the plant, it originates in the axil of a leaf.

Special, interchanged functions

Photosynthetic roots

Many **epiphytic** orchids, orchids which grow attached to the branches of trees, have **aerial roots** which hang down exposed to light. Most of these plants (e.g. *Bulbophyllum* spp., Fig. 4.1) have well-developed leaves but in *Microcoelia* spp. (Fig. 4.2) the leaves are reduced to small brown scales, the stems are small and the aerial roots are the main photosynthetic organs. The aerial roots of epiphytic orchids are white when dry, because the green tissue

4.1 Epiphytic orchid, *Bulbophyllum* sp. Labels: (a) creeping stem attached by adventitious roots, (b) lamina of fleshy leaf, (c) fleshy, green pseudobulb (swollen internode of stem), (d) inflorescence axis from base of pseudobulb

4.2 Epiphytic orchid, *Microcoelia* sp. Labels: (a) attachment by adventitious roots, (b) hanging inflorescence. All the leaves are reduced to scales and the roots and inflorescence stalks are the only green parts

4.3 Prop roots of *Zea mays*. The youngest roots (arrowed) have not yet grown down to the soil

is covered with several layers of air-filled cells (**velamen**). The green colour can be seen when the roots are wet.

Roots which support the plant

The support of the plant in an upright position is usually the function of the stem. However, there are several examples of plants in which roots grow down from overground parts of the stem and give added support. *Zea mays* and related grasses have **prop roots**. These are adventitious roots which grow out from the stem at an angle from immediately above nodes, at successively greater heights above the ground (Fig. 4.3). Once they reach the soil these prop roots form branches and become firmly attached in the soil, giving added support to the stem. Prop roots are formed by *Rhizophora* spp. (Red Mangroves, Fig. 4.4). These trees grow in the muddy soil of mangrove forests along the coast and in estuaries. The mud on which they grow is constantly washed over by salty water. Once the trees are established and have started to branch, prop roots grow out from the undersides of branches, down to the mud. When they reach the mud the prop roots branch profusely, so that each Mangrove tree is supported on a raft of intertwined roots.

Some forest trees (e.g. *Uapaca guineense*) have **stilt roots**. In these trees the trunk may be quite thick at a height of two or three metres above the ground, but it tapers down to a very small diameter at ground level. The tapering part of the trunk bears many woody adventitious roots which grow down to the ground at an angle, so that, partly because of the tapering of the trunk, the whole tree seems to stand on stilts.

Leaves which absorb water

In the family Bromeliaceae, to which the pineapple

4.4 Prop roots of *Rhizophora* sp. (Red Mangrove)

4.5 Supporting leaf bases of *Musa* sp. TS 'stem' of a young plant, which consists entirely of concentric, overlapping leaf bases

(*Ananas sativus*) belongs, roots serve to attach the plant but carry out little or none of the absorbing function. Most such plants are epiphytes or lithophytes (grow on rocks). The absorbing function is carried out by the upper surfaces of the cup-like leaf bases.

Leaves which support the plant

The 'stem' of the Banana (*Musa* spp., Fig. 4.5) is composed entirely of tightly overlapping leaf bases until the plant starts to form an inflorescence. Before this stage is reached, the true stem is all underground and it is the leaf bases which support the leaf blades, often two or three metres above the ground. Later, the inflorescence stalk grows up through the centre of the overlapping leaf bases, appears at the top of the plant, and then produces its flowers and fruits.

Stems which attach the plant

Some underground stems (**rhizomes**) branch profusely through the soil and, with the aid of short adventitious roots, help to attach the plant firmly in the soil (e.g. *Imperata cylindrica*, Spear Grass).

Photosynthetic stems

Most young stems are green because the cells near the surface contain chloroplasts. Therefore, they carry out photosynthesis. In the presence of normal leaves, such young stems cannot be considered to be more than supplementary to the photosynthetic function of the leaves. Many plants however, usually those which live in dry places, have much reduced leaves. The stems are green and usually modified in such a way that they become the principal photosynthetic organs of the plant.

These photosynthetic stems are often flattened and expanded (e.g. many cacti; *Asparagus* spp.) or deeply lobed (e.g. *Euphorbia* spp.) but in some other plants they are merely thick and fleshy (e.g. *Ampelopsis* spp.). When the modified stem is composed of one internode only, it is a **cladode** (e.g. *Asparagus* spp., Fig. 4.6). A cladode is carried in the axil of a scale leaf and does not bear leaves. When the modified stem comprises several nodes and internodes, it is a

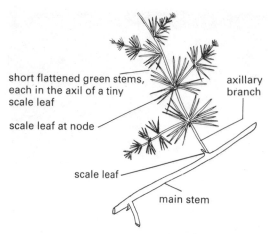

short flattened green stems, each in the axil of a tiny scale leaf

axillary branch

scale leaf at node

scale leaf

main stem

4.6 Photosynthetic stems (cladodes) of *Asparagus* sp. Each flattened, leaf-like structure is a single stem internode

phylloclade (e.g. *Euphorbia* spp., Fig. 4.7). A phylloclade has more or less reduced leaves at each node, but these may drop off soon after they appear and leave only leaf scars.

Special, additional or substitute functions

'Breathing' roots

Avicennia nitida (White Mangrove) grows on soil which is regularly flooded with salty water, if only during the wet season. The root system of *Avicennia* sp. is spreading and many roots run horizontally just below the surface of the mud. Upright (**negatively geotropic**) roots grow upwards from the submerged roots, until their tips are above the surface of the water, at its highest level. These breathing roots (**pneumatophores**) have many lenticels in their outer tissues, through which air may diffuse. Parts of plants which are submerged in water are not able to exchange gases with the atmosphere directly. The lenticels in the pneumatophores of *Avicennia* sp. connect with the intercellular spaces in the main part of the root system, so that exchange of gases can take place. The prop roots of the Red Mangrove also have many lenticels and serve the same function.

Defensive structures

Many plants have sharp structures which certainly

4.7 Photosynthetic stems (phylloclades). The lobed, green stem of *Euphorbia* sp. with small green leaves in the growing region and leaf scars (arrowed) below

assist in protecting them from browsing animals. Similar sharp structures are also possessed by scrambling plants (see below) and it is difficult to distinguish the two functions, except when a plant is definitely not a scrambler.

Prickles are superficial outgrowths (**emergences**) from the surface layers of a stem, leaf or petiole. They are identified as such because their positions are not related to the positions of leaves and axillary shoots. The many examples include *Caesalpinia pulcherrima* (Pride of Barbados), *Solanum macranthum* (Potato Tree) and several other wild species of *Solanum*. Prickles also occur on the trunks of forest trees (e.g. *Ceiba pentandra*, Silk Cotton Tree; *Fagara macrophylla*), in which case they are commonly called 'thorns'.

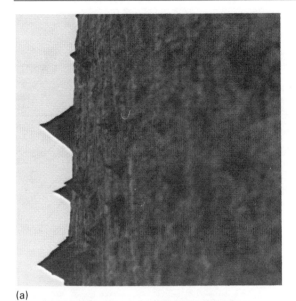

(a)

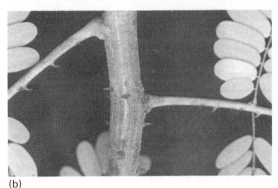

(b)

4.8 Prickles (emergences). (a) large, woody prickles on the trunk of *Ceiba* sp. (b) sharply recurved prickles on the stem and leaf rachis of *Caesalpinia decapetala*, a scrambling shrub

(a)

(b)

4.9 Thorns. (a) *Citrus* sp. with axillary stems modified as thorns. (b) *Bougainvillaea* sp. with thorns from supernumerary axillary buds

True **thorns** are modified stems. Axillary shoot thorns are found in some *Citrus* sp. and in cacti. *Bougainvillaea* sp. also has thorns, derived from supernumerary axillary buds and displaced up the stem, some distance above the axil. However, the thorns on this plant may be related to the scrambling habit rather than to protection, though they undoubtedly serve both functions.

Spines are modified leaves, or parts of leaves. Some *Citrus* spp. have spines which replace leaves and have axillary buds. Several fleshy *Euphorbia* spp. have paired stipular spines at the base of each leaf,

as have *Acacia* spp. common as shrubs in the savanna.

Examples of prickles, thorns and spines are illustrated in Figs. 4.8 to 4.10.

Climbing plants
Climbing plants have weak stems with long internodes and hold themselves up by attaching themselves firmly to supports, usually parts of other plants.

Twining plants (e.g. *Dioscorea* spp., Yams; *Ipomoea* spp.) have stems which twine around any support which is available.

Root climbers (e.g. *Culcasia scandens* and many

4.10 Stipular spines. *Acacia* sp., savanna shrub, with large, woody stipular spines at the base of each compound leaf

4.11 Leaf-tip tendrils of *Gloriosa* sp.

other climbing Araceae) attach themselves by means of adventitious roots. Such roots produce many root hairs which grow into small crevices in the support and then become hardened.

Tendrils are specialized, elongated organs which are sensitive to touch. If a tendril makes contact with a potential support, it quickly coils around the support. In some plants of the pumpkin family (Cucurbitaceae) the tendril is able to coil twice round a support within five minutes of first contact. After attachment has been completed by the tip of the tendril, the lower part of the tendril then shortens by forming spirals. This adds springiness to the connection and avoids the possibility of damage due to sudden stress.

Tendrils are formed by different organs in different plants. **Leaf-tip tendrils** (Fig. 4.11) are formed from the tips of leaves (e.g. *Gloriosa superba*, Flame Lily; *Flagellaria* sp.). **Leaflet tendrils** replace leaflets of compound leaves (e.g. *Bignonia* spp.). *Smilax* sp. has paired **stipular tendrils**, one on each side of the leaf base (Fig. 3.5). *Luffa* spp. and other members of the Cucurbitaceae have branched tendrils which are probably modified leaves, though their nature is difficult to determine according to the rules set out early in the chapter. They arise singly, next to leaf bases and their internal structure resembles that of a petiole at the base, and that of a leaf vein in each of the branches.

Clematis spp. have **petiolar tendrils**. The leaf is compound and the petioles and stalks of leaflets coil around parts of other plants and give support.

Stem tendrils are formed either in the axillary

position or they are leaf-opposed. *Passiflora* spp. (Passion Fruit, Grenadilla) have axillary stem tendrils. In *Cardiospermum* sp. and *Quisqualis* sp. the tendrils are parts of axillary branches forming inflorescences. The leaf opposed tendrils in the family Ampelidaceae (e.g. *Ampelocissus* spp.) are attached to the main stem of the plant exactly opposite the attachment of a leaf. These are formed from the apex of the main stem, which then ceases to grow. The axillary bud then grows into an apparent continuation of the main stem and the tendril is turned into the leaf-opposed position. Examples of stem tendrils are shown in Fig. 4.12.

Scrambling plants

Scrambling plants are different from climbers in that they do not attach themselves to their supports. The main shoots of scramblers grow rapidly upwards between the branches of other plants, and they are prevented from slipping backwards by sharp or recurved (hook-like) structures.

In the garden shrub *Holmskioldia* sp. the petioles are stiff and curved backwards and bear normal leaf blades. In the young growth of the savanna scrambler *Combretum* sp. petiolar hooks have no leaf blades or have very much reduced leaf blades. Other scramblers have recurved lateral branches (e.g. *Canthium* sp.), recurved prickles (e.g. *Caesalpinia bonduc*), thorns (e.g. *Bougainvillaea* sp.), or spines (e.g. *Asparagus* spp.).

(a)

(b)

4.12 Stem tendrils. (a) from axillary bud on the main stem, *Passiflora* sp. (b) leaf-opposed tendrils of *Cissus* sp.

Organs of storage, perennation and vegetative reproduction

Perennation is survival from one growing season, through the adverse season (dry season in the tropics), to the next growing season. Vegetative reproduction is the production of new individual plants by growth and separation into parts, without any sexual process. Most storage organs are organs of perennation and many are also organs of vegetative reproduction.

Roots of most plants act as storage organs, without being modified in any way. In *Daucus carota* (Carrot) the tap root is modified by being greatly swollen. A swollen tap root is also found in *Beta vulgaris* (Beetroot) and *Talinum triangulare* (Water Leaf). In both Carrot and Beetroot, the storage organ starts to form at a very early stage and the top part is the swollen hypocotyl.

Root tubers are formed in several plants of economic importance. A root tuber is a swollen lateral or adventitious root. The edible root tubers of *Manihot* spp. (Cassava) and *Ipomoea batatas* (Sweet Potato) are both swollen adventitious roots. They can be recognized as such by their manner of origin from the stem and the fact that they do not have nodes and internodes. Pieces of the tuber of Sweet Potato will produce shoots, if they are planted, but these come from adventitious buds, not from axillary buds. These root tubers are organs of storage and perennation rather than organs of vegetative reproduction.

Examples of storage roots and root tubers are shown in Figs. 4.13 and 4.14.

Rhizomes are underground stems. Those of many plants are swollen with stored food (e.g. *Canna indica*, Canna Lily, *Zingiber officinale*, Ginger; Fig. 4.15). The rhizome of the Canna Lily grows roughly parallel to the surface of the ground and bears scale leaves and adventitious roots. The tip of each rhizome ultimately grows up and forms a leafy branch, while lateral branches, in the axils of scale leaves, continue to grow underground. This sort of rhizome is a storage organ, an organ of perennation and an organ which effects vegetative reproduction, when old parts of the rhizome die and decay leaving several separate rhizomes where there was only one before.

Stem tubers are the swollen tips of underground stems. *Solanum tuberosum* (Irish Potato) is a good example. In this plant, axillary shoots are formed underground and grow out horizontally or downwards. The tip of each shoot swells with stored food, forming a tuber. If left in the ground until the next growing season then the tuber produces new leafy shoots from buds in the axils of scale leaves.

The tubers of *Dioscorea* spp. are variously described as root tubers and stem tubers. Recent work has shown that they are neither, strictly, modified stems nor modified roots. A Yam is a lateral swelling from a hypocotyl or the internode of a stem

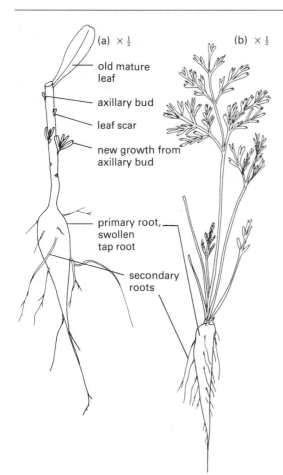

4.13 Swollen tap roots. (a) *Talinum triangulare*, an old plant broken off above the soil, with new axillary shoots developing, using stored food from the swollen tap root. (b) *Daucus carota*, a biennial, during its first season of growth

4.14 Root tubers of *Manihot* sp.

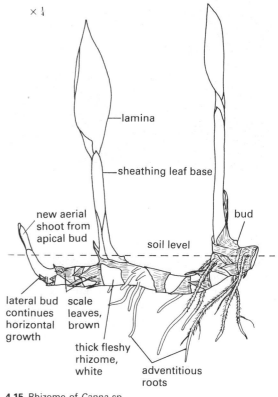

4.15 Rhizome of *Canna* sp.

and is composed entirely of secondary tissue (Chapter 10). Like those of the Sweet Potato, buds which are formed on the surface of a Yam tuber are adventitious.

The tubers of Irish Potato and Yam are organs of perennation and of vegetative reproduction, as well as storage organs. Figure 4.16 shows the structure of the tuber of the Irish Potato.

Corms (Fig. 4.17) are short, greatly swollen underground stems. Typical corms (e.g. *Gladiolus* sp.) are vertical, but the storage organ of *Colocasia* sp. (Cocoyam) may also be interpreted as a corm, although it grows obliquely in the soil. A corm has several nodes, each with a scale leaf. There is an

(a)

(a)

(b) × 1

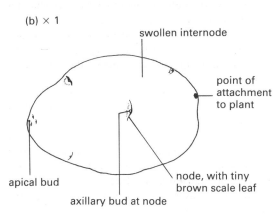

4.16 Stem tubers (a) *Solanum tuberosum* (Irish Potato), tuber exposed at the base of the plant. (b) tuber of *S. tuberosum*

(b) × ½

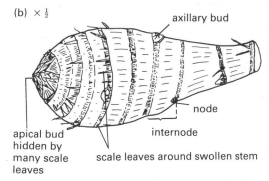

4.17 Corms. (a) *Colocasia* sp., young corms growing from the base of the parent plant. (b) corm of *Colocasia* sp.

apical bud at the end and this grows out, at the start of the growing season, to form a leafy plant. Buds in the axils of scale leaves then form new corms, which will be ready to produce new leafy shoots at the beginning of the next growing season.

The vertical corm of *Gladiolus* sp. shrivels up after it has formed a leafy shoot. A new corm is produced, above the remains of the old one, from the base of the leafy stem. The base of the new corm produces

adventitious roots which attach themselves in the soil and then shorten. These **contractile roots** pull the corm down, deeper into the soil, so that the new corm is at about the same distance below the surface of the soil as the old one was, before it shrivelled. Corms are organs of storage, perennation and vegetative reproduction.

Bulbs have a short upright stem which bears swollen leaf bases or scale leaves in which food is stored. The food in the leaf bases is used up when a leafy shoot is produced, and new bulbs are formed at the base of the new leafy shoot and from axillary

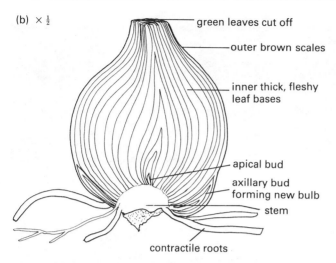

(b) × ½

green leaves cut off

outer brown scales

inner thick, fleshy leaf bases

apical bud

axillary bud forming new bulb

stem

contractile roots

4.18 Bulbs. *Hippeastrum* sp., LS mature bulb

buds. Adventitious roots are formed each year and like those of the corm, may be contractile. Examples of bulbs are *Allium cepa* (Onion) and *Haemanthus* spp. (Blood Lilies). Bulbs are also organs of storage, perennation and vegetative reproduction (Fig. 4.18).

Plants with underground organs of perennation (bulbs, corms, rhizomes) are especially common in the savanna areas of Africa. They produce first flowers, then leafy shoots, each year, before the savanna grass has grown tall. Their underground parts are able to survive the burning of the savanna because they are protected by the soil. Other savanna plants have large underground woody root stocks (**xylopodia**) which are protected from fire and send up new aerial shoots each year.

Organs of vegetative reproduction only

Organs of vegetative reproduction are not always storage or perennating organs.

Runners (**stolons**) are axillary shoots with long internodes which grow along the surface of the ground, and root at the nodes (e.g. *Cynodon dactylon*, Bahama Grass, Fig. 4.19). Axillary buds at some of the nodes develop into new plants, which ultimately become separated from the parent plant by breakage or decay of the internodes of the prostrate stem.

Offsets originate above the soil in the same way as runners. They are usually shorter and thicker than runners. New plants are formed when the apical bud turns upwards, putting down roots and growing into a new plant (e.g. *Sansevieria* spp.)

The lateral, horizontal shoots of *Pistia stratiotes* (Water Lettuce) are usually interpreted as stolons, though they are formed under the surface of water.

Suckers are formed from axillary buds which are under the surface of the soil and grow upwards at an angle. *Musa* spp. (Bananas, Plantains, Fig. 4.20) form suckers, which grow up around the parent plant and take its place when it has fruited and died.

In addition to terminal and axillary buds, some plants form **adventitious buds** on parts of roots, stems and leaves. Adventitious buds grow into adventitious shoots.

Root suckers are adventitious shoots that arise from the roots of a plant. The leguminous tree *Millettia* sp. has a shallow, spreading root system which produces root suckers. They grow into new young trees a short distance from the parent tree.

Adventitious shoots may also arise from broken parts of roots (e.g. Sweet Potato) and from cut or broken woody stems (e.g. *Cassia* spp.).

Adventitious buds are also formed on some succulent plants, on their leaves. In *Kalanchoe* spp. and in some closely related *Bryophyllum* spp. adventitious buds are formed while the leaves are still attached to the plants. The buds form at the leaf edge, producing short shoots which develop adventitious

(a) × ¼

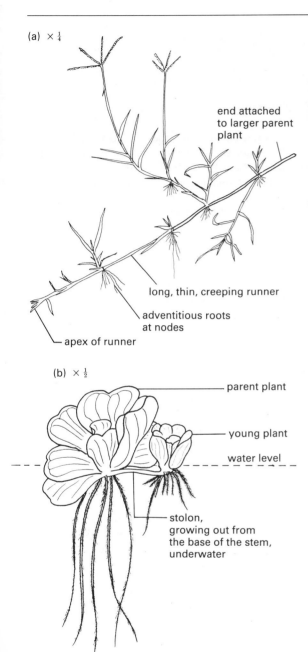

end attached
to larger parent
plant

long, thin, creeping runner

adventitious roots
at nodes

apex of runner

(b) × ½

parent plant

young plant

water level

stolon,
growing out from
the base of the stem,
underwater

4.19 Runners (stolons). (a) *Cynodon dactylon*. (b) *Pistia* sp.,
a floating water plant with stolons growing out under
water

4.20 Stem suckers. *Musa* sp. with a young plant growing
up beside the parent from the corm-like rootstock below

roots at the base of the shoot, after which the young
shoot drops off and roots in the soil. In *Bryophyllum
pinnatum* buds are formed after leaflets have been
shed from the plant. Detached leaves of *Begonia* spp.
form adventitious buds from the surface of the
lamina or from the petiole, usually close to places
where the tissues have been broken.

Bulbils are sometimes produced on aerial parts of
plants. *Agave* sp. (Sisal) has large inflorescences
which produce many flowers. In addition, some of
the tips which would normally develop into flowers
form small bulbs instead (Fig. 4.21). Each small
bulb or bulbil is made up of a short stem and a few
reduced leaves. Similar bulbils are found in the
inflorescences of some varieties of *Allium* sp. (Onion)
and *Crinum* sp. (Crinum Lily).

Bulbils which resemble tubers are formed from
axillary buds on the stems of some yams.

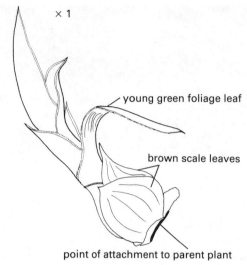

× 1

young green foliage leaf

brown scale leaves

point of attachment to parent plant

4.21 Bulbils: single detached bulbil of *Agave* sp.

Key words

Primary functions: leaf, photosynthesis; stem, support; root, attachment and absorption. **Special functions:** interchanged functions, additional functions, substitute functions. **Interchanged functions:** roots, photosynthetic, supporting; leaves, absorbing, supporting; stems, attaching, photosynthetic. **Additional functions:** ventilation ('breathing'), defence, climbing, scrambling, storage, perennation, vegetative reproduction. **Structures:** roots, aerial root, prop root, stilt root, pneumatophore; leaves, spine, tendril; stems, cladode, phylloclade; thorn,

tendril. **Storage organs:** tuber, rhizome, corm, bulb. **Vegetative reproduction:** runner (stolon), offset, sucker, bulbil.

Summary

The primary functions of leaves, stems and roots are photosynthesis, support and attachment and absorption respectively. Each of the three types of organ can take on functions which normally belong to other organs. Some organs are modified in such a way that they carry out additional functions. The true nature of a modified organ (leaf, stem or root) can be identified by studying morphological and anatomical features of the modified organ. Some organs are modified in relation to storage and vegetative propagation. Vegetative propagation offers a rapid means of increasing numbers of plants.

Questions

1 Suppose you find a plant which has modified organs which do not, at first, look like leaves, stems or roots. What features would you look for, on and around the organ, in order to decide whether it was (a) a modified leaf, (b) a modified stem, or (c) a modified root.

2 Look for a plant with sharp organs or tendrils and apply your answer to the first question to decide the true nature of the modified organs.

3 A Yam or Sweet Potato tuber does not bear buds when it is formed. However, either plant can be propagated by planting parts of tubers. What sort of buds grow to form the green parts of the plant?

5

The plant cell

The **cell** is the smallest structural unit of a living organism that can exist completely or partly independently. An organism is made up of one or more cells. In a one-celled (**unicellular**) organism, the one cell performs all the living functions of the organism. In a many-celled (**multicellular**) organism, there are several or many different types of cells. Each type of cell is **specialized** and carries out only one or a few functions. A group of cells, all of one type or of several types together, forms a **tissue**. Several tissues are grouped together to form an **organ**. Organs make up the **organism**. The cells of all eucaryotic organisms, plants, animals and fungi, are very similar in most respects, but differ in detail. Differences lie in the presence or absence of certain structures, for example cell walls and sap vacuoles, both of which are possessed by plants and fungi, but are absent in animal cells.

Cells and microscopes

The unaided human eye can distinguish objects as small as about 0.1 mm (100 μm), but no smaller. Some plant cells are as large as this, but most are smaller and the structures they contain are smaller still. For this reason, the study of cells has always depended on the quality of microscopes available.

Light microscopes

Until about 1940, the only microscopes in use formed a magnified image by means of glass lenses, using visible light. Microscopes of this type are **light microscopes**, such as we use in the laboratory for most purposes today. The useful magnification which can be obtained depends on the properties of light, especially its wavelength. In fact, it is the **resolving power** of a microscope which determines the highest magnification that can be usefully obtained. The resolving power of a microscope is its ability to form separate images of two or more objects which are close together. The resolving power of a light microscope is about 0.2 μm. Objects separated by less than this distance appear as a single

object. Thus, at a magnification of about 600 times ($\times$ 600), the maximum resolving power is utilized. Magnifications as high as 1000 times are commonly used, however, as the larger image is easier to observe. At greater magnifications, objects become more and more blurred and no greater detail is seen.

Electron microscopes

Electron microscopes, in which an image is formed on a fluorescent screen or on photographic film by a beam of electrons, first came into use in the 1940s. The wavelength of a beam of electrons is less than one thousandth of that of visible light and much greater resolving power is the result. The highest resolving power so far achieved is about 0.3 nm (3 Å). By taking photographs at high magnifications and then making photographic enlargements, useful magnifications of about one million times can be obtained.

Structure observed by use of the electron microscope (EM) is often referred to as **ultrastructure**. In our study of cells, we shall include references to structures visible under the light microscope (e.g. Fig. 5.1) and to ultrastructural features, visible only with EM (e.g. Fig. 5.2).

The generalized plant cell

Since the cells of multicellular green plants are all specialized and differ from each other in structure and function, it is impossible to choose any one cell type as 'typical'. Instead, we must describe a hypothetical, 'generalized' plant cell, constructed to show all the various features of all types of plant cell.

The appearance of cells from a staminal hair of *Setcreasea* sp., as seen through a student microscope, is shown in Fig. 5.1. Since this type of microscope is not ideal and this type of cell could not be described as generalized, only some of the main features are seen. An EM photograph of a section through cells in a root is shown in Fig. 5.2.

When we are describing structures that occur in plant cells, it is necessary to remember that any

Table 5.1 Visibility of plant cell structures with the light microscope (LM) and electron microscope (EM) after suitable staining.

Structure	Visibility	Comments
middle lamella	Visible with LM	
cell wall	Visible with LM	
pits and plasmodesmata	Visible with LM	
plasmalemma	Visible only with EM	Often apparently visible with LM, due to difference in density of materials on the two sides.
endoplasmic reticulum	Visible only with EM	
Golgi apparatus	Visible with LM	Membranes visible only with EM
plastids	Visible with LM	Internal structure visible only with EM
mitochondria	Visible with LM	Internal structure visible only with EM
microbodies	Visible with LM	Structure visible only with EM
ribosomes	Visible only with EM	
tonoplast	Visible only with EM	See comment for plasmalemma above.
nuclear envelope	Visible only with EM	See comment for plasmalemma above.
nucleolus	Visible with LM	
chromosomes	In non-dividing (interphase) nucleus, visible only with EM	Visible with LM in a dividing nucleus, at mitosis or meiosis.

(a)

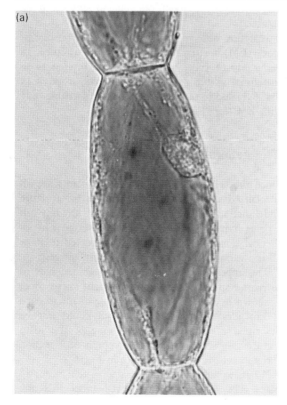

(b)

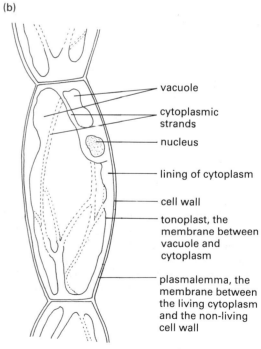

vacuole

cytoplasmic strands

nucleus

lining of cytoplasm

cell wall

tonoplast, the membrane between vacuole and cytoplasm

plasmalemma, the membrane between the living cytoplasm and the non-living cell wall

5.1 Cells from staminal hair of *Setcreasea* sp. (cf. *Tradescantia* sp., Commelinaceae): (a) Photomicrograph (HP); (b) drawing

structure with a diameter of less than about 0.2 μm is visible only with EM and beyond the resolving power of the light microscope. Plant cell structures seen with a light microscope and those visible only with EM are listed in Table 5.1

The cell wall and middle lamella

The cellulose walls of neighbouring cells are separated from each other by a **middle lamella** composed of pectin and calcium pectate. The middle lamella is seen, with the light microscope, as a dark line between the cellulose walls. Its origin, at the time of cell division, is described in Chapter 6.

The **primary cell wall**, which develops continuously until a cell attains its full size, is composed mostly of cellulose microfibrils. Cellulose **microfibrils** are about 25 nm in diameter and are laid down in all directions, without any special orientation. There are many pores between the microfibrils, in the same way as there are pores between the fibres which make up a piece of woven cloth or a piece of filter paper. The primary wall is not of uniform thickness but includes many thin places (**pits**) through which fine strands of cytoplasm (**plasmodesmata**, *sing.* **plasmodesma**) pass from cell to cell. The cell wall, pits and plasmodesmata can be observed with the light microscope.

The cell wall provides some mechanical support for the cell as a whole and determines the shape of the cell. It also prevents the cell from bursting when hydrostatic pressure builds up within the cell. The cell wall is not a physiological barrier. Water and dissolved substances pass freely through the fine pores formed by the network of microfibrils of which the cell wall is composed. Since the wall is more or less impenetrable to larger solid bodies, it probably gives some protection against invading microorganisms. It encloses the plasmalemma with the protoplast or protoplasm (= cytoplasm + nucleus).

The cytoplasm

Because of the presence of plasmodesmata between all living cells, there is an unbroken continuity (**continuum**) of cytoplasm throughout the whole plant body. This continuum is described as the **symplast** of the plant.

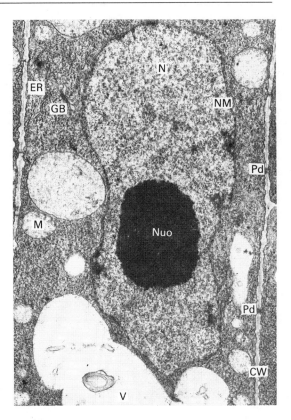

5.2 EM photograph of a cell from the root of a plant. Labels: CW = cell wall; Pd = plasmodesmata; M = mitochondria; ER = endoplasmic reticulum; GB = Golgi body; V = vacuole; N = nucleus; NM = nuclear membrane Nuo = nucleolus. Magnification × 8200. Reproduced from Hall, J. L., Flowers, T. J. and Roberts, R. M. (1982), *Plant Cell Structure and Metabolism*, 2nd edn, Longman, London

Organelles and inclusions

Cytoplasm is the site of most of the physiological activities of the cell. Its detailed structure therefore varies according to the type of cell concerned. Basically, however, there is a ground substance of fluid consistency in which metabolic enzymes are dissolved. Suspended in this ground substance are several types of minute particles and structures. Some people refer to all these cytoplasmic particles, collectively, as cytoplasmic inclusions. Others separate them into two categories:

(a) **Organelles** – specialized parts of the living substance of the cell.

(b) **Inclusions** – accumulations of food materials and stored secretory substances which are not part of the living material of the cell.

We shall make this distinction between cytoplasmic organelles and cytoplasmic inclusions.

Membranes

The cytoplasm is rich in membranes. Several types of organelles are limited by membranes and some of them have membranes as part of their internal structure. In addition, the entire protoplasm is enclosed in a membrane, the **plasmalemma** (**plasma membrane**). Many of the enzymes which control chemical activity within the cell are located on or associated with membranes. Membranes are therefore of central importance in the study of the physiology of the cell.

The concept of **unit membrane** (UM) is based on the fact that, in all organisms, the membranes which form the plasmalemma and other cytoplasmic structures all give a very similar picture when examined with EM. The essence of the UM hypothesis is that all membranes have the same structure and all are composed of protein and lipid. This was proposed by J. F. Danielli in 1934, before practical electron microscopes were available.

High resolution EM photographs of sections through UM show the membrane as a thin, double line. The two dark lines are interpreted as protein layers, together with the water-attracting (**hydrophilic**) ends of phospholipid molecules. The lighter, central region of the membrane is interpreted as a double layer (**bilayer**) of the non-polar, fatty acid components of the phospholipid molecules. Thus, the bilayer consists of phospholipid molecules with their water-repelling (**hydrophobic**) 'tails' pointing to the centre of the membrane and their hydrophilic 'heads' next to the protein, towards the two sides of the membrane. A very simplified diagram of this model of the structure of UM is shown in Figure 5.3. This is also the very structure predicted by Danielli on the basis of biochemical and biophysical evidence.

It is now thought that this postulated structure of UM, although observable in sections which have been fixed and prepared for sectioning in the usual way, is not a true reflection of the structure of UM in the living cell. In 1972, new and sophisticated EM techniques led to its replacement by the **mosaic bilayer** (**fluid mosaic**) model of S. J. Singer and G. L. Nicholson. According to this model, a phos-

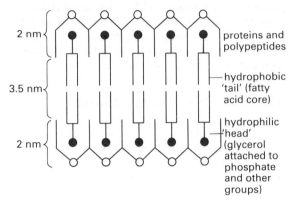

5.3 Interpretation of the structure of the cell membrane

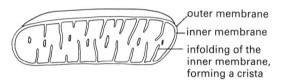

5.4 Mitochondrion: structure as seen with EM

pholipid bilayer forms the basis of UM, exactly as described and illustrated above. However, instead of coating the surfaces of the bilayer, protein molecules are partially embedded in the phospholipid on both sides of the membrane, leaving much of the phospholipid bilayer exposed. Some of the protein molecules extend all the way through the bilayer, from one face to the other, and thus form pathways for the movement of water and hydrophilic substances through the otherwise phospholipid membrane. There is evidence that the proteins embedded in one side of the membrane may differ from those embedded in the other side, and that UM from different cell components have different types and amounts of protein.

The plasmalemma

The plasmalemma separates the cytoplasm from the non-living cell wall and is also continuous with the plasmalemma of neighbouring cells, through plasmodesmata. This membrane regulates or selectively controls molecular movements into and out of the cell.

The mitochondrion

A **mitochondrion** (*pl.* **mitochondria**) is sometimes spherical in shape but is usually elongated, about 2 to 4 μm in length and about 0.5 μm in diameter. It is made up of two membranes. The outer membrane is smooth, the inner membrane is thrown into folds called **cristae** (*sing.* **crista**). The cristae contain dense fluid and their inner surfaces are covered with stalked granules which are involved with the synthesis of the important, energy-containing compound known as ATP. Under the light microscope, mitochondria appear as dots or as short, thin threads. Their internal structure can be seen only in sections examined with EM.

The number of mitochondria contained in a plant cell varies according to the level of metabolic activity in the cell, but it is usually between fifty and 5000.

Mitochondria contain some deoxyribonucleic acid (DNA). DNA is best known as the fundamental genetic material of the nucleus. Nuclear DNA directs the synthesis of proteins in the cytoplasm as a whole. Mitochondrial DNA is believed to control the synthesis of some proteins within the mitochondrion.

Mitochondria are the site of **aerobic respiration** in the cell. They control the breakdown of respiratory substrates (e.g. glucose) in the cell and provide the energy for making ATP. The energy incorporated in ATP is, in turn, used to drive other syntheses and reactions requiring energy. Hence, the mitochondrion is often described as the powerhouse of the cell.

The structural features described above are illustrated in Figure 5.4.

The chloroplast and other plastids

The **chloroplast** of a higher plant is a circular, biconvex structure limited by a double membrane. The average size is about 5 μm in diameter and about 3 μm in thickness. The chloroplast as a whole is therefore well within the resolving power of the light microscope, but details of its internal structure can be observed only with the EM. The cell of a flowering plant may contain up to one hundred chloroplasts.

Within the inner membrane of the chloroplast is a ground substance (**stroma**). The stroma contains proteins, most of which are enzymes. Within the stroma are interconnected piles of sac-like discs (Fig. 5.5). The piles of discs (**grana**, *sing.* **granum**) are interconnected at various levels by pairs of membranes (double lamellae) or single sacs, known as **frets** or **stromal lamellae**. The individual sac-like discs which make up a granum are known as **thylakoids** or **granal lamellae**. Chlorophyll molecules are embedded between the granal lamellae. **Chlorophyll** is the green pigment which is responsible for photosynthesis. Other pigments present in the chloroplast are **carotene** and **xanthophylls**, which do not play a direct role in photosynthesis.

The essential structure of the chloroplast is shown in Figure 5.5.

Like the mitochondrion, the chloroplast contains its own DNA, but nuclear DNA is responsible for the programming of the synthesis of most of the substrate, structural and enzymatic proteins. The light-requiring reactions of photosynthesis take place in the grana, where chlorophyll is present. Other reactions take place in the stroma.

Chloroplasts increase in number by **fission**. When a chloroplast has reached its full size, both membranes become constricted at the centre of the chloroplast, which separates into two parts.

Chloroplasts are **plastids** which contain chlorophyll. Other types of plastids are amyloplasts, leucoplasts and chromoplasts.

Cells from regions of a plant in which cells are constantly dividing do not contain fully developed plastids. Instead, they contain proplastids. **Proplastids** are small bodies, about 1 μm in diameter, limited by a double membrane. The type of plastid formed from a proplastid as a cell matures depends on the position of the cell in the plant. In most tissues which are exposed to light, the proplastid enlarges and the inner membrane is thrown into folds, which form the complex, chlorophyll-containing membrane system which is characteristic of the chloroplast. In the absence of light, the proplastid may develop into a leucoplast. A **leucoplast** is smaller than a chloroplast and has only a rudimentary internal membrane system, without chlorophyll. A leucoplast is capable of developing into a chloroplast, in the presence of light. Alternatively, a proplastid may form an **amyloplast**, with a rudimentary membrane system without chlorophyll and containing one or more grains of starch.

A chloroplast may, under certain conditions, form a different coloured plastid (**chromoplast**). This most frequently happens in parts of flowers and in ripening fruits. The chlorophyll-containing membrane

(a)

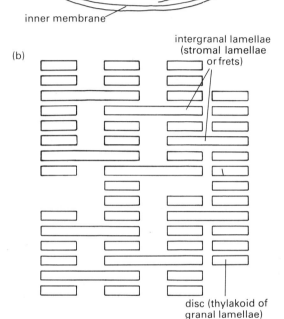

(b)

(c)

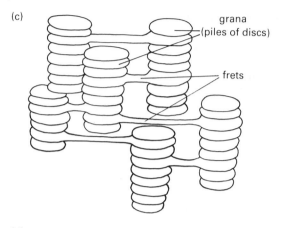

(d)

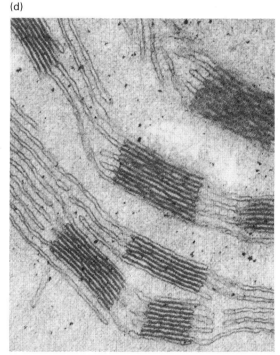

5.5 Chloroplast structure interpreted from EM photographs: (a) whole chloroplast in LS; (b) diagram to show sectional view of granal lamellae and intergranal lamellae (frets); (c) three-dimensional view of grana interconnected by frets; (d) EM photograph of part of a chloroplast of *Bougainvillaea* sp. (× 62 000), showing granal and intergranal lamellae, photography by J. G. Duckett and J. B. Kirkham

system degenerates and large lipid drops are formed in the stroma. The lipid contains a high concentration of carotene and/or xanthophyll, which gives the plastids, and the tissue containing them a bright yellow, orange or red colour. Chromoplasts are responsible for these colours in many flowers and fruits (e.g. *Lycopersicum esculentum*, Tomato). They are also responsible for the yellow or orange colour of the tap roots of *Daucus carota* (Carrot). Chromoplasts may also develop directly from proplastids.

Golgi apparatus

The **Golgi apparatus** or **Golgi body**, named after Camillo Golgi who first described the organelle in 1898, is made up of one or usually more than one unit known as a dictyosome. Each **dictyosome** in turn consists of several layers of flat, saucer-shaped sacs or vesicles called cisternae (Fig. 5.6). The **cisterna** is a membrane-enclosed structure with edges that are usually swollen and perforated (**fenestrated**). Numerous small, approximately spherical

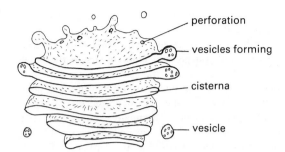

5.6 Golgi body (dictyosome): three-dimensional view of a sectioned dictyosome, interpreted from EM photographs

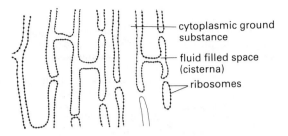

5.7 Endoplasmic reticulum: diagram to show section through several sacs of ER

vesicles arise from the cisternae by being pinched off from the edges.

The whole Golgi body is usually several micrometres in diameter and is therefore visible with the light microscope, after suitable staining.

It is thought that the Golgi body is continuous with the endoplasmic reticulum but, unlike the latter, it is never associated with ribosomes (see below for endoplasmic reticulum and ribosomes).

The Golgi body is responsible for the production of a number of secretions, including cell plate and cell wall material. Secretions leave the Golgi body in vesicles, which travel to the part of the cell where, for example, new cell wall material is being laid down.

Endoplasmic reticulum
The **endoplasmic reticulum** (ER, Fig. 5.7) is a system of paired membranes which spreads to a greater or lesser extent throughout the cytoplasm. The membranes are UM and form flattened sacs, tubes and small vesicles. The space (**cisterna**) between the membranes is fluid filled. Some parts of the ER have many granules called ribosomes (see below) attached to or associated with the outer surface. ER with attached ribosomes is known as **rough ER**. Other parts of the ER have no attached ribosomes and are known as **smooth ER**. ER is known to be continuous with the outer membrane of the nuclear envelopes and the cisternae of ER and nuclear envelope are continuous. The nuclear envelope may be considered as a specialized part of the ER.

The main function of the ER is synthesis. Secretory cells, where much synthesis is taking place, are always especially rich in ER. ER may also play a part in the compartmentalizing of different metabolic activities in different regions of the cytoplasm.

Ribosomes
Ribosomes are small bodies, about 20 to 25 nm in diameter, and are responsible for protein synthesis. They are composed of ribonucleic acid and protein and have no membranes. However, many are attached to the ER (rough ER). Smaller ribosomes, 16 to 17.5 nm, are found in mitochondria and plastids.

Microbodies
Small organelles, each limited by a single layer of UM, are collectively known as microbodies. Two types of microbodies, peroxisomes and glyoxysomes, will be dealt with here. These organelles are more or less spherical and are 0.5 to 1.5 μm in diameter. They appear to function as 'packets' of enzymes which are responsible for specific processes in the cell.

Peroxisomes are found mostly in photosynthetic cells in the leaves of higher plants and contain two enzymes associated with the process of photosynthesis. The first is **glycolate oxidase**, which controls a reaction in which hydrogen peroxide, a toxic substance, is produced as a by-product. Peroxisomes also contain **peroxidases** and **catalase**, enzymes which break down hydrogen peroxide to water and oxygen, both of which are harmless.

Glyoxysomes are structurally similar to peroxisomes, but contain enzymes responsible for reactions involved in the conversion of fats to sugars. Many glyoxysomes are found in the cells of fat-containing seeds like *Ricinus communis* (Castor Oil), in which they are particularly active during germi-

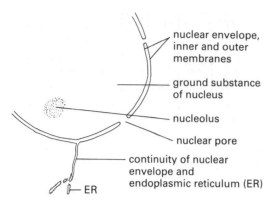

nuclear envelope, inner and outer membranes

ground substance of nucleus

nucleolus

nuclear pore

continuity of nuclear envelope and endoplasmic reticulum (ER)

ER

5.8 The nucleus and ER: diagram to show part of the nucleus, with the (double) nuclear envelope with pores, and the outer membrane of the nuclear envelope continuous with ER

nation, when fat reserves are converted to carbohydrate.

The tonoplast and vacuole

The membrane which separates the cytoplasm from the contents of the vacuole (sap vacuole) is the **tonoplast**. It is a cytoplasmic membrane made up of a single layer of UM, similar to the UM of the plasmalemma and ER but usually slightly thinner.

Immature, developing plant cells usually have several very small vacuoles in the cytoplasm. As a cell matures, these small vacuoles join together and increase in size to form a single, large, central vacuole. Vesicles from the Golgi apparatus and ER may also coalesce with the tonoplast and thus deposit their contents in the vacuole. The sap vacuole usually makes up 80 to 90 per cent of the volume of a mature cell.

The contents of the **sap vacuole** (**cell sap**) are generally regarded as non-living cell inclusions. The vacuole is a water reservoir containing storage materials and waste products of the cell. Enzymes present in the vacuole can change some harmful waste products into simple substances which can be used again in the cytoplasm.

Dissolved substances in the cell sap include mineral salts, various polysaccharides including sugars, and nitrogenous substances including proteins and anthocyanin pigments. Tannin and various acids may also be present. In addition to the dissolved materials, there may be droplets of oil, and crystals

of calcium oxalate are often found in the vacuoles of mature cells.

Anthocyanin pigments, which are water-soluble, give the red, purple and blue colours to the petals of some flowers and to the leaves of some flowering plants. The colour given by anthocyanin pigments varies according to the pH of the cell sap.

The pH of vacuoles frequently differs from that of the cytoplasm. That of the cytoplasm is stable and ranges from about pH 6.8 to about pH 8.0. The pH of the cell sap varies from about pH 1.0 to pH 10.0. The pH of the cell sap is usually on the acid side of neutrality, owing to moderate accumulations of citric, oxalic, tartaric and other organic acids.

Hydrostatic pressure exerted on the cell wall by the vacuolar contents, via the cytoplasm, is very important in maintaining the shape and rigidity of cells and organs.

The nucleus

The **nucleus** is the largest and most conspicuous structure within the cell. It is a spherical or ovoid body, enclosed in the nuclear envelope. The **nuclear envelope** is a double layer of UM, the outer layer continuous with ER. There are many pores (**nuclear pores**) in the nuclear envelope, which allow communication between the nuclear contents and the cytoplasm (Fig. 5.8).

Within the nucleus, there are usually one, two or more nucleoli, which stain deeply with certain stains. A **nucleolus** is a store of ribonucleic acid (RNA) and proteins, which are subsequently used in the formation of cytoplasmic ribosomes. Nucleoli have no membranes.

The ground substance of the nucleus contains threads composed of **deoxyribonucleic acid** (DNA) associated with protein. The association of DNA and protein is known as **deoxyribonucleoprotein** (DNP). These threads, **chromosomes**, are invisible to the light microscope when the nucleus is not dividing but are detectable with the EM. Chromosomes shorten and thicken and become visible to the light microscope when a nucleus starts to divide (Chapter 25). Under the light microscope, the ground substance of the nucleus appears structureless or granular.

DNA and DNP are sometimes referred to as **chromatin**, a name which refers to the fact that they stain

(a) Double ring structure of adenine

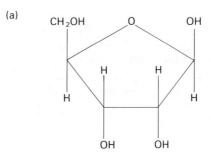

(b) Single ring structure of cytosine

5.9 Nitrogenous bases, molecular structures of (a) a purine, and (b) a pyrimidine

(a)

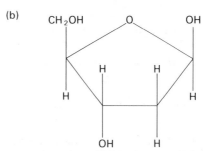

(b)

5.10 Pentose sugars, molecular structures of (a) ribose, and (b) deoxyribose

strongly with basic stains. The remainder of the contents of the nuclear membrane is fluid and may be called **nuclear sap**.

The chromosomes

The number of chromosomes present in the nucleus of a cell is characteristic of the species to which the organism belongs. As has been stated, chromosomes are visible to the light microscope only when the nucleus is in the process of dividing (e.g. Fig. 25.4), but they carry out the most vital part of their activity in the non-dividing nucleus (**interphase** nucleus).

DNA is the hereditary material of the cell. It also controls almost all of the metabolic activities of the cell. The basis of each chromosome is a continuous thread of DNA, which carries the pattern of the cell and all the activities of the cell within its structure, in the form of a molecular code (see below).

As we shall see in the next part of this chapter, DNA is able to reproduce itself exactly (**replicate**) before a nucleus divides. It is also able to transfer parts of its code to **ribonucleic acid (RNA)**, which leaves the nucleus through nuclear pores and passes on its pattern to proteins, which are synthesized in the cytoplasm at ribosomes. The proteins formed are mostly enzymes, though some are merely parts of the structure of the cell (**structural** proteins). Enzymes directly control all the metabolism which takes place in the cell.

DNA, RNA and protein synthesis

DNA and RNA are nucleic acids. Both are polymers of nucleotides. A **nucleotide** consists of a nitrogenous base, a sugar and a phosphate ($-PO_4$) group.

The nitrogenous base may be one of two kinds, a purine or a pyrimidine (Fig. 5.9). **Purines** (adenine and guanine) have molecules with a double ring structure, while **pyrimidines** (thymine, cytosine and uracil) have single ring molecules. The sugar is a **pentose** (5-carbon) sugar, **ribose** or **deoxyribose**. The molecule of deoxyribose is similar to that of ribose, but lacks one oxygen atom (Fig. 5.10). The sugar and the nitrogenous base together make up a **nucleoside**. Nucleoside and phosphate make up the nucleotide.

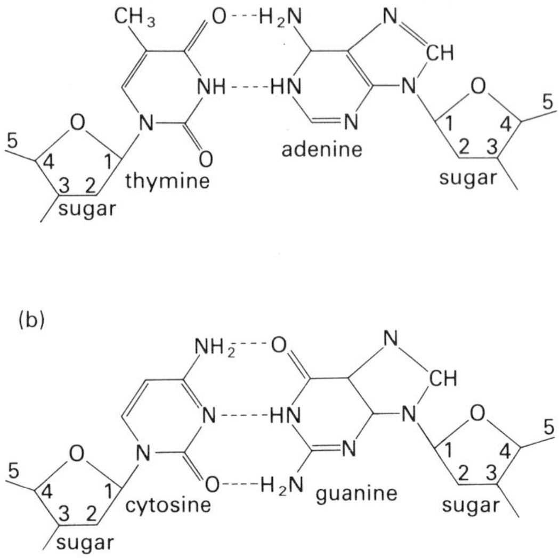

(a)

(b)

5.11 DNA. Diagrams showing hydrogen bonding between (a) thymine and adenine, and (b) cytosine and guanine

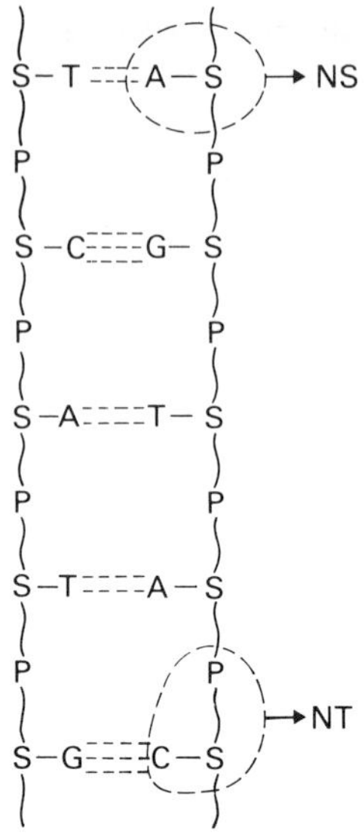

5.12 DNA. Parallel chains of nucleotides in the DNA molecule showing the component sugars (S), phosphate groups (P), bases (T, A, C, G), nucleosides (NS) and nucleotides (NT)

The DNA molecule consists of two parallel chains of nucleotides. In each chain, a phosphate group regularly alternates with a pentose sugar (deoxyribose). A nitrogenous base is attached to the deoxyribose molecule at carbon atom 1 (C_1). The bases attached to the two parallel sugar-phosphate chains face each other and form complementary base pairs by **hydrogen bonding** (Fig. 5.11). Thus, the parallel chains are regularly bridged by the complementary base pairs and are firmly linked together.

Each base pair is made up of a purine and a pyrimidine. Adenine always pairs with thymine and guanine always pairs with cytosine (Fig. 5.12). The whole ladder-like structure is twisted into a right-handed spiral. To continue the analogy of the ladder, the uprights of the ladder represent the sugar-phosphate chains and the steps of the ladder represent the linked base pairs.

There are ten base pairs to each complete turn of the double spiral (**double helix**, Fig. 5.13). The distance across the helix is about 2 nm and the base pairs are separated, on each strand, by about 0.34 nm.

The determination of the structure of DNA, as described above, is probably the most important single discovery in biology to have been made this century. The work was published in 1953 and Francis H. C. Crick, James D. Watson and Maurice

H. F. Wilkins, whose work it was, were later awarded a Nobel Prize (medicine and physiology).

The replication of DNA

The double strand of DNA is precisely self-duplicating because the bases always pair in the same way. When replication is about to take place, the double helix uncoils and the two strands of which it is composed separate, exposing the nitrogenous bases. Each base of each parent strand attracts the complementary base from its immediate environment. The second strand of a double helix is built up on each of the strands which were derived from the original (parent) double helix (Fig. 5.14). As a result, two identical double helices are formed. Replication takes place during what is known as the **S-phase** of the interphase nucleus.

RNA

The structure of **RNA** is similar to that of DNA in many respects, but there are several important differences:

(a) RNA consists of a single polynucleotide chain, rather than a double chain. The single chain may fold upon itself, due to the formation of hydrogen bonds between its components.

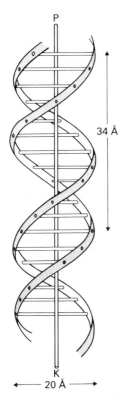

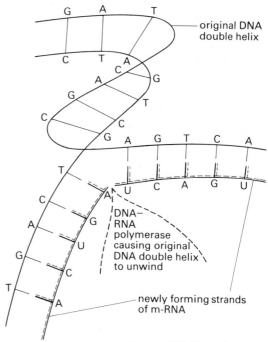

5.15 DNA-dependent formation of m-RNA illustrating the transcription of DNA

5.13 DNA. Diagram showing the right-handed spiral 'ladder' structure (double helix) coiling round an imaginary core (PK)

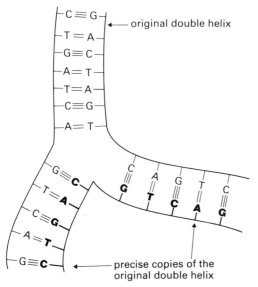

5.14 DNA. A diagram showing the process of replication

(b) Thymine, one of the nitrogenous bases of DNA, is replaced by **uracil**, another pyrimidine.

(c) The deoxyribose of DNA is replaced by **ribose**.

Protein synthesis

The mechanism of protein synthesis is very complicated. The account which follows is a brief and simplified description of the process.

Genetic information is incorporated in the DNA of the chromosome in the form of specific sequences of nitrogenous bases along the molecule. A sequence of three nitrogeneous bases (**codon**) is a code for the inclusion of a particular amino acid in the molecule of a polypeptide. A sequence of codons (**cistron**) determines the sequence of amino acids in the molecule of a polypeptide. Before a polypeptide can be synthesized, the code must first be transferred from the DNA to a molecule of RNA of the type known as **messenger RNA** (m-RNA).

The formation of m-RNA takes place in much the same way as the replication of DNA (Fig. 5.15). The double helix of DNA separates into two single strands and RNA is assembled, from its compenents,

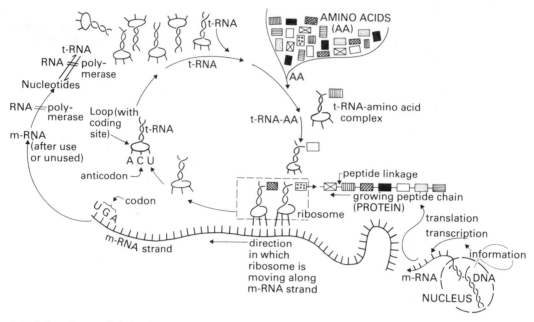

5.16 A flow diagram depicting the hypothesis for the process of protein synthesis

using the exposed bases of one of the two separated DNA strands as the pattern (**template**). Afterwards, the m-RNA separates and the two strands of DNA reunite into a double helix. The m-RNA diffuses from the nucleus into the cytoplasm.

Two other types of RNA are formed in a similar manner, on specialized parts of chromosomes.

Ribosomal RNA (r-RNA) first accumulates, together with protein, in the nucleolus. When the nuclear envelope breaks down at the beginning of nuclear division (Chapter 25), r-RNA and protein are distributed into the cytoplasm and incorporated into ribosomes. **Ribosomes** are the sites of protein synthesis and contain enzymes which are involved in protein synthesis. The exact role of the r-RNA they contain is not known.

Transfer RNA (t-RNA), also known as **soluble RNA** (s-RNA), has a much lower molecular weight than m-RNA. The molecule is folded and, near one end, has a sequence of three nitrogenous bases (**anticodon**) which is complementary to the m-RNA codon for one amino acid. Once it has diffused into the cytoplasm, another part of the molecule attracts and attaches itself to a molecule of the corresponding amino acid.

A **polypeptide** is an amino acid polymer consisting of three or more amino acids linked together by peptide bonds. Polypeptides formed in cells usually contain very large numbers of amino acid units. There is no definite distinction between a polypeptide and a **protein**, but a polypeptide with a molecular weight greater than about 5000 is usually described as a protein.

The synthesis of a polypeptide begins when a molecule of m-RNA becomes attached to a ribosome. Attachment is close to one end of the molecule and the codon close to the point of attachment attracts and bonds with the complementary anticodon of a molecule of t-RNA which is already attached to the corresponding amino acid (Fig. 5.16). The point of attachment of the m-RNA to the ribosome then changes slightly and the second codon of the m-RNA attracts and bonds to another t-RNA molecule with its amino acid. The two amino acid molecules, which have been brought close together, unite by forming a peptide bond and the first molecule of t-RNA separates from its amino acid, leaving it attached to the second amino acid. In the same way, each m-RNA codon in turn adds another amino acid to the chain until the end of the m-RNA molecule is reached and the polypeptide is complete. The polypeptide is set free and the m-RNA separates from the ribosome. The same m-RNA molecule may again associate with a ribosome and repeat the process but, sooner or later, it is destroyed by enzymes in the cytoplasm. Transfer RNA molecules also may act several times but are eventually destroyed.

The molecule of a polypeptide has a spiral structure. If the component amino acids include some which contain sulphur, sulphur atoms join together by forming **disulphide bonds** and the molecule becomes folded. Further folding may result from the formation of hydrogen bonds at various points and the end product may be a molecule of very complex, but constant, shape.

The genetic code

A polypeptide is made up from a pool of about twenty amino acids which exist in living cells. Because of the very large numbers of amino acids in a polypeptide chain, the possible variety of combinations of twenty different amino acids is almost infinite. The properties of an amino acid or protein depend on the number, type and sequence of amino acids which make up the molecule.

The genetic code must code for the type of amino acid at each position on the molecule. Since there are only four nitrogenous bases in DNA, it is clear that there must be more than one nitrogenous base in the code for each amino acid. If there were two in each code, there would be only sixteen possible combinations and this would not give enough codes for the number of amino acids involved. Three nitrogenous bases give the possibility of sixty four combinations, more than enough to code for twenty amino acids. As stated earlier, it is now known that a codon is a sequence of three nitrogenous bases and that some amino acids are coded for by more than one triplet code. Also, some triplets apply to no amino acid (**nonsense codons**) and others signal the beginning or end of a polypeptide molecule.

The cistron, the sequence of codons which codes for one complete polypeptide chain, is equivalent to a **gene** (Chapter 26).

Water relations of plant cells

Since water is of such overriding importance to plants (Chapter 12), the movement of water between the cell and its surroundings is of fundamental significance. What happens at the cellular level affects the gross movement of water and the water balance in the plant. Before we study the movement of water into and out of the cell, we must first look at some more general aspects of water movement.

Water potential

Any movement, including that of water, requires energy. Without energy, no movement can take place. The energy available for movement, or for any other work without change in temperature, is termed **free energy**. The free energy per mole (gram molecular weight) of any substance is the chemical potential of that substance. The chemical potential of water is termed water potential. In a non-uniform medium, water moves from a region of higher energy content (**water potential**) to a region of lower energy content.

Water potential is denoted by the Greek letter psi (ψ) and may be described in terms of (a) atmospheres (atm), (b) dynes per square centimetre (dynes cm^{-2}), or (c) bars, or kilopascals (kP_a). The inter-relationships between the units are as follows:

1 atm (76 cm Hg) = 1.0307 kg cm^{-2} = 1.013 bars;

1 bar = 10^6 dynes cm^{-2} = 75 cm Hg = 0.985 atm = 100 kP_a

The movement of water between two adjacent cells, or two adjacent regions, is governed by the difference in water potential between the two. Such a difference is denoted by $\triangle\psi$ (delta psi). If the two regions are denoted by Y and Z, then the **water potential difference** is given by $\psi Y - \psi Z = \triangle\psi$. If ψY is greater than ψZ, then $\triangle\psi$ is positive and water will move from Y to Z. If ψY is less than ψZ, then $\triangle\psi$ is negative and water will move from Z to Y.

By definition, the potential of pure, unbound water is zero. When water contains any dissolved substance its potential decreases. The water potential of a solution at atmospheric pressure is negative (less than zero). Raising the pressure to which water is subjected raises its potential.

An understanding of the above terms and the use of the terms is essential to an understanding of diffusion and osmosis, both important processes in biology.

Diffusion

Individual molecules free in a fluid medium (e.g. gas molecules; water or solute molecules in a solution) are in constant random motion and tend to distribute themselves uniformly in the entire space occupied by the medium. This tendency means that molecules tend to move from a region of higher concentration (higher potential) to one of lower concentration (lower potential). This is the process of **diffusion**. In the case of a solution, the rule of movement from a region of higher concentration to one of lower concentration applies to both solute and solvent molecules. Wherever there is a difference in water (solvent) potential or solute potential between two regions of a solution, a potential gradient (**concentration gradient**) is set up. Molecules diffuse down their concentration gradients, each type moving from the region of higher potential to the region of lower potential. When a non-uniform solution achieves perfect uniformity throughout, a state of dynamic equilibrium is established. In this state, the continuous random movement of both solute and solvent molecules continues, but the amounts of diffusion in different directions will be equal and there will be no net movement.

Osmosis

A barrier or membrane which will allow the passage of water molecules but not of solute molecules is described as semipermeable. Some biological membranes behave in this way and are called **semipermeable membranes**. However, most biological membranes (e.g. plasmalemma, tonoplast) are freely permeable to water and are also slightly permeable to some solute molecules. These are **differentially permeable membranes**.

Suppose that two aqueous solutions, or pure water and an aqueous solution, are separated by a semipermeable or differentially permeable membrane. Water will diffuse through the membrane down the water potential gradient, from the side of the membrane with higher water potential (pure water, less concentrated solution) to the side with lower water potential ('stronger' solution). This process is called osmosis. **Osmosis** is a special case of diffusion of water from a region of higher potential to a region of lower potential.

Wherever there is water or a weaker solution on one side of a semipermeable or differentially permeable membrane and a stronger solution on the other,

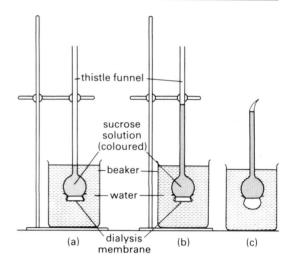

5.17 Apparatus for demonstrating osmosis. The bulb of the thistle funnel, with dialysis membrane tied over the mouth, contains sucrose solution to which a dye has been added so that it can be seen more easily. This is placed, as shown, in distilled water in a beaker. Water molecules can easily diffuse into the solution through the differentially permeable dialysis membrane but the sucrose molecules cannot easily diffuse out. The level of the solution therefore rises in the stem of the funnel (b). If the stem of the funnel is sealed up (c) before equilibrium is established, the membrane becomes distended as more water diffuses into the solution

an **osmotic system** is set up. The living cell, with various membranes (plasmalemma, tonoplast, membranes of organelles), provides such systems in abundance.

Osmosis may be demonstrated if we use an artificial membrane such as dialysis tubing or uncoated cellophane. A piece of membrane is tied firmly over the mouth of a thistle funnel, the funnel inverted and partly filled with a sucrose solution containing dye, to make the level of the liquid easy to see. The funnel is then placed, membrane downwards, in a beaker of water and clamped in such a position that the levels of liquid in the stem of the funnel and in the beaker are the same. In this experiment, the level of the solution rises up the stem of the thistle funnel in a matter of minutes (Fig. 5.17). Theoretically, the level will stop rising when the pressure of the water in the vertical tube balances the force with which the water is entering through the membrane. Alternatively, pressure could be applied to the solution by a sliding piston in the stem of the funnel, to completely prevent the diffusion of water through the membrane into the solution. The amount of pressure which would prevent further entry of water through the membrane would be a measure of the **osmotic potential** of the solution. This would be impracticable with our apparatus, which is too fragile, but more sophisticated apparatus exists for this purpose. Such apparatus is called an **osmometer**.

It is evident from Figure 5.17(b) that the level of the liquid in the stem of the funnel is higher than that in the beaker. The higher level is being maintained as a result of water diffusing from its lower level in the beaker, through the membrane, against the pressure of the liquid at a higher level in the stem of the funnel. This shows that water is doing work, raising the level of the liquid. This can be made more obvious by sealing off the stem of the funnel (Fig. 5.17(c)). Water will still continue to enter. The glass walls will not yield but the membrane becomes distended to accommodate the influx. The distended membrane exerts pressure on the contents of the funnel. This is analogous to what happens in an isolated cell placed in water. When the cell is swollen, with its cell wall distended as a result of water diffusing in by osmosis, it is said to be **turgid**. The pressure exerted by its distended walls on the cell contents is described as **turgor pressure**. Under such conditions, water is doing work by entering the cell against a pressure gradient.

In the artificial setting of the thistle funnel, the membrane will burst if it is distended too far. The intact plant cell does not normally burst, because the delicate, differentially permeable plasmalemma is enclosed by the cell wall, which is fairly rigid and stretches only a little. Pure water will diffuse into a solution through a membrane with a potential called the osmotic potential of the solution. **Osmotic potential** is denoted by ψ_π. Water diffuses from a region of high water potential to a region of lower water potential. Since the water potential of pure water is zero, it follows that osmotic potential must be less than zero. However, despite its negative value, osmotic potential represents the actual pressure generated in a cell as water diffuses in by osmosis.

Since the distended cell wall in a turgid cell presses against the cell contents, it tends to squeeze water out of the cell. Any water leaving the cell under such conditions is doing so down a pressure gradient. This means that the water in the cell has a pressure potential. **Pressure potential** is denoted by ψ_p which, unlike ψ_π, is positive. The pressure potential of the water inside the cell is greater than that of the water outside the cell. By definition, the pressure potential of water at atmospheric pressure is zero.

Referring to Figure 5.17(c), when equilibrium is attained the water potential will be the same in the beaker and in the funnel, that is, in all parts of the system. We reach a steady state in which:

ψ (inside) = ψ (outside).

Water enters a cell in response to water potential gradients between the inside and outside of the cell. In doing so, it generates a pressure potential (turgor potential) which opposes the osmotic potential of the cell contents. Thus, pressure potential and osmotic potential are components of water potential, so that:

$$\psi = \psi_p + \psi_\pi.$$

Rewriting the above equation to represent the steady state, we have:

ψ_p (inside) + ψ_π (inside)
= ψ_p (outside) + ψ_π (outside).

When the external solution is pure water at atmospheric pressure, both ψ_p and ψ_π are zero and, therefore, at equilibrium:

ψ_p (inside) + ψ_π (inside) = 0, or
ψ_p (inside) = $-\psi_\pi$ (inside).

This means that the pressure potential (turgor pressure) developed inside the cell acts in the opposite direction to the osmotic potential of the cell contents, but is numerically equal to it.

If the liquid outside the cell is a solution instead of pure water, the value for turgor pressure is given by the difference between the osmotic potential of the cell contents and that of the external solution:

$$\psi_p = \psi_\pi \text{ (outside)} - \psi_\pi \text{ (inside)} = \triangle\psi_\pi.$$

Osmotic potential and the gas laws
There is a close analogy between the laws governing the pressure-volume relationships of gases and those governing the osmotic potentials of solutions. Gas pressure varies in direct proportion to the concentration of the gas. Similarly, the pressure, P, developed in an osmotic system is proportional to the concentration of the osmotically active solute. If the volume of solution containing a given quantity of

solute is denoted by V, the concentration can be expressed as $\frac{1}{V}$. Thus, $P \propto \frac{1}{V}$, or:

(a) $P = K_1 \left(\frac{1}{V}\right) = \frac{K_1}{V}$,

where K_1 = a constant.

Because the kinetic energy of molecules is proportional to temperature (T), an osmotic system is sensitive to absolute temperature. Thus, $P \propto T$, or:

(b) $P = K_2T$, where K_2 = another constant.

Combining both equations:

(c) $P = \frac{K_1K_2T}{V}$.

From Boyle's Law for gases:

(d) $P_1V_1 = P_2V_2$, or $PV = R$ (a constant).

Allowing for the temperature effect:

(e) $PV = RT$.

For the osmotic system, equation (c) applies to solute molecules free to diffuse in the water, just as if they were gas molecules diffusing freely in their confined space. The constants K_1 and K_2 in equation (c) can therefore replace the gas constant R in equation (e). Equation (c) can then be rewritten:

(f) $P = \frac{RT}{V}$, or $PV = RT$, which is the same as

(e).

In equation (f), the pressure P can be read directly from an osmometer.

In the osmometer, the pressure (P) developed is equal and opposite to the osmotic potential. Therefore, substituting in equation (f), at equilibrium,

when $P = \psi_\pi$, $\psi_\pi = -\frac{RT}{V}$

Continuing the analogy with gases, it will be remembered that one mole of gas occupies a volume of 22.4 litres at standard temperature and pressure (STP), or 0°C and 1 atm. Put another way, one mole of gas occupying 22.4 litres at 0°C exerts a pressure of one atmosphere (1.01 bars). Conversely, one mole of gas confined to a volume of one litre at 0°C exerts

a pressure of 22.4 atmospheres (22.62 bars), since

$P \propto \frac{1}{V}$.

Treating a solution in the same way, one mole of a solute occupies one litre in a molar solution. If the solute molecules were gas molecules, they would exert a pressure of 22.62 bars. It is found that this is true for dilute solutions only. If solute concentration increases, the solvent volume decreases, so there is no strict proportionality between osmotic potential and the molarity of solutions. In order to maintain a reference volume of solvent to varying concentrations of solute, a molal solution is used. A **molar** (M) solution contains one mole of solute in one litre of solution. A **molal** (m) solution contains one mole of solute dissolved in one litre of solvent (making a total volume of more than one litre). The gas law relationships apply more closely to molal solutions than to molar solutions.

The pressure potential of a solution depends on the number of particles in solution and not necessarily on the number of molecules. This means that the relationship $\psi_\pi = \frac{-RT}{V}$ holds only for non-ionizing solutes, in which the number of particles is equal to the number of molecules in solution. Suppose that the solute molecule in solution separates into ions, so that:

(a) $AB \rightarrow A^+ + B^-$,
(b) $AB_2 \rightarrow A^+ + 2B^-$, or
(c) $A_2B \rightarrow 2A^+ + B^-$.

In each of these cases, there is only one solute molecule but the number of particles (ions) varies. In (a), the molecule gives two ions while in (b) and (c) the molecule gives three. So, if a molecule does not ionize in solution (e.g. sucrose) and each particle is one molecule:

$\psi_\pi = -\frac{RT}{V} \times 1 = -\frac{RT}{V}$.

If the molecule ionizes into two ions (e.g. NaCl) and each molecule produces two particles, then:

$\psi_\pi = -\frac{RT}{V} \times 2 = -\frac{2RT}{V}$.

If the molecule ionizes into n ions, then

$\psi_\pi = -\frac{nRT}{V}$.

The above equations assume 100 per cent ionization of the molecules in the solution, but very often ionization is less than 100 per cent. In such a case, the number of particles in the solution will be proportionately reduced. If n particles per molecule at 100 per cent ionization gives

$$\psi_\pi = -\frac{n\,RT}{V} \text{ bars,}$$

then, for x per cent ionization,

$$\psi_\pi = -\frac{n\,RT}{V} \times \frac{x}{100} \text{ bars.}$$

Since concentration (C) varies inversely with volume, it is more meaningful to replace $\frac{1}{V}$ with C, for solutions as opposed to gases. The last equation then becomes:

$$\psi_\pi = -\frac{nx}{100} \times CRT \text{ bars.}$$

Measurement of the osmotic potential of a solution

The osmotic potential of a solution can be measured directly with an osmometer. An indirect method, which can be used in laboratories with modest facilities, is the **freezing point depression** method.

The freezing point of pure water is 0°C. That of a molal solution of a non-ionized substance is −1.86°C. This depression of 1.86°C is produced by a molal solution with a theoretical osmotic potential of 22.62 bars. It is found that there is proportionality between this depression of freezing point and the osmotic potential of a solution, so that:

(a) if the depression produced = 1.86°C,
$\quad \psi_\pi = -22.62$ bars,
(b) if the depression produced = 1.00°C,
$\quad \psi_\pi = -\dfrac{22.62}{1.86}$ bars, and
(c) if the depression produced = x°C,
$\quad \psi_\pi = -\dfrac{22.62}{1.86}\,x$ bars,
$\quad = -12.16x$ bars.

Cells immersed in solutions

If a cell is immersed in a solution which has a solute concentration which is less than that of the cell sap (**hypotonic** solution), then the water potential of the solution is greater than that of the cell sap and water diffuses into the cell. The protoplast exerts pressure on the inside of the cell wall and the cell is said to be **turgid**. If the external liquid is pure water, the cell becomes **fully turgid**, after which no more water can enter.

If a cell is immersed in a solution of higher solute concentration than the cell sap (**hypertonic** solution), the water potential of the cell sap is higher than that of the solution and water diffuses out of the cell. Because of its relative rigidity, the cell wall does not collapse beyond a certain point and the protoplast shrinks away from the inside of the cell wall. Such a cell is said to be **plasmolysed**.

If a turgid cell is immersed in a solution which has the same osmotic potential as the cell sap (**isotonic** solution), water leaves the cell until pressure potential is reduced to zero, after which there is no further net movement of water out of the cell. The protoplast exactly fills the cell wall, without exerting pressure on it. Such a cell is said to be **flaccid**, not plasmolysed and not turgid.

If a cell is severely plasmolysed, the protoplast may be damaged. However, cells can recover from a moderate degree of plasmolysis if they are subsequently placed in a hypotonic solution. Recovery from plasmolysis can also occur automatically if they are left in a hypertonic solution of certain substances for a long time. This happens only when the solute is one to which cell membranes are permeable, though less permeable than to water. For instance, if filaments of the alga *Spirogyra* sp. are mounted on a microscope slide in a molar solution of ethylene glycol, plasmolysis occurs very quickly. If a cell which has been plasmolysed is carefully observed under the microscope, the protoplast is seen to swell until it again fills the cell wall, within a period of about three minutes. The cell has **deplasmolysed**. Plasmolysis first takes place because the solution of ethylene glycol is strongly hypertonic and the cell membranes are completely permeable to water. However, the membranes are slightly permeable to ethylene glycol, which diffuses into the cell. The accumulation of ethylene glycol molecules in the cell raises the osmotic potential of the cell sap and water again enters the cell, by osmosis.

Measuring the osmotic potential of cell sap

Cell sap may be expressed from plant tissues by squeezing the tissue, for example between perspex blocks in a vice. Expression of sap is made easier if the tissue is first frozen, to partially disrupt the cells, and then unfrozen before squeezing. Expressed sap is collected in a tube and its freezing point determined. The osmotic potential can then be calculated, as already described. It is, however, uncertain that the properties of the expressed sap are exactly the same as those of the sap in intact cells.

A simple, direct method of determining the osmotic potential of cells is the **plasmolytic method**. A graded series of solutions (e.g. sucrose or mannitol) is prepared, each of known concentration and osmotic potential. Small pieces of tissue are immersed in each solution and left to equilibrate, which usually takes only a few minutes. Each piece of tissue in turn is then mounted on a microscope slide in a drop of the solution with which it has equilibrated and examined under the microscope. Assuming that the range of concentrations used for the solutions was, for example, from 0.1 M to 1.0 M, the cells in the weakest solution will all be turgid and the cells from the strongest solution will all be plasmolysed. For the tissue from each solution, a count is made of unplasmolysed cells and of cells showing any sign of plasmolysis. The condition in which the protoplast of a cell is just starting to separate from the inside of the cell wall is **incipient plasmolysis**. There is some variation in osmotic potential between cells that make up a tissue. The osmotic potential of the solution which induces plasmolysis in fifty per cent of the cells is taken to be the osmotic potential of the tissue. When such an experiment is carried out, it is unusual for any of the solutions used to cause exactly fifty per cent plasmolysis. The percentages of plasmolysis obtained are therefore graphed against the osmotic potentials of the solutions used and the fifty per cent point read from the graph (Fig. 5.18).

At the point of incipient plasmolysis,

ψ_p (inside) = 0 (i.e. no turgor), and
ψ_π (inside) = ψ_π (outside).

So, the osmotic potential of the cell sap corresponds to the osmotic potential of the external solution.

This method is particularly suited for cells with coloured sap, which makes it easier to observe the withdrawal of the protoplast from the cell wall.

Turgor pressure of cells and tissues
Once the osmotic potential of the cell sap has been determined by the above method, then turgor

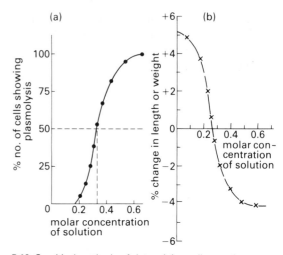

5.18 Graphical methods of determining cell osmotic potential. (a) The plasmolytic method: graphical representation in which the molar strength of the external solution in which fifty per cent of the cells show incipient plasmolysis can be read off. (b) Determination by per cent change in weight or length of tissue: graphical representation showing how the concentration at which there is no change in weight or length is read off (where the curve cuts the horizontal axis at zero per cent change)

pressure (pressure potential, ψ_p) can be determined by the relationship:

(a) ψ_π (outside) = ψ_π (inside) + ψ_p (inside), or
(b) ψ_p (inside) = ψ_π (outside) − ψ_π (inside).

If a state can be established in which there is no net movement of water between cells and the external solution, then:

(c) ψ (external) = ψ_π (internal) + ψ_p (internal).

It is the ψ_p component that is being calculated in equation (b) above. The state in which neither the tissue nor the external solution loses water to the other is established when there is no change in the tissue length or weight when immersed in a solution. Once more, a graded series of solutions is prepared and uniform strips of tissue are measured in length, or weighed, and placed in the solutions. After equilibrium, the measurement or weighing is repeated and the solution in which there is no change in length or weight is noted. Again, graphing may be necessary. The solution in which there was no change in length or weight has an osmotic potential equal to ψ_π (outside) in equation (b) above. Still referring to equation (b), if we have already obtained a value

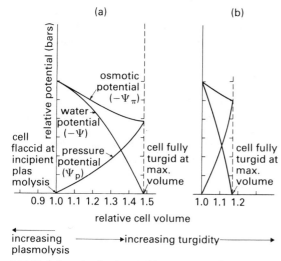

5.19 Variation of cell volume with water, osmotic and pressure potentials in (a) a young cell with elastic walls, and (b) an old cell with relatively inelastic cell walls

for osmotic potential inside the cell (ψ_π inside), we can calculate the value of pressure potential (ψ_p inside) by subtraction. If weighing is used, this method is often called the **gravimetric method**.

Cell volume and turgor

The cell wall is not absolutely rigid. It is elastic to some extent, but elasticity decreases as a tissue becomes older. A cell immersed in water will take up water until it is fully turgid. At full turgor, the volume of a cell with a relatively elastic cell wall may be as much as 1.5 times its volume in the flaccid state. In a very old tissue, the volume increase between the flaccid state and full turgor is very small. In either case, the water potential of the cell sap rises from a negative value, in the flaccid cell, to zero at full turgor. At the same time, its pressure potential rises to a maximum value.

During the change from flaccidity to turgidity, the cell sap is diluted and its osmotic potential becomes less negative (Fig. 5.19).

Difference in water potential – the driving force for water movement

Water in cells and tissues moves in response to differences in water potential ($\triangle\psi$) on either side of a partitioning membrane or membranes. The direction of movement depends on these differences.

Consider two neighbouring cells, cell A and cell B, having water potential x and y respectively. Water will move from A to B if y is more negative then x, that is, if x is greater than y. Furthermore, the rate of movement is proportional to the magnitude of the gradient or water potential difference between the two cells. The rate of diffusion is faster when the water potential difference is greater. The water potential difference actually constitutes the driving force for the movement of water between cells and within tissues. Thus, if we know the components of water potential in two adjacent cells or tissues, we can predict which way and with how much force water will move between them. For instance, suppose that cell A has cell sap with an osmotic potential of -5.5 bars and a pressure potential of 3.5 bars. Suppose also that neighbouring cell B has a cell sap osmotic potential of -11.0 bars and a pressure potential of 3.0 bars. Since water potential equals osmotic potential plus pressure potential, the water potentials of the two cells are:

cell A, $\psi_A = \psi_\pi + \psi_p = -5.5 + 3.5 = -2.0$ bars,
cell B, $\psi_B = \psi_\pi + \psi_p = -11.0 + 3.0 = -8.0$ bars.
$$\triangle\psi = \psi_A - \psi_B \quad \text{or} \quad \psi_B - \psi_A$$
$$= -2 - (-8) \quad \text{or} \quad -8 - (-2)$$
$$= +6 \quad \text{or} \quad -6.$$

Irrespective of the sign for difference in water potential, the value of $\triangle\psi$ shows the magnitude of the driving force, a positive value. However, the sign shows the direction of movement. In the above example, $\triangle\psi$ for A–B is positive, so water will move from cell A to cell B. The value for B–A is negative, so water will move from cell A to cell B, with a force of six bars.

Matric potential

Water movement in tissues does not always involve movement through a differentially permeable membrane. This is true of the process termed **imbibition**. When water or another solvent is imbibed by any material (the **imbibant**), it moves with a force of electrical or electrostatic attraction.

During the course of imbibition (hydration of colloids), water molecules are held quite tightly to the colloidal substances. We say they are **adsorbed**. Therefore, water molecules are packed together more closely than under normal conditions. As a result, the total volume of water plus imbibant before imbibition will be greater than the volume of water plus imbibant after imbibition. Because of the tight packing of the water molecules, some of the kinetic energy (energy of movement) of the water molecules is released as heat. Therefore, a measurable increase in temperature accompanies imbibition. Both the contraction in total volume and temperature increase can be demonstrated quite simply in the laboratory by soaking dry seeds (e.g. *Oryza* sp., Rice) in a flask.

With uptake of the solvent, the imbibant swells and thereby generates pressure which can be quite considerable. This is why the testa of a germinating seed breaks. There is an affinity between the imbibed liquid and the imbibant. This helps colloidal substances in the cell to hold water strongly and thereby to minimize the effects of drought. The forces holding water in a matrix of any kind, including cellulose cell walls and colloidal materials in the cell and in the soil, and other surface forces, are described as matric potential (ψ_M). Thus, if matric forces are involved, matric potential must be included in the calculation of overall water potential:

$$\psi = \psi_\pi + \psi_p + \psi_M.$$

Key words

Organism: unicellular, multicellular; cell, tissue, organ. **Microscope:** light microscope, electron microscope (EM); magnification, resolving power. **Cell:** protoplast, protoplasm, cytoplasm, nucleus. **Plant cell:** cell wall, middle lamella; cytoplasmic organelles; membranes; vacuole; plasmodesmata, symplast; cytoplasmic inclusions. **Membranes:** unit membrane; plasmalemma, endoplasmic reticulum (ER), tonoplast. **Organelles:** mitochondrion, crista; plastid, proplastid, leucoplast, amyloplast, chloroplast, chromoplast; Golgi apparatus (Golgi body); microbody; ribosomes. **Chloroplast:** granum, thylakoid, stroma. **Vacuole:** cell sap, water, mineral salts, enzymes, pigments, storage materials, waste products. **Nucleus:** nuclear envelope, nuclear pores; nucleolus, RNA, chromosomes, DNA, DNP; replication. **Protein synthesis:** m-RNA, r-RNA, t-RNA; amino acid, polypeptide, protein. **Genetic code:** codon, anticodon, cistron, gene. **Water relations:** water potential (ψ), water potential difference ($\triangle\psi$). **Diffusion:** concentration gradient. **Osmosis:** semipermeable, differentially permeable; osmotic potential (ψ_π), osmotic potential difference ($\triangle\psi_\pi$), pressure potential (ψ_p), turgor pressure; turgid; flaccid, plasmolysis, deplasmolysis, incipient plas-

molysis. **Solutions:** hypotonic, isotonic, hypertonic. **Concentration of solutes**: molar solution, molal solution, ionization. **Imbibition**: matric potential (ψ_M).

Summary

The cell is the basic unit of all living things. The tissues and organs of all multicellular animals and plants are made up of cells. The plant cell has a non-living cell wall, inside which the protoplast is contained within a plasmalemma. Within the plasmalemma, the cytoplasm contains several types of organelles, a nucleus and a vacuole. The plasmalemma and other membranes of the cell are formed from unit membrane, which is more or less constant in structure and composition. Within the nuclear envelope, the nucleus contains one or more nucleoli and a number of chromosomes. Chromosomes are composed mostly of DNA.

DNA is the fundamental genetic material and carries the pattern for all the proteins of the cell, most of which are enzymes, in the form of a molecular code. DNA replicates before the nucleus divides. DNA also passes on its pattern to messenger RNA. Molecules of messenger RNA become attached to ribosomes, which contain ribosomal RNA, and interact with transfer RNA to synthesise proteins from amino acids in the cytoplasm.

The force which causes water to move into and out of cells and tissues is water potential difference. The water potential within a cell is the sum of the osmotic potential of the cell sap (negative figure) and the pressure potential of the cell (positive figure). The osmotic potential of any solution is related to solute concentration. Matric potential also influences the uptake of water, by imbibition, where colloidal substances are present in a dehydrated state.

Questions

1 What is unit membrane (UM)? In which structures within the cell is UM present?

2 Describe the structure of (a) a chloroplast, and (b) a mitochondrion, as seen in section with the electron microscope. What is the function of each of these organelles?

3 How do DNA and RNA work together to control the metabolic activities of the cell?

4 Distinguish between water potential, osmotic potential, pressure potential and matric potential.

6

Cells and tissues

The root system and shoot system of a flowering plant grow as a result of cell divisions which take place in cells of an **apical meristem** at the tip of each root and shoot. A **meristem** is a tissue in which the cells are capable of repeated cell division. When meristematic cells divide, some of the new cells remain as part of the meristem while others **differentiate** (change) into cells of the permanent tissues of the plant.

In addition to apical meristems, some plants also have intercalary meristems or lateral meristems. **Intercalary meristems** are characteristic of certain monocots, including grasses. They are located at the base of each internode and are really parts of the apical meristem which have been left behind, below fully developed regions of the internode. They are responsible for lengthening of the internode, for example before flowering, and for the bending upwards of stems which have fallen over. **Lateral meristems** include the vascular cambium, which is responsible for the secondary thickening of woody stems and roots and phellogen, which forms cork (**periderm**) at the surface of woody organs. At this stage, we shall concentrate our attention on apical meristems.

Meristematic cells

Cells of an apical meristem are more or less isodiametric, and are packed closely together without any air spaces between them. They have thin cellulose walls, large nuclei and dense cytoplasm containing only a few small vacuoles. Their cytoplasm is very rich in all types of organelles, including proplastids.

Division of the nucleus is by mitosis, the type of nuclear division that results in the formation of two nuclei, each with exactly the same genetic material as the parent cell (see Chapter 25). Division of the nucleus is followed by the formation of a cell plate between the new nuclei (Fig. 25.9).

This **cell plate** is composed of calcium pectate and gradually extends across the cell, dividing the cyto-

plasm into two parts. The developing cell plate extends through the cell wall and unites with the middle lamella (inter-cellular material) which separated the original cell, which has just divided, from its neighbours. The **middle lamella** between cells is, in fact, derived from the cell plates of previous divisions and is composed of the same material. While the cell plate is completing its growth, a new cellulose wall is laid down on each surface of the cell plate. This unites at the edges with the cell wall of the parent cell and each of the new nuclei with associated cytoplasm is now enclosed in its own cell wall, partly derived from the wall of the parent cell and partly from the new wall laid down on the surface of the cell plate. The cell walls and the cell plate (now middle lamella) do not completely separate the two cells. They are perforated by pores through which pass cytoplasmic connections (**plasmodesmata**, *sing.* **plasmodesma**).

Differentiation of cells

The first stage in the differentiation of cells, the change from a cell similar to a meristematic cell to the mature form, is always **vacuolation**. During this process, vacuoles increase in size and usually coalesce to form a single, large vacuole. The nucleus of the cell comes to lie close to the cell wall within a thin layer of cytoplasm. Vacuolation usually results in a great increase in cell size and it is during this process that the final shape of the cell is determined. As the cell increases in size, more cellulose is added to the cell wall, which retains its continuity and increases in thickness. The cell wall which is laid down during vacuolation is the **primary wall** of the cell. Some cells cease the process of differentiation at this stage. In others, more cellulose is added after vacuolation has ceased. This additional wall material is the **secondary wall** of the cell. Again, in some types of cells, the secondary wall, primary wall and often the middle lamella become impregnated with a different type of material. If this material is lignin, as in

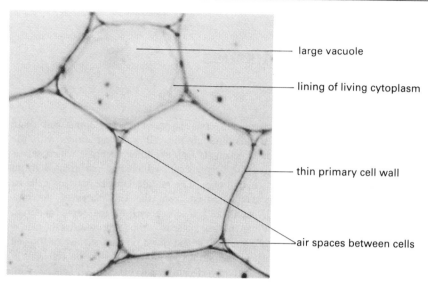

— large vacuole

— lining of living cytoplasm

— thin primary cell wall

air spaces between cells

6.1 Parenchyma: TS parenchyma in the medulla of a dicot (HP)

woody tissues, the wall is said to be **lignified**. In other tissues, notably the outer tissues of woody plants, cell walls may be impregnated with suberin (**suberized**). The cellulose of the cell wall is composed of interwoven fibres and is therefore porous. Lignification or suberization fills in the pores in the cellulose walls, so that they become impermeable to water. Either process therefore results in the virtual sealing off of a cell from its neighbours, except for plasmodesmata between cells.

Plasmodesmata are mostly grouped into small areas of the cell wall where the wall remains thin, even when other parts of the wall have been greatly thickened. These thin parts of the cell wall, perforated by plasmodesmata, are **pits**.

In some types of cells, lignification or suberization of the cell wall is followed by the death of the cell contents, so that the mature cell is dead. The space that was occupied by living contents forms the empty **lumen** of the cell.

Cell types and tissues

All plant organs are made up of a number of tissues. Some tissues are made up of one type of cell only. These are **simple tissues**. Other tissues are made up of a mixture of several different types of cells. These are **complex tissues**.

Simple tissues

Parenchyma tissue is made up of parenchyma cells (Fig. 6.1). Parenchyma cells are the least altered during the process of differentiation, which is limited to vacuolation and the addition of a certain amount of primary wall material. Their main characteristics are as follows:

- Living when mature;
- Not greatly elongated;
- Primary cellulose cell wall only;
- Pits simple (small, with parallel sides);
- Usually rounded at least at the corners so that there are air spaces between the cells.

The functions of parenchyma cells are many. They make up the **ground tissue** of many organs, in which more specialized tissues are embedded. They often store starch and other substances. Their porous walls with living contents allow the movement of water and dissolved substances from cell to cell. When turgid (well supplied with water), they give mechanical support to herbaceous organs (e.g. in seedlings). Because they are alive, they are capable of returning to the meristematic condition. There are also several slightly specialized types of parenchyma. **Chlorenchyma** (Fig. 6.2) contains many chloroplasts and forms the main photosynthetic tissue of the plant. Chlorenchyma also usually has rather large air spaces and the cells are rounded or irregular in shape.

(a)

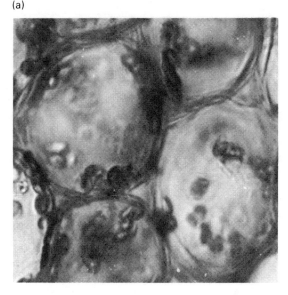

(b)

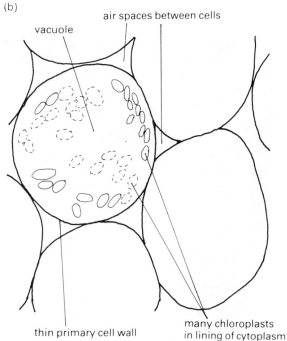

vacuole

air spaces between cells

thin primary cell wall

many chloroplasts in lining of cytoplasm

6.2 Chlorenchyma: (a) TS chlorenchyma (HP); (b) drawing to show details of cells

large air spaces

long 'arms' of irregular shaped cell

thin primary cell wall

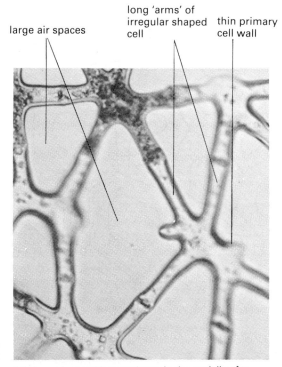

6.3 Aerenchyma: TS aerenchyma in the medulla of a monocot (HP)

Aerenchyma (Fig. 6.3) has especially large air spaces, allowing the free movement of gases. It is often found in submerged organs of water plants.

Collenchyma tissue is made up of collenchyma cells (Fig. 6.4). Collenchyma cells are a little more altered in the process of differentiation than parenchyma cells. Their principal characteristics are as follows:

- Living when mature;
- Elongated along the axis of the organ in which they are situated;
- In addition to the cellulose wall, they have an extra layer of primary wall, composed of hemicellulose and pectic substances, either evenly distributed within the cellulose wall or concentrated in the angles of the cell;
- Pits simple;
- Air spaces usually absent, but small air spaces may be present.

Collenchyma cells are part of the ground tissue of organs but are usually close to the surface, in stems and leaves. They store starch and often contain chloroplasts. Their most important function is mechanical. They add strength, with elasticity, to

tiny air space

Extra layers of
primary wall

(a)

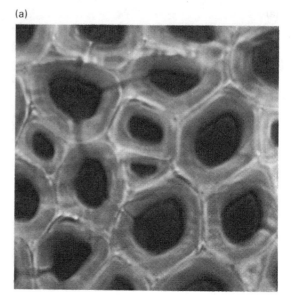

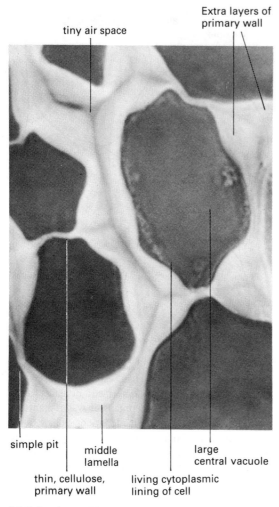

simple pit

middle
lamella

large
central vacuole

thin, cellulose,
primary wall

living cytoplasmic
lining of cell

6.4 Collenchyma: TS collenchyma in the outer cortex of
Talinum sp., from a fresh preparation (HP)

(b)

lumen of
dead cell

smaller cells,
cut ends of
tapering fibres

simple pit

middle lamella

primary
wall

secondary
wall

both lignified

6.5 Sclerenchyma fibres: (a) TS sclerenchyma fibres in leaf
of *Sansevieria* sp. (Bowstring Hemp, HP); (b) drawing to
show details of cells

herbaceous organs. Like parenchyma, they are
capable of returning to the meristematic condition.

Sclerenchyma tissue is made up of sclerenchyma
cells. There are two types of sclerenchyma cells,
sclerenchyma fibres and sclereids. Both types share
the following characteristics:

- Dead when mature;
- Primary and secondary cell walls, both lignified;
- Pits simple.

Sclerenchyma fibres (Fig. 6.5) are greatly elongated and are grouped together to form part of the

(a)

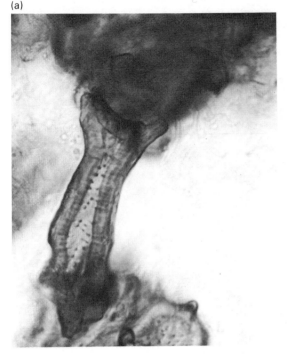

(b)

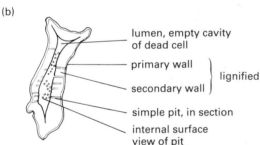

lumen, empty cavity of dead cell

primary wall ⎫
secondary wall ⎭ lignified

simple pit, in section

internal surface view of pit

6.6 Sclereids: (a) sclereid from the flesh of a fruit of *Psidium* sp. (Guava) (HP); (b) drawing to show details of a cell

(a)

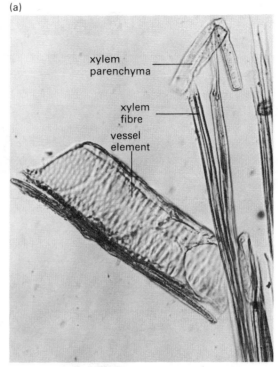

xylem parenchyma

xylem fibre

vessel element

6.7 Xylem: (a) macerated xylem of *Mangifera* sp. (Mango), to show several cell types (HP) (continued overleaf)

ground tissue. They give strength and rigidity to organs. Some fibres are very long, reaching a length of several millimeters. During their differentiation, cells which are to form fibres grow in length rapidly and their tips intrude between other differentiating cells before the secondary wall is laid down.

Sclereids (Fig. 6.6) are variable in shape and may be grouped together to form a continuous tissue, as in the testas of many seeds, to which they give mechanical protection. Star-shaped and H-shaped sclereids give support to parenchyma tissues in some roots, stems and leaves.

Complex tissues
Complex tissues are made up of a mixture of several types of cells.

Xylem (wood) is the tissue which conducts water and dissolved mineral ions from the root to the various parts of the shoot system. It is made up of up to four types of cells, tracheids, vessel elements, xylem fibres and xylem parenchyma (Fig. 6.7).

Tracheids and vessel elements have several features in common and may be referred to jointly as **tracheary elements**. Their common characteristics are as follows:

- Dead when mature;
- Usually elongated along the axis of an organ;
- Lignified secondary cell wall, deposited in any of a number of patterns leaving areas of primary wall uncovered.

(b)

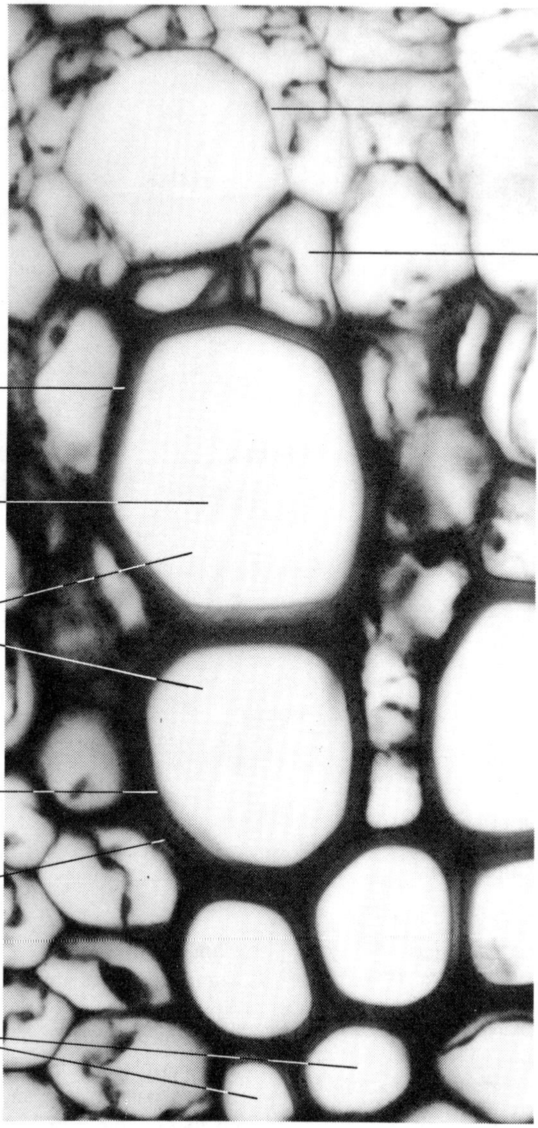

developing
xylem vessel,
with living
contents

living
parenchyma
next to
vessels

secondary
wall with
some lignin

lumen
of dead
cell

metaxylem

thick
lignified
secondary
wall

middle
lamella

protoxylem

6.7 (continued) (b) TS xylem in a vascular bundle of *Tridax* sp. (HP)

Tracheids are elongated and tapering at their ends, arranged together as strands of overlapping cells. Water is able to pass along the strand of tracheids by passing from the lumen of one cell to the lumen of another through the thin areas in the cell wall.

Vessel elements (Fig. 6.8) differ from tracheids in that they are arranged end to end along the axis of an organ and that the cross walls between successive vessel elements are perforated, giving free passage of water from cell to cell. Perforation involves the dissolution of part of the cross wall, often in a characteristic pattern. A series of vessel elements thus forms an open tube along the axis of an organ and a series of interconnected elements is called a **vessel**.

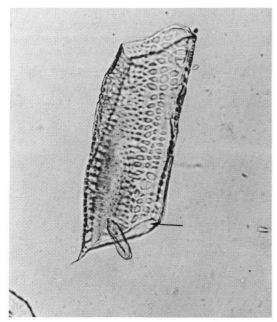

6.8 Xylem vessel element from macerated wood of *Mangifera* sp. showing typical pitting and perforation (arrowed) at end of cell (HP)

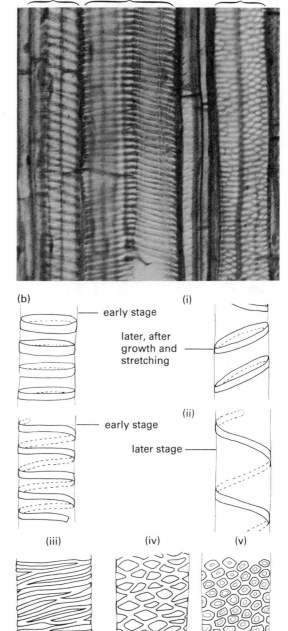

The pattern of the secondary wall laid down in tracheary elements (Fig. 6.9) varies according to the stage of elongation of the organ in which they are situated at the time of differentiation. Tracheary elements which differentiate first, before the organ has finished elongating, have patterns of thickening which will allow stretching. These include **annular** thickening (in separate rings), **spiral** thickening (in a continuous spiral) and **scalariform** thickening (like the rungs of a ladder). Elements in which the secondary wall is laid down after elongation has ceased have **reticulate** (netlike) thickening or thickening with **bordered pits**. Bordered pits are so called because the edge of the secondary wall overhangs the edge of the pit, so that the pit field is wider than the pit aperture and the pit is seen as two concentric circles, when seen in surface view.

When xylem vessels and tracheids become old, or parts of the plant are injured, they are often closed by tyloses (Fig. 6.10). **Tyloses** are parts of neighbouring parenchyma cells which have grown into the lumen of the conducting element through the pits between the cells.

6.9 Patterns of thickening in tracheary elements. (a) photomicrograph showing (i) spiral thickening, (ii) scalariform thickening and (iii) bordered pits (HP). (b) drawings to show (i) annular thickening, (ii) spiral thickening, (iii) scalariform thickening, (iv) reticulate thickening and (v) thickening with bordered pits

In addition to tracheary elements, xylem includes xylem fibres and xylem parenchyma. **Xylem fibres** are similar to sclerenchyma fibres but may have slightly bordered pits. **Xylem parenchyma** is often a little more elongated than the typical parenchyma of ground tissue and the primary wall may be slightly lignified.

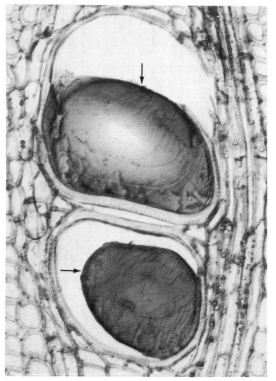

6.10 Tyloses: TS xylem of *Khaya* sp. showing tyloses (arrowed) in vessels (HP)

Phloem is the tissue responsible for the transport of dissolved organic substances (sucrose, amino acids) from the places in which they are formed to the places in which they are used or stored. Phloem contains up to four types of cells, sieve tube elements, companion cells, phloem fibres and phloem parenchyma.

Sieve tube elements and **companion cells** (Fig. 6.11) originate at the same time. During the vacuolation process, differentiating phloem comes to contain rows of rather elongated cells arranged end to end along the axis of the organ. Each of these cells then divides longitudinally into unequal parts. The larger of the two cells produced has a nucleus and

(a)

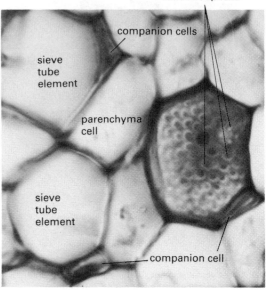

(b)

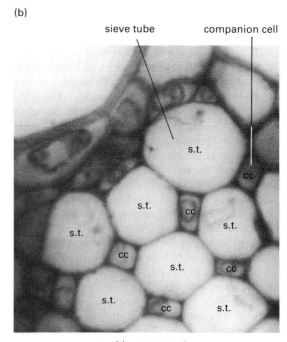

no phloem parenchyma

6.11 Phloem: (a) in a dicot; (b) in a monocot (both HP)

a very thin lining of cytoplasm. It subsequently differentiates into a sieve tube element. The smaller of the two cells contains a nucleus and dense cytoplasm and becomes a companion cell, or may divide transversely and produce two companion cells, end to end, beside the developing sieve tube element. During the further differentiation of the sieve tube element, the nucleus degenerates and disappears. The end walls, adjacent to sieve tube elements in the same row, become perforated with many large, simple pits. The pitted wall is the **sieve plate**. Large, conspicuous cytoplasmic strands pass through the pits in the sieve plate. A series of such sieve tube elements, arranged end to end, is known as a **sieve tube**.

Each sieve tube element has a companion cell, or two companion cells end to end, beside it.

The function of the sieve tube element is the rapid transport of organic substances, especially sucrose and amino acids. The function of the companion cell is believed to be the control of the activities of the sieve tube element, to which it is connected by many plasmodesmata. Sieve tube elements not only lack a nucleus, but have cytoplasm which is almost devoid of organelles. However, both types of cells must be regarded as living when mature.

In monocots, phloem contains only sieve tube elements and companion cells. In most dicots, there is phloem parenchyma and sometimes phloem fibres as well.

Phloem parenchyma is similar to the parenchyma of ground tissue, but the cells are narrower and relatively more elongated.

Phloem fibres are like sclerenchyma fibres.

Surface tissues

The surfaces of the shoot and root systems of herbaceous plants are limited by **epidermis**. Epidermal tissues form the barrier between the tissues of the plant and the immediate environment. Those of the shoot have to cope, usually, with contact with relatively dry air and must provide resistance to desiccation and to mechanical damage, including damage by predators. The epidermis of the root is in contact with the soil and soil water and its functions include the uptake of water and mineral ions from the soil.

The shoot epidermis is derived from the surface layer of the meristem and is almost always one cell thick. It is two or more cells thick in the leaves of some plants (e.g. *Nerium oleander*, Oleander; *Commelina* spp.).

Epidermal cells of stems and elongated leaves are usually elongated along the axis of the organ. Those of broad leaves, including most dicots, are wide and often irregular in outline (Fig. 8.8(a)). The outer surface of the epidermis is covered with a secreted layer of cuticle, continuous from one cell to the next. The **cuticle** is composed of cutin, a hard, waxy substance, and is resistant to the passage of water vapour and other gases.

The epidermis of most above-ground organs includes stomata. A **stoma** is a pore between two specialized epidermal cells, **guard cells**. Usually, there are two or four subsidiary cells associated with the guard cells. The walls of the guard cell are

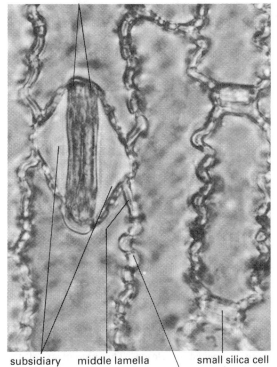

6.12 Leaf epidermis of *Zea* sp. (monocot, HP)

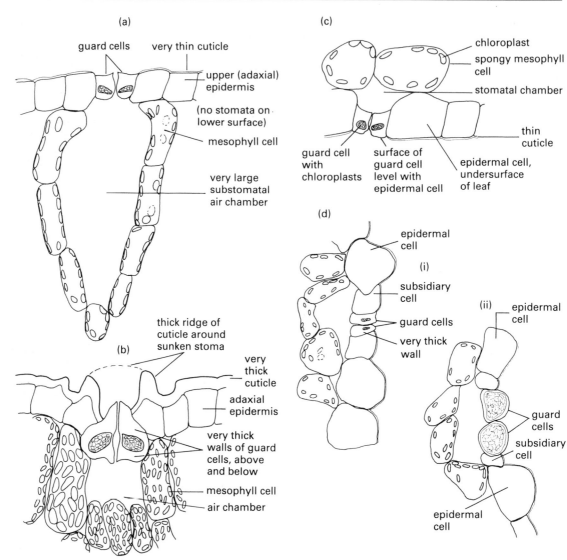

6.13 Stomata in TS: (a) in upper epidermis of a floating (dicot) leaf (see Fig. 8.6); (b) in epidermis of a (monocot) xerophyte; (c) in lower epidermis of a (dicot) mesophyte; (d) in grass epidermis, (i) through the centre of a stoma, (ii) through the end of a stoma (see Fig. 8.8 (b))

unevenly thickened and changes in the water content of the cells cause them to change shape, so that the pore is opened or closed. The physiology of the action of stomata is dealt with in Chapter 8. Examples of different types of stomata are given in Fig. 6.13.

Guard cells have chloroplasts. Other epidermal cells do not.

Many plants have hairs (**trichomes**) on the epidermis of stem and leaf. Trichomes, if numerous, function to trap a layer of moist air on the surface of the organ and thus reduce evaporation. Some trichomes produce sticky secretions, in which case they are said to be glandular.

Trichomes may be unicellular, multicellular and unbranched, or multicellular and branched (Fig. 6.14). Unicellular trichomes are simple outgrowths of epidermal cells. Multicellular trichomes arise from one or more cells of the developing epidermis, by cell division.

(a)

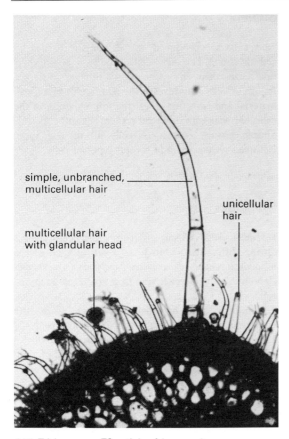

6.14 Trichomes on TS petiole of *Lycopersicum* sp. (Tomato)

Prickles (epidermal emergences, Chap. 4) are formed as more massive outgrowths of groups of epidermal cells.

The root epidermis (Fig. 6.15) is almost always one cell thick. Cells are thin-walled, lack cuticle and are elongated along the axis of the root. A proportion of the epidermal cells form root hairs.

Each **root hair** is an outgrowth of a single epidermal cell. Since most plants have root hairs which are several millimetres long, the root hairs significantly increase the cell surface available for the absorption of water and solutes from the soil.

In epiphytic orchids, the layer of cells which would normally form the epidermis divides several times to produce a velamen, several cells thick. The cells of the velamen have thin cellulose walls and, when mature, lack contents.

(b)

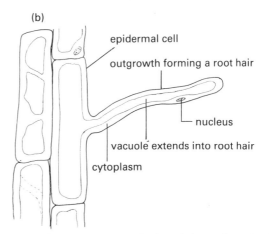

6.15 Root hairs: (a) the growth of root hairs on the roots of seedlings of *Brassica* sp. (Cabbage), 72 hours after the dry seeds were placed on moist filter paper in a covered petridish; (b) diagram to show the structure of a root hair

In woody plants, including woody dicots and some woody monocots, epidermis is eventually replaced by cork (phellem). **Phellem** consists of several layers of cells, the walls of which are suberized (impregnated with suberin). Suberin is a fatty substance. Suberized cell walls are impermeable to water. The cells of phellem are dead when mature. The formation and structure of phellum and associated layers of cells is dealt with in Chapter 10.

Key words

Cells: meristematic cells, mitosis; differentiation, vacuolation, primary wall, secondary wall; pit, simple pit, bordered pit; plasmodesma. **Tissues**: simple tissue, complex tissue, ground tissue, surface tissue. **Simple tissues**: parenchyma, chlorenchyma, aerenchyma; collenchyma; sclerenchyma. **Complex tissues**: xylem, tracheid, vessel element, tracheary element; phloem, sieve tube element, companion cell. **Secondary walls**: lignin, annular, spiral, scalariform, reticulate, with bordered pits. **Surface tissues**; epidermis, cuticle, stomata, guard cells; trichomes; root hairs; phellem, suberin.

Summary

New cells are produced in meristematic tissues by cell division. The new cells produced attain their final size through vacuolation. Further differentiation of cells may involve the addition of more wall material and, in some cases, the impregnation of the cellulose wall with other substances (lignin, suberin). Simple tissues consist of cells all of one type. Complex tissues include several types of cells.

Questions

1 How does a cell cut off from a meristem change during differentiation to form a cell of each of the following types? (a) a xylem tracheid; (b) a vessel element; (c) a sclerenchyma fibre; (d) a parenchyma cell.

2 Compare and contrast trichomes from the epidermis of a stem with root hairs from a root.

3 List all the types of cells which are found in (a) xylem and (b) phloem. State the functions of each cell type in the tissues in which they occur.

7

Stem structure and function

The shoot system is composed of stems and leaves. Leaves are the main photosynthetic organs of the plant. The primary functions of the stem are to support the leaves in the best position for photosynthesis and to transport materials to and from the leaves.

The shoot apex

The **apical meristem** of the shoot is dome-shaped and lies enclosed by the developing leaves of the bud (Fig. 7.1). The apical meristem consists of two distinct layers of cells. The outer layer is the **tunica**, which may be one cell thick, or more than one cell thick, and is characterized by the fact that the cross-walls formed at each cell division are all at right angles to the surface of the meristem (**anticlinal** cell divisions). Within the tunica lies the **corpus**, in which cell divisions occur in all planes (directions). The tunica and corpus together form the **promeristem** of the shoot.

Behind the promeristem lie the **derived meristems**. These are made up of cells which are dividing, but all the products of cell division will eventually differentiate into cells of the primary tissues of the shoot..

Immediately behind the promeristem, the surface develops transverse ridges. These are **leaf primordia**, which gradually elongate as they develop into leaves. As they grow in length, they curve upwards and protect the apical meristem. A little further from the apex, **axillary shoot meristems** develop in the axils of the leaf primordia. Axillary shoot meristems later develop a tunica-corpus structure like that of the main shoot apex.

Derived meristems include protoderm, ground meristem and provascular strands (procambial strands). The **protoderm**, which later forms the epidermis, is derived from the outer layer of the tunica. The outer part of the **ground meristem** is also developed from the tunica, if the tunica is more than one cell thick. The rest of the ground meristem

and the **provascular strands** are derived from the corpus of the promeristem. The distribution of the provascular strands within the ground tissue in dicots is different from that in monocots. In dicots, they usually form a ring as seen in transverse section (TS). In monocots, they rarely form a perfect circle and are always to some extent scattered throughout the thickness of the stem.

Although the promeristem itself is small and narrow, the stem of the shoot produced may be quite wide (e.g. Maize). The thickening of the stem which takes place results from many longitudinal divisions of cells in the region of derived meristems and is known as **primary thickening** of the stem.

Differentiation of tissues in the internode

Differentiation of vascular tissues begins soon after the provascular strands have become distinct. Differentiation of xylem starts at the edge of the provascular strand closest to the centre of the stem and continues centrifugally towards the outside (Fig. 7.2). Differentiation of phloem cells starts at the outside edge of the provascular strand and continues centripetally towards the centre. The first parts of the xylem and phloem to become fully differentiated and all parts which differentiate before the internode has stopped growing in length are the **protoxylem** and **protophloem** respectively. The xylem of the stem is said to be **endarch** because the protoxylem is towards the centre of the stem. Phloem is said to be **exarch** because the first part to differentiate is on the side of the provascular strand nearest the outside of the stem.

Cells of the xylem and phloem which differentiate after the internode has ceased to elongate are **metaxylem** and **metaphloem**. Protoxylem and metaxylem together form the **primary xylem** of the stem. Protophloem and metaphloem form the **primary phloem**.

The extent of differentiation of the cells of the

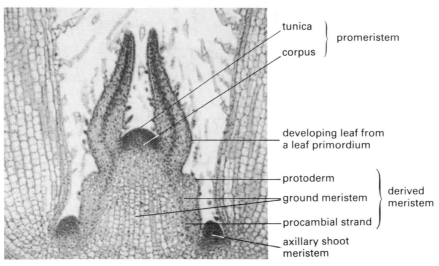

tunica ⎫
         ⎬ promeristem
corpus ⎭

developing leaf from
a leaf primordium

protoderm ⎫
           │ derived
ground meristem ⎬ meristem
           │
procambial strand ⎭

axillary shoot
meristem

7.1 The shoot apex: LS meristematic region,
photomicrograph

(a)

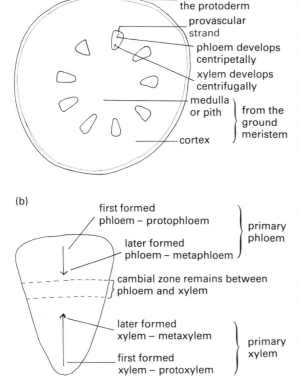

epidermis from
the protoderm

provascular
strand

phloem develops
centripetally

xylem develops
centrifugally

medulla ⎫
or pith │ from the
         ⎬ ground
         │ meristem
cortex ⎭

(b)

first formed
phloem – protophloem ⎫
                      ⎬ primary
later formed   │ phloem
phloem – metaphloem ⎭

cambial zone remains between
phloem and xylem

later formed
xylem – metaxylem ⎫
                   ⎬ primary
first formed   │ xylem
xylem – protoxylem⎭

7.2 Diagrams showing the differentiation of vascular
tissues in the shoot apex, in TS

provascular strands differs in dicots and monocots (Figs. 7.3, 7.4). In dicots, differentiation of xylem and phloem ceases when there is one layer of cells left between xylem and phloem. This row of cells remains meristematic and forms the **fascicular cambium**, in which the cells subsequently divide periclinally (by walls parallel to the surface of the stem) and add to the vascular tissues. Also, the outermost cells of the protophloem, which first differentiated as protophloem parenchyma, often redifferentiate to form fibres. After this has happened, they are known as **protophloem fibres** or **pericycle fibres**. In monocots, differentiation of xylem and phloem continues until all the procambial cells have formed xylem and phloem. There is no cambium. Also, in most monocots, cells immediately around the vascular tissues form a **sheath** of fibres around the vascular tissues.

The strands of xylem and phloem together are **vascular bundles**. Because xylem and phloem lie on the same radius of the stem, their arrangement is said to be **collateral**. The dicot vascular bundle, containing a cambium, is an **open** vascular bundle. That of the monocot, which lacks a cambium and therefore has no possibility of the addition of further vascular tissues, is a **closed** vascular bundle.

At the same time as the tissues of the vascular bundles are differentiating, other tissues are differentiating. The protoderm differentiates into

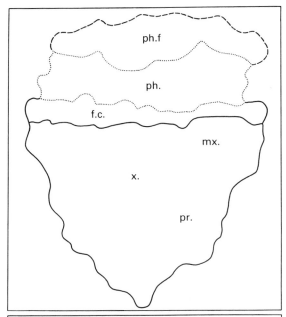

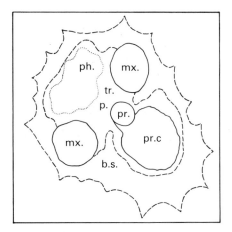

ph.f. protophloem fibres
ph.   phloem (sieve tubes, companion cells,
       parenchyma)
f.c.   fascicular cambium
x.   xylem; mx. metaxylem, pr. protoxylem

ph.   phloem (sieve tubes, companion cells)
mx.   metaxylem
p.   parenchyma
tr.   tracheids, mixed with parenchyma,
       between metaxylem vessels
pr.   protoxylem
pr.c. protoxylem canal
b.s.   bundle sheath

7.3 Open vascular bundle of a dicot, from TS stem of *Tridax* sp. See also Fig. 7.6

7.4 Closed vascular bundle of a monocot, from TS stem of *Zea mays*. See photo Fig. 7.8

epidermis. In dicots, ground tissue outside the vascular bundles forms the **cortex** of the stem. Ground tissue inside the ring of vascular bundles forms the **medulla** (**pith**) of the stem. In most monocots it is impossible to separate cortex and medulla, because the vascular bundles are more or less scattered, and we may only refer to the **ground tissue**.

All the tissues which differentiate from cells derived directly from the apical meristem are **primary tissues**. The primary tissues make up the primary structure or primary body of the stem.

The primary structure of the dicot internode

The epidermis is composed of a layer of cells which are closely attached to each other, with a layer of cuticle on the outside surface. The cells are rectangular in TS and elongated in LS. There are stomata in the epidermis wherever there is chlorenchyma towards the inside.

The outer part of the cortex usually contains alternating longitudinal strips of collenchyma and chlorenchyma. Sclerenchyma replaces collenchyma in some stems. The inner cortex is usually composed of rather large-celled parenchyma. The innermost layer of the cortex, lying next to the protophloem and dipping inwards between the vascular bundles to about the level of the cambium, is more compact than the rest and forms the **starch sheath** or **endodermis** (Fig. 7.5). In most dicots (e.g. *Talinum triangulare*, Water Leaf; *Tridax procumbens*), the cells of this layer contain conspicuously more starch grains than neighbouring parenchyma cells, which is the reason for the first name. In others (e.g. *Helianthus annuus*, Sunflower) the cells have an additional thick-

× 500

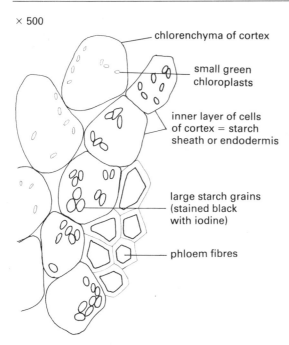

chlorenchyma of cortex

small green
chloroplasts

inner layer of cells
of cortex = starch
sheath or endodermis

large starch grains
(stained black
with iodine)

phloem fibres

7.5 Starch sheath in TS stem of *Tridax* sp.

ening in their radial walls. This thickening is called the **Casparian strip** (**Casparian band**). Where a Casparian strip is present, it is more appropriate to refer to this layer of cells as an endodermis. The structure and function of the endodermis and the Casparian strip are dealt with in Chapter 9.

Between the vascular bundles there is parenchyma forming the **medullary rays**, continuous with the parenchymatous medulla at the centre of the stem.

Primary phloem in dicots is composed of sieve tubes, companion cells and phloem parenchyma, often with a group of protophloem fibres towards the outside (Fig. 7.6). In the **primary xylem**, tracheids and vessels may be scattered or arranged in radial rows. Xylem parenchyma is always present but there are few or no xylem fibres in primary xylem. It is rare for cambium to be seen as a single row of cells that are clearly different from their neighbours. This is because cambial cells usually start to divide and add to xylem and phloem before the metaxylem and metaphloem have completed their differentiation. Cells divided off from the cambium, mostly indistinguishable from cambial cells, are arranged in radial rows each side of the cambium. Because of the

difficulty in identifying the cambium as a definite layer of cells, it is often best to label or refer to the **cambial region**, when interpreting a section.

The primary structure of the monocot internode

The epidermis of the monocot is similar to that of the dicot. The nature of the ground tissue immediately inside the epidermis varies a great deal. It may be composed of collenchyma and chlorenchyma (e.g. *Commelina africana*) or it may be composed of small, closely packed cells which soon become lignified (e.g. Maize and most other grasses). The rest of the ground tissue is parenchyma, except for the fibrous bundle sheaths.

The phloem in the vascular bundle consists of sieve tubes and companion cells only. There is no phloem parenchyma. Adjacent to the phloem, there are usually two large metaxylem vessels, with tracheids between them (Fig. 7.8). Towards the centre of the stem there is a radial row of protoxylem vessels. In the young stem, these are embedded in small-celled xylem parenchyma but this parenchyma is broken up during the final elongation of the stem leaving the vessels lying in a space, the **protoxylem canal**.

The pattern of the monocot vascular bundle, as seen in TS, is very constant. Variation lies in the number of metaxylem vessels and in the extent of the bundle sheath, which may be incomplete (e.g. *Dioscorea* sp., Yam).

The anatomy of the node

At some level within the dicot internode, some of the vascular bundles branch and produce additional bundles, which supply the vascular system of the leaf at the next higher node. These **leaf trace bundles** remain as part of the cylinder of vascular bundles until close to the node, at which point they pass obliquely through the cortex and connect with the vascular bundles of the leaf base. **Axillary branch traces** are formed in the same manner.

In monocots, leaf and branch traces include small bundles from the periphery of the stem and larger bundles from nearer the centre of the stem, all of

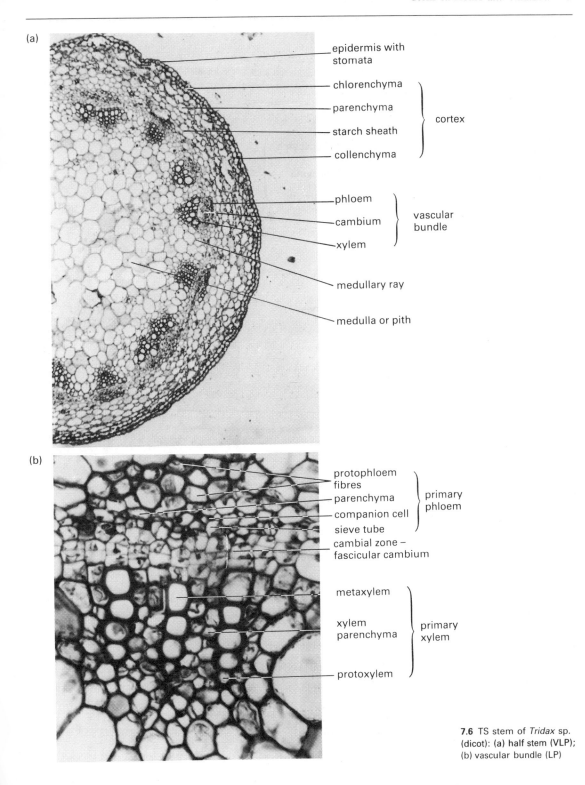

(a)

— epidermis with stomata

— chlorenchyma

— parenchyma   } cortex

— starch sheath

— collenchyma

— phloem

— cambium   } vascular bundle

— xylem

— medullary ray

— medulla or pith

(b)

— protophloem

fibres

— parenchyma   } primary phloem

— companion cell

sieve tube

cambial zone –

fascicular cambium

metaxylem   } primary xylem

xylem

parenchyma

protoxylem

7.6 TS stem of *Tridax* sp. (dicot): (a) half stem (VLP); (b) vascular bundle (LP)

× 20

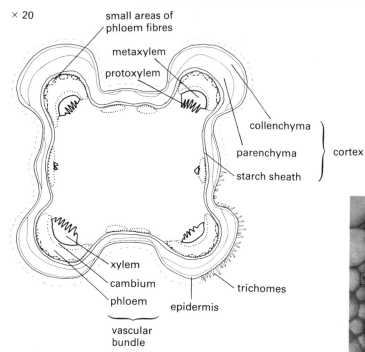

small areas of
phloem fibres

metaxylem

protoxylem

collenchyma

parenchyma   } cortex

starch sheath

xylem

cambium

phloem

trichomes

epidermis

vascular
bundle

7.7 TS stem of *Salvia* sp. (dicot), showing the distribution
of tissues in a 'square' stem

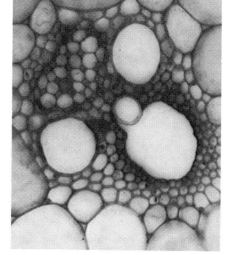

7.8 Vascular bundle of *Zea mays* (monocot).
Labelling as Fig. 7.4

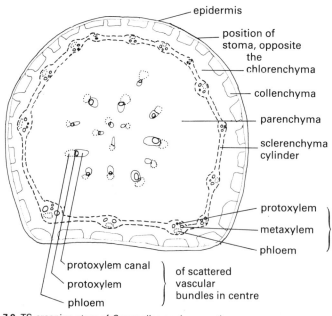

epidermis

position of
stoma, opposite
the

chlorenchyma

collenchyma

parenchyma

sclerenchyma
cylinder

protoxylem   } of vascular
                 bundle embedded
metaxylem   in cylinder of
                 sclerenchyma
phloem

protoxylem canal   } of scattered
                     vascular
protoxylem   bundles in centre
phloem

7.9 TS creeping stem of *Commelina* sp. (monocot)

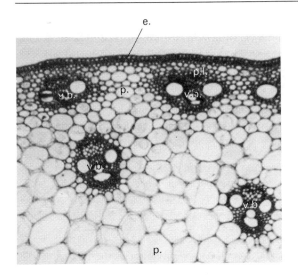

e. epidermis

p.l. parenchyma lignified

p. larger thin-walled parenchyma

v.b. vascular bundle

7.10 TS of outer part of stem of *Zea mays* (monocot) showing thickening and lignification of walls of cells of the ground tissue close to the surface (LP)

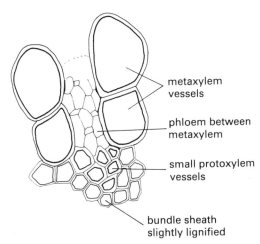

7.11 TS vascular bundle of *Asparagus* sp. (monocot) showing U-shaped xylem

which pass obliquely through the ground tissue in the region below the node.

The origin of leaf and branch trace bundles and

their connection with the leaf is best studied by making a series of sections (serial sections) from the internode and up into the node. In this way, it is possible to follow the course of each vascular bundle and to reconstruct a three-dimensional picture of the pathway followed by leaf and branch traces (Figs. 7.12 and 7.13).

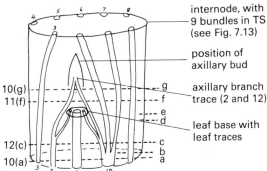

letters (a) to (g) refer to transverse sections in Fig. 7.13
Numbers to left of diagram are the number of vascular bundles in the section

7.12 Diagram of the vascular supply in the node of *Talinum* sp., built up from a study of serial sections, as in Fig. 7.13

Older herbaceous stems

Monocots

Since there is no cambium, the structure of the monocot stem usually undergoes no major changes after the primary tissues have differentiated. Epidermal cells and cells in the outer part of the ground tissue often become lignified and nothing more. Examples of **anomalous secondary thickening** in monocot stems are described in Chapter 10.

Dicots

In dicots, further development usually starts before the primary tissues are fully differentiated. As differentiation proceeds in metaxylem and metaphloem, cells of the fascicular cambium start to divide. Divisions of cambial cells are all **tangential** (parallel to an imaginary tangent to the surface of the stem) and new cells are added towards the xylem and phloem in the ratio of about 3:1. Cells cut off from

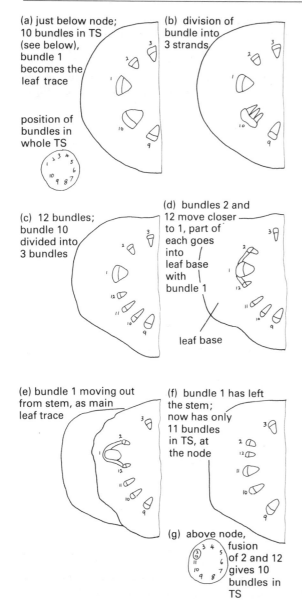

(a) just below node; 10 bundles in TS (see below), bundle 1 becomes the leaf trace

position of bundles in whole TS

(b) division of bundle into 3 strands

(c) 12 bundles; bundle 10 divided into 3 bundles

(d) bundles 2 and 12 move closer to 1, part of each goes into leaf base with bundle 1

leaf base

(e) bundle 1 moving out from stem, as main leaf trace

(f) bundle 1 has left the stem; now has only 11 bundles in TS, at the node

(g) above node, fusion of 2 and 12 gives 10 bundles in TS

7.13 Serial transverse sections of stem of *Talinum* sp., in nodal region, showing origin of leaf trace

the cambium usually divide again, tangentially, once or twice, before they begin to differentiate into cells of the **secondary xylem** and **secondary phloem**.

When the cambium starts to divide, tangential divisions also occur in neighbouring parenchyma cells of the medullary ray (Fig. 7.14). Each parenchyma cell in contact with the fascicular cambium

divides twice. The middle of the three cells so formed becomes a new cambial cell. The formation of cambial cells from medullary ray parenchyma cells continues across the ray until the fascicular cambia are linked by newly formed **interfascicular cambia** so that the two together form a continuous cylinder of cambium around the stem.

Strictly speaking, the fascicular cambium is a remnant of the primary meristem and its cells have remained meristematic since they were derived from the apical meristem. The interfascicular cambium is a **secondary meristem**, since it is formed by **dedifferentiation** of parenchyma cells. However, new tissues added by both fascicular and interfascicular cambia are regarded as secondary tissues.

Division of the cells of the fascicular cambium adds **secondary xylem** towards the centre of the stem and **secondary phloem** towards the outside. Cells cut off from the interfascicular cambium add secondary tissue to the medullary rays. Medullary rays are usually wide in herbaceous plants but are usually narrow in woody plants.

Most herbaceous plants form a limited amount of secondary vascular tissues (Fig. 7.15). This leads to an increase in the thickness of the vascular cylinder, which tends to stretch the cortex and epidermis. Cortical cells accommodate to the increasing diameter of the vascular cylinder by undergoing **anticlinal** cell divisions and thus producing more cells. The epidermis is less able to accommodate to increasing diameter, owing to the rigidity of the cuticle, and may ultimately be replaced by **periderm**, a secondary tissue (Chapter 10).

Secondary xylem and secondary phloem can be distinguished from the primary tissues because their cells are arranged in radial rows. Also, secondary xylem usually contains fibres while primary xylem usually does not.

Variations in primary stem structure

The structures which have so far been described and illustrated are those of 'typical' dicots and monocots. In fact, it is difficult to determine what should be regarded as 'typical' and there is considerable variation in both groups.

Dicots
Nearly all dicots have their vascular bundles arranged in a ring, as seen in TS.

(a)

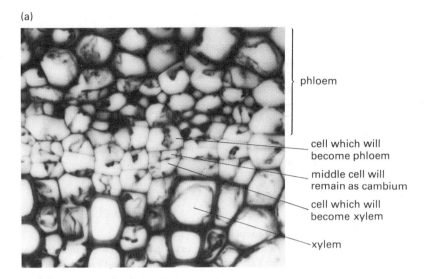

phloem

cell which will
become phloem

middle cell will
remain as cambium

cell which will
become xylem

xylem

(b)

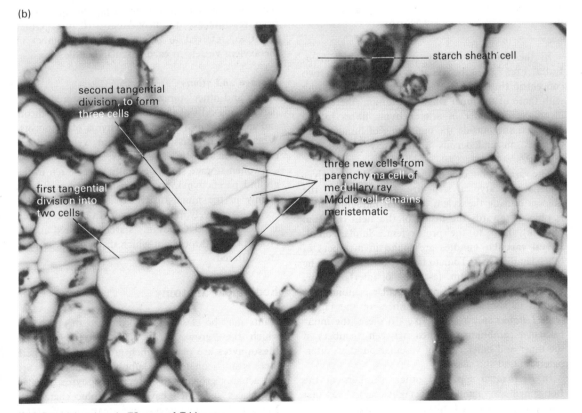

starch sheath cell

second tangential
division, to form
three cells

three new cells from
parenchyma cell of
medullary ray
Middle cell remains
meristematic

first tangential
division into
two cells

7.14 Cambial regions in TS stem of *Tridax* sp.:
(a) fascicular cambium in the vascular bundle;
(b) interfascicular cambium developing in the medullary ray
(both HP)

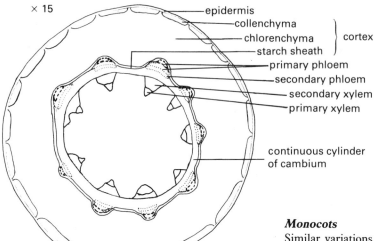

× 15

epidermis
collenchyma
chlorenchyma
starch sheath
} cortex
primary phloem
secondary phloem
secondary xylem
primary xylem

continuous cylinder
of cambium

7.15 TS old (herbaceous) stem of *Talinum* sp. showing the beginnings of secondary thickening

In the pumpkin family (e.g. *Luffa cylindrica*, Luffa, Fig. 7.16), the vascular bundles are in two concentric rings, those of the outer ring alternating with those of the inner ring. In addition, there is an incomplete cylinder of sclerenchyma fibres in the ground tissue outside the outer ring of bundles and there is an endodermis immediately outside the fibres.

The vascular bundles of the Luffa and its relatives also have an atypical structure. In addition to the phloem outside the xylem (**external phloem**), there is phloem towards the inside of the stem (**internal phloem**), next to the protoxylem. Similar **bicollateral vascular bundles** are found in members of other families (e.g. Tomato family, Solanaceae). The Luffa has no cambium, fascicular or interfascicular, but both fascicular and interfascicular cambium are formed in some perennial members of the family (e.g. *Telfairia* sp.; *Sechium* sp.). In these, the interfascicular cambium is formed between members of the inner and outer rings of vascular bundles, alternately, round the stem.

In *Boerhaavia* spp. and other members of the Bougainvillaea family (Nyctaginaceae), vascular bundles are scattered through the ground tissue, rather like those of a monocot, and lack cambium. This family will be referred to later as an example of anomalous secondary thickening (Chapter 10).

Monocots

Similar variations exist in monocots. In members of the Yam family (Dioscoreaceae) and hollow-stemmed grasses (e.g. *Oryza sativa*, Rice), the largest vascular bundles form an almost perfect ring, giving a superficial resemblance to dicot stem structure. Other variation, referred to before, is in the number of metaxylem vessels in each vascular bundle.

Storage and stems

Starch and other storage products are stored in the parenchyma cells of the cortex and medulla of dicots and in the ground tissue parenchyma of monocots. When secondary tissues are present, xylem and ray parenchyma are also storage sites.

Swollen storage stems generally store starch in the ground parenchyma. In the Irish Potato, storage is mainly in the much distended medulla, while the cortex is relatively much narrower. In monocots such as the Yam and Cocoyam, starch is stored in the ground tissue throughout which the vascular bundles are scattered.

Ecological anatomy

Plants may be classified according to the places in which they grow. Most plants are mesophytes. **Mesophytes** are plants which grow naturally where there is an adequate water supply but where they are not actually growing in water. **Xerophytes** grow in dry places and often have xeromorphic characters. Xeromorphic characters are features of their morphology or anatomy which appear to be related to the conservation of water. **Hydrophytes** are water plants and have hydromorphic characters.

× 15

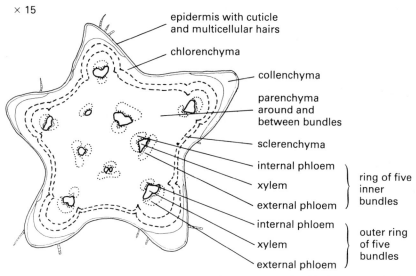

epidermis with cuticle and multicellular hairs

chlorenchyma

collenchyma

parenchyma around and between bundles

sclerenchyma

internal phloem ⎫
xylem ⎬ ring of five inner bundles
external phloem ⎭

internal phloem ⎫
xylem ⎬ outer ring of five bundles
external phloem ⎭

7.16 Variation in dicot stem primary structure, TS *Luffa* sp. showing bicollateral vascular bundles

(a)

flower

fleshy cylindrical leaves

succulent stem

(b) × 15

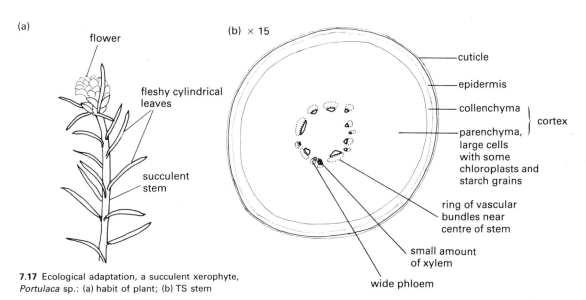

cuticle

epidermis

collenchyma ⎫
⎬ cortex
parenchyma, ⎭ large cells with some chloroplasts and starch grains

ring of vascular bundles near centre of stem

small amount of xylem

wide phloem

7.17 Ecological adaptation, a succulent xerophyte, *Portulaca* sp.: (a) habit of plant; (b) TS stem

Some xerophytes are succulent. **Succulent** plants have much fleshy, water-storing tissue in both stems and leaves. Stems of succulent plants (e.g. *Portulaca* spp.) have large water-storing parenchyma cells in cortex and medulla and often have a wider cortex than mesophytes (Fig. 7.17).

Succulent plants in which the stems are the main photosynthetic organs (e.g. many *Euphorbia* spp.) have much chlorenchyma in the cortex.

Non-succulent xerophytes with photosynthetic stems also have well-developed chlorenchyma in the cortex (e.g. *Casuarina* spp.) or in the outer part of the ground tissue (*Asparagus sp.*). The stems of *Casuarina* spp. are longitudinally grooved and stomata, surrounded by trichomes, are confined to the bases of the grooves (Fig. 7.18).

The stems of hydrophytes (e.g. *Ceratophyllum* sp.) have a small medulla and the vascular bundles are poorly developed. In dicots, the cortex is wide in proportion and contains large air spaces (Fig. 7.19). In monocots, there are air spaces in the ground tissue. There are no stomata in the epidermis of submerged branches and the cuticle is very thin.

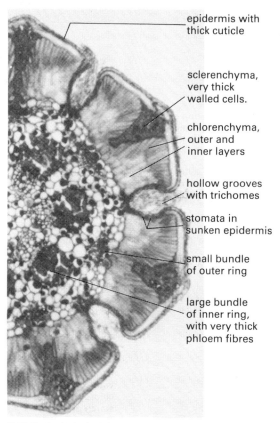

epidermis with thick cuticle

sclerenchyma, very thick walled cells.

chlorenchyma, outer and inner layers

hollow grooves with trichomes

stomata in sunken epidermis

small bundle of outer ring

large bundle of inner ring, with very thick phloem fibres

7.18 Ecological adaptation, a non-succulent xerophyte, *Casuarina* sp., TS stem

Stem function

In all typical plant stems, the mechanical tissue is distributed well away from the centre of the stem, where it gives the best resistance to bending strains. Mechanical tissue includes sub-epidermal collenchyma or sclerenchyma, phloem fibres and, to a lesser extent, xylem. Primary xylem does not contain many mechanical cells.

Stems are always adequately supplied with vascular tissue, which enables them to act as efficient organs for the transport of water to the leaves and the transport of organic substances from the leaves to other parts of the plant.

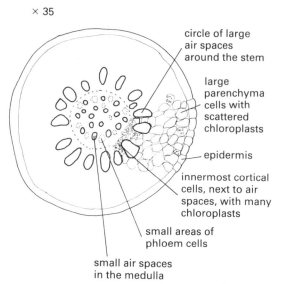

× 35

circle of large air spaces around the stem

large parenchyma cells with scattered chloroplasts

epidermis

innermost cortical cells, next to air spaces, with many chloroplasts

small areas of phloem cells

small air spaces in the medulla

7.19 Ecological adaptation, a hydrophyte stem, TS stem of *Ceratophyllum* sp.

Key words

Apex: promeristem, tunica, corpus; derived meristem, protoderm, ground meristem, provascular (procambial) strands. **Primary xylem**: centrifugal, endarch, protoxylem, metaxylem. **Primary phloem**: centripetal, exarch, protophloem, metaphloem. **Vascular bundle**: xylem, phloem, open (dicot), closed (monocot). **Cambial region**: interfascicular, secondary meristem, dedifferentiation. **Ground tissue**: cortex, medulla (pith), medullary rays; starch sheath (endodermis), Casparian strip.
Secondary tissue: fascicular cambium, interfascicular cambium, tangential divisions, secondary xylem, secondary phloem. **Ecological anatomy**: mesophytes; hydrophytes, hydromorphic characters; xerophytes, xeromorphic characters; succulents.

Summary

The primary structure of the stem develops from cells derived from the apical meristem. The apical meristem consists of a promeristem, from which all other cells are derived, and three derived meristems. Of the three derived meristems, the protoderm gives rise to the epidermis of the stem, the ground meristem gives rise to the ground tissues and the provascular (procambial) strands give rise to the vascular bundles. In dicots, vascular bundles are usually arranged to form a cylinder, with cortex outside, medulla inside and medullary rays between adjacent bundles. In monocots, vascular bundles are usually scattered throughout the ground tissue.

In each bundle, xylem is endarch and phloem is exarch. In dicots, a layer of meristematic tissue, the fascicular cambium, remains between xylem and phloem and soon divides to give rise to secondary xylem and phloem. The cambia of adjacent vascular bundles become joined into a cylinder by the formation of interfascicular cambium, by dedifferentiation of ray parenchyma cells. The interfascicular cambium adds secondary tissues to the rays.

There is no cambium in monocots.

Stem structure is often modified in several characteristic ways in xerophytes and hydrophytes.

Questions

1 What is a primary meristem? Distinguish between (a) promeristem, and (b) derived meristems.

2 In what ways does the development of primary xylem differ from that of primary phloem, in a dicot stem.

3 Suppose you are given a slide of TS stem and asked to decide whether the stem was that of a dicot or a monocot. What characters would you look for and what would you expect to find, if it was (a) a section of a dicot stem, and (b) a section of a monocot stem?

8

Leaf structure and function

The leaf is the main photosynthetic organ of the plant. Photosynthesis takes place mainly in the lamina. The leaf base, petiole, midrib and veins all carry out contributory functions.

The origin and development of the leaf

As has been described in Chapter 7, leaf primordia first appear as transverse ridges on the surface of the apical meristem of the shoot (Fig. 7.1). Cells close to the surface of the corpus divide periclinally and anticlinal divisions take place in the overlying tunica. Once the primordium is established, it increases in length by meristematic activity at the tip and increases in width by meristematic activity at the sides. The final shape and size of the leaf results from further cell divisions throughout the developing lamina.

The parts of the leaf

Leaf base and petiole

The **leaf base** is often slightly swollen. If it is very swollen, it is called a **pulvinus** (e.g. *Mimosa pudica*, Sensitive Plant; *Cassia* spp. Fig. 3.7). A pulvinus is capable of growth movements which turn the leaf so that it is held at right angles to incident light. Stipules may be present or absent and, if present, may drop off at an early stage. When they are persistent, they are sometimes large and leafy and form part of the photosynthetic apparatus of the leaf. The sheathing leaf bases of some monocots (e.g. Maize; Banana) are also important in the positioning of the leaves. They raise the leaf blade well above the level of the node, on what is initially a very short stem (Fig. 4.5, Banana 'stem').

The **petiole** is essentially stem-like in structure. Some petioles are radially symmetrical like a stem (e.g. *Luffa cylindrica; Carica papaya*, Pawpaw). Most are dorsiventral in structure (e.g. *Tridax procumbens; Lycopersicum esculentum*, Tomato). The vascular bundles may be widely separated or may be so close together that they form an arc of vascular tissue with

only very narrow rays. In dicots, cambium may be present or absent but, if present, is almost or totally inactive. There are usually conspicuous strips of sub-epidermal collenchyma. Examples of petiole structure in dicots and monocots are shown in Figs. 8.1 and 8.2.

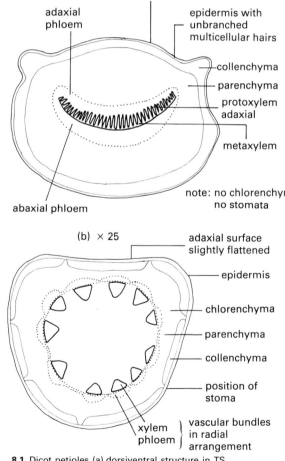

(a) × 50

adaxial surface

adaxial phloem

epidermis with unbranched multicellular hairs

collenchyma

parenchyma

protoxylem adaxial

metaxylem

abaxial phloem

note: no chlorenchyma no stomata

(b) × 25

adaxial surface slightly flattened

epidermis

chlorenchyma

parenchyma

collenchyma

position of stoma

xylem
phloem } vascular bundles in radial arrangement

8.1 Dicot petioles (a) dorsiventral structure in TS *Catharanthus* sp. (formerly *Lochnera* sp.), (b) radial structure in TS *Ricinus* sp.

The structure of the rachis of a pinnate leaf is similar to that of the petiole.

Like the leaf base, the petiole is important in the positioning of the leaf blade. If a plant is viewed from above, its leaf blades form a leaf mosaic. **A leaf mosaic** is an arrangement of the leaves in such a way that they shade each other as little as possible. This is achieved by the combined actions or characteristics of leaf bases and petioles, but may also involve bending of the stem.

much the same even in ferns and lycopods. There is always an upper (**adaxial**) and lower (**abaxial**) epidermis, one or both of which contain stomata. Chlorenchyma forms **mesophyll** between the adaxial and abaxial epidermis. Vascular bundles, each enclosed in a parenchymatous **bundle sheath**, lie at about the centre of the mesophyll. Variations in detailed structure are related to the symmetry of the

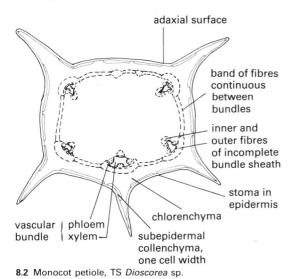

8.2 Monocot petiole, TS *Dioscorea* sp.

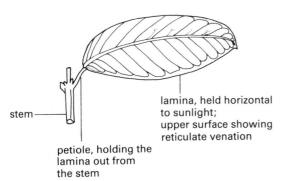

8.3 Dorsiventral symmetry, a simple horizontal leaf of *Psidium* sp. (× ½)

Leaf blade

The **midrib** of the leaf is dorsiventral in structure and generally repeats the pattern of the petiole. The same is true, on a smaller scale, of the lateral veins.

The basic structure of the **lamina** of the leaf is very constant throughout flowering plants and is very

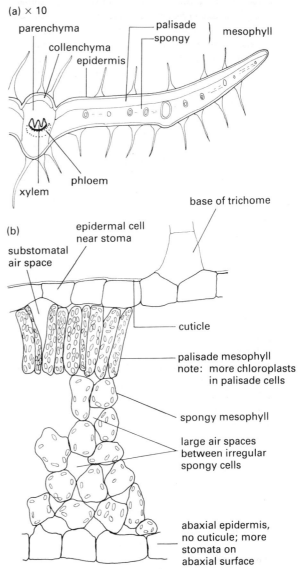

8.4 Dorsiventral leaf structure, *Tridax* sp., (a) TS half leaf (VLP), (b) cellular detail in lamina (HP)

leaf. Leaves which are held in a horizontal position have **dorsiventral** symmmetry. Leaves which stand upright, or hang downwards, have **isobilateral** symmetry. A leaf with isobilateral symmetry has no 'upper' and 'lower' sides and has similar internal structure next to each epidermis.

Dorsiventral leaf structure (Fig. 8.4) is typical of dicots but is found also in some monocots (e.g. *Dioscorea* sp., Yam). In such leaves, the adaxial epidermis has a thicker cuticle and has fewer stomata than the abaxial epidermis. There may be no stomata in the adaxial epidermis. Immediately beneath the adaxial epidermis there are one or several layers of **palisade mesophyll**. The cells of palisade mesophyll are elongated at right angles to the epidermis. Although the cells of palisade mesophyll often appear closely packed when seen in a TS of the leaf, thin sections and sections cut parallel to the surface of the leaf show abundant air spaces between the cells. **Spongy mesophyll** occupies the lower part of the leaf. In this, the cells are rounded or irregularly shaped and air spaces are very conspicuous. The largest air spaces exist next to where there are stomata in the abaxial epidermis. These are **substomatal chambers**. The vascular bundles of small veins are located at the junction of palisade and spongy mesophyll. Each vein contains xylem and phloem, surrounded by a layer of parenchyma cells which form a bundle sheath. The cells of the bundle sheath are closely attached to each other and are larger in plants with C4 photosynthesis (e.g. *Amaranthus* spp.) than in those with the commoner C3 photosynthesis (see Chapter 11).

Isobilateral leaves (Fig. 8.5) are typical of monocots (e.g. *Gladiolus* sp.) but occur in some dicots (e.g. *Protea* spp.; *Eucalyptus* spp.). Here, there is no difference in cuticle thickness or stomatal distribution between the adaxial and abaxial epidermis. There is usually little differentiation of the mesophyll into palisade and spongy layers. In some leaves (e.g. Maize), the mesophyll is all more or less of the spongy type. In others (e.g. *Eucalyptus* spp.), it is all palisade mesophyll.

Ecological anatomy

Xerophytes

The leaves of many xerophytes (e.g. cactus-like *Euphorbia* spp.) are much reduced and the photo-synthetic function is taken over by the stem. Here, the leaves are strongly cuticularised and have few stomata. The mesophyll is rather compact, with small air spaces and small veins. Such leaves are usually short-lived.

Other xerophytes have succulent leaves or are sclerophyllous. A sclerophyllous plant has hard leaves, with thick cuticle and much lignified tissue.

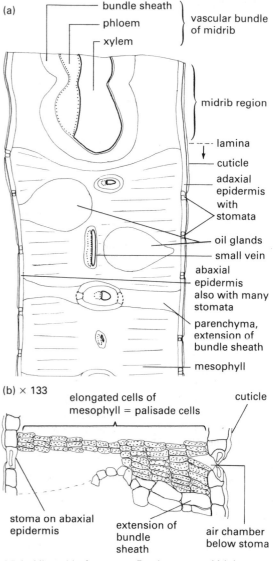

8.5 Isobilateral leaf structure, *Eucalyptus* sp., which has a pendant leaf, (a) TS midrib and lamina (LP), (b) cellular detail in lamina (HP)

Succulent leaves (e.g. *Bryophyllum* spp.; *Portulaca* spp.) have a thick cuticle and the mesophyll is composed of large cells with rather small air spaces. The mesophyll cells have the additional function of water storage, thus enabling the plant to survive dry periods.

Sclerophyllous leaves (e.g. *Butyrospermum* sp., Shea Butter Tree) usually have no stomata in the adaxial epidermis and those in the abaxial epidermis are often sunken below the general level of the epidermis and lie in small pits. In *Nerium oleander*, the lower surface of the leaf is covered with deep, branched pits which are infoldings of the epidermis. The stomata, surrounded by trichomes, are all located at the deepest points of the pits. In addition, the Oleander leaf has a multiple epidermis (more than one layer of cells) on the adaxial surface and on exposed parts of the abaxial surface.

Other xerophytes have a dense covering of trichomes on both adaxial and abaxial epidermis. A dense covering of hairs restricts air movement and thus retains a layer of relatively moist air at the leaf surface.

Eucalyptus spp., introduced trees which are often planted in tropical countries as a source of timber, show several xeromorphic characters in their leaf structure. The leaves of young plants (**juvenile foliage**) are sessile and opposite and have fairly typical dorsiventral structure but the leaves of older plants (**adult foliage**) are petiolate and **isobilateral** and hang downwards. The hanging position is achieved by twisting of the petiole, which can be easily observed. The hanging position is believed to protect the photosynthetic tissue from direct sunlight, when the light is most intense, in the middle of the day.

Hydrophytes

Leaves on totally submerged parts of plants are usually narrow, or divided into several narrow lobes (e.g. *Ceratophyllum* spp.). Their shape protects them from possible damage by strong water currents. They are also of rather simple structure. The epidermis has a very thin cuticle and no stomata. The mesophyll is all spongy, with large air spaces.

Floating leaves (e.g. *Nymphaea* spp., Water Lilies) have normal dorsiventral leaf structure, except that the air spaces in the spongy mesophyll are very large (Fig. 8.6). These large air spaces give buoyancy to the leaf. Stomata are confined to the upper epidermis, which has a waxy cuticle. The waxy cuticle repels water, so that any water that splashes onto the upper surface of the leaf runs off without wetting the surface.

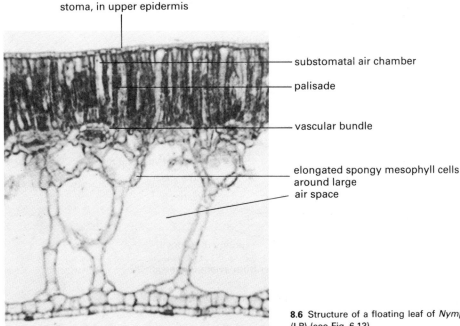

stoma, in upper epidermis

substomatal air chamber

palisade

vascular bundle

elongated spongy mesophyll cells around large air space

8.6 Structure of a floating leaf of *Nymphaea* sp., TS lamina (LP) (see Fig. 6.13)

Leaves of the floating plant *Pistia stratiotes* (Water Lettuce) have a dense covering of hairs on both surfaces. This traps a layer of air against the leaf surface so that, if submerged, the plant quickly surfaces again without wetting the leaf surface. The mesophyll of the Water Lettuce leaf is in the form of air chambers. Chlorenchyma forms the walls of the chambers.

Another floating plant, *Eichhornia crassipes* (Water Hyacinth) has its petioles inflated at the base, where there are large air spaces in the ground tissue.

Leaf function

The process of photosynthesis requires supplies of water, carbon dioxide and light. It also requires a means for the removal of sugar and oxygen, the products of photosynthesis. All these requirements are provided for by the structure of the leaf.

Water

Water is supplied to the mesophyll cells by way of the xylem, with which the leaf is abundantly provided by way of the vascular bundles in petiole, midrib, major veins and the fine veins of the lamina. No mesophyll cell is more than a few cells away from a vein or vein ending. Water passes from the xylem to the mesophyll by way of the living cells of the bundle sheath (Fig. 8.7).

Carbon dioxide

Air contains about 0.03 per cent of carbon dioxide. Carbon dioxide diffuses into the leaf by way of open stomata. The mechanism of stomatal action is dealt with below.

Light

The broad, thin shape of the leaf exposes the mesophyll cells of the leaf to light, to the best advantage. The movements of leaf base and petiole ensure that the leaves are normally held at right angles to incident light.

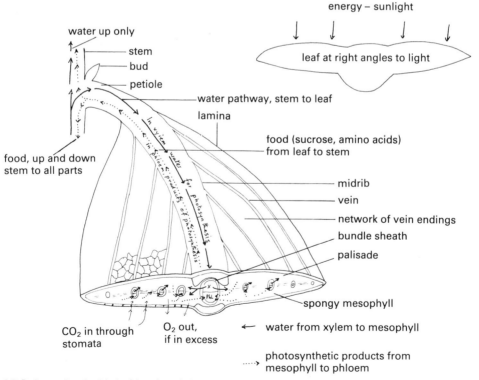

8.7 Pathways involved in leaf function of photosynthesis

Removal of the products of photosynthesis

The first stable carbohydrate product of photosynthesis is glucose. This is usually converted into starch, as a temporary storage product, either in the chloroplasts of mesophyll cells (C_3 plants) or in the chloroplasts of cells of the bundle sheath (C_4 plants). Starch is subsequently converted to sucrose and is removed from the leaf in this form. Sucrose travels from the leaf in the phloem of veins, midrib and petiole. Other substances which are synthesised in the leaf (amino acids) are removed from the leaf by the same pathway.

Oxygen which is produced by photosynthesis diffuses from the cell into the intercellular spaces and thence into the atmosphere by way of open stomata.

Stomatal action

One of the requirements for photosynthesis is free movement of gases (carbon dioxide and oxygen) between the atmosphere and air spaces in the leaf. Water vapour, also, is a gas and free diffusion of gases must involve free movement of water vapour.

In a healthy plant, the walls of the mesophyll cells are constantly wet and the pores in the cellulose walls are always full of water. This facilitates the uptake of carbon dioxide from the air in the intercellular spaces. The air in the intercellular spaces is, therefore, always saturated with water vapour. Since the air outside the leaf is usually less than fully saturated with water vapour, much water vapour is lost, through open stomata, to the atmosphere. This loss of water by evaporation (**transpiration**) is a necessary evil, if photosynthesis is to take place. It is therefore vital that there should be a mechanism by which loss of water vapour is kept to a minimum.

The mechanism of stomatal action

A stoma is a pore in the epidermis between two specialized epidermal cells (**guard cells**). The mechanism by which stomata open and close depends as much on the structure and composition of the guard cells, relative to other epidermal cells, as upon physiological factors.

Guard cells

Guard cells usually contain chloroplasts. Other epidermal cells do not. The shape of guard cells varies from species to species (Fig. 6.13), but there are two basic types, (a) banana-shaped guard cells, found in most plants, and (b) dumb-bell-shaped

guard cells, found in grasses. In either case (Fig. 8.8), stomata are open when the guard cells are turgid and closed when they are flaccid.

In the elliptical type of stoma, with banana-shaped guard cells, opening of the pore is due to the fact that the guard cells curve away from each other as they become more turgid. The wall of each guard cell adjacent to the pore is thicker than that on the side away from the pore. Thinner parts of the wall

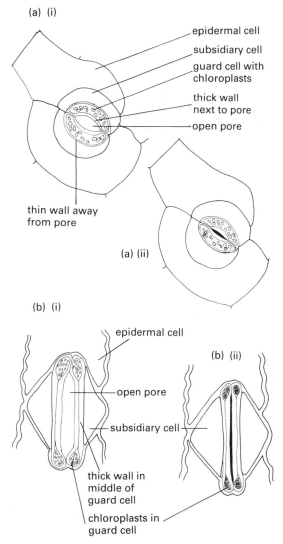

8.8 Stomata, showing different types of guard cells, (a) elliptical stoma of *Talinum* sp., (b) graminaceous stoma of *Zea mays*. In each pair, (i) open pore, (ii) closed pore

stretch more easily than thicker parts and this causes the guard cells to curve away from each other as they take up water.

In the grass (**graminaceous**) type of stoma, with elongated guard cells which are wide and thin-walled at the ends, the walls of the swollen ends become more swollen as water is taken up and thus push the sides of the stoma apart. In both types of stoma, loss of water as the guard cells become flaccid allows the guard cells to close, or almost close, together.

Changes in guard cell turgor

The two main factors which affect guard cell turgor are light, and carbon dioxide concentration within the guard cells.

The classical theory of stomatal control was based on the observation that pH changes affect the reaction of a starch-hydrolysing enzyme, starch phosphorylase. This enzyme is very sensitive to pH. When carbon dioxide is removed by photosynthesis in the presence of light, the pH of the guard cells is increased to about 7. At this pH the enzyme causes starch to be hydrolysed and thus increases the concentration of sugar in the cells. As the starch is hydrolysed to sugar within the guard cells, water enters the cells by osmosis. The guard cells become more turgid and the stoma opens. In the dark, when there is no photosynthesis and respiration is taking place, the concentration of carbon dioxide builds up and the pH falls to about 5. At this pH, the enzyme acts in reverse and causes the polymerization of glucose, which is converted to starch. The osmotic potential of the guard cells is thus lowered. Stomata close whenever the osmotic potential of the guard cells decreases relative to that of the surrounding epidermal cells.

Although this theory is very simple and fits most of the facts, it does not fit all of the facts. For instance, the guard cells of some plants (e.g. *Allium cepa*, Onion) never contain starch, but their stomata open and close in response to light. A hypothesis which is gaining wider acceptance is that changes in guard cell turgor are due to the movement of mineral ions between guard cells and neighbouring epidermal cells by a process which involves the expenditure of energy. Such processes are called '**active ion pumps**'.

It appears that an active ion pump operates the transport of potassium ions (K^+) into and out of the guard cells. This movement of ions, which are osmotically active, results in the formation of an osmotic potential gradient between the guard cells and other cells and causes the stoma to open or close. According to this hypothesis, photosynthesis which takes place in the guard cells in the presence of light reduces the carbon dioxide concentration and thereby increases their pH. These events lead to the synthesis of ATP. An ion pump, powered by ATP, then moves K^+ into the guard cells, either directly or in exchange for H^+ which leave the cell at the same time. It is thought that once K^+ ions move into the guard cells, chloride ions (Cl^-) and other anions follow passively. The resulting increased negative osmotic potential leads to the osmotic uptake of water and opening of the stoma.

In the indirect accumulation of K^+, in exchange for H^+, the H^+ probably comes from malic acid formed by reaction between bicarbonate ion (HCO_3) and phosphoenolpyruvic acid (PEP, see photosynthesis).

This hypothesis is an improvement on the classical hypothesis, but still does not explain all the facts. For instance, guard cells on parts of variegated leaves may contain no chloroplasts, but open and close just the same. Also, the stomata of many plants continue to open and close each day, even though the plants are kept continuously in the dark.

Osmotic potential differences between guard cells and neighbouring cells can sometimes be overridden by water potential differences, which result in the closure of stomata. Leaves exposed to dry air may wilt due to excessive loss of water. This causes their content of the hormone abscisic acid (ABA) to rise sharply. This hormone causes stomata to close and thus has the effect of preventing or reducing further water loss. The mode of action of ABA is not well understood, but it is known to act directly at the level of the plasmalemma and to increase the latter's permeability to ions and water (to cause it to 'leak').

The possible mechanisms of stomatal action given above are hypothetical and much remains to be learned before a full explanation can be given. However, we can be certain that factors affecting stomatal movement are light, carbon dioxide concentration, pH and the water status of the tissues. Stomatal opening is also increased by higher temperatures, as long as water is not a limiting factor.

Key words

Structure: leaf base, petiole, leaf mosaic; lamina; dorsiventral, isobilateral. **Lamina**: epidermis, cuticle, stomata; sub-stomatal cavity, mesophyll, palisade, spongy, bundle sheath. **Function**: photosynthesis, water, carbon dioxide, light. **Stomatal action**: transpiration; guard cells, stoma; chloroplasts, turgor, pH; enzyme, starch phosphorylase; active ion pump.

Summary

The leaf is the main photosynthetic organ of the plant. Combined growth movements of the leaf base, petiole and sometimes the stem keep the lamina in the best position for photosynthesis, thus producing a leaf mosaic.

Leaf structure may be dorsiventral or isobilateral. The cuticle on the epidermal surfaces restricts the loss of water by evaporation (transpiration) mostly to stomata.

Chloroplasts are found in all mesophyll cells and in the guard cells of stomata. The xylem of veins brings water close to all mesophyll cells. Phloem takes away the sugar formed by photosynthesis. Stomata, when open, allow carbon dioxide and oxygen to diffuse into and out of the mesophyll.

Stomata open when it is light and close when it is dark. Stomatal opening and closing depend on the turgor of guard cells.

Questions

1 Which cells of the leaf are actively involved in photosynthesis? How does the structure of the tissues in which these cells occur and the structure of the lamina as a whole enable them to carry out their photosynthetic function efficiently?

2 What is transpiration? Why does it occur and under what conditions is it most active?

3 According to the theories given, what may be the causes of turgor changes in guard cells? Why is it that neither of the theories described is entirely satisfactory?

9

Root structure and function

The primary functions of the root system are the attachment of the plant to the soil, the absorption of water and mineral ions from the soil and the transport of materials to and from the shoot system. A secondary, though very common function, is storage.

The root apex

External features

The tip of a growing root is always protected by a **root cap**. This is a conical cap of tissue which protects the apical meristem which lies within. In Maize, cells are added to a root cap at the rate of about 10 000 a day and are worn away by friction with the soil at about the same rate. Because it is being worn away all the time, the root cap is not easy to see in roots taken from the soil. It is easiest to see on aerial roots (e.g. *Rhizophora* spp., Red Mangroves; epiphytic *Ficus* spp., Figs; *Pandanus* sp., Screw Pine), the roots of water plants (e.g. *Lemna* sp., Duckweed; *Pistia* sp, Water Lettuce) and the roots of seedlings grown on moist filter paper (e.g. *Brassica oleracea*, Cabbage, Cauliflower).

The apical meristem lies within the root cap and cannot be seen in the intact root. Above the root cap, the root has no outgrowths for some distance. This bare part of the root is the **region of elongation**, in which cells derived from the apical meristem are increasing in length by vacuolation. This is also the region in which curvature occurs if the root is laid on its side (Chapter 13). Above this region the root is usually covered with root hairs for a short distance. Root hairs usually have a short life and wither away on the older parts of the roots. Above the region of root hairs, **lateral roots** emerge from within the tissues.

The apical meristem

When seen in LS, the root cap appears as a sheath of parenchyma covering the tip. Within the root cap (Fig. 9.1), are the small, thin walled and densely packed cells of the **promeristem**. Many of these cells

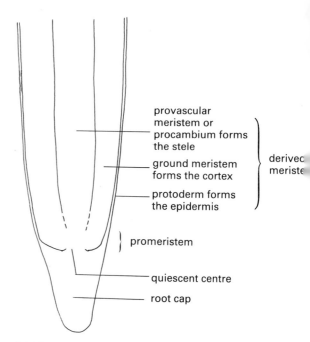

provascular meristem or procambium forms the stele

ground meristem forms the cortex

protoderm forms the epidermis

derived meriste

promeristem

quiescent centre

root cap

9.1 LS root tip showing the meristematic regions and the root cap

are actively dividing, but the most actively dividing cells are grouped around a so-called **quiescent centre**, in which few cell divisions occur. Recent work has shown that the apical meristem of an actively growing Maize root may add as many as 170 000 cells in 24 hours, including those added to the root cap.

As in the shoot apex, cells derived from the promeristem form **derived meristems** which, together with the promeristem, make up the apical meristem of the root. The outermost layer forms the **protoderm**, which is one cell thick. Within the protoderm is the **ground meristem**, which is several cells thick. **Provascular meristem** (procambium) forms the central core. Cells in the three layers differ from each other in shape and in the directions in which they divide. The cells which finally differentiate from

(a)

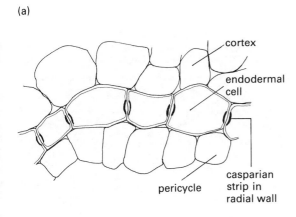

cortex

endodermal cell

casparian strip in radial wall

pericycle

(b)

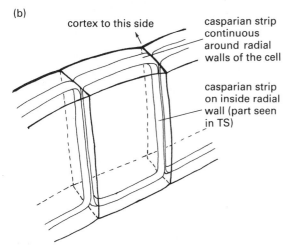

cortex to this side

casparian strip continuous around radial walls of the cell

casparian strip on inside radial wall (part seen in TS)

9.2 Root endodermis: (a) primary stage in a dicot root (HP); (b) diagram to show position of Casparian strip. See Fig. 9.7 for later stage

them are determined by their planes of division and subsequent vacuolation.

Differentiation of primary tissues

Vacuolation of cells begins before cell division has ceased but is most active after divisions have ceased. This is when the greatest elongation of tissues takes place. Cells from the protoderm form the **epidermis** (**piliferous layer**) of the root. Differentiation of epidermal cells involves only elongation and the formation of root hairs from some of them (Chapter 6). The ground meristem produces only the parenchyma of the **cortex** of the root. During the maturation of most of the cortical parenchyma cells, some

of the intercellular material softens so that they become rounded and intercellular spaces appear. The layer of cells closest to the epidermis (**hypodermis**) usually remains more compact, as do the cells of the innermost layer of the cortex (**endodermis**). The endodermis is of special significance in roots. Its cells develop a distinctive thickening of the radial and transverse walls (Fig. 9.2), the **Casparian strip** (**Casparian band**). Suberin is deposited in a continuous strip all round each cell and each strip coincides with and is fused with the strips around neighbouring endodermal cells. This thickening involves the impregnation of the middle lamella and primary walls with suberin and sometimes becomes overlaid with cellulose which becomes lignified. The strip is also intimately associated with the protoplast of each cell.

In some plants (e.g. epiphytic orchids), the cells of the hypodermis also develop Casparian strips, after which they are known as the **exodermis**.

It is in the stele, derived from the procambium, that the greatest variety of differentiation occurs. The layer immediately inside the endodermis remains more or less unchanged except for lengthening of the cells during vacuolation. This is the pericycle. The **pericycle** is usually formed from a single layer of closely packed parenchyma cells but it may be more than one cell thick (e.g. Luffa).

The first cells of the vascular system to differentiate are patches of protophloem. Protophloem cells are difficult to recognise with certainty because they are simply narrow, elongated cells without clearly distinguishable features. Sieve tube elements and companion cells appear only later. The first cells to become lignified are the vessel elements of the protoxylem, formed in groups alternating with the patches of protophloem, immediately inside the pericycle.

Differentiation of both xylem and phloem proceeds towards the centre of the root. The last vessel elements to become lignified after that part of the root has stopped elongating are the largest vessel elements of the metaxylem. There is always a layer of parenchyma left between adjacent groups of primary xylem and primary phloem. There is no cambium at this stage, even in dicots (Fig. 9.3).

A source of variation in the appearance of roots in TS is the number of protoxylem and protophloem groups. Dicots (Fig. 9.4) usually have the vascular tissues gathered closely together at the centre of the

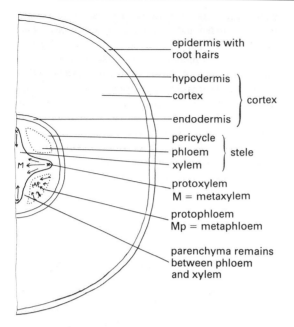

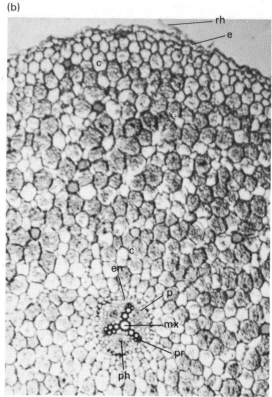

arrows show direction of
development from protophloem
to metaphloem, and from
protoxylem to metaxylem

9.3 Diagram of TS dicot root to show direction of
differentiation in the stele

rh	root hair	p	pericycle
e	epidermis	mx	metaxylem
c	cortex	pr	protoxylem
en	endodermis	ph	phloem

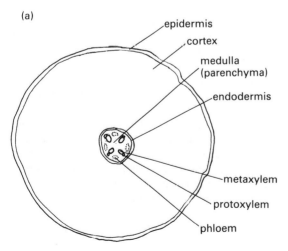

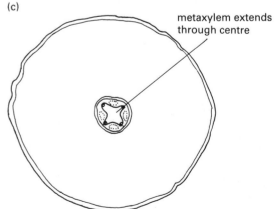

9.4 Variation on dicot root stucture as seen in TS:
(a) drawing of a tetrarch root with medulla;
(b) photomicrograph of a triarch root; (c) drawing of a
tetrarch root without medulla. All VLP

root with the xylem either united by a central mass of metaxylem, or with a small medulla. There are usually four or five groups of protoxylem. Such roots are said to be **tetrarch** or **pentarch**, according to the number of protoxylem groups. Monocots (Fig. 9.5) usually have thicker roots with more protoxylem groups and a larger medulla at the centre. There are rarely fewer than 6 protoxylem groups (**hexarch**

(a)

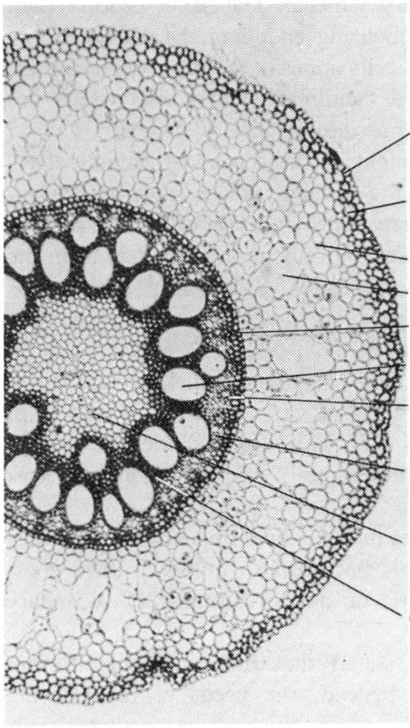

- epidermis
- hypodermis, slightly lignified
- cortex
- air space
- endodermis
- metaxylem
- protoxylem
- phloem groups alternating with protoxylem
- parenchyma of medulla
- lignified cells around metaxylem

(b)

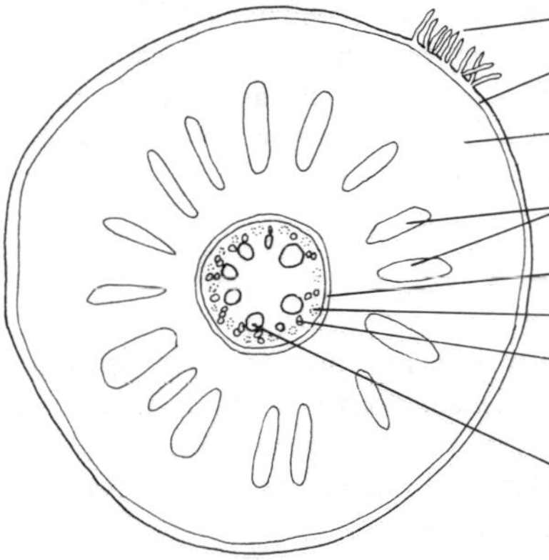

- root hairs, over whole epidermal surface
- epidermis
- cortex, parenchyma
- air spaces
- endodermis
- phloem alternating with protoxylem
- metaxylem, vessels very large and few in number

9.5 Monocot root structure: (a) TS young root of *Zea mays*; (b) TS root of *Canna* sp. Both VLP

roots) and often many more (**polyarch** roots). However, some dicots (e.g. *Rhizophora* spp., Red Mangroves) have polyarch roots and some monocots (e.g. *Crinum* sp., Crinum Lily) may be pentarch.

It is in small dicot roots that differentiation of metaxylem extends to the centre of the root, so that the xylem is star-shaped when seen in TS. Thicker dicot roots and all monocot roots have a central medulla and alternating, but unconnected, groups of primary xylem and phloem.

Note that the position of the protoxylem relative to the metaxylem is different in the root from that in the stem. In roots, the protoxylem is outside the metaxylem and we describe the primary xylem as **exarch**. In stems, the primary xylem is endarch, as previously described. Also, primary xylem and phloem are on different radii of the root. This **radial** arrangement of the two vascular tissues differs from the collateral, or bicollateral, arrangement of the tissues in the stem. The true nature of an organ of doubtful identity can be ascertained by an examination of the vascular tissues, to discover whether the xylem is exarch or endarch and whether the xylem and phloem are radially or collaterally arranged. Note also that the term 'vascular bundle' is used only to describe the groups of xylem and phloem found in a stem. There is no comparable juxtaposition of xylem and phloem in the root.

The origin of lateral roots

Both dicot and monocot roots form lateral branches but dicot roots branch more than monocot roots. Lateral roots begin to form just above the region of root hairs (Fig. 9.6). The development of lateral roots commences with the division of cells of the pericycle, usually opposite the protoxylem but sometimes opposite the protophloem. Cell division leads first to the formation of a group of meristematic cells in which the layering typical of a normal root apical meristem soon appears. Growth of the new apical meristem leads to penetration of the endodermis and the rest of the cortex, until the new lateral root bursts through the epidermis and out into the soil. Lateral roots are said to be **endogenous** in origin, because they grow out from within the tissues. This is in contrast to the superficial origin of shoot branches.

Once the vascular system of the lateral root starts to differentiate, it becomes connected to the vascular system of the main root.

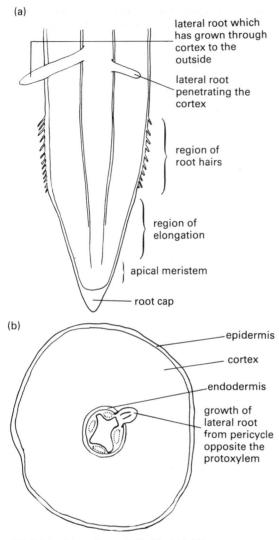

(a)

lateral root which has grown through cortex to the outside

lateral root penetrating the cortex

region of root hairs

region of elongation

apical meristem

root cap

(b)

epidermis

cortex

endodermis

growth of lateral root from pericycle opposite the protoxylem

9.6 Origin of lateral roots (a) in LS; (b) in TS

Older roots

Monocots

In monocots, no cambium is formed and the primary structure remains more or less unchanged throughout the life of the root. Changes are restricted to the outer layers of the cortex and to the endodermis.

Root hairs disintegrate after they have ceased to carry out their absorbing function. This results in disruption of a large part of the epidermis and the loss of its protective function. This is compensated for by the lignification of the hypodermis and neighbouring layers of the cortex. In very old monocot roots, lignification of the walls of cortical cells extends as far as the endodermis.

At about the same time as root hairs cease to function, changes take place in the endodermis. The inner and radial walls of endodermal cells opposite the protophloem become covered with a **secondary lamella** of suberin. There may also be lignification of the secondary lamella. The same process then occurs in neighbouring endodermal cells until only the endodermal cells opposite the protoxylem are left in their original condition. Those cells which are unchanged are **passage cells**.

The development of the secondary lamella effectively seals off the vascular tissues from the cortex, except for the gaps left by passage cells. The passage cells allow outward movement of substances from the inner tissues to the cortex. By the time this stage is reached, this part of the root is no longer an absorbing organ.

In yet older roots, a tertiary lamella is laid down inside all the walls of each endodermal cell (e.g. *Smilax* sp., Fig. 9.7). This often includes the passage cells and completely seals off the vascular tissues from the cortex.

These changes in the endodermis are characteristic of monocot roots but are also found in dicot roots which have little or no development of secondary tissues.

It has been stated that monocot roots do not branch much. Instead, the needs of the growing plant are met by the addition of more adventitious roots to the fibrous root system. These facts can be related to the absence of secondary vascular tissue. Older roots cannot expand their carrying capacity by the addition of secondary tissues and the increasing needs of the growing shoot system can be met only by the addition of more roots.

Dicots

A dicot root with primary structure has no cambium but a cambium usually forms at an early stage. It arises in the cells of the parenchyma which lie between the xylem and phloem, and in the pericycle outside the protoxylem.

The cambium is formed by a series of cell divisions similar to those by which interfascicular cambium is formed in stems. Divisions start next to

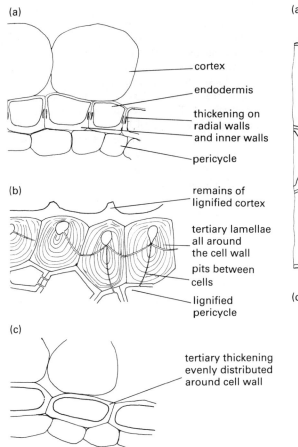

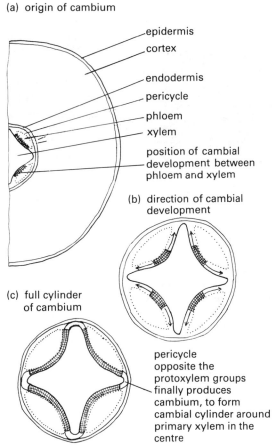

9.7 Later developments in the endodermis: (a) TS monocot endodermis showing secondary lamella, note that Casparian strips are still visible; (b) TS *Smilax* sp. (monocot), endodermis in old root with tertiary lamella; (c) TS endodermis with tertiary lamella in dicot root in which there is no secondary thickening in the stele. All HP

9.8 Diagrams to show the origin of cambium in a dicot root

the metaphloem and development of cambium progresses each way, towards the protoxylem (Fig. 9.8). The full cylinder is completed when the separate arcs, as seen in TS, are united by the formation of cambial cells in the pericycle, where it passes around the protoxylem.

The first activity of the cambium, next to the metaphloem, adds secondary xylem and phloem so that the cambium soon forms an almost perfect circle, as seen in TS. Primary rays are formed from the part of the cambium next to the protoxylem groups. Subsequent formation of secondary vascular tissues and of periderm follows the same pattern as in stems (Chapter 10).

Variation in the primary structure of roots

Variation in the structure of primary roots is centred on relative proportions of cortex and medulla, associated with the number of protoxylem groups, and on the presence or absence of large air spaces in the cortex.

Ecological anatomy

There is little difference between the structure of the roots of mesophytes and those of xerophytes and hydrophytes, except that most plants which have roots which are submerged in water, even in wet soil, develop large air spaces in the cortex (Fig. 9.9).

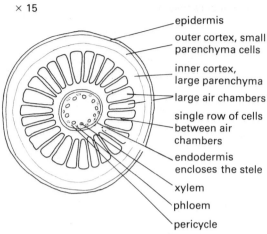

× 15

- epidermis
- outer cortex, small parenchyma cells
- inner cortex, large parenchyma
- large air chambers
- single row of cells between air chambers
- endodermis encloses the stele
- xylem
- phloem
- pericycle

9.9 Structure of a hydrophyte root, TS root of *Pistia* sp.

Aerial roots

Aerial roots which are also prop roots or stilt roots (e.g. Maize; Red Mangrove) differ from the normal roots of the same plants in having many more protoxylem and protophloem groups distributed around a large pith (Fig. 9.10). As a result, the distribution of vascular tissues superficially resembles that in a stem. The radial arrangement of xylem and phloem and their exarch xylem make them readily distinguishable from stems.

The photosynthetic aerial roots of many epiphytic orchids (Figs. 4.1, 4.2) have a specialized type of multiple epidermis known as velamen. The **velamen** may be from two to several cells thick. The cells of which it is composed are dead when mature, with narrow thickenings and slits in their walls. Inside the velamen, the outermost layer of cortical cells forms an exodermis. Cells of the **exodermis** have Casparian bands and the walls of all but a few (passage cells) become thickened and lignified. Only the passage cells of the exodermis keep their living contents. Although the velamen absorbs water when it rains, it is probable that little of this water passes into the cortex through the passage cells. The main function of the velamen is said to be to protect the inner tissues of the cortex from desiccation.

Storage roots

Almost all dicot and monocot roots store starch or other storage substances in the cortical parenchyma. The main site of storage in swollen storage roots is in the parenchyma cells of secondary tissues.

The swollen tap roots of *Daucus carota* (Carrot) and *Beta vulgaris* (Beetroot) store sugar in secondary xylem and phloem parenchyma and in ray parenchyma. In Cassava and Sweet Potato, storage is almost entirely restricted to secondary xylem parenchyma and ray parenchyma. The large amount of

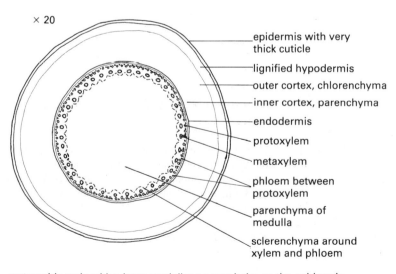

× 20

- epidermis with very thick cuticle
- lignified hypodermis
- outer cortex, chlorenchyma
- inner cortex, parenchyma
- endodermis
- protoxylem
- metaxylem
- phloem between protoxylem
- parenchyma of medulla
- sclerenchyma around xylem and phloem

note: wide stele with a large medulla; no root hairs on the epidermis

9.10 Structure of an aerial root, TS prop root of *Sorghum* sp. Compare with Fig 9.5

secondary xylem formed in these roots contains only a few small areas of lignified cells. The greatest part is made up of unlignified parenchyma cells.

Root–stem transition

The junction of the vascular system of the primary root with that of the primary shoot of a plant is located in a transition zone in the hypocotyl. The

(ii)
stele enlarges, medulla of parenchyma in centre, metaxylem moving to outer position

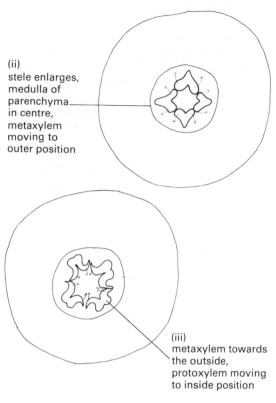

(iii)
metaxylem towards the outside, protoxylem moving to inside position

(iv) region of lateral roots, vascular traces from sides of bundles into roots

(a)

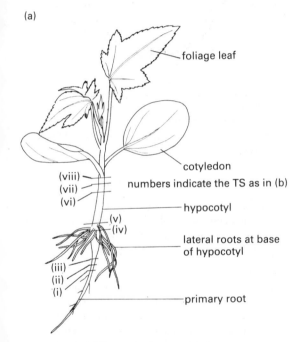

foliage leaf

cotyledon

numbers indicate the TS as in (b)

hypocotyl

lateral roots at base of hypocotyl

(viii)
(vii)
(vi)
(v)
(iv)
(iii)
(ii)
(i)

primary root

sections taken from root (i), up through hypocotyl, to the base of the cotyledons (viii)

(b)

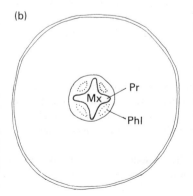

(i)

root, tetrarch Small stele with metaxylem (Mx) in centre, protoxylem outside (Pr), and phloem (Phl) alternating with the protoxylem

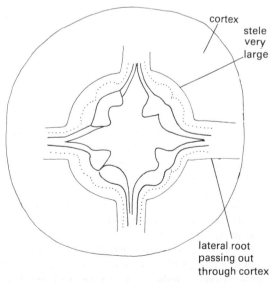

cortex

stele very large

lateral root passing out through cortex

9.11 Root–stem transition: (a) young plant of *Ricinus* sp. to show region of serial sections (continued overleaf)

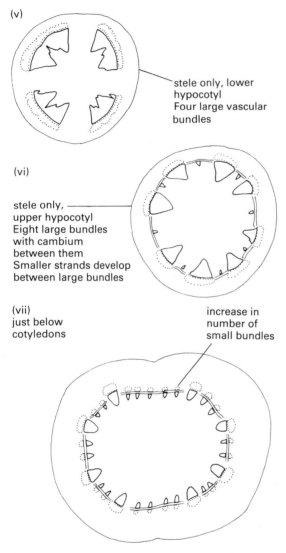

(v)

stele only, lower
hypocotyl
Four large vascular
bundles

(vi)

stele only,
upper hypocotyl
Eight large bundles
with cambium
between them
Smaller strands develop
between large bundles

(vii)
just below
cotyledons

increase in
number of
small bundles

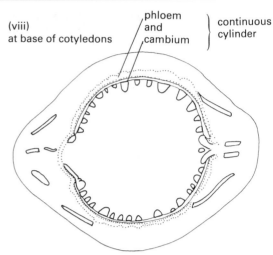

(viii)
at base of cotyledons

phloem
and
cambium

} continuous
cylinder

Above the cotyledons TS shows bundles
of various sizes with continuous cylinder
of phloem and cambium

9.11 (continued) (b) serial sections from root up through
the hypocotyl

Root structure and function

As stated, the primary functions of a typical root are
the attachment of the plant, absorption of water and
mineral ions and the conduction of water and
dissolved substances to and from the shoot system.

Attachment

Roots form a branching system which usually
penetrates deeply and widely into the soil. They
therefore provide firm attachment for the plant. The
concentration of lignified tissues at the centre of the
root gives mechanical strength which can resist
pulling strains. Resistance to bending strains, such
as is provided for in the stem by the wide spacing
of the vascular bundles near the surface, is not
needed in a typical root. Roots which must resist
bending strains always have a wide medulla, compar-
able with that of a stem.

Absorption

The root system is spread through, and can absorb
materials from, a very large volume of soil. Because
of their shape, root hairs provide an extremely large
surface area through which water and mineral ions
can be absorbed.

The role of the endodermis in controlling the
movement of water and mineral ions is very
important and is dealt with in detail in Chapter 12.

change required from root structure (exarch xylem,
radial arrangement of vascular tissues) to stem structure (endarch xylem, collateral or bicollateral
vascular bundles) is quite a large one and varies from
species to species.

Transition from root structure to stem structure
usually involves radial division of groups of xylem
and groups of phloem, the inversion of the relative
positions of protoxylem and metaxylem and the
union of groups of xylem and groups of phloem to
form vascular bundles. Serial sections of the transition zone in the hypocotyl of a tropical species are
shown in Fig. 9.11.

Conduction

The large vessels of the metaxylem provide a ready passage for the movement of water and dissolved substances. Phloem provides for the transport of sucrose and amino acids.

Key words

Root apex: root cap; apical meristem, promeristem, quiescent centre. **Derived meristems**: protoderm, ground meristem, provascular meristem (procambium). **Primary structure**: epidermis (piliferous layer); cortex, hypodermis, endodermis. **Endodermis**: Casparian strip (Casparian band), secondary lamella, tertiary lamella, passage cell. **Stele**: pericycle; exarch xylem, tetrarch, pentarch, hexarch, polyarch; phloem; radial arrangement of vascular tissues. **Functions**: attachment, absorption, conduction, storage.

Summary

The root apex has the same types of derived meristems as the stem apex, but procambium forms a solid core. Both xylem and phloem are exarch. Variation in root structure is mostly in the number of protophloem and protoxylem groups. Neither dicot nor monocot roots have cambium at first, but a cambium develops in dicot roots by dedifferentiation of parenchyma cells between xylem and phloem. Dicot roots have a solid xylem core, or separate groups of xylem and a small medulla. Monocot roots are usually polyarch, with a larger medulla. The branching system of roots provides for attachment of the plant, root hairs provide for absorption of water and mineral salts and the vascular system provides for the conducting function. In storage roots, starch and other substances are usually stored in secondary vascular tissues.

Questions

1 As seen in TS, how does the structure of a pentarch dicot root differ from that of a dicot stem with five vascular bundles?

2 How and where do secondary roots develop in the tap root system of a dicot? Describe their origin and explain how this differs from the origin of branches in a shoot system.

3 How do the various tissues of a root show adaptation to the principal root functions of attachment, absorption and conduction?

4 Explain how differences between dicot and monocot root systems are correlated with the presence or absence of secondary thickening.

10

Woody plants

Woody plants are here taken to include shrubs and trees (Figs. 3.10, 3.11). They include dicots which continue to add secondary vascular tissues from year to year and monocots which have the habit and dimensions of shrubs and trees. Most woody dicots start to form secondary vascular tissues in the same way as has been described for herbaceous dicots. This is regarded as normal secondary thickening. Others (e.g. *Bougainvillaea* sp.) form additional hard, wood-like secondary tissues in an entirely different way (**anomalous** secondary thickening). Some tree-like monocots also have anomalous secondary thickening (e.g. *Dracaena* spp., Fig. 10.20). Others develop thick, hard stems containing much wood-like tissue at the stage of primary growth and do not add secondary vascular tissues (e.g. *Bambusa* sp., Bamboo; *Cocos nucifera*, Coconut Palm; *Elaeis guineensis*, Oil Palm).

Secondary thickening in woody dicots

As has been shown, most herbaceous dicots have some secondary thickening (Fig. 7.15). The origin of the cambial ring (Figs. 7.14, 9.8) is the same in woody dicots as in herbaceous species. A superficial difference in woody dicot stems is that their primary structure usually includes very many, narrow vascular bundles with very narrow rays between them (Fig. 10.1). However, some trees have larger primary vascular bundles with wide rays, and almost resemble overgrown herbs in structure (e.g. *Carica papaya*, Pawpaw, Fig. 10.2; *Cussonia* sp.).

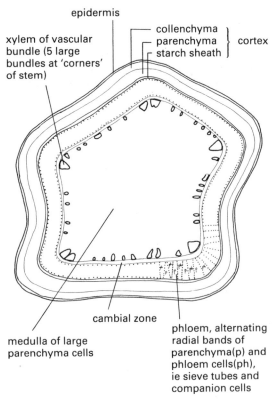

10.2 TS young stem of *Carica* sp. with wide rays and much secondary phloem

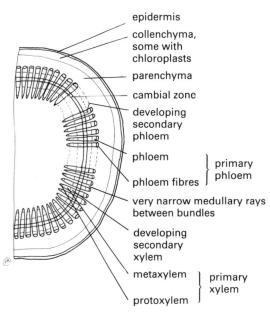

10.1 Woody dicot, TS stem of *Hibiscus* sp. showing many narrow vascular bundles separated by narrow medullary rays

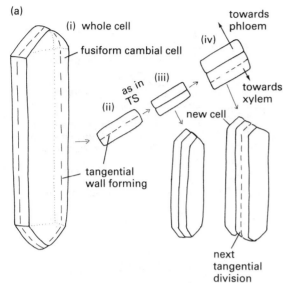

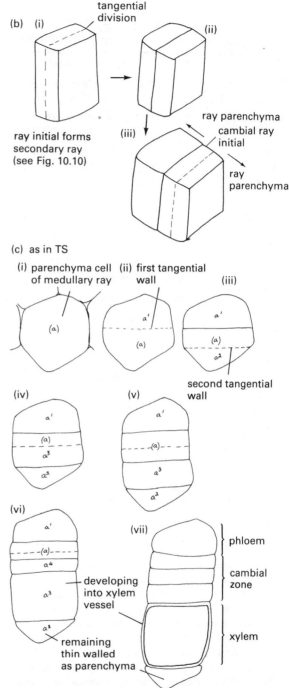

(a)

(i) whole cell

fusiform cambial cell

(iv) towards phloem

as in TS

(iii)

(ii)

tangential wall forming

new cell

towards xylem

next tangential division

10.3 Cambial development: (a) fusiform cambial initial; (b) ray initial; (c) origin of interfascicular cambium in TS stem. See also Fig. 7.14

(b) (i)

tangential division

(ii)

ray initial forms secondary ray (see Fig. 10.10)

(iii)

ray parenchyma

cambial ray initial

ray parenchyma

(c) as in TS

(i) parenchyma cell of medullary ray

(ii) first tangential wall

(iii)

second tangential wall

(iv)

(v)

(vi)

(vii)

developing into xylem vessel

remaining thin walled as parenchyma

phloem

cambial zone

xylem

cell (a) remains meristematic, new cells a^3 and a^4 look similar to (a), these are together referred to as the cambial zone

The cambium and cambial action

Cells of the vascular cambium which add to the xylem and phloem are elongated along the axis of the organ. Those of the rays are not elongated, or are only slightly elongated. All divide by tangential walls, seldom by radial walls. When an elongated (**fusiform**) cambial cell divides, it cuts off a derivative towards the xylem or towards the phloem. Cells are added to the xylem at about three times the rate of those added to the phloem. Each cambial derivative usually undergoes one or more further tangential divisions before all of the products differentiate into xylem or phloem cells. Cells of the cambium located in the rays divide to produce ray parenchyma, in both directions.

Secondary vascular tissues

Secondary xylem includes vessels, tracheids, xylem parenchyma and xylem fibres. Vessels and tracheids are always in contact with living xylem parenchyma cells, and there is always a series of xylem parenchyma cells between each vessel or tracheid and the nearest ray. The arrangement of the different kinds of xylem cells varies from species to species, as also do the proportions of the different cell types. Hard

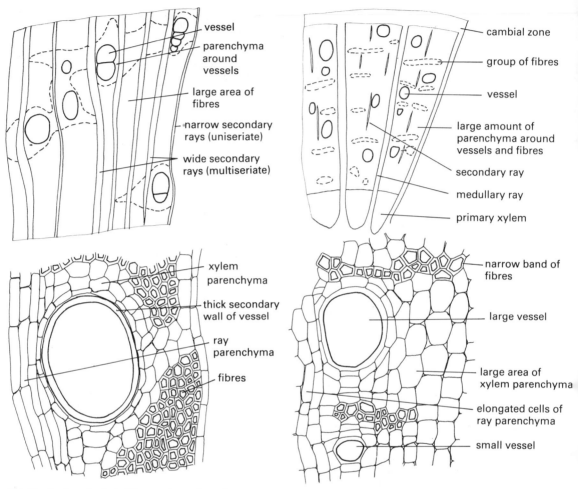

10.4 Distribution of cells in TS secondary xylem of *Chlorophora* sp. (Iroko), a hard wood with many fibres, LP and HP

10.5 Distribution of cells in TS secondary xylem of *Erythrina* sp., a soft wood with much xylem parenchyma and a few fibres in tangential bands, LP and HP

woods contain many fibres (e.g. *Chlorophora excelsa*, Iroko). Soft woods contain a great deal of xylem parenchyma and few fibres (e.g. *Triplochiton scleroxylon*, Obeche; *Erythrina* spp.). Differing proportions and arrangements of xylem cells are illustrated in Figures 10.4 to 10.6.

Secondary phloem is composed of sieve tubes, companion cells, phloem parenchyma and phloem fibres. Fibres usually occur in tangential or radial bands. Parenchyma is mixed with sieve tubes and companion cells. Some examples of the arrangement of cell types are shown in Figs. 10.7 and 10.8.

Secondary ray parenchyma cells are usually elongated along the ray. They do not have secondary cell walls but parenchyma cells in the secondary xylem sometimes have their primary walls lignified.

As secondary xylem is added, the perimeter of the xylem occupied by the cambium increases. The distance between neighbouring rays also increases. Both these changes are compensated for by further developments.

Increase in the number of cambial cells is brought about by transverse or oblique division of fusiform cambial cells. This produces two new cambial cells,

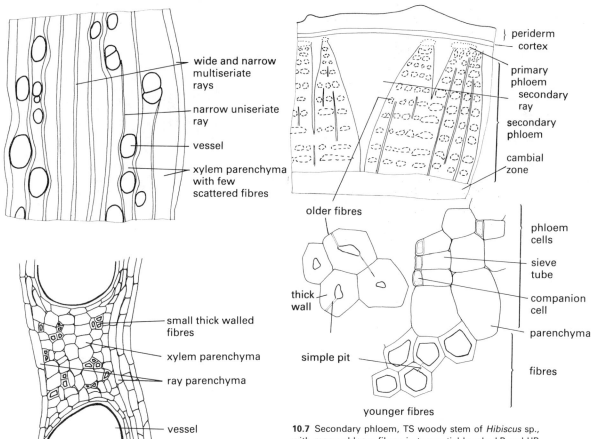

wide and narrow
multiseriate
rays

narrow uniseriate
ray

vessel

xylem parenchyma
with few
scattered fibres

periderm
cortex

primary
phloem
secondary
ray

secondary
phloem

cambial
zone

older fibres

thick
wall

simple pit

younger fibres

phloem
cells

sieve
tube

companion
cell

parenchyma

fibres

10.7 Secondary phloem, TS woody stem of *Hibiscus* sp.,
with many phloem fibres in tangential bands, LP and HP

small thick walled
fibres

xylem parenchyma

ray parenchyma

vessel

10.6 Distribution of cells in TS secondary xylem of
Triplochiton sp. (Obeche), a soft wood with a few scattered
fibres, LP and HP. See also Fig. 10.12

one above the other. The new cambial cells then
gradually change in shape in such a way that they
come to lie more or less side by side (Fig. 10.9). The
number of cells around the circumference of the
cambium is thus increased.

When the increasing diameter of the stem causes
existing rays to separate more than a certain amount,
a new ray is formed equidistant between them. A
new ray starts to form when a fusiform cambial cell

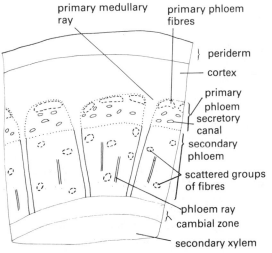

primary medullary
ray

primary phloem
fibres

periderm

cortex

primary
phloem

secretory
canal

secondary
phloem

scattered groups
of fibres

phloem ray

cambial zone

secondary xylem

10.8 Secondary phloem in TS woody stem of *Erythrina* sp.,
with scattered groups of phloem fibres, LP

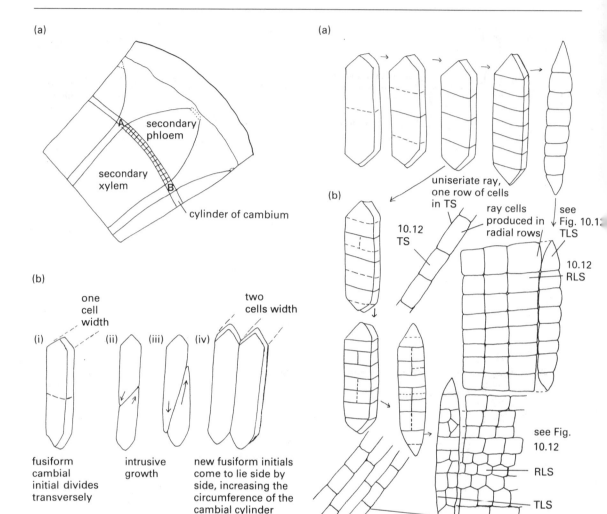

(a)

(b)

one cell width

two cells width

(i) (ii) (iii) (iv)

fusiform cambial initial divides transversely

intrusive growth

new fusiform initials come to lie side by side, increasing the circumference of the cambial cylinder

10.9 Diagrams to show increase in circumference of the cambium: (a) position of cambium in TS woody stem; (b) division of a cambial cell in region A–B of (a)

10.10 Origin of a secondary ray: (a) division of fusiform cambial initial to uniseriate ray; (b) multiseriate ray. Compare with Fig. 10.12

divides several times, transversely. In this case, the new cambial cells become ray initials and subsequently divide tangentially to produce a new ray. The original rays (**primary rays**) can be recognized when seen in TS by the fact that they extend from the outer edge of the phloem to the medulla (in the stem) or to the protoxylem (in the root). Rays which are formed as described above (**secondary rays**) extend only part of the way towards the centre of the stem or root (Fig. 10.10).

Just as the properties of a type of wood are affected by the proportions of different types of cells, they are also affected by the proportions and distribution of the rays. A TS of a woody stem shows the arrangement and relative amounts of different types of cells in the secondary tissues and the widths and

(a) Direction of cutting section

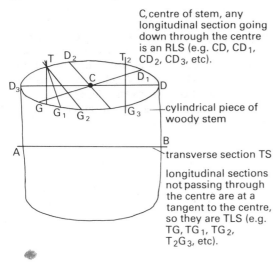

C, centre of stem, any longitudinal section going down through the centre is an RLS (e.g. CD, CD_1, CD_2, CD_3, etc).

—cylindrical piece of woody stem

—transverse section TS

longitudinal sections not passing through the centre are at a tangent to the centre, so they are TLS (e.g. TG, TG_1, TG_2, T_2G_3, etc).

(b) Woody stem in TS, RLS and TLS

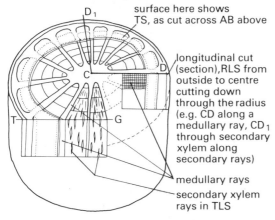

surface here shows TS, as cut across AB above

longitudinal cut (section), RLS from outside to centre cutting down through the radius (e.g. CD along a medullary ray, CD_1 through secondary xylem along secondary rays)

medullary rays
secondary xylem rays in TLS

10.11 Diagram to show the planes of various sections: (a) directions of cutting sections; (b) woody stem in TS, RLS and TLS

(a)

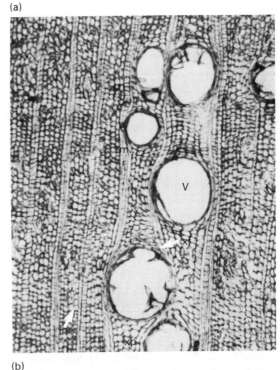

(b)

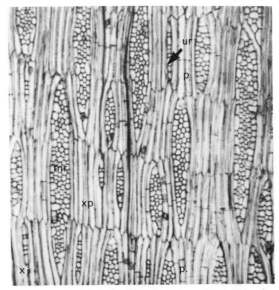

10.12 Secondary xylem of *Triplochiton* sp. (Obeche), a soft wood: (a) TS, compare with Fig. 10.6, scattered fibres arrowed (HP); (b) TLS, showing multiseriate and uniseriate rays, elongated xylem parenchyma, some cells of which (p) have divided transversely, and fibres (HP) (continued overleaf)

radial extent of the rays. A **radial longitudinal section** (RLS, a longitudinal section cut along a radius of the organ) shows the longitudinal and radial extent of the rays. A **tangential longitudinal section** (TLS, a longitudinal section cut at right angles to a radius) shows the vertical extent and width of the rays. Figure 10.11 shows the relationship between the three types of sections. Figure 10.12 shows secondary wood, as seen in TS, RLS and TLS.

(c)

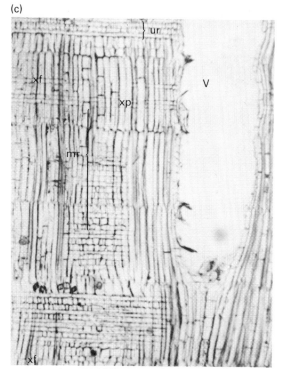

10.12 (continued) (c) RLS elongated xylem parenchyma near the vessel, dark thick-walled fibres and both multiseriate (mr) and smaller uniseriate (ur) rays (HP)

Stems and roots

Development of secondary xylem and phloem is the same in stems and roots. The only difference between a secondarily thickened stem and a secondarily thickened root is that the stem has a medulla, around which are several or many endarch primary xylem groups, while a root usually has no medulla but has a central core of exarch primary xylem with the protoxylem groups at the inner ends of the primary rays (Fig. 10.13).

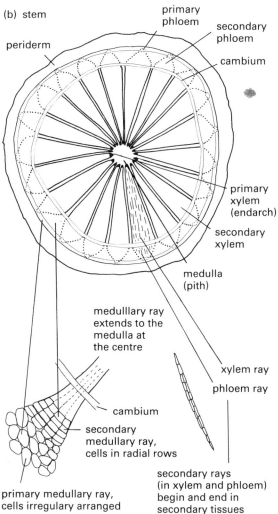

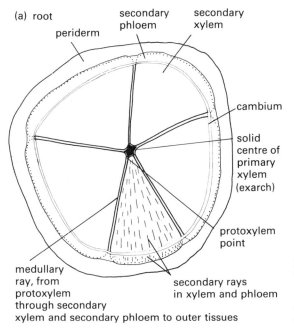

10.13 Diagrams to compare secondary thickening in root and stem: (a) TS woody dicot root; (b) TS woody dicot stem

Growth rings

Trees which grow in temperate climates go through an annual cycle which includes a period of active growth when temperatures are relatively high, and a period of total dormancy, when they have lost their leaves and the temperature is very low. This annual cycle is reflected in the structure of their secondary wood. Usually, the onset of the warm growing season is marked by the formation of secondary xylem which is composed of rather wide, thin-walled cells. As the growing season progresses, the xylem formed has narrower cells with thicker walls. Since the xylem formed at the end of the season is succeeded by the very different xylem formed at the start of the next growing season, the change of seasons is easily seen when a TS is looked at under the microscope or even with the naked eye. The result is the formation of **growth rings**, each of which corresponds to one year of growth. In this case, each growth ring is an **annual ring**.

Cambial action in tropical trees is less related to the seasons and more affected by variation in rainfall throughout the year. Trees in the wettest rain forest may show no growth rings at all. Growth rings observed in the wood of trees growing in drier places are not necessarily annual rings. The interruption of a dry season by a heavy fall of rain may lead to a burst of growth but the resulting growth ring represents only a few weeks of cambial activity (Fig. 10.14).

Heart wood and sap wood

A large tree may have a large volume of secondary wood, but only the youngest, outermost part of the secondary wood, the **sap wood**, is alive and functioning. In the sap wood, the xylem and ray parenchyma cells are living and contain stored starch and the vessels are transporting water from the roots to the leaves. The older, central part of the wood is dead. This is the **heart wood**. As wood becomes older, stored materials are removed from the parenchyma, the vessels are blocked by tyloses and cell walls become impregnated with a variety of substances, including oils, tannins and pigments. These substances make the heart wood more resistant to decay than the· sap wood. They also change the appearance of the wood but do not affect its mechanical strength. It is the heart wood that makes the most valuable timber, where durability and appearance are important.

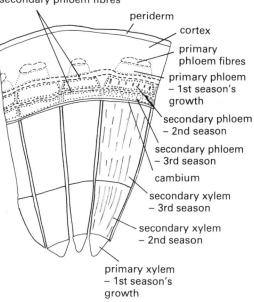

tangential bands of
secondary phloem fibres

periderm

cortex

primary
phloem fibres

primary phloem
– 1st season's
growth

secondary phloem
– 2nd season

secondary phloem
– 3rd season

cambium

secondary xylem
– 3rd season

secondary xylem
– 2nd season

primary xylem
– 1st season's
growth

10.14 Growth rings in a woody dicot stem, TS *Jacaranda* sp. (LP)

Periderm

Increase in the diameter of the vascular tissues due to secondary thickening has a great effect on the outer tissues, which sooner or later must be disrupted.

Stems

As in the case of herbaceous plants, the cortex of a woody stem is able to accommodate to the increasing thickness of the vascular tissues, to a small degree. However, the epidermis can stretch only a little. The function of the epidermis is soon replaced by the formation of **periderm**.

The formation of periderm is initiated by the formation of a new, secondary meristem, the **phellogen** (cork cambium). The phellogen is commonly formed in the outer part of the cortex, often in the layer of cells immediately inside the epidermis (e.g. *Polyscias* sp., African Holly). In some plants it is formed in the epidermis (e.g. *Nerium oleander*, Oleander). In either case, the cells divide twice periclinally, in the same way as cells of the medullary ray divide during the formation of interfascicular cambium. The middle layer of cells produced remains meristematic and is the phellogen.

Cells of the phellogen divide periclinally, to produce radial rows of cells towards the outside only (e.g. *Nerium* sp.) or to both outside and inside (e.g. *Polyscias* sp.). Cells cut off towards the outside form the **phellem** (cork) (Fig. 10.15). During their differentiation, their walls become suberized. They are air-filled and dead when mature. The cells are closely packed together, without air spaces, so the main part of the phellem is impermeable to water and gases. Certain parts of the phellogen form **lenticels**, which take over the function of stomata in so far as they allow gaseous diffusion between the atmosphere and the inner tissues. In fact, periderm formation often starts under stomata. In the region of the lenticel, the cells of the phellem round off, leaving large air spaces, before they become suberized.

The **phelloderm**, formed of cells cut off from the phellogen towards the inside of the stem, consists of radial rows of living, unsuberized, parenchyma cells. The cells have small air spaces between them.

Phellogen is active for only a short time. Soon, the meristematic cells themselves differentiate into phellem. After this has happened, a new phellogen is formed deeper within the tissues of the plant, ultimately in the secondary phloem parenchyma and in corresponding cells of the ray parenchyma. Each

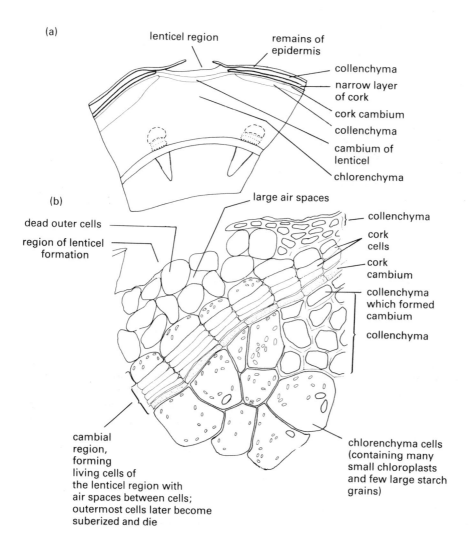

(a)

lenticel region

remains of epidermis

collenchyma

narrow layer of cork

cork cambium

collenchyma

cambium of lenticel

chlorenchyma

(b)

dead outer cells

region of lenticel formation

large air spaces

collenchyma

cork cells

cork cambium

collenchyma which formed cambium

collenchyma

cambial region, forming living cells of the lenticel region with air spaces between cells; outermost cells later become suberized and die

chlorenchyma cells (containing many small chloroplasts and few large starch grains)

(c)

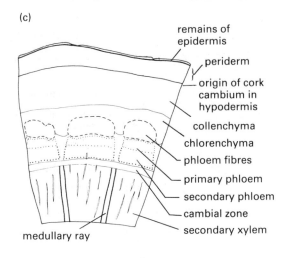

- remains of epidermis
- periderm
- origin of cork cambium in hypodermis
- collenchyma
- chlorenchyma
- phloem fibres
- primary phloem
- secondary phloem
- cambial zone
- secondary xylem
- medullary ray

(d)

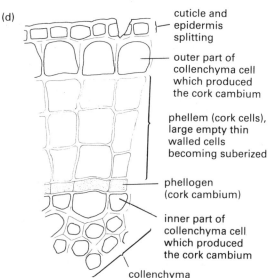

- cuticle and epidermis splitting
- outer part of collenchyma cell which produced the cork cambium
- phellem (cork cells), large empty thin walled cells becoming suberized
- phellogen (cork cambium)
- inner part of collenchyma cell which produced the cork cambium
- collenchyma

10.15 Periderm formation in dicots: (a) TS old stem of *Talinum* sp. showing lenticel formation opposite regions of chlorenchyma in the cortex (LP); (b) details of the lenticel region in *Talinum* sp. (HP); (c) TS outer part of young woody stem of *Gmelina* sp., LP; (d) details of cork formation in outer collenchyma of *Gmelina* sp., HP

time a new phellogen is formed and starts to form phellem, some outer tissues are isolated from the vascular tissues and die. In some cases, they are sloughed off and the surface of the stem is covered only with the most recently formed layer of periderm. This is often the case in forest trees. In other

shrubs and trees, the dead tissues, including cortex and secondary phloem, remain attached, sandwiched between successive layers of periderm. These dead tissues, the **rhytidome**, give further protection to the living tissues of the stem. The thick, fissured 'bark' of many savanna trees gives protection against fire (e.g. *Cussonia* sp.). Note that the word 'bark' is a non-scientific term and may be used in different ways. It is sometimes used to describe the exposed surface of a woody plant. It is usually used to describe all those tissues of a woody plant outside the vascular cambium. If you take the bark off a tree, you remove the living phloem as well as layers of periderm and rhytidome.

There is some evidence that some forest trees with very thin bark retain the first formed periderm indefinitely. In such cases, the original cortex must also be retained, though greatly increased in diameter by radial cell divisions. If such trees bear 'thorns', each of these is tipped by the prickle which was first formed as an epidermal emergence in the young plant. Such thorns are composed of periderm of an especially compact type (e.g. *Ceiba pentandra*, Silk Cotton Tree, Fig. 4.8; *Fagara macrophylla*).

Roots

Woody roots form periderm in the same way as woody stems. The first phellogen is usually formed in the pericycle. Succeeding layers of phellogen are formed in the secondary phloem. Aerial roots (e.g. *Rhizophora* spp., Red Mangroves; *Chasmanthera dependens*) and some roots with a restricted amount of secondary thickening (e.g. some *Ipomoea* spp.) form the first phellogen in the epidermis or hypodermis.

Anomalous secondary thickening.

Dicots

Anomalous types of secondary thickening are found in the roots and stems of many dicots, belonging to several different families.

In stems of members of the Nyctaginaceae (e.g. *Boerhaavia diffusa; Bougainvillaea* sp.; *Mirabilis jalapa*, Four O'Clock), the vascular bundles contain no cambium and are scattered throughout a central ground tissue. Secondary thickening begins when a complete ring of cambium forms in the cortex, immediately outside the vascular bundles. This cambium cuts off secondary tissues towards the centre of the stem. Most of the secondary tissue forms lignified ground tissue, in which are scattered secondary vascular bundles (Fig. 10.16). At first sight, the secondary tissue resembles xylem containing isolated islands of phloem.

The storage roots of the *Beta vulgaris* (Beetroot) have an anomalous form of secondary thickening. The first cambium is formed in the usual way but acts for only a short time. When the first cambium becomes inactive, a new cambium is formed outside the previous one in the pericycle or secondary phloem. The process is repeated several times and the mature beetroot contains many alternating layers of secondary phloem and secondary xylem. In the root tubers of *Ipomoea batatas* (Sweet Potato), secondary xylem and phloem are formed in the usual way. Both tissues contain a very high proportion of parenchyma. Subsequently, cambia are formed in the secondary xylem, each surrounding a large vessel, or a group of large vessels. These produce secondary xylem towards the original vessel and secondary phloem towards the outside. Both secondary tissues include a high proportion of parenchyma, but the end result is that small patches of phloem appear within the original secondary xylem.

Other types of anomalous secondary thickening are found in woody *Bauhinia* spp., members of the Bignoniaceae and in many woody forest climbers (lianes).

Monocots

Bamboos (e.g. *Bambusa* spp.; *Arundinaria* sp.) and most palms (e.g. *Cocos nucifera*; *Elaeis guineensis*) have thick stems and grow to tree-like dimensions but do not have secondary thickening. Their stems become thick by primary growth (**primary thickening**), which takes place before the part of the stem concerned has finished elongating, while the primary meristems are still active.

Bamboos have no periderm and the toughness of the outer part of the stem is due to lignification of the ground tissues.

Variation in the thickness or increase in the thickness of the trunks of some palm trees is not due to the addition of secondary tissues. It is due to the increase in size of cells of ground tissues and the air spaces between them. It does not involve the addition of secondary tissues. Increase in thickness

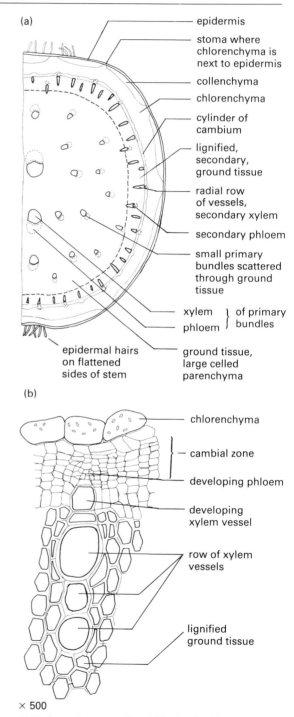

(a)
- epidermis
- stoma where chlorenchyma is next to epidermis
- collenchyma
- chlorenchyma
- cylinder of cambium
- lignified, secondary, ground tissue
- radial row of vessels, secondary xylem
- secondary phloem
- small primary bundles scattered through ground tissue
- xylem ⎫ of primary
- phloem ⎭ bundles
- epidermal hairs on flattened sides of stem
- ground tissue, large celled parenchyma

(b)
- chlorenchyma
- cambial zone
- developing phloem
- developing xylem vessel
- row of xylem vessels
- lignified ground tissue

× 500

10.16 Anomalous secondary thickening in old stem of *Mirabilis* sp. (dicot): (a) TS half stem (LP); (b) details of cambium and formation of secondary tissues (HP)

of this type is known as **secondary enlargement**.

Some palms (e.g. *Cocos* sp.) form periderm like that of dicots. Others have storied cork. **Storied cork** is formed by repeated periclinal divisions in cells of the ground tissue close to the surface of the stem, followed by suberization of the cells produced. There is no layer of cells which can be recognized as a phellogen.

Anomalous secondary thickening similar to that in the Nyctaginaceae occurs in species of *Dracaena*, *Aloe* and *Sansevieria*. In *Dracaena* spp., a cambium forms in the cortex, immediately outside the part of the ground tissue occupied by scattered vascular bundles. The cambium cuts off radial rows of cells towards the inside and the outside. Most of the new cells remain as ground tissue, but some divide several times lengthways and form secondary vascular bundles. The primary vascular bundles are of normal monocot structure. Secondary vascular bundles have a central core of phloem surrounded by tracheids.

Dracaena spp. also form storied cork.

Abscission of plant parts

Abscission means cutting off. Leaves, floral parts and fruits are amongst the parts of a plant which may be cut off during the normal life of the plant.

Leaf abscission
The older leaves of even a herbaceous annual plant may undergo abscission before the life of the plant is over. Leaf abscission occurs annually in deciduous plants and after longer periods in evergreen plants. In addition, leaves may be cut off after they have been severely damaged, perhaps by insect attack or disease.

At some point in the leaf base there is a region in which there are no strengthening tissues except the conducting tissues of the xylem. It is in this region that an abscission layer is formed. It is formed by the softening of the intercellular material and sometimes also of cell walls in a plane across the leaf base. At the same time, a protective layer usually develops between the abscission layer and the stem. Cells of the protective layer become suberised and **wound gum** accumulates between the cells. Tyloses also form in the vessels and tracheids at about the same point, thus sealing off the conducting system. The leaf finally falls off when the cuticle and conducting cells break, due to some strain, normally from wind.

In woody plants, phellogen develops across the wound, below the protective layer, and the leaf scar is finally sealed with a layer of periderm, continuous with the periderm of the rest of the stem. Thus, the severe wound is totally healed.

Other plant parts, including fruits, are separated by abscission layers in a similar way. Some trees form abscission layers across small woody branches and are thus said to be 'self pruning' (e.g. *Antiaris* sp.).

Wound healing

Plants have a considerable ability to heal wounds, whether the wounds are superficial scratches or cuts or major damage such as the loss of a whole branch.

Wounding of herbaceous parts
When a soft tissue is damaged, the immediate result is the formation of a **cicatrice**. In this process, exposed living cells become suberized and wound gum is deposited in intercellular spaces. This process is usually followed by the formation of a wound phellogen similar to normal phellogen, parallel to the surface of the wound.

Wounding of woody organs
When a woody branch is cut or broken off, the first reaction is the formation of callus. **Callus** is a tissue composed of parenchyma cells in an undifferentiated mass. It is formed by cell division from cells of the cambial region and rays. Callus forms a pad over the living tissues. Periderm is then formed close to the surface of the callus and a vascular cambium also forms, continuous with the cambium of the broken surface.

Callus and cambium subsequently spread gradually over the dead heart wood exposed by the wound until the broken end is completely enclosed in living tissues and the vascular cambium is continuous from side to side.

Callus formation, following wounding of woody organs, is very important in the techniques of budding and grafting. In these processes, both stock and scion produce callus. Callus from the two sources merge and new vascular tissues differentiate within, connecting stock and scion.

Key words

Woody plants: shrubs, trees; secondary thickening, anomalous secondary thickening; primary thickening, secondary enlargement. **Secondary vascular tissues**: secondary xylem, secondary phloem, secondary rays. **Secondary xylem**: vessel elements, tracheids, xylem parenchyma, xylem fibres; growth rings, annual rings; sap wood, heart wood. **Secondary phloem**: sieve tube elements, companion cells, phloem parenchyma, phloem fibres. **Periderm**: phellogen (cork cambium), phellem (cork), phelloderm; lenticels. **Monocots**: primary thickening; secondary enlargement; storied cork. **Abscission**: abscission layer. **Wound healing**: cicatrice, wound gum; callus; periderm.

Summary

Dicot trees and shrubs usually have secondary thickening which originates from the vascular cambium of stems and roots. In tree-like monocots, a thick stem usually results from primary thickening, close to the apex, but some palms also have secondary enlargement. Some dicots and woody monocots have anomalous secondary thickening. The secondary xylem (wood) formed by dicots varies in quality according to the proportions and distribution of xylem parenchyma and xylem fibres. Xylem includes living sap wood and dead heart wood. In woody dicots and some woody monocots, the outer tissues of stems and roots are replaced by periderm. Leaves, flowers and fruits fall from a plant when an abscission layer breaks, after which the scar is usually sealed by periderm. Wounds are healed by wound gum or callus, usually followed by the formation of periderm.

Questions

1 What is secondary thickening? Give named examples of a dicot and a monocot with secondary thickening and list the similarities and differences in their secondary structure.
2 What is commonly known as 'wood'? Show how differences in structure between different woods affect the uses to which they are put.
3 Describe what happens to tissues outside the vascular system of a woody plant, after secondary thickening starts.
4 What happens to the end of a branch of a woody dicot after the end of the branch has been cut off? How is the wound sealed and finally healed?

11

The plant in action – I Respiration and photosynthesis

This chapter deals with two biochemical processes which are fundamental to all plant life, and, indeed to animal life. Respiration (cell respiration) takes place all the time in all living cells and involves the release of chemical energy, most of which is then used to support all types of living processes. Photosynthesis takes place in all green cells exposed to light and, with the exception of chemosynthesis as carried out by some bacteria, is the only source of new organic compounds and of the oxygen in the air we breathe. An understanding of these topics requires some knowledge of biochemical processes and of the enzymes that govern them.

Biochemical processes and enzymes
As will be seen, most of the chemical processes which take place in a cell are very complex. Every stage of every complex process is governed by an enzyme. An **enzyme** is a protein which acts as a catalyst in a chemical reaction.

A **catalyst** is a substance which facilitates a chemical reaction, without itself being changed. It is easy to demonstrate the action of a simple, inorganic catalyst. For instance, manganese dioxide catalyses the breakdown of hydrogen peroxide to water and oxygen:

$$2 H_2O_2 \rightarrow 2 H_2O + O_2$$

If a small amount of manganese dioxide is added to some hydrogen peroxide, the mixture immediately froths up as oxygen bubbles are released, until all of the hydrogen peroxide has broken down. More hydrogen peroxide may then be added and will be broken down in the same way. The manganese dioxide is unaffected and may be filtered off and used again indefinitely.

The action of the enzyme peroxidase, which also causes the breakdown of hydrogen peroxide to water and oxygen, can also be easily demonstrated. All living cells contain peroxidase. If a small piece of any living tissue is crushed and added to hydrogen peroxide, oxygen bubbles immediately form as

hydrogen peroxide is broken down. This works equally well with animal and plant tissue.

Some enzymes act alone. Other enzymes can only operate in the presence of another substance, a **coenzyme**. Some coenzymes are simple (e.g. magnesium ion), others are complex substances (e.g. nicotinamide adenine dinucleotide, NAD, see 'Respiration' below).

The names of enzymes
Each enzyme catalyses only one chemical change, or only one type of chemical change. For instance, peroxidase acts only in the breakdown of peroxide groups, while peptidases cause the hydrolysis of the peptide linkages between amino acids in many different peptides and proteins. Except for a few cases, in which old ('traditional') names are used, the name of an enzyme is built up from the name of the substrate on which it acts, or the name of the type of reaction which it catalyses, plus the suffix '-ase'. Thus, peroxidase acts on peroxide groups, while dehydrogenases catalyse reactions in which hydrogen atoms are removed from substrates. The names of enzymes are, therefore, descriptive.

The nature of enzyme action
Inorganic catalysts are thought to act by adsorbing reactant molecules on their surfaces, thus bringing them closer together than they would otherwise be, or by inducing strains within substrate molecules so that they break down. Enzymes appear to act similarly.

All enzymes are proteins. A protein is composed of one or more polypeptides. The polypeptide chains of proteins have a complex tertiary structure, brought about by the formation of hydrogen bonds and disulphide linkages between different parts of the molecule. Each protein molecule therefore has its own characteristic and often complex shape. Parts of the surface of an enzyme molecule form **active sites**, to which substrate molecules are attracted and to which they become temporarily attached. Once

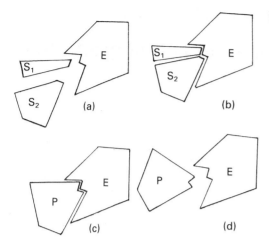

11.1 A diagrammatic representation of enzyme action:
(a) enzyme molecule E attracts two reactant molecules, S1 and S2 to its active site, with which they form hydrogen bonds; (b) the two substrate molecules fit closely into the shape of the active site and are thus brought very close together; (c) the substrate molecules react with each other and (in this case) form a single molecule of a product, P; (d) the union of the two substrate molecules changes the hydrogen bonding relationships, so that the product is set free from the active site and moves away. The enzyme is now ready to accept a new pair of substrate molecules

attachment is complete, the stresses in chemical bonds within the enzyme molecule and in the substrate or substrates are redistributed in such a way that chemical change is facilitated. Once the chemical change has taken place, the product or products of the change are no longer attracted to the active site and move away, leaving the enzyme molecule in its original state and ready to receive more substrate. This process is illustrated diagrammatically in Fig. 11.1.

Factors affecting enzyme action
The rate at which an enzyme-catalysed reaction takes place is affected by several factors.

Substrate concentration and **enzyme concentration** both affect the rate of reaction. In either case, an increase in concentration up to a certain level leads to more rapid activity. This is because substrate and enzyme come together by diffusion, so that higher concentrations make close contact between molecules more frequent. In the case of substrate concentration, reaction rate increases only

up to the optimal level at which all the enzyme molecules are working at their maximum rate.

Temperature affects the rate of all chemical reactions, including enzyme-catalysed reactions. In general, the higher the temperature, the faster the reaction. This is true for enzyme-catalysed reactions only within the range of temperatures at which life processes normally take place. We may refer to 'physiological temperatures', which range from about 0 °C to about 35 °C, depending on the organism and tissue involved. Above this temperature range, enzymes start to break down (become **denatured**) and the rate of reaction falls off.

The effect of temperature on the rate of a reaction is usually described by stating the **temperature coefficient** (Q_{10}) of the reaction:

$$Q_{10} = \frac{\text{the rate of reaction at } (t + 10 \text{ °C})}{\text{the rate of reaction at } t \text{ °C}}$$

Simply expressed, Q_{10} is a ratio which indicates the extent to which a rise in temperature of 10 °C affects the rate of reaction. For a process that is insensitive to temperature (e.g. the light reaction of photosynthesis, see below), the Q_{10} is 1.0. Most enzyme-catalysed reactions have a Q_{10} between 2.0 and 3.0 within the range of physiological temperatures.

Hydrogen ion concentration (pH) also affects the rate of enzyme-catalysed reactions. As proteins, enzymes are composed of amino acids. Every amino acid has a basic (amino) group and an acidic group. Changes in pH affect these groups and so affect the shape of the whole enzyme molecule. All enzymes act most rapidly at an optimal pH, usually close to neutrality (pH 7.0). Certain enzymes, notably some of those which are digestive enzymes of animals, have an optimal pH far from neutrality. For example, the mammalian stomach protease known as trypsin acts most rapidly at pH 2.0.

Inorganic ions may also affect the rate of enzyme action. If a solution of sodium chloride is added to a mixture of amylase solution and starch solution, the rate of breakdown of starch increases greatly.

Measurement of the rate of enzyme-catalysed reactions
There are hundreds of enzymes in every living cell. However, for simple experiments to demonstrate features of enzyme activity it is best to use digestive enzymes which can be obtained commercially, or which can be easily extracted from living tissues.

The rate of activity of a solution of an enzyme is measured by the rate at which a substrate is used up, or the rate at which the product or products of a reaction are formed.

The amylase which is extracted from yeast or from germinating grain is, in fact, a mixture of enzymes. This mixture attacks starch (amylum) molecules and breaks them down to the hexose sugars of which they are composed. We may estimate the rate of this reaction most conveniently by measuring the time taken for a solution of the enzyme to break down all the starch in a given solution (e.g. 0.1 per cent), using an aqueous solution of iodine in potassium iodide as an indicator.

A solution of amylase may be made from commercially produced enzyme (e.g. 0.01 per cent), or by grinding up recently germinated maize grains in water and filtering off the solid matter. If we mix equal volumes of the enzyme solution and 0.1 per cent starch solution, the starch is all broken down in a time measurable in minutes, at room temperature. The mixture of enzyme and substrate solutions is placed in a test tube, shaken occasionally and sampled at one minute intervals. Each sample is a drop of the mixture, which is immediately placed on a white tile (**spotting plate**) and mixed with a drop of iodine solution. As long as starch is present, a blue-black colour immediately appears. When all the starch has been broken down, the blue–black colour no longer appears. When this point is reached, the time taken (minutes) for the total breakdown of the starch is recorded. This is then converted into a 'rate', in arbitrary units:

$$\text{rate} = \frac{1}{t}.$$

The experiment may now be repeated at different temperatures, to determine Q_{10}, at different pHs, with different concentrations of starch and enzyme, and in the presence of sodium chloride. In every case, the time taken is converted to rate, as above, and the rates may then be directly compared.

The above experiment is both simple and crude, but helps to illustrate some of the properties of enzymes. As has been stated, there are hundreds of enzymes acting continuously within every living cell, some in the ground substance of the cytoplasm and nucleus and some within organelles such as chloroplasts and mitochondria. Knowing something of the action of enzymes, we may now proceed to study some of the complex processes which they catalyse.

Respiration

In the process of respiration, energy, which was acquired during photosynthesis and stored in the chemical bonds of carbohydrates and other molecules, is liberated. Some of this liberated energy is converted to heat, in which form it is no longer available for the biochemical work of the cell and may be regarded as waste energy. However, most of the energy liberated in respiration is made available for the immediate use of the cell by being transferred to and stored in **adenosine triphosphate** (ATP). This energy transfer is achieved when a molecule of ATP is synthesized from a molecule of adenosine diphosphate (ADP) and inorganic phosphate (P_i). This synthesis of ATP requires an input of energy of about 7000 calories per molecule:

$$\text{ADP} + P_i + \text{Energy (7000 cals)} \rightarrow \text{ATP}.$$

ATP

The energy required to make ATP from ADP is much more than that required to make ADP from adenosine monophosphate (AMP). Binding the extra phosphate group on to ADP uses a great deal of energy. It follows that, when ATP breaks down to ADP + P_i, the energy with which the terminal bond is held will be released and made available for the work of the cell. In general, reactions which yield energy which can be used by the cell are coupled with the synthesis of ATP. Similarly, reactions which require energy are coupled with the breakdown of ATP to ADP and P_i. Thus:

In the scheme represented above, the forward reaction (A to B + C) is a breakdown reaction which releases energy, which is utilized in the formation of ATP. The reverse reaction (B + C to A) is synthetic and requires energy, which is provided by ATP.

ATP, like other adenosine phosphates (Fig. 11.2), is a nucleotide. By the nature of its reactions, it

11.2 The configuration of the ATP molecule. The terminal phosphate bond is represented differently from other bonds connecting phosphate groups and yields more energy on breakdown

serves as a temporary energy store which provides for the immediate needs of reactions in the cell. ATP serves to couple energy-yielding and energy-consuming reactions in the cell, even when they are not taking place at the same time. The chemical energy in the high-energy terminal bond of the ATP (ADP$\sim$P) is readily transferred from one compound to another.

In man-made machines, large amounts of energy are lost as heat. In contrast, ATP synthesis is an efficient energy transfer system and oxidation of substances within the cell occurs with relatively little loss of energy. As a result, the energy stored in a biological compound may be transferred over and over again. In the dynamic system of the living cell, the energy from one source transfers to ATP and to different compounds at different times.

Substrates of respiration
Respiration is concerned with the breakdown of substances in the cell, with the liberation of energy. The substrate used in respiration is usually a carbohydrate, but lipids and proteins are also used. The substrate used directly in respiration is the simple hexose sugar glucose, which forms the building blocks for starch and which, together with fructose, forms the disaccharide sucrose. More complex carbohydrates undergo hydrolysis to glucose prior to serving as respiratory material (**respiratory substrate**). For this reason, glucose is traditionally regarded as the respiratory substrate.

The complete breakdown of glucose molecules in respiration takes place in three stages. The initial stage is **glycolysis**. The second stage is the **Krebs cycle**, also known as the tricarboxylic acid (TCA) cycle or the citric acid cycle. The third and final stage is the **electron transfer chain**.

Glycolysis
Three important events take place during glycolysis. First, the simple sugar is phosphorylated. **Phosphorylation** involves the breakdown of ATP to ADP + P_i + energy. The energy released is utilized in converting the simple sugar to sugar phosphate. The process of phosphorylation is, therefore, energy-consuming and two molecules of ATP are used up for every molecule of glucose involved. This phosphorylation is referred to as **substrate phosphorylation**. In the process, the six-sided glucose ring is also changed to a five-sided fructose ring, but this rearrangement does not change the number of carbon atoms in the molecule (Fig. 11.3).

Specific enzymes regulate the different stages in the process, which is summarized below. Note that in these and other equations and diagrams that follow, words in brackets are the names of the enzymes which catalyse each change.

$$\text{ATP} \qquad \text{ADP} + P_i$$
(i)   glucose $\longrightarrow$ glucose-6-phosphate
(hexokinase)

(ii) glucose-6-phosphate $\longrightarrow$ fructose-6-phosphate
(phosphogluco-isomerase)

$$\text{ATP} \qquad \text{ADP} + P_i$$
(iii) fructose-6-phosphate $\longrightarrow$ fructose-1, 6-diphosphate
(phosphofructo-kinase)

Thus far, the process has required an input of energy in the form of energy-yielding hydrolysis of ATP.

The second stage of glycolysis is the break-down of the phosphorylated 6-carbon sugar molecule into two molecules of a 3-carbon carboxylic acid, pyruvic acid. This is a complicated process involving six reaction steps. During these reactions, hydrogen ions

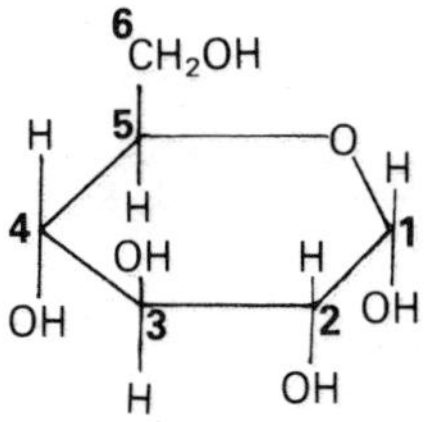

6 CH$_2$OH
H **5** O H
H **4** OH H **1**
OH **3** **2** OH
H OH

(a) glucose molecule with the carbon atoms numbered

6 CH$_2$OH O **1** CH$_2$OH
5 **2**
H H OH H
H **4** **3** OH
OH H

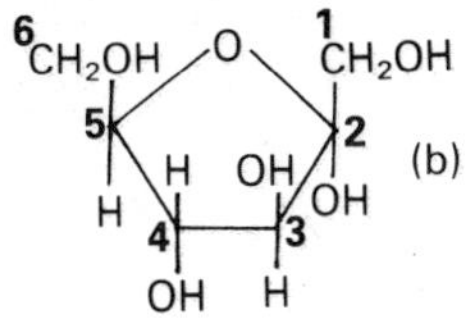

(b) fructose molecule with the carbon atoms numbered

11.3 Structure of the glucose and fructose molecules: (a) glucose; (b) fructose

and electrons are given up by the substrate and transferred to a substance called **nicotinamide adenine dinucleotide** (NAD). NAD is an **electron acceptor** and is reduced or oxidized as it accepts or loses electrons. When it gains an electron it can also combine with protons (hydrogen ions), the presence of which (NADH$_2$) also indicates a reduced condition of the NAD.

Although oxygen is commonly an electron acceptor from an **electron donor**, which thereby becomes oxidized, oxygen is not necessarily involved in oxidation–reduction reactions. Oxidation and reduction are concerned with **electron transport**, in which there is an electron donor and an electron acceptor. When electron transfer takes place, the electron donor is oxidized and the electron acceptor is reduced. In biochemical reactions, the oxidation and reduction of **electron carriers** (donors and acceptors) are commonly indicated by the presence or absence, respectively, of a hydrogen ion. This is illustrated in the equation below, in which AH$_2$ is a substrate and B is an electron carrier.

AH$_2$ +   B
(reduced)   (oxidized)

↓

A   +   BH$_2$
(oxidized)   (reduced)

Associated with the splitting of hexose to pyruvic

acid is a release of energy into the system. This energy is conserved (utilized) in making ATP from ADP and P$_i$. It is also utilized when NAD is reduced to NADH$_2$. Thus, the third main event in glycolysis is the acceptance by NAD of protons and electrons originating from the substrate. The rest of the process of glycolysis can be summarized as follows, (iv) to (vi) below:

(iv) Fructose-1, 6-diphosphate
↓ (aldolase)
dihydroxyacetone phosphate
↓ ↑
glyceraldehyde phosphate

The interconversion of glyceraldehyde phosphate and dihydroxyacetone phosphate is regulated by the enzyme isomerase.

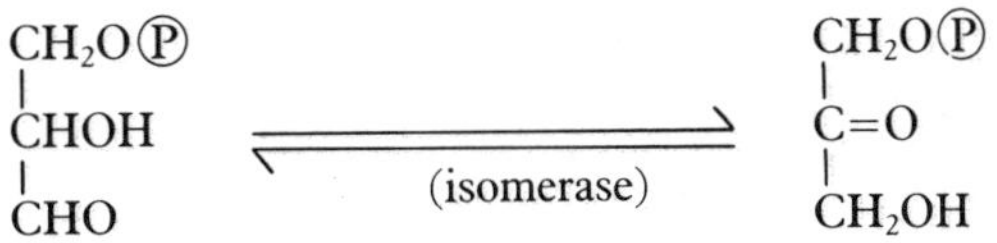

CH$_2$O℗   CH$_2$O℗
| |
CHOH ⇌   C=O
| (isomerase) |
CHO   CH$_2$OH

Note that in the above and in later equations and diagrams, the symbol ℗ represents a 'phosphate group' ($-PO_4$).

(v) Glyceraldehyde phosphate → 1,3-diphosphoglyceric acid

CH$_2$O℗   CH$_2$O℗
| (triose-phosphate dehydrogenase) |
CHOH ——————————————→ CHOH
| |
CHO   P$_i$ NAD   NADH$_2$ COO℗

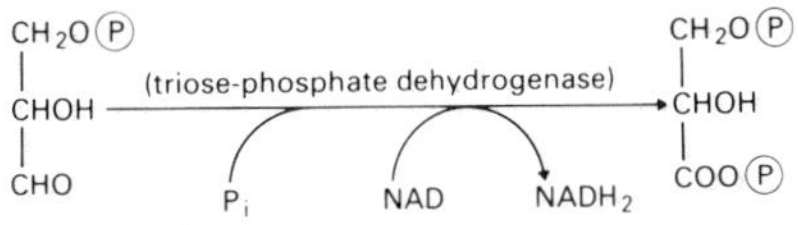

(vi) 1,3-diphosphoglyceric acid → 3-phosphoglyceric acid

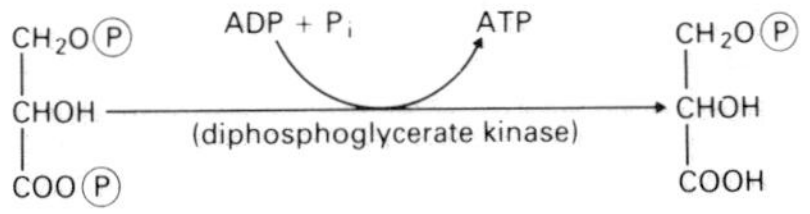

CH$_2$O℗   ADP + P$_i$ ATP   CH$_2$O℗
| |
CHOH ——————————————→ CHOH
| (diphosphoglycerate kinase) |
COO℗   COOH

Two further reactions, omitted here, which involve intramolecular rearrangement and loss of a water molecule lead to the formation of phosphoenolpyruvic acid, from which pyruvic acid is formed:

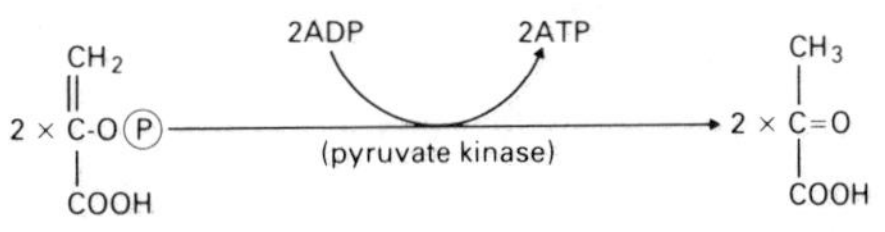

CH$_2$ 2ADP   2ATP   CH$_3$
‖ |
2 × C-O℗ ——————————→ 2 × C=O
| (pyruvate kinase) |
COOH   COOH

The overall reactions of glycolysis are summarized in Fig. 11.4.

$$C_6H_{12}O_6 \xrightarrow{\begin{array}{c} 2\text{ ATP} \quad 2\text{ ADP }(*-2) \\ \\ 2\text{ NAD} \quad 4(\text{ADP} + P_i) \quad 4\text{ ATP} \quad 2\text{ NADH}_2 \\ (*+4) \quad (*+(2\times3)= +6) \end{array}} 2\times \begin{array}{c} CH_3 \\ | \\ C=O \\ | \\ COOH \end{array}$$

*total gain + 8

11.4 The overall reactions of glycolysis and the energy (*) involved

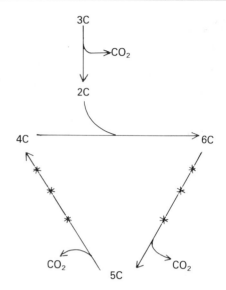

11.5 Very much simplified outline of the Krebs cycle in terms of the carbon atoms involved. The starting substrate is a 3-carbon (3-C) compound, which is oxidized to three molecules of carbon dioxide in one complete cycle. Starred positions represent omitted steps

This summary represents a net gain of 2 ATP molecules in terms of energy conservation. The formation of $2NADH_2$ in aerobic respiration also means that when this is finally oxidized the energy gain will be equivalent to six further ATP molecules, as in the following scheme:

The above is an illustration of an **electron transfer chain**, in which $NADH_2$ is reoxidized to NAD by handing over two electrons and hydrogen ions to an oxygen atom (= half-molecule of oxygen). Three points are worthy of special note. These are (a) the handing-over of electrons to oxygen is not direct, but takes place through a chain of three electron carriers, (b) a total of three ATP molecules are produced per atom of oxygen reduced, or per pair of hydrogen atoms oxidized to water, which conserves the energy used in attaching the high-energy phosphate to ADP in the reaction $ADP + P_i \rightarrow ATP$ in three stages, and (c) the reoxidation of the reduced coenzyme, **flavine adenine dinucleotide** ($FADH_2$), to FAD, which entails the synthesis of two ATP molecules. Thus, in summing up, account must be taken of both the direct ATP synthesis and the ATP equivalents of $NADH_2$ and $FADH_2$.

The electron transfer chain outlined above is actually the third and final stage of respiration. It has been necessary to mention it at this point in order to discuss the fate of the protons and electrons eliminated from the substrate during glycolysis.

The Krebs cycle

The second stage of aerobic respiration is the Krebs cycle, which starts where glycolysis ends. Pyruvic acid, produced by glycolysis, is the substrate in the Krebs cycle. The essential feature of the cycle is that the 3-carbon substrate is completely oxidized to CO_2 and H_2O in one round of the cycle. It therefore requires two turns of the cycle to oxidize the six atoms of carbon of the original glucose molecule.

Before going into details, the following greatly simplified description may be helpful. Both the substrate and the carboxylic acids in the cycle may be considered simply in terms of the number of carbon atoms in each. The substrate, a 3-C (3-carbon) compound, first oxidizes one carbon atom to CO_2 and the remaining 2-C fragment becomes bonded to a 4-C acid to form a 6-C acid. This loses one carbon by oxidation to CO_2, at two different stages in the cycle. In this way, the initial 4-C acid is regenerated and ready to accept another 2-C fragment from pyruvic acid and the cycle is repeated. This is illustrated in Fig. 11.5. It should be emphasized that this oxidation process releases energy by the formation of ATP and in the reduction of NAD to $NADH_2$. The products of the cycle are therefore ATP,

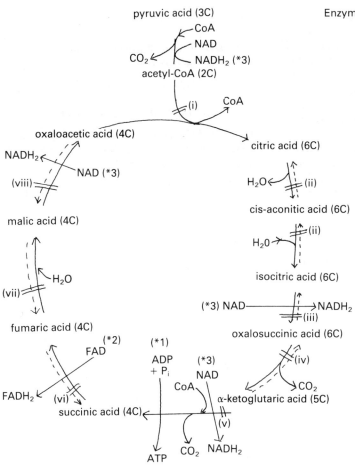

Enzymes:
(i) citrate condensing enzyme
(ii) aconitase
(iii) isocitric dehydrogenase
(iv) carboxylase
(v) α-ketoglutaric dehydrogenase
(vi) succinic dehydrogenase
(vii) fumarase
(viii) malic dehydrogenase

11.6 A more detailed diagram of the Krebs cycle, showing the points (*) at which energy is released. Dotted arrows show possibly reversible reactions while solid arrows represent the main direction of reaction. The enzymes involved are numbered (i) to (viii)

$NADH_2$ and CO_2. It is also important to understand that many more steps are involved than are shown in Fig. 11.5. There are four 6-C acids and at least four 4-C acids in the full scheme (Fig. 11.6).

This is because molecular rearrangements are involved in the scheme. Particularly obvious is the involvement of dehydrogenases such as NAD dehydrogenase and FAD dehydrogenase. These are enzymes which eliminate pairs of hydrogen atoms from the substrates. Sometimes, also, the incorporation or removal of a molecule of water results in the rearrangement of chemical bonds within the

substrates, for example, in the change from citric acid to cis-aconitic acid:

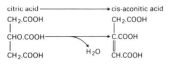

or from fumaric acid to malic acid:

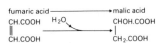

We can now do the arithmetic of the energy involved in the aerobic respiration of one molecule of glucose.

During glycolysis, two ATP molecules are used while four are produced, giving a net gain of two molecules of ATP. This is referred to as substrate level phosphorylation. Also, two $NADH_2$ molecules are produced and their eventual re-oxidation gives rise to six molecules of ATP from oxidative phosphorylation.

From the conversion of pyruvic acid to acetyl CoA, one $NADH_2$ is produced. Again, the reoxidation of this through the electron transport chain leads to the synthesis of three molecules of ATP, or six ATP molecules from the two molecules of pyruvate derived from one glucose molecule.

From the Krebs cycle oxidation of acetyl CoA are produced three $NADH_2$, giving nine ATP, one ATP directly and one $FADH_2$, giving two ATP. This makes a total of twelve ATP from one turn of the cycle, or a total of twenty-four ATP for each molecule of glucose (two turns of the cycle).

In summary, one molecule of glucose gives eight ATP from glycolysis, six ATP from the conversion of pyruvate to acetyl CoA and twenty-four ATP from the Krebs cycle, a total of thirty-eight ATP.

The complete oxidation of one molecule of glucose to carbon dioxide and water is represented by the equation:

$$C_6H_{12}O_6 + 6O_2 \rightarrow 6CO_2 + 6H_2O + 686\,000 \text{ calories.}$$

Only 266 kilocalories (kcal) or about forty per cent of this liberated energy is converted to chemical energy, and thus conserved, in the thirty-eight ATP molecules. The hydrolysis of each ATP molecule ($ATP \rightarrow ADP + P_i$) will therefore liberate seven kcal of energy for the metabolism of the cell.

Anaerobic respiration

In anaerobic respiration, which takes place in the absence of oxygen, the energy produced is from glycolysis only. Glycolysis results in a net gain of two molecules of ATP only, when glucose is broken down to pyruvic acid.

In some bacteria, yeasts and other fungi and in green plants, pyruvic acid is further converted to ethanol (ethyl alcohol) by the process of **alcoholic fermentation**, which may be summarized as follows:

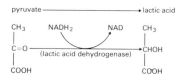

In this conversion, there is no further gain in ATP. In fact, the process is energy-consuming and utilizes the $NADH_2$ formed in gylcolysis. The net reaction, starting from glucose, is:

$$C_6H_{12}O_6 + 2ADP + 2P_i$$
$$\downarrow$$
$$2CH_3CH_2OH + 2CO_2 + 2ATP$$

Since very little energy (about seven per cent) is released from the glucose in this process, most of the energy is still locked up in the ethanol.

In some other bacteria, fungi and in animal cells, the pyruvic acid is converted to lactic acid, by **lactic acid fermentation**, as follows:

It can be seen that this type of respiration, in the absence of oxygen (anaerobic respiration or fermentation), is quite inefficient in releasing energy from the substrate.

An alternative pathway of respiration

The main pathway of respiration is glycolysis followed by the Krebs cycle. However, in many organisms an alternative pathway exists. The essential sequence of reactions in this alternative pathway may be summed up in the diagram below. Note that NADP (nicotinamide adenine dinucleotide phosphate) is another hydrogen carrier and has the reduced form $NADPH_2$.

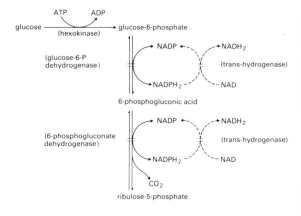

Thus, a hexose (6-C sugar) molecule is directly oxidized to a pentose (5-C sugar) molecule and CO_2. By the action of the enzyme transhydrogenase, the hydrogens of $NADPH_2$ are transferred to NAD to produce $NADH_2$. Two molecules of $NADH_2$ are produced in this reaction and their oxidation to NAD and H_2O through the electron transport chain is accompanied by the synthesis of six molecules of ATP in the usual way. This pathway is oxygen-requiring and is variously termed the **hexose-monophosphate shunt**, the direct oxidation pathway, the pentose–phosphate shunt or, simply, the pentose shunt.

Only one molecule of CO_2 is produced from one molecule of glucose, since only one of the six carbon atoms in the glucose molecule is oxidized. Therefore, six turns of the cycle are needed to oxidize the equivalent of one molecule of glucose. This gives twelve molecules of $NADH_2$ and, by the electron transport chain, these give thirty-six molecules of ATP. By comparison, the energy production of this pathway is therefore almost as efficient as the glycolysis–Krebs cycle pathway. Since one phosphate is necessary to phosphorylate the substrate in the first place, the net gain is thirty-five molecules of ATP, compared with the thirty-eight molecules produced by way of the glycolytic-Krebs cycle. In fact, if triose phosphate, which is invariably produced by the shunt, joins the glycolytic pathway, the net energy yield adds up to thirty-seven molecules of ATP per glucose molecule oxidized.

This mention of triose phosphate above is important. When one carbon atom from the hexose substrate is oxidized, the remaining five carbons of the resulting pentose undergo a complex sequence of rearrangements. Some of the enzymes involved in the process are the same as those involved in the glycolytic pathway and are also involved in the Calvin cycle of photosynthesis (see below). It is beyond the scope of this book to deal with all the details of the shunt. It is enough to note that the reorganization of carbon atoms following the elimination of CO_2 from the hexose gives rise to a variety of 3-C, 4-C, 5-C, 6-C and 7-C compounds. Most of the reactions are reversible. Triose phosphate is one of the intermediate substances in the cycle. The reactions catalysed by two enzymes, transketolase and transaldolase, make up the greater part of the scheme and account for most of the carbon rearrangements. This may be summarized as follows:

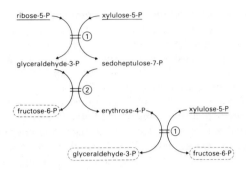

In the above scheme, ① represents transketolase, ② represents transaldolase and –P represents phosphate. Other figures preceding 'P' indicate the terminal carbon atoms to which the phosphate is attached. The three pentose molecules which are underlined may be regarded as the reactants in the scheme and the three ringed molecules (two hexoses and one triose) may be regarded as the products. The existence of this pathway was first indicated when it was found that iodoacetate and fluoride, the classical inhibitors of glycolysis, were found to have no effect on glucose utilization in some tissues.

The leaves of higher plants metabolize glucose largely by this pathway, which also has a role in photosynthesis. The pathway is involved in the interconversion of compounds produced in the initial stages of photosynthesis. It also seems to be more important in mature tissues than in meristematic tissues.

The site of respiration

Respiration as a whole takes place in the cytoplasm of the cell. Glycolysis takes place in the ground substance of the cytoplasm. The Krebs cycle and the electron transport chain are located within the mitochondria.

Within the inner mitochondrial membrane, the cristae are surrounded by a dense solution containing enzymes and coenzymes, water, phosphates and other substances which are involved in respiration. The enzymes and other substances involved in the electron transport chain are lodged in the surfaces of the cristae. The enzymes of the Krebs cycle are in solution in the inner compartment, which permits free passage of certain molecules, pyruvic acid and ATP for instance, but limits the passage of other substances.

Factors affecting the rate of respiration

Temperature affects the rate of respiration. All chemical reactions are sensitive to temperature and enzyme-catalysed reactions such as respiration are particularly sensitive. There is a rather narrow range of temperatures in which respiration can take place. The rate is very slow at temperatures near 0 °C and increases to a maximum between about 35 to 40 °C. Enzymes tend to be denatured at high (**non-physiological**) temperatures. Therefore, above a certain optimal temperature, respiration rate drops as enzyme activity slows, due to the destruction of some of the enzymes. **Exposure time** is a factor of importance in the temperature effect. At higher and inhibiting temperatures, above 30 °C, the period of time before the respiration rate drops becomes shorter. The higher the temperature, the shorter the period.

Molecular oxygen is required for the reactions of the Krebs cycle and the electron transport chain. Oxygen is the final acceptor of electrons in the electron transport chain and is indispensable if aerobic respiration is to take place. As oxygen concentration increases from zero, the rate of aerobic respiration increases. With most plants, the rate of increase in respiration rate becomes less with greater oxygen concentration. This is to say that the increase in respiration rate is hyperbolic. In germinating grains of *Oryza sativa* (Rice) and some other plants, however, the increase in rate is linear over a certain range of oxygen concentrations.

In practice, the atmospheric oxygen effect is slight unless the concentration falls by at least five per cent of the usual atmospheric concentration, and so great a fall is rare. However, variations in oxygen concentration in the soil may be appreciable. Whereas the rise in respiration rate at concentrations below atmospheric concentration is small, there is a steady fall in rate with decrease in concentration below the normal atmospheric level. The effect is less drastic in tissues with a capacity for anaerobic respiration.

Carbon dioxide concentration, like oxygen concentration, varies little in the air but may vary very much in the soil, where oxygen concentration may approach zero and carbon dioxide concentration may rise to as high as ten per cent. The effect of increasing CO_2 concentration is to retard the rate of respiration and consequently to retard or inhibit processes which require respiratory activity. These include water and mineral absorption by roots, root growth and seed germination.

The **state of hydration** of tissues has a general effect, but it is particularly striking in seed germination. Seeds will not germinate unless their tissues are hydrated. When hydration reaches a certain level, depending on the type of seed, there is a marked increase in respiration rate leading to germination.

Inorganic salts can also affect respiration rate. In 1933, Lundegärdh and Burstrom found that the rate of respiration increases when a plant tissue is transferred from water to a solution of a salt. This increase in respiration rate over normal rate is termed 'salt respiration'. Active uptake of mineral ions through the action of 'ion pumps' requires the expenditure of energy and this is supplied by increased respiration.

Some **other chemicals**, often in very low concentrations, retard or completely inhibit respiration by acting, more or less, as enzyme inhibitors. These include cyanides, carbon monoxide, malonates and iodoacetate. This inhibition may be reversible or may cause permanent injury.

Wounding a tissue causes an increase in the respiration rate of cells close to the wound. This has been correlated with increased sugar content or the availability of respiratory substrates. Generally, wounding initiates meristematic activity in the neighbourhood of the wound but its stimulatory effect is not fully understood.

Mechanical stimulation of some plant organs (e.g. leaves) has been reported to increase their respiration rate. This stimulation may be brought about by handling, stroking or bending the organ and seems to take place only in the presence of oxygen. Furthermore, the effect is reported to decrease if stimulation is continued over a period of time. The effect of mechanical stimulation cannot, so far, be explained.

The **age** and **type** of tissue also affects respiration rate. Generally, young and developing tissues respire more vigorously than mature tissues. In this case, as in other cases described above, we can conclude that respiration liberates energy for all cell activities and that there is a higher rate of respiration wherever there is greater activity.

The measurement of respiration

When living tissue is incubated in a dilute, colourless solution of tetrazolium chloride, a red precipitate coats the surface of the tissue within a few minutes.

This is a qualitative test for respiration. The appearance of the colouration indicates that the tissue is alive and respiring. In addition, variation in the density of colouration indicates where respiration is most active. Tetrazolium salt is an artificial hydrogen acceptor and the tetrazolium part of the salt is reduced to an insoluble red compound when dehydrogenase transfers hydrogen to it.

The obvious way to make quantitative measurements of respiration is to measure the quantity of respiratory substrates remaining after a period of respiration, or to measure the quantity of respiratory products over a period of time. The amount of oxygen used can also be measured. It is most convenient to measure the amount of oxygen used, or the amount of carbon dioxide produced.

A simple technique is to pass the carbon dioxide produced by respiration through a solution of sodium hydroxide and afterwards to titrate a sample of the original sodium hydroxide solution and the experimental solution against standard hydrochloric acid, using phenolphthalein as indicator. The difference between the two titrations represents the amount of carbon dioxide absorbed. Alternatively, barium hydroxide may be used to absorb the carbon dioxide. In this case, insoluble barium carbonate is precipitated and the solution titrated against acid, to find the amount of barium hydroxide remaining. This is then compared with a titration of the original barium hydroxide and, again, the difference is a measure of the amount of carbon dioxide absorbed. In either case, the gas passed over the tissue must first be passed through barium or sodium hydroxide solution, to free it from carbon dioxide, before it passes over the tissue, and a known weight of tissue (e.g. seeds of *Vigna unguiculata*, Cow Pea) must be used.

More accurate determinations of respiration rate are done by a **manometric technique** which entails measuring changes in pressure of gas in a closed system. Several simple versions of such apparatus are in use, but the standard one is a manometer called the Warburg apparatus or Warburg respirometer (Fig. 11.7). This comprises a manometer attached to a special flask which has a centre well, and usually a side-arm with a glass stopper. The capillary bore of the manometer is connected to a reservoir at the base, with a screw clip for adjusting the level of the liquid in the manometer. This fluid is the blue Brodie's solution, the density of which is adjusted

to a known value, to facilitate calculations. Tissue under test is placed in the flask and ten per cent potassium hydroxide solution is placed in the centre well, to absorb all the carbon dioxide produced. Thus, volume changes in the closed system result from oxygen uptake. The effects of respiratory stimulants, inhibitors and poisons can be measured by placing them in the side-arm and adding them to the tissue by tipping the flask, after measuring the respiration rate in their absence. Since gas pressures are involved, temperature is kept constant by means of a thermostatically controlled water bath, in which the flasks are immersed. A shaking device keeps the contents of the flasks thoroughly mixed. Some versions are also provided with lighting, for the study of photosynthesis, and cooling devices, for the study of respiration and photosynthesis at lower temperatures.

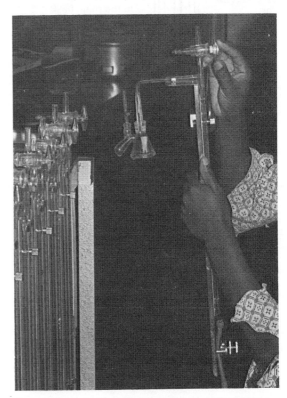

11.7 Warburg apparatus for measuring respiration rate: (a) a student about to put the apparatus illustrated into the water bath (left) (continued overleaf)

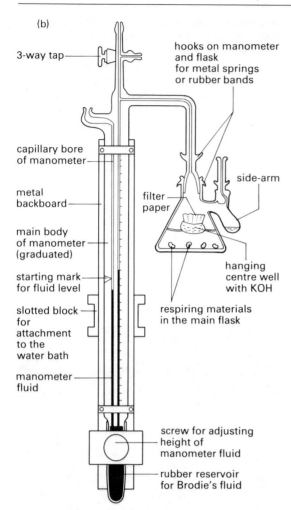

(b)

3-way tap

hooks on manometer
and flask
for metal springs
or rubber bands

capillary bore
of manometer

side-arm

metal
backboard

filter
paper

main body
of manometer
(graduated)

starting mark
for fluid level

hanging
centre well
with KOH

slotted block
for
attachment
to the
water bath

respiring materials
in the main flask

manometer
fluid

screw for adjusting
height of
manometer fluid

rubber reservoir
for Brodie's fluid

11.7 (continued) (b) the essential features of the apparatus, excluding the thermostatically controlled water bath (not to scale)

Respiratory quotient

Respiratory quotient (RQ) is calculated from the equation:

$$RQ = \frac{\text{Volume of } CO_2 \text{ produced by respiration}}{\text{Volume of } O_2 \text{ consumed by respiration}}$$

From the equation

$$C_6H_{12}O_6 + 6O_2 \rightarrow 6CO_2 + 6H_2O,$$

it follows that in the complete respiration of one molecule of glucose, six molecules of carbon dioxide are produced for every six molecules of oxygen

consumed. Therefore, the RQ for carbohydrate respiration is $6/6 = 1.0$. For respiratory substrates in which the ratio of oxygen to carbon is lower than in carbohydrates, the RQ is less than 1.0. For instance, it is about 0.6 for fats and 0.9 when proteins are the substrate for respiration.

The oxidation of the fat, glyceryl tristearate, requires 163 oxygen molecules for the liberation of 114 carbon dioxide molecules, as follows:

$$\begin{array}{l} CH_2O.CO.C_{17}H_{35} \\ | \\ 2 \times CHO.CO.C_{17}H_{35} + \mathbf{163}\ O_2 \rightarrow \mathbf{114}\ CO_2 \\ | \qquad\qquad\qquad\qquad\qquad\qquad + 110\ H_2O \\ CH_2O.CO.C_{17}H_{35} \end{array}$$

So, the $RQ = \dfrac{114}{163} = 0.699 = 0.7$ approximately.

The oxidation of malic acid, a common substrate in succulent plants, gives an RQ greater than 1.0:

$$\begin{array}{l} CH_2.COOH \\ | \qquad\qquad + 3\ O_2 \rightarrow 4\ CO_2 + 3\ H_2O. \\ CHOH.COOH \end{array}$$

In this case, the $RQ = \dfrac{4}{3} = 1.33$.

In addition to the nature of the substrate, various internal and external factors can affect the RQ value. The incomplete oxidation of sugars, for example, gives an RQ lower than 1.0. During the germination of seeds of *Ricinus communis* (Castor Oil) and other fat-containing seeds, oxygen is used first in converting the fats to fatty acids and simple carbohydrates. It is only after this that oxidation of the fatty acids and carbohydrates results in carbon dioxide production as more oxygen is consumed. The ratio of carbon dioxide produced to oxygen consumed is small and, in germinating seeds of *R. communis*, may be as low as 0.3.

Photosynthesis

Respiration takes place in all living cells and results in the release of energy by the breakdown of organic molecules. Photosynthesis takes place in all green plant cells exposed to light and results in the synthesis of organic molecules. Photosynthesis is a light-dependent **anabolic** (synthetic) process, unlike the **catabolic** (breakdown) process of respiration. Materials which are produced by photosynthesis are taken apart in respiration.

Briefly, we shall be looking at photosynthesis as a process of energy conversion in which the initial form and source of energy is light and the final form is in chemical bonds. This conversion is made possible by the special attributes of the light-absorbing green pigment chlorophyll. The process can be summarized as follows. A **photon** (discrete unit) of light is absorbed by chlorophyll. At the same time, a molecule of water is split up in the presence of chlorophyll by the energy from the photon of light. This action gives rise to electrons, hydrogen atoms and molecular (gaseous) oxygen. The energy of the absorbed light is also used to raise one electron (of a lone pair in the chlorophyll molecule) to a higher energy level. The excess energy of this electron is used to synthesize ATP from ADP and P_i. The displaced electron is ultimately accepted by an electron acceptor, nicotinamide adenine dinucleotide phosphate (NADP) in the chloroplast. As it accepts the electron, NADP becomes reduced to $NADPH_2$, taking on the hydrogen from water. In the meantime, the electrons from the water can replace those being lost from the chlorophyll and being donated to the NADP. The products of this reaction (the **light reaction** of photosynthesis) are thus ATP and $NADPH_2$. Both of these products are used in subsequent reactions (the **dark reaction** of photosynthesis) in which carbon dioxide from the air is reduced to carbohydrate.

Thus, the primary event in photosynthesis is the conversion of light energy to electrochemical energy, which goes into the formation of the high energy bonds of ATP and the chemical bonds of the reduced hydrogen acceptor, $NADPH_2$. This energy is then transferred to, and locked up in, the carbohydrate to which the CO_2 is reduced in the energy-absorbing reaction. The sum total of the process is:

$$CO_2 + H_2O + energy \rightarrow (CH_2O) + O_2.$$

The nature of light

Visible light makes up a small part of a continuous spectrum known as the **electromagnetic spectrum**. The electromagnetic spectrum is a spectrum of electromagnetic radiation, the rays or waves of which have varying wavelengths (Fig. 11.8). The total range of the spectrum is from less than one Ångstrom unit (10^{-10}m) to thousands of m. The wavelength of visible light, 380 to 750 nm, represents a very small part of the electromagnetic

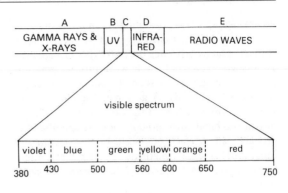

11.8 The visible spectrum in relation to the entire electromagnetic spectrum (not to scale). The ranges of wavelength are:
A = < 1 Å to 100 nm;
B = 100 to 380 nm (ultraviolet);
C = 380 to 750 nm;
D = 750 nm to < 1 m (> 10^9 nm);
E = 1 m to thousands of metres.

spectrum. Physicists have shown that white light is a mixture of light of different wavelengths (colours). Thus, when a beam of white light is passed through a glass prism, it is split up into its component colours.

It is now known that light is transmitted and absorbed not in a continuous stream but in discrete units called photons or quanta. The energy (E) of a single photon of any wavelength is given by the formula $E = hk/\lambda$, where h = a constant (Planck's constant), k = the speed of light (virtually another constant) and λ = the wavelength of light. Since h and k are constants, $E \propto 1/\lambda$, so the energy of a photon is inversely proportional to its wavelength. Light of shorter wavelength is therefore more energetic than light of longer wavelength.

Chloroplast pigments

The chloroplasts of plants belonging to different divisions of algae (Chapter 17) contain different mixtures of pigments. Only chlorophyll a is common to all algae and all land plants. Green algae and land plants have essentially the same pigments in their chloroplasts, namely, chlorophyll a, chlorophyll b, xanthophylls and carotene. The presence of this mixture of pigments can be demonstrated by making an acetone extract of fresh or dry leaves (e.g. of

Amaranthus sp.) and separating the mixture into its components by paper chromatography. Simple paper chromatography separates the four types of pigments as orange (carotene), yellowish–brown (xanthophylls), blue–green (chlorophyll *a*) and yellow–green (chlorophyll *b*) bands. A grey band of phaeophytin sometimes appears behind the carotene band. Phaeophytin is a breakdown product of chlorophyll and is probably an artefact produced by the technique. More sophisticated methods of chromatography separate the xanthophylls into several separate bands.

The role of light in photosynthesis

Light must be absorbed before it can cause a photochemical reaction. Characteristically, light is absorbed by pigments (coloured substances). The effect of light absorption by the photosynthetic pigment is to cause instability in the pigment molecule. An unstable molecule tends to react in such a way that it assumes a state of greater stability. The unstable molecule is said to be in an 'excited' state.

With the absorption of light energy, the molecule of chlorophyll assumes a higher energy level and is therefore more reactive. The destabilization of the chlorophyll molecule causes one member of a lone pair of outermost electrons to go into a higher energy orbit. The molecule then remains unstable until it regains an electron. The displaced electron may return to 'ground' state and restore the stability of the molecule by rejoining its partner in the molecule. Alternatively, it may join a molecule of a different substance. In this latter case, the pigment from which an electron is lost is serving as an electron donor and the receiving substance is an electron acceptor. In the chloroplast, there is a system of electron donors and electron acceptors and the light reaction of photosynthesis is associated with electron transport.

Any flow of electrons is electrical current and can be measured in terms of volts. When a substance receives an electron it becomes 'reduced' and when it loses an electron it becomes 'oxidized'. An electron donor is thus a reducing agent and an electron acceptor is an oxidizing agent. Reactions involving electron transfer are termed oxidation/reduction reactions. The tendency of a reducing agent (or an oxidizing agent) to donate electrons, that is to be oxidized (or to accept electrons and be reduced) is referred to as its **redox potential**. Electrons move

from a more negative electrode to a less negative electrode. Thus, strong reducing agents are associated with large and negative redox potentials. Any compound more electronegative than hydrogen (as standard) has a negative redox potential. So the transfer of electrons to, as well as the consequent movement of, electron acceptors to a higher energy level is associated with increase in electronegativity of redox potentials. The more reducing substances are at higher energy levels than less reducing ones. Electron transfer reactions can therefore be represented on a scale of redox potential (volts, V). The role of light is to create an activated form of chlorophyll by raising some electrons in its molecule to an excited state. In this state the redox potential of chlorophyll *a* changes from about plus 0.4 V to minus 0.6 V, which is more electronegative than the other electron carriers in the chloroplast electron transport chain.

The electron transport chain is illustrated in Fig. 11.9, where it is obvious that at least two forms of chlorophyll and two light reactions are involved. One chlorophyll, absorbing light maximally at about 680 nm wavelength, operates one photochemical system (PS II). The other absorbs at 700 nm wavelength and operates the other photochemical system (PS I). One system enhances the other for maximum photosynthesis.

From the figure the following summary can be made. When a chlorophyll molecule, P683, with a redox potential of plus 0.8 V is struck by light, its redox potential changes to minus 0.2 V. In this excited state it donates electrons directly to an unidentified acceptor, Q. At the same time, the photon energy in the presence of the chlorophyll causes the splitting (**photolysis**) of water to produce oxygen and electrons. It is thought that the electron from the water molecule replaces the one lost by the chlorophyll. The electrons accepted by Q are successively transferred through three or more electron carriers to another chlorophyll molecule, P700, at a redox potential of plus 0.4 V. A second light reaction takes place when this chlorophyll absorbs light of longer wavelength. Its redox potential changes to minus 0.6 V. It donates the electrons directly to another unknown acceptor, X. From this the electrons are transferred to NADP through ferredoxin, in the presence of NADP reductase. The reduced NADP (NADPH$_2$) is reoxidized to NADP when it donates its electrons to the system in which carbon dioxide is reduced.

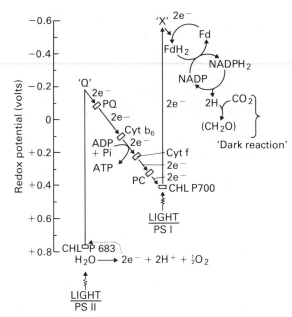

11.9 The electron transport scheme in photosynthesis. Abbreviations used above are:
CHL = chlorophyll molecule;
PQ = plastoquinone;
Cyt b_6 = cytochrome b_6;
Cyt f = cytochrome f;
PC = plastocyanin;
Fd = ferredoxin;
PS = photochemical system;
e = electron

Photophosphorylation

Another aspect of the light-requiring or light-dependent reaction of photosynthesis is very important. At some stage of the electron transfer reaction, probably between cytochrome b_6 and cytochrome f, the excess energy of the electrons is utilized in the synthesis of ATP. This is termed **photosynthetic phosphorylation** or, more simply, photophosphorylation. When the ATP synthesis is coupled to the electron transfer from chlorophyll through the various electron carriers to NADP, this is described as **non-cyclic photophosphorylation**. This produces both ATP and reducing power, both needed in the carbon dioxide reductive fixation. Sometimes the electrons, instead of reaching NADP from ferredoxin, get back to one of the carriers between plastoquinone and plastocyanin. Such a route is to some extent cyclic and the ATP synthesis

associated with it is termed **cyclic photophosphorylation**. This results in ATP production but no reducing power. Under some other conditions the electrons may rejoin oxygen or even water. Again, no reducing power is produced and the phosphorylation in this case is termed **pseudo-cyclic photophosphorylation**.

Light absorption and photosynthesis

Light is absorbed in each of the two photochemical systems by a different form of chlorophyll, possibly a different form of chlorophyll *a*. Chlorophyll *a* is thus in effect the main photosynthetic pigment. However, not all chlorophyll *a* molecules absorb light at the same wavelength. There is experimental evidence that groups of chlorophyll molecules absorbing at different wavelengths act together for maximum efficiency in light energy 'harvesting' in each photosystem. This group is referred to as the **photosynthetic unit**. In higher plants and green algae, the unit comprises 250 to 300 chlorophyll molecules. The molecules absorbing at short wavelengths outnumber those absorbing at longer wavelengths. In PS I, for instance, all the light quanta absorbed in different parts of the unit are transferred to one reaction centre with a single chlorophyll *a* molecule called P700 absorbing at the longest wavelength. The energy in the longer wavelengths is less than that at shorter wavelengths. As a result, energy is transferred preferentially and spontaneously from the shorter wave absorption band to the longer one, with only very small losses of energy in the form of heat at each stage.

The cooperation of pigment molecules in light absorption is not limited to chlorophyll *a*. Energy may also be transferred from chlorophyll *b* to chlorophyll *a*. In fact, non-chlorophyllous pigments such as carotene and xanthophylls also absorb light, the energy of which is subsequently used in photosynthesis. Since these pigments are not the site of photosynthesis they are termed **accessory pigments**. The light they absorb is used only indirectly and must be transferred to the main photosynthetic pigment. In the intact plant, transfer of energy from one pigment to another is possible, provided that the pigment particles are close enough together. Energy transfer is always from short wave absorber to long wave absorber. Accessory pigments are also believed to function protectively against the photo-oxidation of chlorophyll in intense light.

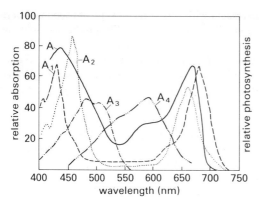

11.10 Comparison of the photosynthetic action spectrum (A) with absorption spectra of chlorophyll *a* (A₁), chlorophyll *b* (A₂), carotenes (A₃) and phycobilins (A₄). Adapted from R.G.S. Bidwell, 1979. *Plant Physiology*, 2nd edn, Macmillan Publishing Co. and Collier Macmillan Publishers, London

A graphical representation showing the relative amounts of light of different wavelengths absorbed by a pigment is termed the **light absorption spectrum** for that pigment. On the other hand, a representation of the relative amounts of photosynthesis carried out with the same amount (intensity) of light of different wavelengths is the **photosynthetic action spectrum**. When a generalized photosynthetic action spectrum is compared with the light absorption spectra for various photosynthetic pigments, the role of chlorophyll *a* as the principal photosynthetic pigment becomes obvious. As shown in Fig. 11.10, the absorption spectrum showing the closest fit or nearest coincidence with the action spectrum is that of chlorophyll *a*. The best interpretation of this is that the light absorbed by chlorophyll *a* is that most directly utilized in photosynthesis.

The absorption of light by two different forms of chlorophyll with different wavelength absorption peaks is represented by the two photosystems mentioned above. We previously referred to the group of co-operating pigment molecules in the absorption of light within the photosynthetic unit. There is also co-operation between the two photosystems. As a result of, and as evidence of this co-operation, the net photosynthesis carried out when the two systems are operating simultaneously exceeds the sum of the photosynthesis from both systems operating separately. This kind of effect, which is more than additive, is termed **synergistic**. It is also

known as the **Emerson enhancement effect**, after its discoverer. Emerson found that:

(i) With 700 nm wavelength light alone, the (photosynthetic) quantum efficiency was less than one per cent;

(ii) with 680 nm wavelength light alone, the quantum efficiency slightly exceeded nine per cent;

(iii) the net efficiency when (i) and (ii) are added slightly exceeds ten per cent;

(iv) with the two different wavelengths (i) and (ii) given simultaneously, the quantum efficiency was of the order of fifteen per cent.

Thus, the two co-operating photochemical systems are necessary for maximum photosynthetic efficiency.

The Calvin cycle

The reactions discussed so far require light and are therefore light-dependent. They are usually referred to as the light reaction of photosynthesis. The ATP and reducing power obtained in the light reaction are required for carbon dioxide fixation. Carbon dioxide fixation does not require light, but it can take place in the light as well as in darkness. It is therefore a light-independent reaction and is often termed the **dark reaction** or formerly, Blackman's reaction. It includes a carbon reaction cycle known as the **Calvin cycle**, named after Professor M. Calvin of the University of California. In the Calvin cycle, as in the Krebs cycle of respiration, the starter (acceptor) molecule is again regenerated at each turn of the cycle and a molecule of carbon dioxide is assimilated at the same time (Fig. 11.11). Thus, to produce one molecule of hexose, six revolutions of the cycle are required, while six molecules of carbon dioxide are fixed. The general equation is:

$$6CO_2 + 6H_2O \rightarrow C_6H_{12}O_6 + 6O_2.$$

Two significant events can be isolated in the Calvin cycle. First, carbon dioxide reacts with ribulose biphosphate to produce two molecules of phosphoglyceric acid. This acid is reduced to a carbohydrate, triose–phosphate, by the products of the light reaction. It is from triose–phosphate (glyceraldehyde phosphate) that the main photosynthetic products (sugars, polysaccharides) are produced. Secondly, the Calvin cycle includes a complex series of reactions involving 3-, 4-, 5-, 6- and 7-carbon sugar phosphates. The net effect of the rearrangement of

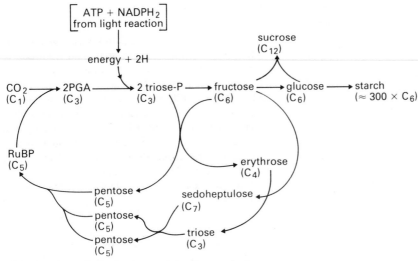

11.11 The Calvin Cycle. A scheme of the photosynthetic carbon cycle. The number of carbon atoms in each molecule is indicated in brackets and each sugar contains either one or two phosphate groups. Triose-P = glyceraldehyde phosphate (GAP)

carbon atoms involved is the regeneration of the original pentose molecule which is the initial carbon dioxide acceptor molecule.

The arrangements of carbon atoms (**carbon skeletons**) formed in the dark reaction of photosynthesis are the basis on which all the other organic molecules of the living system are produced. It should also be noted that the triose–phosphate (glyceraldehyde phosphate) produced by reduction of phosphoglyceric acid (PGA) is common to photosynthesis and the glycolysis stage of respiration. The very steps leading to its production in glycolysis can lead to the production of fructose and glucose in more or less reverse order in photosynthesis, using the high energy bond of ATP.

The immediate products of carbon dioxide fixation

When carbon dioxide is absorbed into the stroma of the chloroplast at the start of the Calvin cycle, it reacts with ribulose biphosphate to produce an organic acid, phosphoglyceric acid (PGA). This is then reduced to the carbohydrate triose-phosphate. The **carboxylation** reaction is catalysed by the enzyme ribulose biphosphate carboxylase, abbreviated here to RuBisCo.

ribulose-1,5-biphosphate(RuBP) $\longrightarrow$ PGA $\longrightarrow$ triose-phosphate

$$
\begin{array}{c}
CH_2O\ \text{(P)} \\
| \\
C = O \\
| \\
CHOH \\
| \\
CHOH \\
| \\
CH_2O\ \text{(P)}
\end{array}
+ CO_2 + H_2O \longrightarrow
\underset{(RuBisCo)}{}
2 \times
\begin{array}{c}
CH_2O\ \text{(P)} \\
| \\
CHOH \\
| \\
COOH
\end{array}
\xrightarrow{2H}
2 \times
\begin{array}{c}
CH_2O\ \text{(P)} \\
| \\
CHOH \\
| \\
CHO
\end{array}
$$

The fixation of carbon dioxide described so far is known as the **C_3 pathway**. In some species of plants, however, the first product of carbon dioxide fixation is a 4-carbon compound. This was first reported from Hawaii in the early 1960s by researchers on sugar cane. This was later confirmed and the mechanism elucidated by M. D. Hatch and C. R. Slack of Australia, who showed that the first product of carbon dioxide fixation in this pathway was oxaloacetic acid. This **C_4 pathway** is alternatively known as the Hatch–Slack pathway. The carbon dioxide acceptor in this case is a 3-carbon compound, phosphoenolpyruvic acid (PEP) and the enzyme concerned is PEP carboxylase. The oxaloacetate formed is then either reduced to malic acid or reacts with alanine

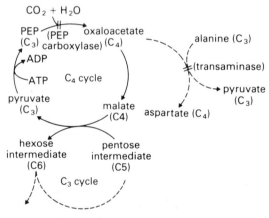

$$COOH - COP + CO_2 + H_2O \rightarrow C=O$$

(PEP)   (oxaloacetate)

(malate)

(aspartate)

(alanine)

(pyruvate)

11.12 The C_4 (Hatch–Slack) pathway, details of reactions involved

11.13 Outline of the C_4 pathway and its connection with the C_3 cycle

(by transamination) to form aspartic acid and pyruvic acid (Fig. 11.12). However, the C_4 pathway and the C_3 pathway are linked because the immediate products of carbon dioxide fixation by the C_4 pathway, malate and aspartate, are no substitute for PGA. The acids are broken down to carbon dioxide, which joins the Calvin cycle, and pyruvic acid. The pyruvic acid then reacts with ATP to regenerate the carbon dioxide acceptor molecule, PEP. The C_4 cycle is summarized in Fig. 11.13.

Kranz anatomy – special leaf anatomy of C_4 plants

Plants which utilize the C_4 pathway (**C_4 plants**) have a leaf anatomy which differs somewhat from that of plants which utilize the C_3 pathway (**C_3 plants**). In C_3 plants, the mesophyll tissue contains chloroplasts which are of uniform appearance and distribution. Carbon dioxide fixation and the rest of the Calvin cycle reactions take place in all the mesophyll cells. In the C_4 leaf, the parenchymatous bundle sheaths are more prominent than in the C_3 leaf. These bundle sheath cells contain chloroplasts with poorly developed grana, while mesophyll cells contain chloroplasts with well-developed grana. The mesophyll cells carry out the fixation of atmospheric carbon dioxide by the C_4 pathway, and lack the Calvin cycle. Malate and aspartate formed in the mesophyll cells are then transported to the bundle sheath cells, which have the Calvin cycle. The acids break down here to give carbon dioxide, which is then fixed by the C_3 mechanism. So, the biochemical events of photosynthesis (C_3 and C_4) are separated in space in C_4 leaves but not in C_3 leaves. In C_4 leaves, the chloroplasts of mesophyll cells usually lack starch grains, but there are numerous starch grains in the chloroplasts of the bundle sheath cells.

The significance of C_4 photosynthesis

Since it eventually ends up in the C_3 mechanism, the C_4 mechanism seems unwieldy and possibly unnecessary. However, the PEP carboxylase which mediates C_4 carboxylation has a much higher affinity for carbon dioxide than RuBisCo, of the C_3 cycle. As a result, under low carbon dioxide concentrations, the PEP carboxylase enables C_4 leaves to absorb carbon dioxide efficiently. High light intensities and high temperatures prevail in the tropics and, under these conditions, plants tend to close their stomata due to the loss of much water. Diffusion of carbon dioxide into the leaves is therefore restricted. As a result, C_4 plants are better adapted to live in the tropics or elsewhere under dry conditions and high temperature conditions. The significance of this is that C_4 plants not only survive but show relatively high rates of photosynthesis under conditions that are counter-productive and possibly lethal to C_3 plants.

Both C_4 and C_3 species are represented in many families and more than a dozen genera have both C_3 and C_4 species. Families, genera and species with C_3

Table 11.1 Examples of families and genera of flowering plants which include C_3 and C_4 species.

Taxa		C_3 Plants	C_4 Plants
DICOTS	families	Chenopodiaceae, Compositae, Cruciferae, Leguminosae, Malvaceae, Solanaceae.	Amaranthaceae, Chenopodiaceae, Euphorbiaceae, Portulacaceae.
	genera	*Helianthus, Lactuca, Gossypium, Arachis.*	*Amaranthus, Gomphrena, Euphorbia, Portulaca.*
	species	*Helianthus annuus, Lactuca sativa, Gossypium hirsutum, Arachis hypogaea.*	*Amaranthus hybridus, Gomphrena globosa, Euphorbia maculata, Portulaca oleracea.*
MONOCOTS	families	Gramineae, Cyperaceae.	Gramineae, Cyperaceae.
	genera	*Avena, Dactylis, Panicum, Triticum, Cyperus.*	*Cynodon, Oryza, Panicum, Pennisetum, Sporobolus, Cyperus.*
	species	*Avena sativa, Dactylis glomerata, Hordeum vulgare, Triticum sativum.*	*Oryza sativa, Panicum maximum, Saccharum officinale, Zea mays.*

and C_4 photosynthesis are listed in Table 11.1.

Crassulacean acid metabolism and photosynthesis

Crassulacean acid metabolism (CAM) is yet another pathway of carbon dioxide fixation and is found in most succulent plants. In these plants, there is a diurnal variation in organic acid concentration. Titratable acidity increases in the dark and decreases when it is light. This phenomenon was first studied in members of the family Crassulaceae, which gave the process its name. The essential feature of the process is that carbon dioxide is fixed into 4-carbon organic acids during the night, when the stomata of the plants concerned are open. At the same time, carbon dioxide produced by respiration is also fixed. Malic acid, citric acid and isocitric acid are the principal products. The acids break down and release carbon dioxide again when it is light.

Under conditions of high light intensity and dryness, in terms of water economy, it is advantageous for a plant to have its stomata closed. This limits water loss, but also limits the entry of carbon dioxide for photosynthesis. Under these conditions, however, carbon dioxide from the organic acids and from respiration is available for photosynthesis. Thus, the ability to carry on with photosynthesis with stomata closed which is conferred by CAM gives succulents the ability to survive in arid and semi-arid environments. **CAM plants** usually possess xeromorphic features such as sunken stomata, thick cuticle and reduced leaves. However, not all CAM plants are succulents and CAM is not

an obligate pathway. If stomata are open during the day, carbon dioxide is absorbed and fixed in the normal way and CAM is bypassed.

CAM photosynthesis is similar to C_4 photosynthesis in the sense that carbon dioxide fixation into C_4 acids is separated from C_3 photosynthesis. In the case of C_4 plants, the separation is spatial. In CAM plants, the two processes are separated in time only. In both cases, the initial fixation of carbon dioxide is referred to as β-carboxylation, because the absorbed carbon dioxide is fixed in the β-carbon position of the resulting oxaloacetate:

$$PEP \longrightarrow oxaloacetate \longrightarrow malate$$

$$H_2O + {}^*CO_2 + \begin{matrix} CH_2 \\ | \\ C\text{-}O\text{-}P \\ | \\ COOH \end{matrix} \xrightarrow[P_i]{\text{(PEP carboxylase)}} \begin{matrix} \alpha\text{-COOH} \\ | \\ C=O \\ | \\ CH_2 \\ | \\ \beta\text{-}^*COOH \end{matrix} \xrightarrow[2H]{\substack{\text{(malic} \\ \text{dehydrogenase)}}} \begin{matrix} COOH \\ | \\ CHOH \\ | \\ CH_2 \\ | \\ COOH \end{matrix}$$

Families in which CAM occurs include the Crassulaceae, Cactaceae, Orchidaceae, Bromeliaceae, Liliaceae and Euphorbiaceae. A summary of the process of carbon dioxide fixation by CAM is given in Fig. 11.14. The CAM probably represents an ecological adaptation. CAM species studied in as many as eighteen families are found to live along the tropics of Cancer and Capricorn. Those found in the wet tropics occupy physiologically dry and epiphytic habitats. The gymnosperm *Welwitschia bainesii* (*W. mirabilis*) of the Namibian desert and some ferns are also known to have CAM.

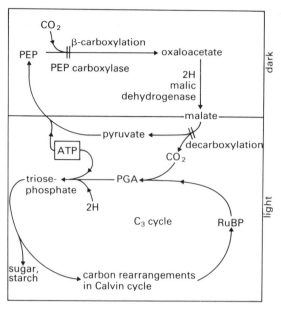

11.14 Carbon dioxide fixation in CAM

Photorespiration

Photorepiration is another pathway of carbon metabolism. It is a special carbon oxidation pathway. In this sense, it is unlike the C_3 or C_4 pathway, where there is a net fixation of carbon into a C_3 or C_4 compound. In the C_3 pathway, we have $C_5 + C_1 \rightarrow 2C_3$. In the C_4 pathway, we have $C_3 + C_1 \rightarrow C_4$. In photorespiration, we have $C_5 + O_2 \rightarrow C_3 + C_2$. Further reactions of the C_2 compound lead to the production of glycine, which is then decarboxylated to serine. The net effect of the sequence of reactions involved is that oxygen is absorbed and carbon dioxide is produced. This is a real oxidation reaction but there is an important difference between this and Krebs cycle oxidation. ATP and $NADPH_2$ are produced in the Krebs reaction but neither is produced in photorespiration, which is, therefore, an energy-wasteful oxidation pathway. The reactions are summarized in Fig. 11.15. This C_2 carbon oxidation cycle is another accessory carbon cycle coupled to the reductive C_3 (Calvin) cycle in photosynthesis. It should be noted that RUBP (RUDP) is the common acceptor molecule for CO_2 in the C_3 cycle as well as for oxygen in the C_2 cycle. PGA is produced via either route but the C_2 acid phosphoglycollate is also produced via the oxygen pathway. As shown in the

diagram, the process involves three sites, chloroplasts, peroxisomes and mitochondria. Carbon dioxide is produced in the mitochondria. This light-dependent carbon dioxide production is photorespiration. At high oxygen and low carbon dioxide concentrations, the C_2 pathway (**glycollate pathway**) predominates in species with glycollate oxidase and photosynthetic efficiency decreases. Conversely, the C_2 pathway (or the oxygenase activity of RUBP carboxylase) is inhibited at high carbon dioxide concentrations and low oxygen concentrations. In this case, the C_3 cycle is favoured.

The phenomenon of photorespiration is compared with both photosynthesis and 'dark' (normal) respiration in Table 11.2. It is characteristic of C_3 plants and virtually absent in C_4 and CAM plants. On balance, photorespiration is undesirable and can be a serious cause of reduced efficiency in photosynthesis. The famous German biochemist Otto Warburg reported in 1920 that oxygen inhibits photosynthesis in algae. This inhibition is common to all C_3 plants and is known as the **Warburg effect**. This inhibition is due to photorespiratory loss (non-energy-conserving oxidation) of products of photosynthesis in species in which photorespiration occurs.

The site of photosynthesis

The chlorophyll a and accessory pigments involved in the light-dependent reactions of photosynthesis are embedded within the grana lamellae of the chloroplast. The light-independent reactions of photosynthesis take place in the stroma of the chloroplast. However, organelles other than chloroplasts are involved in the glycollate metabolism, which is truly a photosynthetic phenomenon. Chloroplasts, peroxisomes and mitochondria are involved in that special photo-metabolic pathway, in which oxygen is absorbed and carbon dioxide is released.

Factors affecting rate of photosynthesis

The principal requirements of photosynthesis are the presence of chlorophyll and supplies of light, water and carbon dioxide.

Chloroplasts containing chlorophyll a and accessory pigments are present in abundance in most chlorenchyma.

Photosynthesis takes place as long as **light** is available. On a sunny day, light is not the **limiting factor**

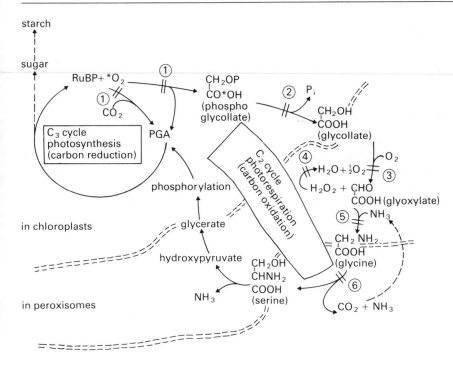

11.15 The pathway of photorespiration. Some of the important enzymes are numbered: 1 = ribulose diphosphate carboxylase/oxygenase; 2 = phosphatase; 3 = glycollate oxidase; 4 = catalase; 5 = transaminase; 6 = glycine decarboxylase

Table 11.2 Comparison of some features of photorespiration, photosynthesis and dark respiration.

Feature	Photorespiration	Photosynthesis	Dark respiration
process occurring only in the light	yes	yes	no
site of process	Chloroplasts, peroxisomes, mitochondria.	chloroplasts	mitochondria
energy in the form of carbon compounds, ATP or NADPH$_2$	decreased	increased	increased
temperature increase within the physiological range	rate increases up to 30 °C	light phase unaffected, dark phase increases	increase continuously
O$_2$ increase from 2 to 40 per cent, CO$_2$ decrease from 0.2 to 0.0 per cent	increased	little effect	little effect
larger increases in O$_2$ concentration	increases	decreases	increases
increases in CO$_2$ concentration	decreases	increases	decreases
CO$_2$ gas	released	absorbed	released
O$_2$ gas	absorbed	released	absorbed

for photosynthesis, that is, it is not the factor which determines the rate of photosynthesis and increasing the light supply will not change the rate. However, simple experiments using a 'bubbler' (Fig. 11.16) show clear decreases in rate of photosynthesis as, for instance, a cloud obscures the sun. The colour of incident light can also be shown to be important, under experimental conditions. It can be shown, again using a bubbler, that those wavebands which are most strongly absorbed by chlorophyll (e.g. red light) cause the most rapid photosynthesis.

Water is always present in healthy cells and it is difficult to demonstrate its necessity for photosynthesis, beyond the fact that photosynthesis slows down when leaves become wilted. However, wilting has many effects on a leaf, including causing the closure of stomata, and the lower rate of photosynthesis in wilted leaves is not a simple matter of water shortage.

Air contains about 0.03 per cent of **carbon dioxide** and is the source of carbon dioxide for the plant. It can be shown that this low percentage of the gas is usually the limiting factor for photosynthesis. Under well-illuminated conditions, the rate of photosynthesis can be increased by keeping a plant in an atmosphere in which the proportion of carbon dioxide is artificially augmented. This can be demonstrated quite simply, using a bubbler once again. If a solution of sodium hydrogencarbonate (sodium bicarbonate) is added to the water in which the plant and its bubbler are immersed, the rate of oxygen production immediately rises. Sodium hydrogencarbonate provides carbon dioxide thus:

$$2 \text{ NaHCO}_3 \rightarrow \text{Na}_2\text{CO}_3 + \text{CO}_2 + \text{H}_2\text{O}.$$

The rate at which the light reaction of photosynthesis proceeds is dependent on light intensity but not on temperature, since only photochemical reactions are involved. However, the dark reactions are ordinary enzyme-catalysed reactions and would be expected to have a Q_{10} between 2.0 and 3.0. In practice, this cannot be demonstrated because temperature is not a limiting factor for the rate of photosynthesis at physiological temperatures. Dark reactions keep pace with light reactions throughout.

The measurement of photosynthesis

As with respiration, the rate at which photosynthesis proceeds is measured by observing changes in the concentrations of substances which are consumed in

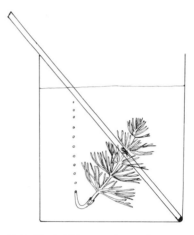

11.16 The bubbler experiment. One end of the bubbler, a short length of glass tube, fits closely around the end of the stem of the water plant, the other end is drawn out into a capillary tube, which ensures uniform bubble size. The rate of photosynthesis can then be measured by counting the number of bubbles released per minute, under different conditions of lighting and carbon dioxide concentration (using potassium hydrogen carbonate (potassium bicarbonate) solutions)

the process or substances which are produced by photosynthesis, usually the latter.

In C_3 plants, carbohydrate which is produced by photosynthesis is incorporated in starch grains which form in the stroma of the chloroplast. It is constantly being removed and **translocated** (transported) to other parts of the plant for more permanent storage, but starch accumulates in chloroplasts while it is light. This is the basis for simple qualitative experiments which can be used to demonstrate some of the requirements of photosynthesis. In such experiments, a leafy branch is kept in darkness for twenty-four hours or more, during which time the starch in the leaves is either respired or translocated into the stem. The leaves are said to be 'destarched'. Leaves, or different parts of the leafy shoot, are then exposed to different conditions. They may be partially masked with an opaque material so that only part of each leaf is exposed to light, exposed to different wavelengths of light, or kept in an atmosphere which lacks carbon dioxide. Subsequently, the leaves are detached, the chlorophyll removed with a suitable solvent (ethanol, acetone) and, after washing in water, placed in iodine solution. The dark parts of the 'starch print' formed indicate where photosynthesis has taken place. In experiments such as these, it is possible to demonstrate that photosynthesis requires light, light of certain wavelengths and carbon dioxide. The necessity for chlorophyll can also be demonstrated if variegated leaves (leaves with white and green patches, e.g. *Breynia* sp.) are used.

The rate at which oxygen is evolved can be measured, in water plants, by use of a bubbler, referred to above, and plants exposed to different conditions. This method is particularly useful for comparing the photosynthetic values of different intensities and wavelengths of light and different concentrations of carbon dioxide.

More sophisticated experiments require the use of a Warburg or similar apparatus, the same as that used for studies of respiration rate but provided with a controlled lighting system. In such an apparatus, it is possible to study, for instance, the relative rates of respiration and photosynthesis and to determine the intensity of light which is sufficient to allow photosynthesis to exactly compensate for the loss of carbohydrate by respiration (**compensation point**). Compensation point varies much in different plants. It is usually high in plants which normally live exposed to full sunlight (**sun plants**) and low in plants which are adapted to living in shade (**shade plants**).

Key words

Catalysts: inorganic catalyst; organic catalyst, enzyme, coenzyme; protein molecule, active site.
Factors: enzyme concentration, substrate concentration, pH (hydrogen ion concentration), inorganic ions, temperature; physiological temperatures.
Processes: respiration, photosynthesis; catabolic (breakdown, energy release), anabolic (synthesis, energy conversion). **Respiration**: cellular, aerobic; respiratory substrate, glucose; glycolysis, Krebs cycle, electron transfer chain; respiratory quotient (RQ). **Glycolysis**: cytoplasmic matrix; substrate phosphorylation. **Krebs cycle**: mitochondria; molecular rearrangement. **Electron transfer chain**: oxidation, reduction, electron donor, electron acceptor. **Anaerobic respiration**: fermentation; ethanol, lactic acid. **Alternative pathway**: pentose shunt. **Photosynthesis**: light energy to chemical energy; light reaction, dark reaction. **Light reaction**: photolysis of water, excitation of chlorophyll mole-cule, redox potential; pigments, chlorophyll, light absorption spectrum, photosynthetic action spectrum; synergistic effect. **Dark reaction**: C_3 pathway, Calvin cycle; C_4 pathway, Kranz anatomy; crassulacean acid metabolism (CAM), ecological adaptation; photorespiration (C_2 carbon oxidation cycle), glycollate pathway.

Summary

Chemical reactions within the cell are catalysed by enzymes. Respiration and photosynthesis are both very complex processes and involve many enzymes. In respiration, a catabolic process, carbohydrates and other substrates are oxidized and the energy which is released is transferred to ATP. The energy stored in ATP is immediately available for energy-consuming reactions which take place in the cell. The first part of respiration, glycolysis, takes place in the cytoplasmic matrix. The other reactions of aerobic respiration, the Krebs cycle and the electron transfer chain, take place in mitochondria. Photosynthesis is an anabolic process in which hydrogen from water is combined with carbon dioxide from the atmosphere, using the energy of light, to produce carbohydrates. Oxygen is produced in the process. The first part of photosynthesis is the light reaction, in which water molecules are broken down and oxygen is released. This is followed by the dark reaction, in which energy obtained in the light reaction is used in the fixation of carbon dioxide. The light reaction takes place in the grana lamellae, the dark reaction takes place in the stroma of the chloroplast.

Questions

1 What are enzymes? Give examples of the involvement of enzymes in photosynthesis and respiration in a green plant.
2 What is the importance of ATP in the processes of respiration and photosynthesis in a green plant?
3 Outline the process of aerobic respiration in a plant cell. Where does the process take place and why are enzymes important? How is respiration affected by outside (environmental) factors?
4 Photosynthesis comprises a 'light reaction' and a 'dark reaction'. Give an outline of each of these two stages of photosynthesis. What factors affect the rate of photosynthesis?

12

The plant in action – II   Water relations, translocation and mineral nutrition

The term **water relations** describes all aspects of the movement of water into, through and out of the plant. We have already looked at the water relations of plant cells (Chapter 5).

The term **translocation** describes the longitudinal movement of water and dissolved substances along the parts of a plant, through the xylem and phloem. Water and mineral ions derived from the water in the soil are absorbed by the roots and translocated through the xylem to all parts of the plant. Water is subsequently lost by evaporation (**transpiration**), mostly from the leaves. Organic products of synthesis (e.g. sucrose, amino acids) are translocated within the plant through the phloem.

In this chapter we shall concentrate our attention first on the uptake and translocation of water and mineral ions, transpiration, and phloem translocation. We shall then look at the requirements of plants for mineral ions (**mineral nutrition**).

The uptake of water and mineral ions

As shown in Chapter 5, the movement of water between two regions of living tissue, or between two adjacent cells, depends on the difference between their water potentials. Most of the water uptake into a rooted plant is through the root epidermis, especially in the region of root hairs. As long as the water potential of the soil solution is greater than that of the cell sap, water will enter the root. Osmosis, a passive (non-energy-requiring) mechanism is responsible for water uptake. There is no known active (energy-requiring) mechanism directly involved in water uptake. However, active uptake of mineral ions into the root from the soil solution results in the accumulation of mineral ions, against a concentration gradient, in the cells of the root. Since this results in an increase in the water potential difference between the root and the water in the soil, this in turn results in increased osmotic water uptake

from the soil. In this way, an active mechanism is indirectly responsible for increased water uptake.

A typical soil, suitable for the growth of most plants, contains both air and water. The soil water contains dissolved substances derived from the soil and forms the **soil solution**. The soil solution is not continuous, but is held around particles of soil and in capillary spaces. Plant roots are in contact with the soil solution and from it absorb both water and mineral ions. Like water, ions are absorbed mostly by root hairs.

Active uptake of mineral ions into the root hairs and the outer cell layers of the cortex results in accumulation of ions in the cells. Similarly, the mineral ion ('salt') content of the inner cortex of the root becomes higher than that of the outer cortex. A gradient of increasing salt concentration starts from the epidermis and continues into the cortex as far as the endodermis. The endodermal cylinder constitutes a physiological barrier. Any movement of ions through intercellular spaces or through the capillary spaces in cell walls is blocked by the Casparian strips and the only pathway into the stele is through the living protoplasts of the endodermal cells. Passage of ions into the endodermal cells and subsequently into the living cells within the stele is again an active process. Since active ion uptake requires respiratory energy, it requires the presence of oxygen. High carbon dioxide concentration is inhibitory. The concentration of oxygen decreases towards the centre of the root and the concentration of carbon dioxide increases in the same direction. The overall effect is that the living cells within the stele cannot expend enough energy to accumulate and hold salt against a concentration gradient. Salt is therefore lost from the living cells. It cannot pass out of the stele, because of the endodermis, and loss (excretion) into the lumina of xylem vessels is the result. In this way, mineral ions accumulate in the xylem vessels.

As stated above, there is a gradient of salt concentration in the root which increases unidirectionally towards the interior. Water absorption and movement from cell to cell follow the same path. Water diffuses from the soil solution, which has low osmotic potential (high water potential), through the cortical cells and into the sap of the xylem vessels, which have high osmotic potential (low water potential).

Entry of water and ions is through root hairs and neighbouring epidermal cells. The particular significance of root hairs in relation to water and ion uptake is that the extension of the wall of the epidermal cell into a hair results in the exposure of a very large surface area in contact with the soil solution. It is estimated that a single small plant possesses several billion root hairs, with a total surface area of three to four million square centimetres. This estimate puts the total surface area of root hairs and other root surfaces at much more than 100 times the total surface area of leaves and other above-ground parts of the plant.

Translocation of water and mineral ions

Xylem is the pathway of water movement throughout the plant. In angiosperms, water movement takes place through the lumina of the xylem vessels. Each xylem vessel is composed of many dead, empty vessel elements arranged end to end. The cross-walls originally present between the vessel elements are extensively perforated or almost entirely removed as the cells differentiate and the resulting tubes (**xylem ducts**) form a network which extends to all parts of the plant. No living cell is far from a source of water for the maintenance of turgor and other functions, or from a source of mineral ions dissolved in the water contained in the vessels.

Almost all angiosperms possess vessels and also possess tracheids, which probably assist in the translocation process. Most gymnosperms have tracheids only and these tracheids are the only pathway for the translocation of water and mineral ions. Unlike vessels, tracheids do not form continuous tubes, since their walls are not perforated. However, their end walls are tapered and overlap each other and a continuous pathway is provided through pits in the walls between tracheids. Although vessels are certainly more efficient conductors of water, the disadvantages of tracheids relative to vessels do not seem to be very serious. Some of the tallest trees in the world are conifers and they lack vessels.

The long axes of vessel elements and tracheids are arranged parallel to the long axes of plant organs and water conduction is predominantly in this direction. However, lateral transport also takes place through the pits in the side walls of vessels and tracheids. Water also passes laterally from vessels and/or tracheids to neighbouring parenchyma cells.

Forces involved in water translocation

The ultimate source of water for all parts of the plant is soil water. As described, this is first passed into the tracheary elements (vessels and tracheids) of the absorbing region of roots. From there, the water is translocated to the stems and leaves through the xylem.

The height of some emergent forest trees such as *Khaya* sp. (Benin Mahogany), *Lophira procera* (Ironwood) and *Chlorophora excelsa* (Iroko) may exceed 40 m. The deepest absorbing tips of the roots may be 10 m below the surface of the ground. The height separating the water-absorbing organs and the transpiring leaves at the top of a tree may, therefore, be as much as fifty metres. The two main forces which may be responsible for raising water up this great distance, and some other possible forces, are our next topic.

Root pressure is the first of these forces. The concept of root pressure is that a root system absorbs water from the soil, as already described, and 'pushes' it up the stem. The existence of root pressure can be demonstrated by a simple experiment (Fig. 12.1). The shoot of a well-watered plant (e.g. Tomato) is cut off, leaving a stump a few centimetres high. A vertical glass tube is connected to the stump by a watertight connection of rubber tubing and a little water added to the tube until the top of the water is visible above the connection. The water in the tube rises slowly, due to the exudation of xylem sap from the detopped plant. This sap is exuded by root pressure. That it is exuded under pressure can be demonstrated by replacing the vertical tube with a mercury manometer, when a pressure of one to two bars may be obtained.

A root pressure of two bars could, in theory, support a water column about twenty metres high. However, it would probably not be as effective as this in the intact plant, since vessels are not completely clear, smooth-sided tubes and must offer more resistance

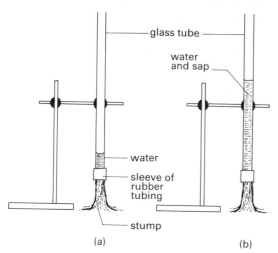

glass tube

water and sap

water

sleeve of rubber tubing

stump

(a)   (b)

12.1 Demonstration of root pressure. The stump of a detopped plant, (a) soon after detopping; (b) long after detopping

to water movement than the glass tube in the experiment. However, due to the evaporation of water from leaves (transpiration), which draws water from the xylem, the water in the xylem is under tension rather than pressure for most of the time. Root pressure cannot be sufficient to raise water to the tops of tall trees. It is also entirely absent in conifers. Under conditions of low transpiration, root pressure does however account for sap exudation (bleeding) in cut stems, for the loss of drops of liquid water from leaf margins (guttation, below) and possibly for forcing air out of vessels in which the water column has been broken.

Capillary rise of water in vessels has been suggested as a mechanism by which water might be moved up the xylem. However, the diameter of a vessel is too large to cause a rise of as much as a metre. Also, a xylem vessel is not open at the top like the glass tubes used to demonstrate capillary rise. The closed tube formed by the vessel would offer too much resistance to allow any appreciable rise by this mechanism.

The **vital theory** of the movement of water in the xylem was proposed in the late nineteenth century. When a stem was killed, for example by heat, the leaves soon died because water did not reach them. This was taken as evidence that the living cells of the xylem played a part in water movement. However, later work showed that if the living cells of the xylem were killed with poisons, water movement through

the xylem continued. The failure of translocation in the earlier experiments was probably due to the blockage of vessels following the death of the living cells.

The **cohesion-tension** theory of xylem translocation was first proposed by H. H. Dixon in 1914. The general principle of the theory can be demonstrated in several ways. Take the familiar experience of drinking a soft drink through one or two straws. Initially, our sucking reduces the pressure of the air in the straws and causes them to fill up with liquid. Thereafter, continued sucking imposes tension on the liquid in the straws, which continues to rise into the mouth in unbroken columns through the straws. If we raise the straws from the liquid, bubbles of air enter and the liquid runs out.

At a more sophisticated level, we may take fine glass capillary tubes of such a length and diameter that they immediately fill to the top when their lower ends are submerged in water. If we now place a very absorbent sponge across the tops of the tubes, in contact with the water in the tubes, the sponge draws water from the tubes, also by capillarity. As water then evaporates from the sponge, more water is drawn up through the capillary tubes, from the water container, as long as the lower ends of the capillary tubes are under the surface of the water. This experiment is possible because water resists being compressed (squashed together) or being pulled apart easily. This is because water molecules **cohere** (stick together) and also **adhere**, for instance, to the walls of a vessel. As a result, the water columns remain unbroken as long as the cohesive and adhesive forces are greater than the downward gravitational pull on the water column.

By analogy, water from the soil is sucked up to the tops of trees through the tracheary elements of the xylem. Evaporation of water from the mesophyll cells in the leaf provides the sucking force. This causes a decrease in the water potential of these cells, so that water from deeper-lying cells moves to replace the water lost. The innermost mesophyll cells draw water from the cells of the bundle sheath and these, in turn, draw water from the xylem. The water in the xylem is thus put in a state of tension. The tension originating in the leaf is transmitted through the unbroken column of water in the xylem, as far as the root system. The name **transpiration stream** is given to water moving by this mechanism (see 'Transpiration' below).

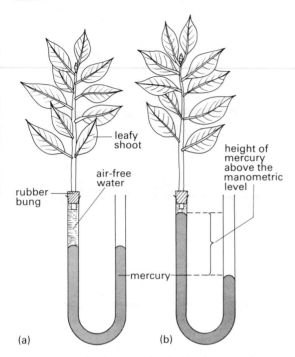

rubber
bung

leafy
shoot

air-free
water

height of
mercury
above the
manometric
level

mercury

(a)   (b)

12.2 Apparatus for demonstrating that tensions are set up by the transpiring shoot, (a) initial setting; (b) later situation, after the mercury has risen above the original manometric level

Direct measurement of tension in the xylem is not yet possible, as all methods cause the breakage of the water columns and therefore release any tension that may exist. However, there are indirect observations on the subject. First, an instrument called a dendrograph continuously records changes in the diameter of a tree trunk over a period of time. Measurements with a dendrograph show that there is a decrease in the diameter of a tree trunk during a period of high transpiration and an increase when transpiration rate is low. It appears that the 'shrinkage' of the tree trunk during periods of great water loss may be due to tension in the xylem. Secondly, if a leafy shoot is cut under the surface of water and connected to one arm of a mercury manometer (Fig. 12.2), the mercury will rise in that arm. The water column above the mercury in that arm must be under tension as a result of water being drawn away by the plant, from which it is lost by evaporation. There is no reason to suppose that the situation in an intact plant should be any different from that in the cut shoot.

Thirdly, when dyes are injected into the xylem of a plant, they usually move rapidly both upwards and downwards. This might result from simultaneous upward and downward movement in the xylem, but is more easily explained by the existence of tension within the xylem.

One of the most serious doubts about the tenability of the cohesion – tension theory has been the tensile strength of water. As has been stated, measurements of 300 bars have been given for the experimental situation in which pure water has been used. The presence of solutes in the water must reduce this figure, but since the tallest tree would require a tension of no more than 13 to 14 bars, the tensile strength of water must be more than sufficient.

Pathways of water movement across living tissues

What has been described so far is water movement along a water potential gradient based on osmosis. There are, however, three pathways of water movement across living tissues, all along water potential gradients.

Water which moves osmotically from cell to cell is, in effect, passing from vacuole to vacuole. This is known as the **vacuolar pathway** of water movement. Water also moves, along a water potential gradient, from cell to cell, through plasmodesmata. All the living cells of a plant are in protoplasmic continuity through plasmodesmata and form the symplast of the plant. Movement through plasmodesmata is said to be by the **symplastic pathway**. Water may also cross a tissue through the capillary spaces in the cellulose walls of cells, without passing into or out of living protoplasts. This is the **apoplastic pathway**.

The walls of living cells, composed of cellulose fibrils, contain very many fine capillary spaces with very powerful water-retaining capacity. The walls of neighbouring cells are in close contact and, as a result, the capillary spaces of all living cells which are in contact form a continuous system of fine capillary tubes.

When water evaporates from the walls of mesophyll cells, it evaporates from the exposed ends of many capillaries. This reduces the water potential in the capillaries by causing tension in the contents. Water is therefore drawn from the protoplasts, and also from the capillary system in the walls of neighbouring cells. If the rate of evaporation is high and the stress is great enough, water is drawn along the walls of cells directly from the tracheids in nearby veins.

Similarly, under stress conditions caused by rapid transpiration and high tension in the xylem, water is drawn from the walls of living cells in the stele of the root in the same way. However, movement of water through cell-wall capillaries cannot operate through the endodermis, where the radial and transverse walls are sealed by the Casparian strip. Here, water can pass only through the protoplasts of the endodermal cells. Outside the endodermis, in the cortex, water may again pass by the apoplastic pathway.

The concept of water movement along the capillary system in cellulose cell walls highlights the importance of the endodermis. The apoplastic pathway is continuous from the walls of root hairs and water entering the capillary system directly from the soil water is not pure, but contains dissolved ions. Thus, since the soil solution does not at first pass through the plasma membrane and cytoplasm of a cell, the ions present have not been selected, as they have when they have been actively absorbed. In this case, selection must occur at the endodermis.

Under the stress conditions described, movement by the apoplastic pathway is most rapid. Movement by the symplastic pathway is next in order of efficiency and the vacuolar pathway is slowest.

12.3 Diffusion through small pores into still air. (a) hemispherical 'diffusion shell' around a single pore; (b) diffusion through several pores in a multiperforate membrane; (c) mutual interference between neighbouring pores due to overlapping of diffusion shells

Transpiration

Transpiration is the loss of water from the plant by evaporation. Plants lose a great deal of water in this way. For instance, it has been shown that a single plant of *Helianthus annuus* (Sunflower) takes up and loses about seventeen times as much water each day as a man.

Transpiration takes place from the surface of the plant. **Cuticular transpiration** takes place through the cuticle. **Lenticular transpiration** takes place, in woody plants, through lenticels in periderm. **Stomatal transpiration**, transpiration through stomata, accounts for about ninety per cent of the water loss from an average plant, whereas cuticular transpiration (especially in a plant with a thick cuticle) and lenticular transpiration are often negligible. Also, if a plant normally has no leaves or has lost its leaves, stomatal transpiration may be very low. The few stomata in the epidermis of the stem make little contribution to the magnitude of transpiration.

The leaves of most angiosperms are amphistomatic (have stomata on both surfaces). Some plants are hypostomatic or epistomatic (have stomata only on the lower or upper surface respectively).

Diffusion through stomata

The diffusion path of molecules of water vapour or other gas molecules through a pore is not always straight. Some molecules come out perpendicularly relative to the surface of the leaf but most describe paths curving towards the surface in which the pore is located (Fig. 12.3). The result of this, in the case of the diffusion of water vapour, is the formation of hemispherical 'shells' of water vapour around each stoma. The air closest to the stoma becomes more and more saturated with water vapour. The humidity within the diffusion shell is intermediate between that of air within the stomatal pore and that of freely circulating air farther from the pore. These diffusion shells offer appreciable resistance to continued diffusion of water vapour from the leaf. The thicker the diffusion shell, the greater is the resistance to further diffusion. The build-up of a thick diffusion shell leads to a decrease in transpiration rate. The thickness of the layer of air which is relatively stationary around the stoma may be up to ten millimetres, in still air. This may be reduced to three millimetres in slow wind and to less than half a millimetre in a strong wind. However, because of the cooling effect of strong wind, the net effect of wind is that the rate of increase in transpiration rate becomes smaller as the wind velocity increases.

The control of transpiration

Transpiration is a diffusive process and the rate of diffusion of a substance between two points depends on two factors. First, it depends on the concentration difference between the two points. This also determines the direction of diffusion. Secondly, the rate is affected by the resistance to diffusion of the intervening pathway.

The situation represented in the formation of diffusion shells, described above, is a model operative only in still air. Since air is never completely still, the hemispherical diffusion shells of the model do not form, and although resistance to diffusion is still present and measurable, it is greatly reduced. Instead of hemispherical diffusion shells, there are flattened boundary layers in which particles move parallel to the leaf surface. Further away from the surface, the air flow is turbulent and entails free mixing of molecules. This is in contrast to the flow of the layer closest to the leaf, where little mixing occurs.

The resistance to flow through the plant has three components, the resistance offered by roots, stems and leaves. Resistance to movement from leaves has been the object of most investigation. The most critical factor is diffusion between the air spaces of the mesophyll and out of the leaf through the stomata. A simple instrument for measuring internal resistance to diffusion in the leaf is the **porometer**.

In the simple porometer (Fig. 12.4), a small cup is attached to a leaf surface by an airtight gelatine washer. This is attached to a horizontal arm of a T-shaped tube, the vertical arm of which is a glass

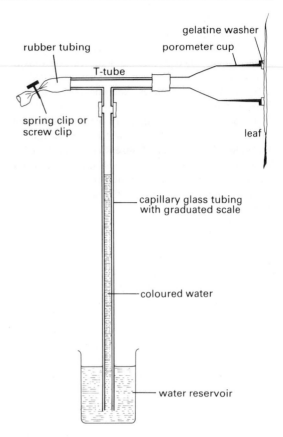

12.4 A simple porometer in use

capillary tube with a graduated scale, dipping into a beaker of coloured water. The other end of the top of the T-tube is closed by a spring clip on a short length of rubber tubing. Holding open the spring clip, water is sucked up the vertical tube and the clip closed again. The reduced air pressure in the vertical tube caused by the falling column of water causes air to be drawn into the leaf through stomata outside the area covered by the cup, through the mesophyll of the leaf and into the cup through the stomata inside the cup. As it does so, the level of the water in the vertical tube gradually falls and is timed doing so. The rate of fall is used to estimate the mean size (degree of opening) of the stomatal pores.

The porometer has several disadvantages which it is difficult to completely overcome. First, the resistance measured is both that of the stomata and that of the mesophyll, which is especially serious if the leaf has stomata only on one side. Also, the presence of the porometer cup may alter the environment of the stomata in any of several different ways. It reduces light intensity, changes the carbon dioxide concentration and increases the humidity. Because of these effects, porometer experiments should be as brief as possible.

There are two self-regulatory aspects of transpiration. First, under limited water supply, transpiration leads to stomatal closure. Secondly, the onset of wilting as a result of water loss leads to a change in the angle of the leaves, so that they are no longer at right angles to the rays of the sun. This reduces the effective intensity of the light and checks transpiration.

The role of transpiration

In most ways, transpiration seems to be a 'necessary evil' which results from the adaptation of the leaf to its primary function as a photosynthetic organ. Thus, the positioning of the leaf at right angles to incident light and the broad, flat shape of the leaf ensure that it intercepts the maximum amount of sunlight. Carbon dioxide must enter mesophyll cells in solution, first through the walls of the cells, which have their capillary spaces constantly full of water. Oxygen produced by photosynthesis goes out of cells the same way. Intercellular spaces ensure the maximum exposure of moist cell walls. Open stomata allow for the rapid diffusion of carbon dioxide into the leaf and of oxygen out of the leaf. All these conditions promote rapid transpiration. The open stomata make the air within the leaf continuous with that outside the leaf and the free exposure of the moist mesophyll cell walls to the intercellular spaces leads to rapid evaporation within the leaf and rapid diffusion of water vapour out of the leaf. Because of its exposure to sunlight, the temperature within the leaf may be higher than that of the surrounding air and this, too, promotes rapid outward diffusion of water vapour.

As a result of rapid transpiration, an average land plant utilizes only about one per cent of the water it absorbs. The rest is lost to the atmosphere. Of all the stresses to which a plant is subjected, including shortage of essential mineral ions, insufficient light and insufficient water, the ill effects of water stress are greater than all the others combined.

On the positive side, transpiration has been questionably associated with three roles, mineral salt absorption, cooling, and more general effects on growth and development.

Circumstantial evidence indicates that the growth of plants is reduced by up to fifty per cent under prolonged conditions of high humidity. Cooling certainly results from transpiration, but the significance of this has not yet been proved. The temperature of leaves in very hot habitats may be up to 15 °C below that of the air as a result of evaporative cooling. It is argued that, under such extreme conditions, this cooling may make the difference between survival and heat damage, or even death. But this is under extreme conditions.

Since mineral salts and water are together in the soil, it was once thought that, since transpiration helped water absorption by the roots, it also encouraged salt absorption. However, it is now known that salt absorption is almost entirely an active process. Only a small proportion is absorbed passively as a result of water absorption but once mineral ions have reached the xylem of the root, their translocation and distribution are certainly positively affected by transpiration.

Xylem sap

The liquid contents of xylem vessels (**xylem sap**) is a clear liquid with an acid pH of around 5.0, with a solute concentration which never exceeds about one per cent and an osmotic potential which rarely exceeds -0.2 bars. Of the dissolved material present, mineral matter, including organic chelates (organic-metal compounds), makes up twenty-five to forty per cent. Organic matter, including amino acids, amides and organic acids, makes up between sixty and seventy-five per cent. Carbohydrates are either undetectable or negligible. The sap, therefore is not merely a solution of mineral salts but contains compounds derived from living cells of the plant.

Guttation

Guttation, the loss of liquid water from above-ground parts of the plant, is not transpiration, but is best considered at this point.

If potted Maize seedlings are well watered and placed under a cool bell-jar, droplets of water appear along leaf margins, especially at the tips, within a few hours. This demonstrates the loss of water by guttation. Under the conditions described, the bell-jar provides an atmosphere of cool air with high relative humidity. The high relative humidity checks transpiration, but water absorption continues. Root pressure forces water up the plant and excess water is forced out by way of specialized structures known as **hydathodes**. Guttation can also be observed on many other plants early in the morning, especially after a humid and relatively cool night.

It appears that water is also actively secreted by gland-like structures of some plants. These structures are called **water glands** or **active hydathodes**. The amount of water lost by guttation is usually small but occurs on a larger scale in some tropical plants. It is common in grasses but many dicots also exude water in this way (e.g. *Ficus capensis*).

Guttation fluid is not pure water but often contains both organic and inorganic compounds. These compounds may include carbohydrates, amino acids and inorganic salts such as nitrates, sulphates and chlorides of calcium, potassium and magnesium. Occasionally, the drying of guttation fluids leaves behind salt particles (e.g. the 'salt glands' of *Avicennia* sp., White Mangrove) or concentrated solutions on the leaf surface. These can damage young leaves through salt stress. Sometimes, such particles are redissolved and reabsorbed into the leaf.

Guttation is of very little significance in the water economy of plants.

Factors affecting transpiration rate

Transpiration rate is affected by both internal and external factors. When these factors are discussed, below, it is assumed in each case that all the other factors are such that they are not limiting transpiration. We shall start with the environmental factors which affect transpiration rate.

Light affects stomatal movement. The stomata of most plants close in the dark and open when it is light. Thus, light is the commonest limiting factor, since stomata must be open for any appreciable amount of transpiration to occur. It has also been reported that light increases the permeability of protoplasm and this could also cause an increase in transpiration.

Temperature affects transpiration rate in several ways. Stomata tend to close at low temperatures (e.g. 0°C) and the degree of opening increases up to 30°C. Therefore, within a certain temperature range, increasing temperature causes an increase in transpiration owing to its effect on stomatal transpiration. A rise in the temperature of leaves above that of the surrounding air also increases the vapour pressure gradient between leaves and the surrounding air and increases the rate of water loss.

Relative humidity (RH) of the air is the amount of water vapour present in a given volume of air, expressed as a percentage of the maximum amount of water vapour that the air can hold at a given temperature. Leaves transpire much faster into dry air than into air with a high RH. This is because the rate of transpiration depends on the water potential gradient between the evaporating surface and the air. The steepness of this gradient is drastically lowered when the RH of the air is high.

Wind affects transpiration in two opposite ways. By sweeping away or reducing the thickness of the layer of water vapour or diffusion shells on the leaf, wind steepens the vapour pressure gradient and thus tends to increase the rate of water loss. On the other hand, the cooling effect of wind tends to reduce the vapour pressure gradient and thus to lower the rate of transpiration. Also, high wind velocities cause stomata to close and thus reduce transpiration.

If the soil water status decreases and water becomes less easily available to the roots, less water is absorbed by the plant. The resulting increase in tension in the xylem causes a decrease in water potential of the cells of the leaf. Water is less readily lost from mesophyll cells by evaporation and transpiration decreases. At some threshold value, further water stress causes stomata to close and transpiration is further checked.

We now come to internal (structural) factors which affect the rate of transpiration.

Stomata are the main avenues of the loss of water vapour and it would therefore be expected that the degree of opening of stomata would be very important. However, the extent to which variations in stomatal aperture control the rate of transpiration has been a matter of some controversy. Here, we summarize some general findings on the matter.

(a) The rate of diffusion from a large, open surface of water is proportional to the area exposed. The rate of diffusion from small, circular areas is, however, proportional to their perimeters. As a result, the degree to which stomata are open has little effect until the stomata are nearly closed. Estimates of the degree of opening at which stomata exert control vary from about two per cent to fifty per cent of their maximum extent.

(b) The 'dead areas' (Fig. 12.3(b)) between diffusion shells provide spaces for the accommodation of

Table 12.1 Variation in stomatal size and stomatal distribution in some legumes and cucurbits which grow in Nigeria. (From Gill *et al.*, 1982 and Gill & Karatela, 1982.)

plant	stomatal frequency		stomatal index	length of stomatal pore (μm)	
	lower surface	upper surface	lower surface	lower surface	upper surface
Caesalpiniaceae					
Bauhinia monandra	22	—	0.20	15	—
B. tomentosa	22	—	0.25	15	—
Cassia alata	28	12	0.36	25	22
C. fistula	—	18	0.50	—	15
Mimosaceae					
Albizia zygia	20	16	0.33	15	10
Pentaclethra macrophylla	30	—	0.23	17	—
Papilionaceae					
Baphia nitida	22	8	0.25	15	17
Crotalaria retusa	15	8	0.36	38	30
Vigna unguiculata	35	28	0.25	15	20
Cucurbitaceae					
Lagenaria siceraria	95	65		22.2	16.5
Luffa acutangula	85	60		12.95	20.35
L. aegyptica	105	102		22.2	22.2
Momordica charantia	102	80		11.1	11.1

water vapour from the shells. At a given distance from the leaf surface near the edge of a pore, the air is drier than at the centre of the pore. This explains, or is related to, the observation that diffusion rate through small pores is proportional to the pore perimeter. This is known as the **edge effect** on diffusion through a pore.

(c) When the distance between neighbouring pores is small, diffusion shells overlap and mutual interference occurs. The rate of evaporation between interfering pores decreases. Diffusion shells overlap if the distance between pores is about ten pore diameters or less. This means that considerable interference does occur, since stomata are usually closer together than this.

In spite of the above qualifications, the influence of stomata on transpiration is far from negligible and open stomata allow a great deal of transpiration. Brown and Escombe's classic work showed that a given area of a leaf of *Helianthus annuus* (Sunflower) may transpire half as much water vapour as that lost from the same area of exposed water surface. This is in spite of the fact that the total area of the stomatal pores when fully open is less than one per cent of the total leaf area.

Ninety per cent or more of a plant's total water loss is through the stomata. Therefore, if stomata are absent, almost completely closed, or hidden in pits or depressions (e.g. *Nerium oleander*, Oleander; *Casuarina* sp., She-Oak), transpiration is reduced in proportion if other factors are constant.

Stomatal frequency, the number of stomata in one square cm of the epidermis, may be measured directly. The distribution of stomata may also be expressed by calculating the **stomatal index** (**stomatal ratio**). Where S is the number of stomata in a given area of epidermis and E is the number of other epidermal cells in the same area:

$$\text{stomatal index} = \frac{S}{S + E}$$

The size of the stomatal pore, as well as stomatal frequency, vary considerably between the two surfaces of the same leaf (Table 12.1).

The **total leaf area** of a plant is another factor to be taken into account. The larger the leaf area, the greater the amount of transpiration. However, small plants lose water at a greater rate per unit surface area than larger plants. When two plants of different sizes but with similar leaf characteristics are compared, it is usually found that the plant with smaller leaf area has the higher intensity of transpiration per unit leaf area.

Since it is the roots which absorb the water which is subsequently lost by the leaves, the **root-shoot ratio** is important. The root-shoot ratio is the ratio of the absorbing surface to the total evaporating surface of the plant. In general, transpiration takes place at a higher rate in plants with a greater root-shoot ratio, as long as water is freely available in the soil. If absorption by the roots is less than water loss from the leaves, transpiration is checked by tensions which build up in the xylem.

Leaf surface structure affects rates of transpiration. The leaves of xerophytes illustrate these types of structural modification best. Characteristic features of the leaves of xerophytes include thick cell walls, deep layers of palisade parenchyma, a thick cuticle, sunken stomata, a covering of epidermal hairs, reduction of the surface to volume ratio of the leaf (i.e. succulent leaves) and reduced leaf size. These are usually called xeromorphic characters and enable xerophytes to survive under dry conditions. However, it has often been found that, with a plentiful water supply, the leaves of xerophytes transpire more rapidly per unit surface area than those of mesophytes. This is probably because most xerophytes have more extensive leaf venation, more stomata per unit area and a higher root-shoot ratio than mesophytes.

With regard to **vascular structure**, leaves with extensive venation tend to transpire more rapidly than leaves with fewer veins. Also, vessels of large diameter (e.g. in many lianes) offer less resistance to water movement than smaller vessels and tracheids. Gymnosperms, most of which lack vessels and have only tracheids in their xylem, are characterized by low transpiration rates.

Measuring transpiration

A simple comparison of rates of transpiration from different leaves, or from different surfaces of the same leaf, may be made by using cobalt chloride or cobalt thiocyanate paper. These papers are blue when dry (e.g. in a desiccator) but rapidly turn pink when exposed to humidity. For comparison purposes, a piece of blue paper is secured to a leaf surface with a pair of microscope slides held together with elastic bands at both ends. The edges of the slides are sealed against the leaf surface with petro-

leum jelly (vaseline). The time taken for the paper to change from a standard shade of blue to a standard shade of pink is taken to be inversely proportional to the rate of transpiration from the surface of the leaf with which the paper was in contact. This method has serious disadvantages. The sealed area of the leaf is transpiring into completely dry air, since both cobalt compounds are strongly hygroscopic, and, if the experiment lasts for more than a few minutes, this will cause closure of the stomata.

Several other, more accurate methods, depend on measuring the amount of water transpired.

If a well-watered, potted plant has its pot and soil completely enclosed in a plastic bag tied at the top round the base of the stem to prevent evaporation from those surfaces, water loss by transpiration may be measured by weighing the plant, in its pot, after various time intervals. Results are best expressed as **transpiration intensity**, which is equal to the weight of water transpired per unit leaf area, measuring one side only of each leaf. Results are expressed in milligrams or grams of water per square centimetre per hour. This method ignores possible weight changes resulting from photosynthesis or respiration, but is a good simple method for use over a short period of time.

Another method is to measure directly the amount of water transpired. A whole plant or part of a plant is enclosed in a glass container connected to a tube of water absorbent (calcium chloride or phosphorus pentoxide). A stream of air is drawn through this apparatus and also through a similar (control) apparatus. Air should be drawn at the same rate through both. The difference between the increases in weight of the two drying tubes gives the weight of water transpired by the plant. The main weakness of this method is that the plant is enclosed under abnormal conditions of temperature, light intensity and relative humidity.

Alternatively, a shoot or leaf is detached from a well-watered plant, quickly placed on a sensitive balance and the time noted. The weight is recorded at one minute or two minute intervals for five to ten minutes and the results plotted on a graph (Fig. 12.5). If there is any delay at all between cutting of the part and the first weighing, the curve is extrapolated to zero time. The initial slope of the line represents the transpiration rate immediately before the leaf was cut off. If the graph is curved, the slope is obtained from a tangent to the line at

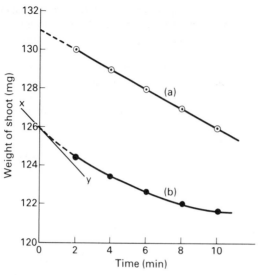

12.5 Determining the transpiration rate of cut shoots (a) and (b) by rapid weighing at short time intervals from zero time of excision. For (b) the graph is curved and the initial slope is obtained by drawing a tangent (x–y) to the curve at zero time

zero time. Using this method, it is especially important that the plant should be well watered. If there is any tension in the xylem when the part is cut off, the rate of water loss will be exaggerated.

It is also possible to measure the rate at which a shoot takes up water in response to transpiration, by use of a **potometer** (Fig. 12.6). A potometer is a glass apparatus with two arms, one carrying a cut shoot and the other a water reservoir with a tap at the base. To this is attached a horizontal arm of capillary tubing of known bore, on which there is a millimeter scale. The apparatus is set up, as illustrated, and an air bubble is introduced into the capillary tube, the end of which is then replaced under the surface of water in the beaker. The rate of movement of the bubble against the scale, with the knowledge of the exact size of the bore of the tube, gives a direct measure of the rate of water uptake by the shoot. After each reading, the bubble can be pushed back along the graduated tube by briefly opening the tap below the reservoir.

Again, this method disregards the small amount of water used in metabolism during the experiment.

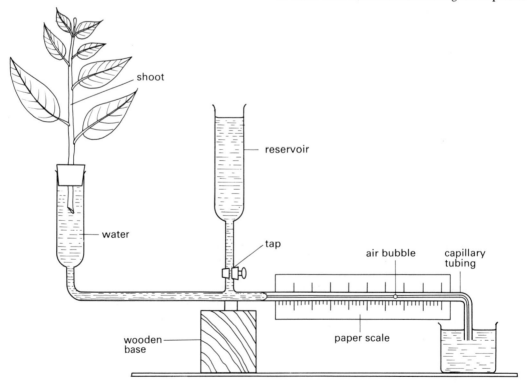

12.6 The essential features of a potometer in use

It measures the rate of water uptake by the shoot and, if used as a measure of transpiration, it assumes that water uptake and water loss are equal. This is true, except during and immediately after a period of very rapid transpiration. When transpiration is very rapid, water uptake may lag behind. After a period of rapid transpiration, water uptake may be greater than transpiration, as deficits are compensated for.

There are several different models of potometer, but whichever is used there should be a second identical apparatus set up, without a shoot, as a control. This records changes in water volume which result from temperature changes during the experiment. Any fluctuation in the position of the bubble in the control may then be used to correct readings of bubble movement in the main apparatus. Use of a cut shoot is artificial in that cutting eliminates root resistance in the intact plant. It is possible to use an intact plant in a potometer, but lack of aeration of the roots again introduces an unnatural condition.

Results of potometer experiments are expressed as grams of water taken up per hour, calculated from the distance travelled during the period of each experiment, the known diameter of the bore of the tube and the density of water.

Sometimes, the rate of transpiration, T, is compared to that of evaporation, E, from an **evaporimeter** or **atmometer** exposed to similar conditions. An atmometer is an apparatus which consists of a standard evaporating surface, usually made of porous pot, connected to a water reservoir. Loss of water is measured by weighing and the results calculated in terms of grams of water lost per hour. The **transpiration coefficient** (**relative transpiration**) is then calculated from the formula:

$$\text{Transpiration coefficient} = \frac{T}{E}.$$

The transpiration coefficient will be found to vary according to conditions of lighting, temperature and air movement.

Translocation of organic solutes

We have seen that movement of water and inorganic solutes in the xylem is essentially a one-way process, from the roots to all the above-ground parts of the plant. Movement of organic solutes in the phloem is not a one-way process. In this respect, it is convenient to see a plant as composed of a number of sources of organic substances, from which the substances originate, and a number of sinks, to which they are translocated. **Sources** include leaves and other green parts, and storage organs, at the time when new growth is taking place. The leaves are the site of photosynthesis and storage organs provide the materials needed for new growth. **Sinks** include active meristems, including the apical meristems of shoots and roots, flowers, developing fruits, storage organs at the time when they are forming, and, to a lesser extent, living non-green cells throughout the plant. Vegetative storage organs thus start as sinks, when they are developing, and later become sources, when new growth starts.

Evidence for the phloem pathway

The first experiments to determine the pathway of transport of organic materials in plants were carried out as long ago as 1675, when Malpighi carried out the first **ringing** (girdling) experiments. In a ringing experiment, a continuous strip of tissue, including periderm and phloem, is removed from around the stem of a leafy woody plant, leaving the cambium exposed. This type of treatment interrupts the continuity of the phloem but leaves the xylem intact. After ringing, it is usually found that the outer tissues above the ring gradually become enlarged, due to the accumulation of organic reserves. No such enlargement takes place below the ring. Further evidence that the phloem is responsible for the transport of organic materials comes from a number of sources, some of which are described below.

Chemical analysis of the phloem shows that it contains large amounts of sugar, mostly sucrose, and nitrogenous compounds, including amino acids.

When **mining caterpillars**, which eat up the inner tissues of leaves, destroy the xylem within a leaf, starch continues to be removed from the leaf. When they destroy the phloem, starch accumulates in the leaf.

When **aphids** attack a plant, their mouthparts are inserted into sieve tube elements of the phloem, from which they extract sap which is rich in organic materials.

In experiments with **stem surgery**, the removal of all but one large vascular bundle from a stem was found to reduce the removal of nitrogenous substances from leaves above the surgery to about seventy-five per cent of the original rate. When the xylem of the remaining bundle was removed, the rate decreased further to sixty-five per cent. When the phloem of the remaining bundle was removed, instead of the xylem, the rate dropped to five per cent.

Growth experiments confirm the importance of phloem in the translocation of organic materials. In 1920, O. F. Curtis carried out experiments with growing shoots, which he subjected to four different treatments and measured growth in length after nine days. The treatments and results were as follows:

(a) A leafy shoot was left intact, with all its leaves, as control. It grew 341 mm during the experiment.
(b) A leafy shoot was ringed at the base but otherwise left intact. It grew 118 mm.
(c) A shoot had all its leaves removed (was **defoliated**) but was not ringed. It grew 270 mm.
(d) A shoot was ringed and defoliated. It grew five mm.

With regard to this experiment, the shoot apex would normally receive organic nutrients from its own leaves, which photosynthesize, and from other parts of the plant. Compared with the control, the plant in (b) had its phloem interrupted by ringing but possessed leaves. The plant in (c) depended for organic nutrients on other parts of the plant. The plant in (d) had no leaves and also had its phloem interrupted.

These and other experiments confirm that assimilatory products (**assimilates**), including the products of photosynthesis, move primarily through the phloem. Movement may be in either direction and may change from one direction to the other with changing circumstances.

The phloem pathway

As has been described in Chapter 6, phloem consists of four types of cells. Sieve tube elements, with perforated sieve plates, are arranged in longitudinal rows to form sieve tubes (**phloem ducts**). Companion cells are closely associated with the sieve tubes. Living phloem parenchyma and dead phloem fibres are also often present. Sieve tubes, unlike xylem vessels, are living cells and have cytoplasm but no nuclei.

Experiments using radioactive tracer elements have provided much useful information about the pathway of translocation in the phloem and elsewhere. In such experiments, a suitable compound containing a radioactive isotope (**tracer**) is fed into a plant and its movements followed. Thus, carbon dioxide containing radioactive carbon (^{14}C) may be fed into a leaf and subsequent movement of radioactive products of photosynthesis (**photosynthates**) followed. Rates of movement and rough locations may be determined by using a Geiger counter. A Geiger counter is an instrument which detects the presence and amount of radiation directly. Precise location of radioisotopes is by the use of autoradiography. After introduction of a radioisotope and after allowing the necessary time for translocation to start, the plant material is treated to immobilize the radioisotope. This may be done by rapid drying, or by freezing and then drying. The material is then placed in close contact with X-ray film for a period of days, weeks or months. The period required depends on the type and intensity of radiation present in the tissues. When the film is developed, regions of tissue containing radioactivity can be identified as those which cause darkening of the film. Using a refinement of this technique, radioactive isotopes may be located at the cellular and even subcellular level. Thus, it is found that photosynthates are translocated from leaves, in the phloem sieve tubes. It is also found that phosphates fed directly into the leaves can be translocated away from the leaves, always in the sieve tubes. Phosphate fed to the roots travels upwards in the xylem, but some is laterally transported into the adjacent phloem and travels upwards that way. Some ions (**mobile** ions) may be translocated out of old leaves, in the phloem, and into younger leaves, or back to the roots and even through the roots back into the soil.

To summarize, organic compounds move up and down in the phloem. If any movement of organic compounds occurs in the xylem, it is always upwards. Mineral ions move upwards mainly in the xylem and any downward movement is in the phloem.

Phloem sap

Some of the best evidence of the contents of phloem sieve tubes (**phloem sap**) has been obtained by the very ingenious use of aphids. Sap obtained by cutting across the phloem and other tissues is likely to be contaminated with material from other cells. However, when an aphid attacks a plant it inserts its sucking mouthparts (**stylets**) through the outer layers of cells and directly into a single sieve tube element. If an aphid is anaesthetized while it is feeding and its stylets carefully cut, close to the head, sap from the sieve tube flows from the top of the cut stump of the stylets. The **exudate** is thought to be very similar to the sap in the intact sieve tube. This liquid has been used for chemical analysis and in experiments involving the use of radioactive tracers.

Phloem sap is far more viscous than xylem sap and has an alkaline pH of about 8.0. Solute concentration is also much higher than that in xylem sap. Sugar content ranges from two to twenty-five per cent and accounts for about ninety per cent of solutes. Sucrose is the main sugar present in most species but there may be traces of other non-reducing sugars, generally oligosaccharides such as raffinose. An oligosaccharide consists of a sucrose molecule with one (raffinose) or more D-galactose units attached. In some plants, these oligosaccharides predominate over other sugars. Reducing sugars are rare in the phloem, but glucose and fructose may occur as their phosphate derivatives. The concentration of nitrogenous compounds such as amino acids and amides varies from 0.03 per cent to twelve per cent, rarely more. Mineral ions include fairly high concentrations of potassium and phosphate. Other substances, which occur in small amounts, include organic acids, vitamins, alkaloids and viruses. The osmotic potential of phloem sap ranges from about -6 to -34 bars.

Rate of phloem translocation

Estimates of the rate at which substances are translocated in the phloem vary under different conditions and from species to species. Rates of movement of most photosynthates lie between twenty and one hundred and fifty cm per hour (cm h^{-1}). In a few species, much higher rates have been reported, for example, 270 cm h^{-1} in *Saccharum officinarum* (Sugar Cane).

The maximum speed of sap movement in phloem is only about one hundredth of that in xylem. However, the rate in phloem is of the order of 50 000 times faster than that of diffusion. It is because of this fact that simple diffusion must be ruled out as a possible mechanism of phloem translocation.

Phloem loading

Sugars must move into the phloem from photosynthetic cells, or other source cells, before translocation can start. High concentrations of sugar build up in the receiving sieve tubes close to these tissues. This is **phloem loading**. Use of radioactive tracer techniques in which $^{14}CO_2$ is introduced into leaves shows that radioactivity first appears in photosynthates. These photosynthates then move out of the cell through the cell wall rather than directly into the phloem. They therefore move from the symplast system into the apoplast system. Subsequently, they enter the symplast of the phloem, possibly by way of the companion cells.

Materials may also move into the phloem through specialized transfer cells. **Transfer cells** have protuberances on the inside of the cell wall, which greatly increase the surface area of the cell membrane. They also have dense cytoplasm. These cells are thought to be specialized in relation to the collection and transfer of solutes to and from the phloem. Similar transfer cells also occur in plant glands (e.g. nectaries) and haustorial organs. Although it has been suggested that parenchyma cells of the bundle sheath found in many leaves play a part in phloem loading, this has never been proved.

Phloem loading is an energy-dependent process. Sugars and other substances are transferred across membranes into sieve tubes against a concentration gradient. Phloem loading is also selective and, in different species, only specific substances are loaded. It is not known why sucrose is the preferred sugar but this selectivity is extended to ions and even to certain synthetic substances such as herbicides. Ions including those of potassium and those containing phosphorus or nitrogen are readily translocated in the phloem and are correspondingly easily loaded. However, passive entry into the phloem is not ruled out and, once inside, any organic or inorganic solute is translocated, just as it would be in the xylem.

Unloading of the phloem occurs at the consumer cell level, within the sinks, where respiration, growth or storage is taking place.

The mechanism of phloem translocation

Theories about the mechanism of phloem translocation fall into two categories. Some propose active mechanisms, others propose passive mechanisms.

Protoplasmic streaming (cyclosis) has been proposed as the mechanism of phloem transport and

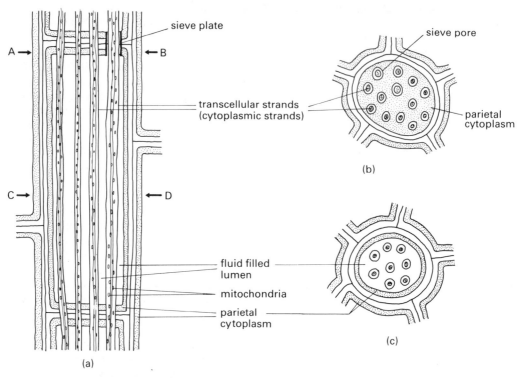

12.7 The essential anatomy of a sieve tube element according to Thaine's theory of the protoplasmic streaming mechanism of phloem transport: (a) vertical section of a sieve tube element; (b) TS at the level A–B; (c) TS at the level C–D. (After Thaine, R., 1964, *J. exp. Bot.* 15: 470–84.)

this falls into the first category. Cyclosis requires the expenditure of ATP. It is easily observed in the cells of staminal hairs of *Setcreasea* sp. (Fig. 5.1) and other members of the Commelinaceae such as *Tradescantia* sp. and *Zebrina* sp.

Since movement of particles by cyclosis takes place in both directions, this would be compatible with simultaneous bidirectional translocation within a single cell. It is clear that when streaming and solute movement from cell to cell take place, the latter would be speeded up by the former. However, it is not generally accepted that streaming could be an important mechanism for phloem translocation. One of the most significant objections to the theory is based on the reported absence of streaming in mature, functioning sieve tube elements. Evidence on this point is conflicting. However, there are other objections. For example, cyclosis is too slow to account for the rates of movement involved and

diffusion from cell to cell required for this process would further reduce the rate.

The strongest advocate for the streaming mechanism, R. Thaine, described cytoplasmic strands passing from cell to cell through the pores in the sieve plate (Fig. 12.7). Particles were observed moving in opposite directions in adjacent strands in the same cell, which could explain simultaneous bidirectional transport. There are four objections to Thaine's version of the cyclosis theory. Firstly, evidence for the existence of streaming transcellular strands is scanty and controversial. Secondly, the concept of membrane-enclosed strands would separate the source and sink from the moving column of solutes and the permeability of the membranes of the strands would be highly critical. Thirdly, the streaming particles appear to move faster than the surrounding solution and the mechanism of 'turn-round' at the ends of the cell is

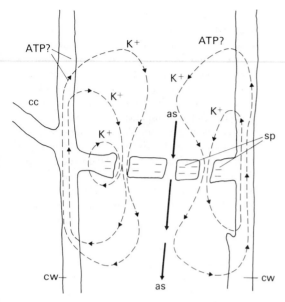

12.8 The electro-osmotic mechanism of phloem transport, showing sieve plate (sp) with negatively charged microchannels between the cytoplasmic P-protein filaments (see Fig. 12.7) in the pores. The movement of K⁺ ions is indicated by the dotted arrows while the solid arrows show the assimilates (as) streaming past. In the sieve tube lumen downstream the K⁺ ions 'leak' into the cell wall (cw) and, upstream, either the companion cell (cc) or sieve tube itself supplies ATP to pump the K⁺ ions back into the lumen

of the sieve plate, potassium ions leak into the walls of the sieve tube unit and are actively pumped back into the sieve tube element on the 'upstream' side of the sieve plate. This maintains a gradient of potassium ions on the two sides of the sieve plate. The ATP driving the potassium pump is supplied by the sieve tube element or by the adjacent companion cell, which is much more active metabolically. The principle of the theory is illustrated in Figure 12.8.

There is some evidence for the existence of electrical potentials across sieve plates. However, more convincing evidence in support of this theory is still lacking. Moreover, the phloem protein (**P-protein**) fibrils which are supposed to ensheathe the transcellular strands have not yet been observed in the sieve cells of gymnosperms. Since potassium ions are required to move through the protein fibres which extend through the sieve plates, these fibres are vital to the mechanism.

Apart from diffusion, which we have already ruled out, the only passive mechanism proposed to explain phloem translocation is the **mass flow** (**bulk flow**) hypothesis, which is the most important of all hypotheses. This, proposed by E. Munch of Germany in about 1930, is based on the assumption that there is a turgor potential gradient in the sieve tube between source and sink. As a result, metabolites move passively, in solution, in the positive direction of the gradient. A physical model of the process is illustrated in Figure 12.9.

The above physical model may be applied to the situation in the plant. Osmometer A represents the phloem in a source and osmometer C represents the phloem in a sink. The connecting link B represents columns of sieve tube elements while the connection D represents a column of xylem elements. Water from the dilute xylem sap is absorbed into the loaded cells of the phloem in the source. Solutes are unloaded into neighbouring cells in the sink.

Observations in support of the mass flow hypothesis include:

(a) When an aphid stylet has pierced a sieve tube element, exuding sap may flow for several days and exceed the total volume of the sieve tube by more than 100 000 times. Also, flow under pressure can be observed under the microscope.

(b) The absence of a nucleus and almost total absence of mitochondria in mature sieve tube

obscure. Lastly, the rate and mass of movement of solutes observed cannot be fully accounted for by the cross-sectional areas of the strands and their flow rate.

The second proposed active mechanism for phloem transport is set out in the **activated mass flow** hypothesis, and involves electro-osmosis. **Electro-osmosis** involves the movement of water or other non-ionized hydrophilic molecules (e.g. sucrose), brought about when polarized ions flow through a charged membrane along a maintained electrical potential gradient.

The theory is that high potassium ion concentration is maintained on one side of the sieve plate and a lower concentration on the other side. These ions, along with phloem sap, flow from the less positive side, through the negatively-charged microchannels between the cytoplasmic filaments in the pores of the sieve plate. On the 'downstream' side

(a)

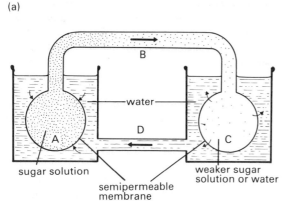

sugar solution

semipermeable
membrane

weaker sugar
solution or water

(b)

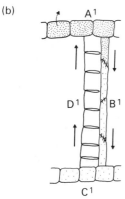

12.9 The pressure flow concept of phloem translocation:
(a) A physical model showing the movement of sugar
molecules from an osmometer, A, filled with concentrated
sugar solution to another osmometer, C, containing a
weaker solution; (b) a diagram showing the application of
the concept to representative essential parts of a plant. A^1
= photosynthesizing leaf cells; B^1 = columns of sieve tube
elements receiving photosynthates; C^1 = consuming root
cells or storage parenchyma; D^1 = columns of xylem
vessel elements containing an ascending column of water,
most of which is later transpired from the leaves. Solute
movement from A^1 to C^1 depends on the maintenance of
hydrostatic pressure

elements support the view that metabolic energy
is not essential for translocation.

(c) When a plant is defoliated, sugar concentration
gradients which previously existed between
source and sink disappear.

(d) Sugar flow along sieve tubes is largely along
gradients of decreasing total sugar.

(e) Pressures are known to exist in the vascular
system as a whole except under conditions which

cause high transpiration rate. It is common
knowledge that before piped water was generally
available, for instance in Nigeria, dry season
'water' could be obtained by tapping from the
wood at the base of the tree *Musanga cecropioides*
(Umbrella Tree). A large water-pot could be
filled overnight when a large tree was tapped.
That the phloem contains fluids under pressure
is even more economically important to palm-
wine tappers and drinkers. When the inflorescence of *Elaeis guineensis* (Oil Palm) or the top of
Raphia vinifera (Raphia Palm) is tapped, from
two to ten litres or more of sugary sap is exuded
from the sieve tubes in a single day. The largest
amount of the resulting palm wine is produced
at night. The palm wine undoubtedly represents
the sugary sap of the phloem diluted with water
which has moved osmotically from the xylem. As
in other cases of injury, sieve plates become
plugged with P-protein callus within a few
hours. Therefore, the wine tapper has to slice
away a thin layer of tissue about three times a
day to maintain the full rate of flow.

Observations seemingly opposing the mass flow
hypothesis, most of which can be explained away,
are as follows:

(a) Sieve tube elements must be living if they are to
function. Phloem transport is also sensitive to
temperature and to metabolic inhibitors. Both of
these facts seem to indicate a physiological rather
than a physical process. However, if basal
metabolism is interfered with, the structure of
the sieve tube and other factors vital to its functioning may also be affected. Such an effect
would be only indirect relative to translocation.

(b) The movement of photosynthates from mesophyll cells into the phloem and other indications
of an active process seem to rule out the passive
flow of translocates. However, most of the active
processes which have been identified have to do
with phloem loading and unloading and not
directly with translocation.

(c) Different substances moving at different rates
and the movement of metabolites in two directions at the same time seem to be contrary to the
mass flow hypothesis. However, different metabolites may leave the source in different sieve
tubes (**phloem ducts**), in which case different
rates of lateral transport out of the phloem might

produce different rates of translocation. It has also been suggested that substances being translocated may separate out by a chromatography-like process during flow, thus giving rise to different rates of movement.

During the last ten years or so, new theories have been put forward. Some researchers have been attracted by the concept of an active, peristaltic type of movement of metabolites within the fibrils of sieve tubes. This, again, would allow simultaneous bi-directional movement within the same cell, since movement could be in different directions in different fibrils.

It should be clear from the above that the problem of the mechanism of phloem translocation is not solved. Of all the present options, the mass flow hypothesis is the simplest and, when set up as a model, it seems to be the most workable.

Factors affecting translocation

It is possible that phytohormones and growth regulators influence the direction and rate of translocation in the phloem. Light is known to increase the export of sugars from photosynthesizing leaves to shoots and shoot tips, while translocation to roots is favoured in the dark. Other factors affecting the process include temperature, metabolic inhibitors and mineral deficiency. With regard to the last, it is reported that boron deficiency makes translocation in the phloem less efficient. Phosphorus deficiency is also known to inhibit the translocation, in the phloem, of the weed-killer, 2,4-D (2,4-dichlorophenoxyacetic acid), when it has been applied to the leaves of a plant.

Mineral nutrition

As we have seen, the roots of plants take in water and dissolved mineral ions from the soil. Soil is generally quite complex. Its main constituent is usually solid **mineral matter**, derived by weathering from the underlying rock or from rocks elsewhere. Most soil also contains a certain amount of more or less decayed **organic matter**. Both mineral matter and organic matter provide the mineral nutrients required for the healthy growth of plants. In addition, there is soil water, in which mineral ions are dissolved to form the **soil solution**, and a certain

amount of air, forming the **soil atmosphere**. The air in the soil atmosphere may be similar to the atmosphere above, if the soil is well drained and porous, but usually has a higher concentration of carbon dioxide and a lower concentration of oxygen. Soil also contains organisms which live there more or less permanently. These comprise the **soil microflora** and the **soil fauna**. It is with the mineral nutrients, dissolved in the soil solution, that we are now concerned.

Mineral nutrients

Some mineral nutrients provide elements which are required for the synthesis of important components of the living material of the cell. Others have less obvious functions. Most of the metallic elements which are necessary act as coenzymes (**enzyme activators**), whether or not they have a more obvious function as well. Other nutrients help to maintain the osmotic and anion/cation balance in cells.

Mineral nutrients are usually divided into two categories, macronutrients and micronutrients. **Macronutrients** are required in relatively large amounts. **Micronutrients** (trace elements) are required in smaller, sometimes very small, quantities. As we shall see, the exact dividing line between macronutrients and micronutrients is to some extent arbitrary.

Macronutrients

The six elements required as macronutrients are nitrogen (usually as nitrate), potassium, calcium, phosphorus (as orthophosphate, $H_2PO_4^-$), magnesium and sulphur (as sulphate). They are listed here in order of the quantity necessary for the plant, the first, nitrogen, being required in greater quantity than any of the others.

Nitrogen is necessary for the manufacture of proteins, nucleic acids and the many other nitrogen-containing components of the cell. Most plants take in nitrogen as nitrate, but some plants are able to absorb ammonium ions. Plants which are deficient in nitrogen are stunted, their leaves become yellow (**chlorotic**) and they die prematurely. Nitrate in the soil is derived from the breakdown of dead organic matter and is held only in the soil water. It is very easily lost by leaching.

Potassium is a coenzyme in several enzyme-catalysed reactions, for instance the action of ATP-ases. It is also important in the formation of cell membranes. Potassium deficiency results in increased cell size and leaf size in some plants. Leaves become chlorotic around the edges and plants suffer premature death. Like other cations, potassium is held in the soil, adsorbed on the surfaces of the smallest mineral particles (clay particles) and is not easily leached away.

Calcium is important as a constituent of the middle lamella (calcium pectate) and is also a co-enzyme. Calcium deficiency causes the disruption of shoot and root apices. Growth is stunted and often deformed and the plant dies early.

Phosphorus is required for the synthesis of many important compounds, including nucleic acids and adenosine phosphates. Deficiency leads to darkening of the leaves and stunting of the whole plant, followed by death. Most of the phosphorus in the soil is held in insoluble form and is released gradually into the soil water, as soluble ions.

Magnesium is a constituent of the chlorophyll molecule and a coenzyme. If a plant is deficient in magnesium, young leaves are green but older leaves become chlorotic and retain green colour only close to major veins.

Sulphur is required in small quantities and is a constituent of some proteins. Deficiency results in severe stunting, general chlorosis and death. Sulphate is easily lost from the soil by leaching.

Micronutrients

A mineral substance can be shown to be essential for normal plant growth only by comparing plants grown in the presence of that substance with plants grown in its absence. The six macronutrients listed above were recognized as essential in the last century. At that time, chemicals available to scientists were mostly very impure by modern standards and techniques for testing their purity were crude. It is probable that the required micronutrient elements were present as impurities in the major nutrients in quantities sufficient for normal growth. As purer chemicals and better analytical techniques became available, more and more elements, in small or very small amounts, were found to be necessary for normal plant growth. As stated above, the dividing line between macronutrients and micronutrients is arbitrary and iron and boron (as borate) are required in quantities only a little less than, for instance, sulphur. Chlorine, copper, manganese, zinc and molybdenum are necessary in only very small

amounts and a trace of cobalt is probably also required.

Iron, which was recognized as an essential element before the end of the last century, is necessary for the synthesis of chlorophyll, although it is not a constituent of chlorophyll. Iron is also a constituent of cytochromes and acts as a coenzyme. An iron-deficient plant has chlorosis of young leaves, which become green only close to main veins, while old leaves remain green. This pattern of chlorosis is, therefore, different from chlorosis caused by deficiency of magnesium or other elements.

Boron has several roles within the plant. Amongst these are effects on calcium metabolism, cell wall formation and cell differentiation. Symptoms of deficiency are also various, usually involving the formation of necrotic (dead) patches of tissue in roots, stems and leaves.

Chlorine, as chloride ion, is involved in the regulation of anion/cation balance and osmosis. Absence of chlorine leads to severe stunting and death.

Copper is a coenzyme in oxidase reactions and protein metabolism. Copper deficiency results in the dying back of shoots.

Manganese is a coenzyme for part of the Krebs cycle and for enzymes involved in photosynthesis. Deficiency is recognized by a number of symptoms, including the presence of chlorotic spots on leaves, turning to brown spots as the tissues die.

Zinc is a coenzyme in several reactions and also seems to be related to auxin metabolism. Deficiency results in a failure of stem elongation.

Molybdenum is involved in nitrogen metabolism and deficiency affects the uptake of nitrate from the soil. Deficiency usually leads to a chlorotic mottling of the leaf, the yellow patches later turning brown. Molybdenum is also necessary for nitrogen fixation in leguminous root nodules and in nitrogen-fixing organisms (bacteria, blue-green algae) in the soil.

The status of **cobalt** as a micronutrient is not certain. However, it has been shown that deficiency leads to chlorosis and leaf-drop in *Theobroma cacao* (Cocoa). Cobalt is certainly necessary for nitrogen fixation in leguminous root nodules.

Several other elements are taken up and utilized by plants, though not necessarily essential.

Sodium is absorbed and is involved in maintaining anion/cation and osmotic balance. Plants which grow in salt water (**halophytes**) usually take up both sodium and chloride in quantity and some halophytes subsequently excrete them as salt, through glands on their leaves (e.g. *Rhizophora* spp., Red Mangroves).

Silicon, in the form of silica (SiO_2) is taken up by many plants (e.g. grasses; *Lantana camara*). In grasses, cluster-crystals of silica form a normal part of the epidermal cell wall, where they may give protection against attack by small herbivores (e.g. snails).

The relationship between mineral nutrients is not simple and not only absolute quantities but also relative quantities of different nutrients are often important. Also, several of the micronutrients are very toxic in excess, even when the total amount in the soil is quite small (e.g. copper). Other elements are also toxic to most plants (e.g. aluminium).

Water culture experiments

In order to demonstrate a plant's requirements for mineral nutrients, it is usual to use water culture experiments. In order to do this, seeds of a selected plant are germinated to the stage at which an adequate root system has formed and the plants are then transferred to jars containing different solutions of nutrients. One jar contains the full nutrient solution and is the control experiment. Others contain solutions which are deficient in one nutrient or another. Growth of the plants is compared over a period of weeks.

The selection of a suitable plant is important. It is best to use a plant with a fairly small seed, as a large seed may contain a store of nutrients in the endosperm or cotyledons and the effects of nutrient deficiency may be delayed in their appearance. Satisfactory results are usually obtained with *Zea mays*, *Sorghum* sp. or another of the smaller grains, or *Helianthus annuus* (Sunflower). Seeds are germinated in a seed germinator, on moist (thoroughly washed) blotting paper or in sand which has been thoroughly washed with acid and distilled water to remove all possible nutrients.

Jars of nutrient are then prepared. Jars of a capacity of 250 ml are usually sufficient. These should be cleaned carefully and wrapped in black paper or plastic to exclude light from the culture solution and thus prevent the growth of algae. Nutrient solutions are added and the jars labelled clearly. Plants are placed between the two halves of a split cork lined with cotton wool and placed in the

mouths of the jars with their root systems immersed in the culture solution. For good results, the culture medium must be continuously aerated, using an aquarium pump with connections to a glass tube extending into each jar. Jars should be checked daily and topped up with appropriate nutrient solution if any noticeable loss of water by evaporation has occurred. If possible, nutrient solutions should be emptied away and replaced with fresh solutions not less than every two weeks.

Suitable nutrient solutions for this experiment can be obtained in powder or tablet form from suppliers or made up from pure chemicals according to formulae given in more specialized books (e.g. Baron, W. M. M., *Organization in Plants*; Thomas, M., Ranson, S. L. and Richardson, J. A., *Plant Physiology*: see booklist at the end of Section B).

Key words

Water uptake: water potential, osmosis, passive uptake. **Ion uptake**: soil solution, active uptake, xylem sap. **Translocation**: root pressure, cohesion-tension, transpiration stream. **Water pathway**: vacuolar pathway, symplastic pathway, apoplastic pathway; endodermis, Casparian strip, xylem vessel (xylem duct), mesophyll, stoma. **Transpiration**: stomatal, cuticular, lenticular; transpiration rate, transpiration intensity, transpiration coefficient. **Stoma**: stomatal frequency, stomatal index (stomatal ratio); relative humidity, diffusion shell, edge effect. **Guttation**: hydathode, active hydathode (water gland). **Phloem translocation**: source, sink; simultaneous bidirectional translocation. **Experiments**: ringing (girdling), defoliation, surgery, aphid stylets; radioisotopes (tracer elements), Geiger counter, autoradiography. **Phloem sap**: assimilates, photosynthates, sucrose, oligosaccharides, amino acids, amides. **Pathway**: sieve tubes (phloem ducts). **Phloem loading** and **unloading**: transfer cells. **Mechanism**: active, protoplasmic streaming (cytoplasmic streaming, cyclosis), transcellular strands; activated mass flow, electro-osmosis, P-protein fibrils; passive, mass flow (bulk flow). **Soil**: mineral matter, organic matter, soil water, soil atmosphere, soil fauna, soil microflora. **Mineral nutrients**: macronutrients, micronutrients (trace elements). **Deficiency symptoms**: stunting, chlorosis, leaf drop, necrotic tissues. Water cultures.

Summary

Water uptake into the root depends on osmosis. The active uptake of mineral ions from the soil water increases water potential difference and thus tends to increase the osmotic movement of water into the plant and into the stele. The excretion of mineral ions into the vessels of the xylem results in osmotic movement of water into the vessels. This results in the development of hydrostatic pressure in the xylem of the root, which produces root pressure, which tends to force water, with dissolved salts, up the xylem. Root pressure is not strong enough to account for all water movement up the stem, but accounts for guttation in some plants. Loss of water vapour from the leaves draws water from the xylem and water moves up in response to the tension produced, due to the adhesive and cohesive powers of water. Transpiration rate is affected by many factors, including the opening and closing of stomata. Most transpiration occurs through open stomata. Translocation of organic compounds and some ions is through the sieve tubes of the phloem and is always from a source (e.g. photosynthesizing leaves) to a sink (e.g. growing apex). The mechanism by which phloem translocation takes place is very controversial. Sugar and other substances are loaded into the phloem by an active process. Subsequent movement is probably passive, as described by the mass flow hypothesis, but some research workers support hypotheses which require the expenditure of metabolic energy. Soil usually contains more mineral matter than anything else but usually also contains organic matter. It also contains soil water and there is a soil atmosphere. The mineral requirements of plants are supplied from the mineral and organic matter, by way of the soil water. Mineral nutrients include a total of about fourteen essential elements, six macronutrients and about eight micronutrients. Some of these elements are required as components of important compounds in the cell, others are coenzymes. Some other elements are taken up by plants and utilized, but are not essential for normal growth.

Questions

1 How do active uptake of mineral ions and passive (osmotic) uptake of water into a root act together to produce root pressure? Name one visible consequence of root pressure.

2 Explain how temperature, wind, relative humidity and stomatal aperture affect transpiration rate.

3 Describe how (a) ringing and (b) experiments using radioisotopes have been used in research on phloem translocation.

4 To what extent is the movement of sugars from the photosynthetic cells of a leaf to, say, the cells of a developing storage root an active or a passive process? Distinguish between phloem loading and unloading and phloem translocation.

5 Name the six macronutrient elements and explain why they are essential to the plant.

6 Describe how you would set up a water culture experiment to demonstrate the need for micronutrients.

13

The plant in action – III Growth and development

In this chapter, we are going to look at the life history of the sporophyte generation of a flowering plant from a physiological point of view. We have already dealt with the germination of seeds (Chapter 2). After the seedling is established, every plant goes through a number of phases, which involve overall growth and lead eventually to the production of flowers, fruits and seeds. Development during all these phases is governed by a number of internal and external factors, which control and modify the course of events.

Growth

Every organism starts its life as a single cell. In animals and in the sporophyte generation of plants, this is a zygote. From this cell, through processes of growth and development, the new organism gradually takes on its adult form, as determined by its genetic material and influenced to a greater or lesser extent by the environment. In all multicellular organisms, growth depends first on cell division, followed by cell growth and the differentiation of cells and organs until the characteristic structure of the mature organisms has been attained. The way in which organs attain their final form is referred to as **morphogenesis.**

The nature of growth
It is not easy to give a simple definition of growth, but it is necessary to arrive at a working definition. We can start by stating that growth results from an increase in the quantity of protoplasm. This is basic and can be applied most obviously at the level of the cell. At the level of the organism, growth results in an increase in size, which can be measured as increase in mass, weight, height, length or volume of the organism. This results from the fact that growth is accompanied by development, so that there is some organization or reorganization of the accumulating mass of protoplasm. An increase in size resulting from growth must also be irreversible. When a seed imbibes water it becomes larger and

heavier but it cannot be said to have grown. The imbibed water can be lost quite easily by drying, so the change is reversible. Also, from the time that water is imbibed until the seedling starts to carry out photosynthesis, the overall weight (measured as dry weight) of the seedling decreases, but the seedling is growing. The protoplasm of the embryo increases at the expense of stored nutrients in the cotyledons or in the endosperm. The net decrease in weight results from the mobilization of materials from the non-growing parts and their utilization, for instance in respiration. For the same reason, an etiolated seedling, grown in the dark, continues to grow but loses weight (measured as dry weight) all the time. Therefore, when we consider growth, emphasis is placed on the increase in the amount of protoplasm in the growing system.

As growth continues, organs are differentiated and are arranged in space to produce forms characteristic of the species. In animals, some cells migrate from their original positions during growth and development. The shapes of plant organs are produced mainly by differential growth in different parts and/or in different directions.

Plant cell growth
We have already described (Chapter 6) how cells differentiate after cell division. A cell derived from an apical meristem at first has very dense cytoplasm with a few very small vacuoles. There are no intercellular spaces. As such a cell starts to differentiate, the cell wall gradually thickens and the small vacuoles become larger and run together to form a single, central vacuole. As the cell attains its final size and shape, the amount of cytoplasm increases, but not in proportion to the increase in cell size. The cytoplasm finally forms a thin lining to the cell, with the embedded nucleus at one side (Fig. 5.1). In some tissues, for example, parenchyma, intercellular spaces appear.

These changes during the differentiation of a cell have certain physiological implications. The first stage of differentiation involves the formation of a relatively large amount of cytoplasm and a smaller

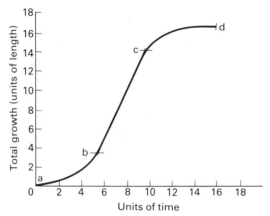

13.1 An idealized growth curve showing the exponential phase (a–b), the linear growth phase (b–c) and the phase of declining growth (c–d)

amount of cell wall material. Therefore, the cell at this time assimilates more proteinaceous substances, for the production of cytoplasm, than carbohydrates, for the production of cell walls. Salt uptake is sufficient for assimilatory needs but, since the vacuoles are small, very little accumulation of salts takes place. As cells enlarge, the rate of increase in the quantity of cytoplasm falls and that of cell wall substances increases. Therefore, smaller quantities of proteins and larger quantities of carbohydrates are assimilated. Much water is taken in for the hydration of the protoplasm and for the enlargement of vacuoles. As the vacuoles enlarge, uptake of salts increases beyond the amounts required for assimilation as they accumulate in the vacuoles.

The growth curve

However we measure growth, it proceeds in a characteristic manner. The rate of growth is always slow at first, then increases until it reaches a maximum rate and subsequently decreases until the cell, organ or organism attains its maximum size. A time-course of such a growth pattern gives an S-shaped (sigmoid) curve (Fig. 13.1). This idealized curve can be divided into various phases, as shown. These phases differ according to the growth pattern and the proportions of the various phases also vary. For instance, the growth curve of a germinating seed, using dry weight as an index of growth, starts by decreasing due to loss of weight. The phase of senescence preceding death may also involve loss of weight. A growth pattern showing all these phases may exhibit a truly S-shaped growth curve. In

others, the 'S' may be less distinct but the shape is always recognizable. The two phases that seem to be represented in all cases are the **exponential** pattern (a–b in Fig. 13.1) and the phase of decline or decreasing growth rate (c–d). In most cases the exponential phase leads into a phase of linear growth (b–c). The whole exponential phase represents what is variously referred to as '**the grand curve of growth**' or '**the grand phase of growth**'. If the graph is plotted as growth rate (size increase per unit time) against time, instead of in absolute units of size against time, the curve obtained is rather like a normal distribution curve (Fig. 13.2). The expansion

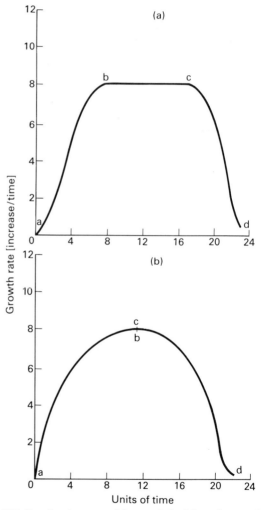

13.2 Growth rate curves: (a) curve derived from the growth curve in Fig. 13.1; (b) as above, but lacking the linear growth phase

growth rate of the first trifoliate leaf of *Vigna* sp. (Cow Pea) plotted against time is shown in Figure 13.3. Two leaves of different sizes are represented and each curve resembles the idealized curve shown in Figure 13.1.

Distribution of growth in space and time

As mentioned earlier, growth is not uniform throughout the entire plant, but is concentrated in specific and characteristic growing regions. This is very easy to demonstrate with root growth. A germinating seed of *Vigna unguiculata* (Cow Pea) is allowed to produce a radicle about 25 mm long (about 36 h after soaking). The entire length of the root is marked, with indian ink, at equal intervals (e.g. 2.0 mm) and the marked root is left to grow for another day or so, suitably suspended in a specimen tube and kept moist. After a period of growth, it is found that the rate of elongation (growth) is slow at the base of the root and close to the tip, and fastest in the intervening region. If measurements are made, they may be plotted on a graph (Fig. 13.4).

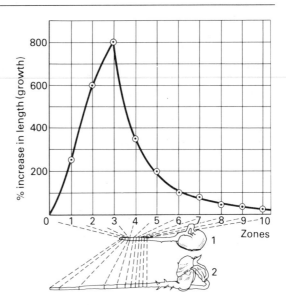

1 seedling with original marking numbered 1–10, equally spaced on the radicle

2 same seedling later, exhibiting differential growth rate at various zones

13.4 Growth distribution in the growing root of a seedling of *Vigna unguiculata* (Cow Pea). Initially, the root was marked into ten equal zones. Distribution of elongation growth is determined from the elongation of each zone. The curve of this growth pattern is shown in the graph

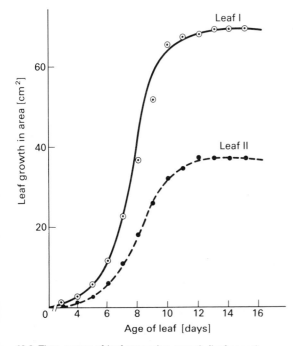

13.3 Time-course of leaf expansion growth (leaf growth curve). From: Fred Asiegbu, *Changes in water status, and their implications, in detached and intact first trifoliate leaf of black-eye bean* (Vigna unguiculata), University of Benin B.Sc. Honours Project Dissertation, June, 1983)

A convenient device for measuring extension growth in stems is the **auxanometer** (Fig. 13.5). In essence, this is a pivoted lever with unequal arms. The end of the short arm is connected to the shoot apex with thread. The thread is kept taut by, if necessary, adding weights to the long arm. The smallest extension of the stem below the attachment of the thread is reflected in a downward movement of the long arm. The upward movement of the short arm, resulting from growth, is magnified in the long arm by the ratio of the length of the long arm to the length of the short arm. The tip of the long arm is pointed and so arranged that it scratches a vertical mark on the surface of a piece of paper which has been blackened by smoke. If the smoked paper is attached to the surface of a metal drum and the drum is periodically turned to one side and back by a clock mechanism, the mark on the paper also records time, as shown in the illustration. With this arrangement,

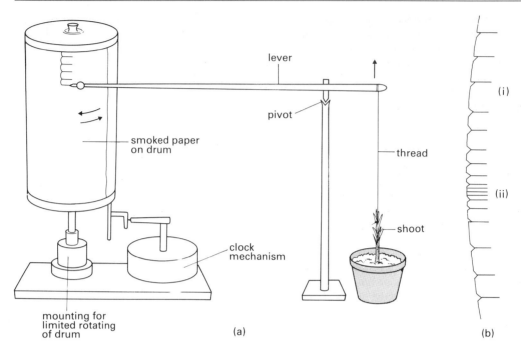

13.5 Auxanometer (a) Diagram showing the principle on which the auxanometer works. (b) Illustration of the kind of tracing made on smoked paper mounted on a drum. The region (i) shows a relatively fast growth rate and (ii) shows a very slow rate

growth in time can be recorded. It is found, for instance, that most extension growth takes place at night.

Measurement of growth

Growth may be measured in many different ways. The parameter chosen depends to some extent on the particular interest of the worker concerned.

Growth in height (length) of the above-ground parts of a plant is a simple index of growth. The advantages of using this parameter are that its use is simple, speedy and non-destructive. The organ or organism used remains undamaged and can be used again for a continuous assessment of growth. However, the measurements obtained do not necessarily represent a picture of growth of the whole organism. Linear growth may not be proportional to overall growth. For instance, dark-grown (**etiolated**) plants may be much taller than plants grown in light but their overall rate of growth is much lower. Also, linear measurements take no account of growth in girth, degree of branching or lateral spread.

Growth in volume is appropriate for fruits and similar organs but is not easy to measure. One serious disadvantage is that the volume measured may include a lot of intercellular spaces, which are not truly part of the organism.

Growth in area is a parameter suitable only for the study of growth in expanded organs such as leaves. Again, leaf area may not be increasing at the same average rate as the rest of the plant and it is not easy to measure. Leaf area meters, constructed to give immediate measurements of leaf area, are very expensive. They operate by measuring the amount of light interrupted as detached leaves are fed in on a moving belt.

The simplest method of measuring the area of a leaf is to trace its outline on to a piece of graph paper and then to count the number of squares (area) covered. Estimation of fractional squares leads to some inaccuracy but, if there are many fractional squares, it is accurate enough to count the number of fractional squares included, whether they are almost complete squares or very small parts of squares, and divide the number by two, to give the number of whole squares they represent.

Measurement of leaf area is very important to a plant physiologist who is interested in the primary productivity of flowering plants, since total leaf area is related to the total amount of photosynthesis which can take place.

Measurements of fresh weight, which are very easy to make, may be used to estimate growth. However, the plant parts measured may have to be sacrificed and are subject to errors arising from variations in the amount of water in a tissue. Fresh weight is not, therefore, a constant and reliable index of growth.

Dry weight is a better measure of growth which has taken place, since it is a measure of dry matter, mostly representing materials which have been accumulated during growth. Interpretation may not be straightforward, for instance, because of the decreasing dry weight of germinating seeds or etiolated plants. The most serious disadvantage is that the organ or organism must be destroyed in the process, which involves drying in an oven to constant weight, at a temperature of between 70 and 105°C.

Growth substances

The term **growth substance** is used to describe both natural and synthetic substances which affect and control growth, development and metabolic processes. They are effective at very low concentrations and are neither intermediate substances nor products of the pathways which they control. Thus, they are not metabolites. Plant growth substances may also be referred to as **phytohormones** or **plant hormones**. Hormones are substances which are synthesized in one part of an organism and translocated to another part, where they act to regulate growth. There are several classes of plant growth substances, the chemical structures of which are illustrated in Figure 13.6.

Auxins

The existence of **auxins** was first suspected by Charles Darwin in the late 1880s but the definitive discovery was made by Fritz W. Went of Holland in 1928. Within two years of its discovery the chemical nature and structure of auxin were known. Auxin is the name given to a class of substances typified by indole-3-acetic acid (**IAA**).

One of the experiments which led Went to his discovery is illustrated in Figure 13.7. In this experiment, he found that substances that could diffuse into an agar block from the cut tip of an Oat coleoptile could cause extension growth of the coleoptile, while a plain agar block could not.

Auxin is typically produced in the apical meristem of shoots and in other tissues in which much synthetic activity is going on, for instance, in expanding and photosynthesizing leaves and in developing fruits. It characteristically moves **basipetally** in the shoot, from the apex to the morphological base, even if the shoot is turned upside-down. Thus, its transport is strictly **polar**. It moves through parenchymatous and similar tissues, at a rate which is greater than can be explained by diffusion alone. As it moves down the shoot, its concentration becomes less, partly due to the action of an enzyme, **IAA** oxidase, also by combining with other compounds. The resulting **auxin gradient** within the shoot has several important effects on development.

The main biological effect of auxin is to promote or inhibit cell elongation, according to its concentration in the tissue concerned. The concentrations which cause these two opposite effects vary in different parts and organs. Thus, maximum stimulation of growth is achieved in stems at about 10^{-4} molar, in buds at about 10^{-8} molar and in roots at about 10^{-10} molar. On the other hand, growth is inhibited in stems at concentrations above about 10^{-2} molar, in buds at about 10^{-6} molar and in roots at about 10^{-8} molar.

As well as its effects on cell elongation, auxin gradients affect cambial activity, the development of new vascular tissues and are largely responsible for the phenomenon of apical dominance (see below).

Auxin is also formed in root apices, but it is thought that much of the auxin found in roots has its origin in the shoot.

Since it is present in such small amounts, auxin is not easy to quantify (**assay**). As an alternative to its chemical detection and quantification, the relative amount of auxin present is assessed by the magnitude of the biological effect produced. This is known as **bioassay**. Many different bioassays have been developed for auxin. Auxin stimulates the elongation of cut lengths of young stem and the curling-apart of the ends of split lengths of stem. The stems of etiolated seedlings of *Pisum sativum* (Garden Pea) are used as standard practice but it can also be standard-

13.6 The structure of some growth substances: (a) naturally occurring auxins; (b) synthetic auxins; (c) gibberellic acid and cytokinins; (d) abscisic acid and ethylene

ized with other seedlings, such as those of *Vigna* sp. (Cow Pea). The percentage increase in length of pieces of stems or the degree of curvature apart of split ends is plotted against known concentrations of auxin (IAA). The amount of auxin in a tissue extract can be found by comparing its effects with those of standard solutions of IAA, plotted on a graph. Bio-assays are non-specific and do not distinguish between auxins proper and other substances with auxin-like activity. Therefore, it is preferable to separate auxins chromatographically and to carry out bioassays on extracts from pieces of the chromatogram.

Gibberellins

The association between the *bakanae* disease of rice in Japan, caused by the fungus *Gibberella fujikuroi*, and excessive growth of shoots of infected plants led to the discovery of **gibberellin**. The first suggestion that such a growth substance existed was given experimental support in the middle 1920s and, in

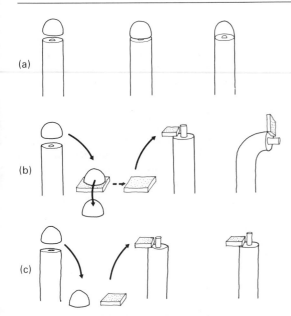

(b)

(c)

13.7 An illustration of the kind of experiment which led to the discovery of auxin by F. W. Went. (a) *Avena* sp. (Oat) coleoptile with the tip cut off and symmetrically replaced. No growth curvature takes place. (b) Coleoptile tip cut off and placed on a small agar block, so that diffusate from the tip accumulates in the block. When the block containing the diffusate is placed asymmetrically on the coleoptile, growth curvature follows. Curvature was due to increased growth on the side of the coleoptile receiving diffusate from the agar block. (c) As last, but plain agar (not containing diffusate) was placed asymmetrically on the cut coleoptile. No growth curvature follows

1938, gibberellin was isolated in crystalline form.

The main action of gibberellins is to promote elongation. When supplied with minute amounts of the hormone, many dwarf plants grow tall. It also plays a part in the characteristic elongation of the floral axis in semicompressed or rosette types of plants. This phenomenon, common in biennial plants in temperate climates, is called **bolting**. Bolting can also be seen, in the tropics, in *Lactuca* sp. (Lettuce) and often occurs in *Brassica chinensis* (Chinese Cabbage) after transplanting. Gibberellin is also involved in some phases of seed germination, in the breaking of dormancy and in some formative processes.

Like auxins, gibberellins interact with other hormones to produce various effects. Unlike auxins, their transport is non-polar. Gibberellins move about freely in the plant.

A typical gibberellin is gibberellic acid (**GA$_3$**, Fig. 13.6), but about fifty-two gibberellic acid and gibberellic acid-like substances are now known. All have the same basic chemical structure but differ in the nature of the various sidechains or substitutions. Different gibberellins occur in different plants but produce similar effects. However, a few species-specific effects are also known.

Cytokinins

Cytokinins are growth substances specifically concerned with the promotion of cell division. Kinetin (6-furfurylaminopurine, Fig. 13.6) was first separated from an old sample of yeast DNA in 1955, by C. O. Miller and his co-workers. It was found to induce cell division and was the first such substance isolated. Other synthetic compounds with similar effects were called kinins, to reflect their kinetin-like nature. Because this could cause confusion with a similar term in animal physiology, all such hormones, which stimulate cell division (cytokinesis) were renamed cytokinins. One of the most effective naturally-occurring cytokinins was isolated from *Zea mays* and named zeatin.

In addition to stimulating cell division, cytokinins also have other effects, either alone or when interacting with other hormones. For instance, in experiments in which cultures of Tobacco pith cells are grown as an unorganized mass of cells (**callus**), the addition of different mixtures of auxin and kinetin can be used to make the callus produce roots, or shoots, or continue growth as callus. Cytokinins also prevent the onset of the ageing process in senescent tissues, perhaps by stimulating cell division. They also prevent the loss of chlorophyll from detached leaves. Some bioassays of kinetin are based on this last property, which is easily measurable.

Unlike auxins and gibberellins, cytokinins are relatively immobile and diffuse only over short distances.

Abscisic acid

Different lines of research led to the naming of certain hormones as abscisin I, abscisin II and dormin. These lines converged in 1965, when P. F. Wareing and his co-workers found that all three substances were one compound and had the same biological effects. The name **abscisic acid** was subsequently officially abbreviated to **ABA** at the sixth International Conference on Plant Growth

Substances held in Ottawa, Canada, in 1967.

ABA is a hormone with growth inhibiting action, very effective in inducing and maintaining dormancy in many seeds and in buds of perennial plants. It also induces stomatal closure and causes leaf abscission. In biosynthesis, it follows the same initial route as gibberellin and is similar to gibberellin in structure (Fig. 13.6). It seems to counteract the effects of gibberellin in some plants. It is thought that ABA is the major and perhaps the only active component of a previously proposed 'β-inhibitor complex' found in many plant extracts.

Ethene (Ethylene)

Ethene is a simple, gaseous compound (Fig. 13.6), produced in leaves, flowers and fruits. Its production is strongly stimulated by auxin. Perhaps its most striking effect is the induction or promotion of senescence. Hence, it strongly affects fruit ripening and the 'vase life' of cut flowers which have been placed in water. Some of its effects resemble the formative effects of auxin. It has even been suggested that some of the formative effects attributed to auxin, particularly in the roots, are really produced by ethene synthesized in response to auxin stimulus. The two hormones probably interact to produce their effects.

Another common effect of ethene is **epinasty**, in which there is more growth on the upper side of an organ (e.g. a leaf) than on the lower side, causing the organ to curve downwards.

Apical dominance

The term **apical dominance** is used to describe the situation in many plants in which the apical bud grows ahead rapidly while axillary buds, especially those close behind the apex, remain dormant. Early physiologists attributed apical dominance to the concentration of nutrients in the apex (**nutritive theory**). Later, it was shown that the auxin gradient in the shoot was responsible. Auxin diffusing down the shoot from the apex was thought to be in sufficient concentration to inhibit the growth of axillary buds. Removal of the growing tip of the shoot cut down the level of auxin and axillary buds immediately started to develop. The application of auxin to the decapitated stump, thus replacing the apex physiologically but not morphologically, maintains

apical dominance. However, this is too simple an explanation. The present concept is that apical dominance is maintained by the interaction of two growth substances, auxin and cytokinin. Application of cytokinin to dormant buds counteracts apical dominance, so that the buds grow out into lateral branches, even when the apex of the shoot is still intact.

It is now known that hormones affect nutrient translocation. According to the **nutrient-diversion theory**, nutrient traffic up the shoot goes as a mainstream directly to the apex, following the gradient of increasing auxin concentration up the shoot. For this reason, there is very little diversion of nutrients to lateral buds, which do not produce auxin while they are dormant. Once they are released from apical dominance by the removal of the shoot apex, or their cells are stimulated to divide by cytokinin, they become sources of auxin. Each produces its own auxin gradient so that nutrients are redirected into the developing lateral branches. It has recently been suggested that ethene, also, may be involved in apical dominance.

The intensity of apical dominance varies in different plants. It is strong in trees which have a characteristic triangular profile, with very short branches close to the top and progressively longer branches towards the base (e.g. *Casuarina* spp.; also many conifers). It is virtually absent in plants with a bushy profile.

Apical dominance is also seen in roots. Where it is strong, it is responsible for strong tap root development. Weak apical dominance in the root system leads to the development of a fibrous root system.

Environmental factors affecting growth and development

Many environmental factors affect growth and development, sometimes to the extent of over-riding internal regulating mechanisms. Such factors may be physical or chemical but physical factors are the more evident. The major external factors which affect growth are:

(a) Light, including wavelength, intensity, duration and periodicity (the lengths of alternating periods of light and dark).
(b) Gravity.

(c) Temperature, including its absolute value and periodicity.
(d) Relative humidity.
(e) Mechanical factors, including wind and other types of mechanical disturbance.
(f) Nutrient status.

Aspects of the effects of wind, humidity and nutrient status have been dealt with already (Chapter 12). We shall now deal with growth and other movements in response to external stimuli and with the effects of light and temperature on flowering.

Tropisms

Tropisms are changes in the orientation of parts of plants (curvature, twisting) brought about by growth in response to external stimuli. When tropic movements occur in response to a stimulus from one side (unidirectional stimulus), they result in the plant part assuming a new orientation related to the direction from which the stimulus is received.

Tropic movements are classified first according to the direction in which the growth movement occurs relative to the source of the stimulus. When the response is directly towards the source of the stimulus, the tropic movement is **positively orthotropic**. If the response is directly away from the stimulus, it is **negatively orthotropic**. The direction of the response may be at an angle to the stimulus, in which case it is **plagiotropic**. If at right angles to the stimulus, it is **diatropic**. Each tropism is also named after the factor providing the stimulus to which the organ is responding and there is a long list of them (Table 13.1). In each of the tropisms named, the response may be positively orthotropic, negatively orthotropic, plagiotropic or diatropic.

In addition to all of the above, higher plants also show autotropism. **Autotropism** refers to the observation that, in any given species, some organs always have a fixed orientation relative to other organs. Thus, for example, leaves have a constant orientation relative to stems, and branches have a constant orientation to the main stem axis. Lateral roots also are borne at definite angles to the main root axis.

All tropic movements are brought about by different growth rates on two sides of a responding organ. Thus, if growth rate is enhanced on one side and/or reduced on the other side, then a **growth curvature** results. For this reason, only a growing organ can exhibit a growth response. Therefore, the

Table 13.1 Tropisms

Tropism	Directional stimulus
geotropism	gravity
phototropism	light
haptotropism (thigmotropism)	touch (rubbing)
hydrotropism	relative humidity
chemotropism	concentration gradient of a specific chemical
aerotropism	concentration gradient of oxygen
thermotropism	temperature gradient
traumatotropism	injury
rheotropism	current of water
galvanotropism	electricity

sensitivity of a reacting organ is high while it is growing but falls to zero when it reaches its maximum size.

The regions in which growth curvature takes place are regions of cell expansion, not regions of cell division. Thus, curvature almost always takes place in sub-apical rather than apical regions. In grasses, curvature may also take place in otherwise mature stems, in the vicinity of the intercalary meristem which is situated at the bottom of each internode.

The tropic response of an organ may be different at different times. For example, the flower stalk of *Arachis hypogaea* (Groundnut) is negatively geotropic when the flower first opens but becomes positively geotropic after the fruit starts to develop, as the fruit is buried in the ground. Some tropic stimuli also interact with each other. Light, for instance, seems to modify the sensitivity of plant organs to gravity. Thus, runners of a grass such as *Cynodon dactylon* (Bahama Grass) are diageotropic when strongly illuminated, but become negatively geotropic when strongly shaded or grown in the dark.

Geotropism

Gravity is a force which acts as a directional stimulus. The stimulus is received only in certain parts of the plant, mainly in the tips of roots, shoots and coleoptiles. In roots, it has been shown that removal of the extreme tip is enough to prevent a geotropic response, although geotropic curvature normally takes place at some distance behind the tip. In *Zea mays*, removal of the root cap alone is enough to prevent geotropic curvature when the root is placed in a horizontal position, but ability to respond returns as a new root cap is formed. Similar experiments with shoots, and with grass coleoptiles, also show that the directional stimulus of gravity is received in the tip. In most plants, primary shoots and grass coleoptiles are negatively geotropic and primary roots are positively geotropic.

Plants react to centrifugal force in the same way as to gravity and centrifuges can be used to test the effects of enhanced 'gravitational' force. The effects of gravity can also be neutralized, experimentally, by the use of a klinostat. A **klinostat** is a machine in which a disc to which plants can be attached is slowly rotated by an electric motor. Seedlings or more mature plants attached horizontally to the disc then receive the stimulus of gravity from constantly changing positions and both root and shoot continue to grow horizontally.

If a seedling is placed in a horizontal position for a short time (not on a klinostat) and is then placed vertically, the root first bends to one side, in response to the stimulus it received when it was horizontal, and then bends vertically downwards again. This shows that some time passes between the perception of gravity and response to the stimulus.

Gravity affects the growth rates of roots and shoots in different ways. In most shoots, the overall growth rate is unchanged when reacting to gravity, since there is an inhibition of growth on the side which becomes concave that is roughly compensated for by a stimulation of growth on the convex side. In roots, the overall growth rate is reduced because either the inhibition of extension on one side is not matched by the stimulation on the opposite side, or because it is accompanied merely by a lower degree of inhibition on the other side.

There is a minimum time for which the stimulus must act before an organ will respond. This is the **presentation time**. The presentation time for the shoots and roots of flowering plants varies from about three minutes to about thirty minutes in different species and under different conditions. The length of presentation time is greater at lower temperatures. Experiments with centrifuges show that the effect of gravity is quantity related. When centrifugal forces greater than normal gravity are applied, the necessary presentation time is correspondingly reduced.

Response to the stimulus is not immediate. The duration of the period between the end of the presentation time to the first signs of response (curvature) is known as the **reaction time**. Since the stimulus is received in the tips of roots and shoots, the stimulus must be transmitted to the region in which the reaction is to take place and this is always at some distance from the tip. We can therefore recognize four stages in a geotropic reaction sequence, (a) the perception of the external stimulus, (b) the induction of metabolic changes in the receiving area, (c) the conduction of the physiological influence to the responding region, and (d) geotropic growth curvature (response).

Previously, a rather simple explanation for the different responses of roots and shoots to gravity was widely accepted. This was based on the fact that auxin regulates extension growth. A different concentration of auxin on two sides of an organ would therefore produce different rates of growth on the two sides. As a result of gravity, auxin accumulates in the lower half of an organ placed horizontally. A relatively high concentration of auxin stimulates elongation in shoots but inhibits elongation in roots. Therefore, a shoot bends upwards and a root bends downwards. However, this simple explanation is not enough to explain all the facts, especially where roots are concerned.

When auxin content is bioassayed, it is found that changes in growth rate are greater than can be explained by auxin concentrations. So, further hypotheses have been proposed. The most important of these is that gravity is perceived by the movement within the cell of microscopic bodies of high specific gravity, which sink to the bottom of the cell under the effect of gravity when the orientation of an organ is changed. These bodies are called statoliths (literally 'position stones'). The **statolith theory** was first put forward in 1900. The theory proposed that statoliths, by moving to the bottom of a cell, 'polarize' the upper and lower parts of the cell. However, it is not known how the position of the statoliths brings about a redistribution of auxin. One suggestion is that as they settle at the lower side they displace metabolites and thus cause reduced metabolism. It is suggested that this, in turn, brings about increased auxin synthesis in, or preferential transport of auxin to, the lower side.

When the statolith theory was first proposed, it was suggested that starch grains acted as statoliths. The present view is that it is not starch grains but amyloplasts that carry out this function. However, the roots of certain seedlings lack any detectable statoliths but still react to gravity.

Apart from auxin, it is now thought that a redistribution of the growth inhibitor ABA, which is present in root caps, may be involved in geotropic response in roots.

Phototropism

Most shoots (including all coleoptiles) are positively phototropic. The primary roots of some seedlings are negatively phototropic but most roots do not respond to unilateral illumination. The leaves of most plants are diaphototropic.

As in geotropism, there are presentation times and reaction times in phototropism and the response is in proportion to the magnitude of the stimulus. At high light intensities the presentation time may be less than one second and reaction time may be only a few seconds. Again, it is the tips of organs that are the sites of light perception. In *Avena* sp. (Oat), it is the terminal 50 μm which is most sensitive. In epigeal seedlings, it is the hypocotyl which is the principal photoreceptor. Cotyledons also play a role in the phototropic response but seem to affect only the intensity of the response.

Since the work of F. W. Went, it has been known that unilateral illumination of coleoptiles of *Avena* sp. results in an asymmetrical distribution of auxin. Auxin accumulates on the side of the coleoptile away from the light. This causes faster cell elongation on the shaded side and the resulting differential growth makes the coleoptile bend towards the light. The unbalanced distribution of auxin was at first attributed to three processes, (a) more photodestruction of auxin on the lighted side, (b) increased auxin synthesis on the darker side, and (c) lateral transport of auxin from the lighted side to the dark side.

From the work of W. R. Briggs, it is now known that the important factor is the lateral transport of auxin. In experiments carried out more recently (Fig. 13.8), it has been shown that when radioactive ^{14}C-IAA is applied to a coleoptile of *Zea mays*, it is laterally transported away from the lighted side to the shaded side, thus confirming the conclusion of

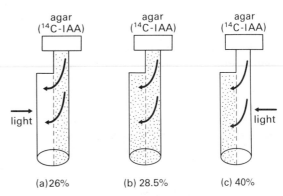

(a) 26%   (b) 28.5%   (c) 40%

13.8 Lateral transport of auxin in *Avena* sp. coleoptiles under different lighting conditions. Radioactive auxin (^{14}C-IAA) was applied to lengths of coleoptile from infused agar blocks. The total transport is indicated as a percentage of the total applied auxin moving from (a) the shaded (stippled) side to the illuminated (clear) side, (b) dark side to dark side (no light), and (c) the illuminated side to the shaded side. (After R. K. de la Fuente and A. C. Leopold, 1968. Lateral movement of auxin in phototropism, *Plant Physiol.* 43: 1031–36)

Briggs. It can be seen from the illustration that lateral transport of auxin from the lighted side to the shaded side is increased by unilateral illumination, while transport in the opposite direction is slightly inhibited.

How the absorption of light causes the distribution of auxin is uncertain. It has been suggested that light absorbed by carotenoids can make plastids take up a specific position in the cell relative to incident light. Chloroplasts certainly migrate in this way, in relation both to direction and intensity of light. Such movements may regulate the movement of auxin to the shaded side, rather as amyloplasts are thought to do, according to the statolith theory of geotropism.

As in the case of geotropism, the exact role of auxin and the possible involvement of other factors in bringing about a phototropic response are not yet fully understood.

Haptotropism

Haptotropism (thigmotropism) is characteristic of climbing plants. The attaching roots of *Culcasia scandens* and similar plants are **positively haptotropic**. When in contact with a tree trunk, for instance, they curve closely around the trunk as they grow. Roots of *C. scandens* and certain other Araceae also produce tough, root-hair like outgrowths where they are in contact with their support and it is these which attach them to the support. Such outgrowths are not formed on parts of the root which are not in contact with the support.

When not attached to any support, the growing tip of a twining plant does not grow straight upwards but moves round and round, describing a spiral as it grows upwards. This is not a response to any external stimulus but is an **autonomous movement**. The movement (**circumnutation**) may be clockwise or anticlockwise, according to the species concerned. When such a shoot comes into contact with a support, it is **positively haptotropic**. It grows round and round in a spiral tightly wound around its support in the same direction as its previous circumnutation.

Tendrils also are positively haptotropic. Rapid curvature of tendrils of *Passiflora* sp. or of many members of the Cucurbitaceae is easy to observe. Stimulation of the surface of a tendril close to the tip, for instance by stroking with a piece of wood, causes the tip of the tendril to curve almost immediately. In several Cucurbitaceae, continuous stimulation leads to the formation of a complete loop within five minutes. The formation of a second loop is completed within twelve to fifteen minutes of the commencement of stimulation. Under natural conditions, this would lead to permanent attachment of the tendril to its support. If stroking is stopped within a few minutes of the beginning of the experiment, the tendril again becomes straight. Stimulation with a liquid, for instance a jet of water, causes no curvature.

It is probable that the initial curvature of a tendril is not truly a growth movement but results from a rapid shortening of the epidermal cells of the tendril at the site of stimulation. However, growth on the convex side of a stimulated tendril subsequently increases to forty to two hundred times its previous rate, while growth on the concave side ceases altogether.

Tendril curvatures can be stimulated by the application of auxin but the extent to which auxin is involved in the natural curvature of tendrils is not known.

Hydrotropism

Hydrotropism is the growth of roots towards a region of high water content. It is doubtful, however, if moisture provides a stimulus to curvature compar-able to that provided by light direction or gravity. In fact, hydrotropism is difficult to distinguish from the greater growth of roots in regions of soil where adequate water is present, compared with that in dryer soil.

Chemotropism and aerotropism

In flowering plants, chemotropism and aerotropism are best seen in the growth of pollen tubes. If pollen grains (e.g. *Carica papaya*, Pawpaw) are germinated in water, under a coverslip, tubes tend to grow towards the centre of the coverslip, away from the edges. They are **negatively aerotropic**. If a small piece of stigmatic tissue is included with the pollen grains under the coverslip, they grow towards the piece of stigma, in response to a diffusion gradient of a substance contained in the stigmatic tissue. They are **positively chemotropic** in relation to the stigmatic secretion.

Nastic movements

Nastic movements are movements of parts of plants which are non-directional with respect to the stimuli which bring them about. They cannot be classified in entirely the same way as tropic movements, since they are non-directional, but are described in a similar manner (e.g. photonastic, thermonastic etc.).

Nyctinastic movements of leaves and floral parts, once known as sleep movements, probably combine photonasty and thermonasty. They are reactions to daily cycles of light intensity and temperature. The leaves of many plants droop slightly downwards from the leaf base as night begins to fall and rise upwards again in the morning. These movements are more obvious in some plants. In many *Cassia* spp., pinnae fold together and leaves hang down at night. In *Bauhinia* spp., the two halves of the leaf fold together, upwards, and the leaf hangs down. In *Mimosa pudica* (Sensitive Plant), which has bipinnate leaves, leaflets fold obliquely upwards, pinnae move upwards towards the apex of the leaf and leaves hang down. Reverse movements occur in the morning. The movements result from turgor changes in the pulvini at the bases of leaflets and leaves. Apart from the stimuli of light and temperature, nyctinastic movements of leaves include an autonomous element. It has been shown in some plants that sleep movements continue for some days after a plant is transferred to a continuously dark place and occur at approximately the same times as in plants left outside. Thus, plants have their own **autonomous** (**endogenous**) rhythms, usually geared to the diurnal (daily) cycle but not entirely dependent on it. Some flowers also show nyctinastic movements, closing in the evening and opening again the next morning.

Nyctinastic movements are brought about by the movement of potassium ions between the upper and lower sides of pulvini which result in turgor changes. The involvement of auxin cannot be ruled out.

A number of plants show rapid responses to touch (**haptonastic** movements), to shaking (**seismonastic** movements) or to injury (**traumatonastic** movements). In the case of the common tropical wasteland plant, *Mimosa pudica* (Sensitive Plant), any of these three types of stimulation causes the leaves to behave exactly as they do in the evening, as described above. In this species, the folding and collapse of a single leaf after stimulation takes only a few seconds. If the tip of a leaflet is very strongly stimulated, perhaps with a flame, the reaction spreads first throughout the whole leaf and then to successive leaves both up and down the stem. The mechanism by which the stimulus is transmitted has been the subject of much investigation. It is now known that the pathway is along series of specialized cells in the phloem and that it involves changes in the electric charges on cell membranes. After the collapse of several leaves following stimulation, all leaves and leaflets return to their normal positions in a period of between ten and fifteen minutes, at normal tropical temperatures.

Similar but slower movements occur in *Schrankia* sp., another member of the Mimosaceae rather similar in appearance to *M. pudica*. The common African weed *Biophytum petersianum* (Sensitive Legs) is also sensitive to touch or shaking, which cause it to fold its leaves upwards, rather slowly, after being stimulated.

The rapid folding of the leaves of the (American) insectivorous plant *Dionaea muscipula* (Venus Fly Trap) is haptonastic and occurs when contact is made with one or more of a row of hairs each side of the midrib on the upper side of the leaf. The inward curving of the tentacle-like appendages on the leaves of *Drosera* spp. is brought about by directional growth movements (haptotropism), not by nastic movements.

Flowering

Most plants undergo distinct phases of growth. A phase of vegetative growth follows germination of the seed. This is followed by a flowering phase. In annuals, these phases both occur in a single year. In biennials (e.g. some *Bryophyllum* spp.), vegetative growth takes place in the first year and the flowering phase follows the year after. Perennials, once established, usually have phases of vegetative growth and flowering each year, or each growing season, though not necessarily in that order.

The factor which stimulates a plant to flower, usually at a specific time of year, varies from species to species. In some annuals, flowering seems to occur when a plant has reached a certain size, or perhaps when it has achieved a certain nutritional status (has a sufficient photosynthetic income, or enough stored food). The stage at which a plant reaches the stage of 'ripeness to flower', before which it will not flower under any circumstances, varies a great deal. Some plants are able to flower at the seedling stage (e.g. *Pharbitis nil*; *Chenopodium* sp.), some must be at least a few weeks old (e.g. *Xanthium* sp.), while some trees must be at least several years old.

Once a plant has reached the stage of ripeness to flower, the time at which flower buds start to form may be affected by external factors. For instance, some annual plants start to flower when they are very small and behave as ephemerals if they have an insufficient supply of water, mineral nutrients, or light (e.g. *Zinnia* sp.; *Cosmia* sp.; *Zea mays*). In many plants, flowering is initiated only under specific conditions of light duration or temperature.

Photoperiodism

Plants are affected by light intensity and quality (colour), for instance in photosynthesis and in the etiolating effects of low light intensity. Many plants are also affected by variations in the relative lengths of light periods (days) and dark periods (nights) throughout the year.

In temperate parts of the world, there is very great variation in the relative lengths of day and night throughout the year. In the summer, there may be eighteen hours or more of daylight and only six hours of darkness in every twenty-four hours. In the winter, the position is reversed and days are very short. Many temperate plants are very sensitive to 'length of day' (**photoperiod**). Some plants start to flower after the end of the summer, when days are becoming short, or after the winter, while days are still short. Other plants start to flower only when the photoperiod is long, and flower in the middle of the summer. Plants which start to form flower buds only when they are subjected to a photoperiod of less than a critical maximum number of hours per day are **short day plants** (e.g. *Ipomoea batatas*, Sweet Potato; *Euphorbia pulcherrima*, Poinsettia). Plants which flower only when the photoperiod is longer than a certain critical minimum number of hours per day are **long day plants** (e.g. *Raphanus sativus*, Radish; *Beta vulgaris*, Beetroot). Plants in which flowering is not related to day length are **day-neutral plants** (e.g. *Cucumis sativus*, Cucumber; *Daucus carota*, Carrot). In addition to these three simple categories there are also species which flower only at times of intermediate day length. Other species flower only when short days are followed by longer days (spring) or when long days are followed by short days (autumn). Yet other species will flower under any day length but flower most profusely under either long days or short days.

In the tropics, there are no such extreme variations in length of day. As in temperate regions, variation depends on latitude. In parts of Nigeria, for instance, day length varies by only about one hour throughout the year. However, the flowering of some tropical plants is affected by day length. In some varieties of *Cajanus cajan* (Pigeon Pea), a difference in day length of only fifteen minutes is sufficient to determine whether or not a plant flowers.

The effects of photoperiod

The term 'short day plant' is, in fact, misleading. A short day plant can be prevented from flowering by artificially extending the length of day with electric light. The extra period of light need not be very intense. A 100 watt bulb at a distance of about 2 m is enough and this is much less than daylight. It is also found that red light is as effective as white light in preventing flowering. Also, a short day plant may be prevented from flowering during short days by interrupting the night with a short period of light, thus dividing the dark period into two parts. Here again, light need not be intense and red light is the most effective. Different plants need different lengths of illumination in the middle of the night to have this effect. Some require several minutes of light, others only a few seconds ('a flash'). Thus, 'short day plants' might better be called 'long night plants'.

Induction of flowering in long day plants during

a season when natural days are short can be initiated by artificially extending the day. As long as the intensity of light during the normal day is sufficient to meet the nutritional needs of the plant, only a low intensity of illumination is required to prolong the photoperiod. Again, red light is as effective as white light.

Plants vary a great deal in their sensitivity to day length and in their rate of response to exposure to a length of day which stimulates flowering. The American weed *Xanthium strumarium* (Cocklebur) is especially sensitive and, for this reason, has been favoured as experimental material. In this case, exposure to a single cycle of short day and long night causes the irreversible initiation of flowering, even if the plant is otherwise exposed to long days. Other short day and long day plants require several days of exposure to the appropriate photoperiod before flowering is irreversibly initiated.

After exposure to a suitable day length, it takes several days for structural evidence of the formation of flower primordia to appear.

Light, phytochrome and florigen

The action spectrum of light with respect to photo-periodism (Fig. 13.9) is similar to that which induces germination in light-sensitive seeds (Chapter 2). Red light (wavelength 660 nm) is most effective. As in the case of light-sensitive seeds, the effects of red

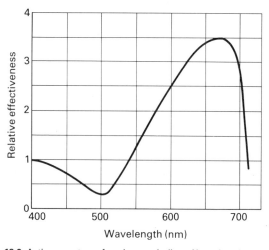

13.9 Action spectrum for photoperiodism. Note that the peak corresponds to the main peak in the absorption spectrum of phytochrome

light can be reversed by exposure to far-red 'light' (wavelength 730 nm). Plants of *Xanthium* sp. exposed to a single short day, followed by a long dark period, are stimulated to flower. If the night is interrupted by a flash of red light, they do not flower. If, however, the flash of red light is followed by a flash of far-red, flowering is again induced. If alternating flashes of red and far-red are used, the plant flowers if the last flash is far-red and fails to flower if the last flash is red.

This reversible red/far-red reaction is characteristic of the pigment **phytochrome** and is consistent with the involvement of phytochrome in the plant's perception of the day-length stimulus.

Many other experiments have been carried out using *Xanthium* sp. Some of the experimental observations are:

(a) Exposure of a single leaf to short days, while the rest of the plant is kept under long day conditions, is enough to induce flowering.
(b) If a single leaf from a plant exposed to short days is grafted onto a plant exposed only to long days, flowering is induced in the plant which receives the graft.
(c) If all the leaves are removed from a plant immediately after the plant has been exposed to a short day sequence, flowering does not follow.
(d) If a plant is defoliated before being exposed to a short day sequence, flowering does not follow.

These and other experimental observations all lead us to the conclusion that the stimulus is received in the leaves and transmitted to shoot apices by a hormone.

As long ago as 1936, M. C. Chailakhyan of Russia proposed the existence of a flowering hormone and also proposed the name **florigen** for the hormone. Since that time, much work has been done to isolate and identify the hormone, so far with only limited success. In this respect, we may give the work of R. G. Lincoln and his associates at Long Beach State College, and K. C. Hamner and his associates at the University of California as examples. Both groups of workers obtained extracts from leaves and buds of flowering *Xanthium* sp. These extracts were found to stimulate flowering in the short day plant *Lemna perpusilla* while growing under long day conditions. They also produced limited induction of flowering in vegetative plants of *Xanthium* sp. exposed to long day conditions, if gibberellic acid was also applied.

The yet unidentified factor or factors in the extract were given the name **florigenic acid**.

Flower initiation is certainly controlled by hormones, but it is possible that both flowering-inducing florigens and flowering inhibitors may be involved. It is also probable that other hormones (e.g. gibberellic acid) are involved.

Temperature and flowering

In some plants, temperature interacts with photoperiod in the induction of flowering. Low temperatures tend to delay flowering, or to reduce the sensitivity of a plant to a photoperiod that would immediately induce flowering at a higher temperature.

Temperature also has more direct effects. In *Oryza sativa* (Rice), flowering does not occur until a plant has been exposed to a temperature of 24°C or above, within the physiological range of temperatures. However, the most striking demonstration of a temperature effect on flowering is the requirement of many temperate biennial plants for exposure to low temperatures before they can start a flowering cycle. This effect is known as **vernalization**. In this case, the stimulus is received in the apical bud, although flower primordia may not appear until a long time after the stimulus is received.

The effect of exposure to low temperatures on flowering in several biennial varieties of cereals (e.g. varieties of: *Avena* sp., Oats; *Secale* sp., Rye; *Triticum* sp., Wheat) have been the subject of much investigation. Normally, these plants make only vegetative growth in their first season and flower in their second growing season, after they have lived through the winter. If, however, they are exposed to temperatures around or just above 0°C during their first year of growth, they will flower and fruit in the first season.

Annual varieties of these cereals, which are planted in the spring and produce a crop the same year, are not as productive as biennial varieties. This observation led to research to find treatments by which biennial varieties could be stimulated to produce their crop in the first season. To be commercially useful, such treatments needed to be applicable on a large scale. The best known and probably the most successful method is that of T. D. Lysenko of Russia, whose process was subsequently applied commercially.

In the vernalization of Winter Wheat, seed grain was first moistened, to initiate germination, and was almost immediately exposed to a temperature of about 0°C for a short time, simply by opening the sides of the building in which it was contained and exposing it to the winter temperature outside. The partially germinated seed grain was then kept at a temperature low enough to inhibit further growth. When warmer weather allowed, the grain could then be sown in the normal way and subsequently produced its crop in one growing season.

It is clear that some kind of chemical change must take place as a result of vernalization. As with 'florigen', hormonal processes are suspected, but have not been proved to exist.

Growth, environment and stress

The growth of a plant is governed by both internal (hereditary) factors and the total factors present in the environment. Environmental factors which affect growth include the availability of (a) water, (b) carbon dioxide, (c) oxygen and (d) mineral nutrients. Other factors are temperature and light. Water is both a medium and a metabolite and water supply has very far-reaching effects. Favourable water balance within the plant depends not only on the presence of water in the soil, but also on the salinity of the soil water and on the relative humidity of the air. In the case of light, the intensity, quality and photoperiod are all important. All these factors, which act directly on the plant, are themselves governed by more general factors such as latitude, altitude and, in the case of the availability of soil water and mineral nutrients, by the structure and overall fertility of the soil.

Optimal growth and development of a plant can occur only if all the environmental factors are themselves optimal for the species of plant concerned. With this knowledge, it can be seen that a very large proportion of the ecological problems in tropical Africa and elsewhere are, in truth, problems in plant physiology.

Stress

If any essential factor is in short supply or in excess, a stress is imposed on the plant. Thus, for example, stress can arise from the presence of too much water in the soil or from a shortage of water in the soil. The same is true for mineral nutrients. Stress can also be caused by the presence of substances which

the plant does not require (e.g. pollutants) in the soil or in the atmosphere.

In all parts of the world, some of the most important stresses to which plants are subjected are associated with seasons. In a temperate climate, the most obvious seasonal factor is temperature. High temperatures produce stress in summer and low temperatures produce stress in winter. In our tropical climate, the most obvious seasonal variation is in water supply rather than temperature. In West Africa, there is a wet season which extends, approximately, from May to October, with the heaviest rainfall in July. There is little or no rain in the dry season, which extends throughout the rest of the year. Moisture stress, excess and deficiency respectively, is associated with this seasonality. Less obvious but still important is seasonality in respect of light energy. This is illustrated in Fig. 13.10, which shows the results of an experiment carried out on *Helianthus annuus* (Sunflower). Here, growth in terms of dry weight and total leaf area correlates well with sunshine hours, when water is available. If the

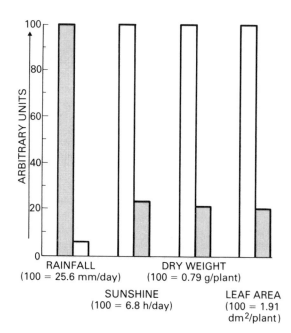

13.10 Meteorological and growth parameters during the wet season (shaded) and dry season (unshaded) for Sunflower grown in Freetown, Sierra Leone. The lower histogram for each function is expressed as a percentage of the higher one, for which the absolute value is indicated in brackets. Adapted from Eze (1973), *Ann. Bot.* 37: 315–29

plants had not been watered, their growth would probably have correlated equally closely with rainfall, as long as light energy was non-limiting.

Dust haze in the atmosphere can reduce the energy of light reaching the surface of the earth, in the same way as cloud cover. The very dry north-easterly harmattan wind carries much dust to West Africa seasonally, between November and February each year, and thus both reduces light intensity and increases the desiccating power of the air.

In the course of evolution, plants have developed various means of accommodating or resisting stresses of different kinds. These may be broadly classified into three categories, as set out below.

Stress avoidance is essentially an escape mechanism, in which active growth practically ceases while the stress is acting. For example, most deciduous trees shed their leaves at the end of the wet season and produce new leaves at, or just before, the beginning of the next wet season. Perennial herbs mostly die back to ground level, leaving their living underground perennating organs to produce new aerial growth when the next wet season starts. The vegetative parts of annuals die completely at the beginning of the dry season and only the seeds, which are drought resistant, remain alive during the unfavourable season.

Stress amelioration invariably involves adaptive modifications which protect metabolic processes from the effects of stress and allow the plant to grow normally under stressful environmental conditions. Thus, cacti and other succulent plants store water in the colloid-rich cells of their extensive parenchyma tissues. Other plants which normally tolerate dry conditions have deep, wide-spreading root systems which ensure the continuous availability of water, or one or more other modifications which reduce transpiration. At the other end of the scale, hydrophytes have modifications (e.g. large air channels in their parenchyma) which compensate for the stress of total submergence in water.

Stress tolerance includes cases in which physiological processes are adapted to operate under stress, or are temporarily more or less suspended under stressful conditions, with no apparent after-effects when the stress is removed. This is commonest in organisms with a simple structure, with little structural basis for amelioration mechanisms. However, similar mechanisms are found in some more complex plants. Some blue-green algae apparently thrive in hot springs and others grow in icy waters. In either case, temperatures are outside the normal range for most plants. Plants which are capable of undergoing a high degree of desiccation and subsequently making a total recovery after water becomes available include algae belonging to several divisions (e.g. *Scytonema* spp., blue-green algae), many mosses (e.g. *Bryum* spp.), many epiphytic liverworts (e.g. *Frullania* spp.), the sporophytes of some ferns (e.g. *Trichomanes* spp., Filmy Fern; *Pellaea* spp.) and a few flowering plants (e.g. *Anastatica* sp., the Resurrection Plant of the Middle East).

The three categories described above are not mutually exclusive, but commonly occur in combination. The avoidance mechanism has serious disadvantages; for instance, stress may be so severe that it damages the dormant stage (e.g. seed). Some degree of resistance combined with avoidance extends the growing season and thus increases overall productivity and makes survival more certain. Similarly, stress tolerant vascular plants (ferns, angiosperms) also have xeromorphic characters which greatly reduce transpiration and therefore ameliorate the stress and defer the time when stress tolerance must come into operation.

Key words

Growth and development: morphogenesis. **Growth:** growth rate, grand curve of growth (grand phase of growth), exponential growth, senescence. **Parameters of growth**: height (length), auxanometer; volume; area; fresh weight, dry weight. **Growth substance:** phytohormone (plant hormone), bioassay; auxin (IAA), extension growth, apical dominance, polar movement, auxin gradient; gibberellin (gibberellic acid, GA), bolting, non-polar movement; cytokinin, cell division, callus; abscisin (abscisic acid, ABA), abscission; ethene, epinasty. **Tropism**: directional stimulus; orthotropic, plagiotropic, diatropic; presentation time, reaction time; geotropism, statolith, centrifuge, klinostat; phototropism, photoreceptor; haptotropism (thigmotropism); aerotropism; chemotropism; traumatotropism. **Nastic movements**: nyctinastic, photonastic, thermonastic, haptonastic, traumatonastic. **Autonomous movements**: circumnutation. **Flowering**: photoperiodism, photoperiod; short day plant, long day plant, day-neutral plant; phytochrome, 'florigen', florigenic acid; temperature, vernalization. **Stress:** stress avoidance; stress amelioration; stress tolerance.

Summary

Growth can be measured in a number of ways including comparative measurements of height (length), area (of leaves), fresh weight and dry weight. Whatever valid parameter is used, a graph of the size of an organ or organism against time gives an S-shaped curve which shows a very rapid increase at first, followed by a period of arithmetical growth and finally a period of decreasing growth rate as the organ or organism approaches its maximum size or becomes senescent. Growth is stimulated, inhibited or regulated by growth substances. Each of the known growth substances has specific effects and growth substances also interact to produce other effects. Auxins promote or inhibit elongation of organs, according to their concentration and the nature of the organ. Gibberellins promote elongation. Cytokinins promote cell division. Abscisic acid inhibits growth. Ethene speeds senescence. Unequal distribution of auxin in a shoot or root is responsible for growth curvatures such as those produced in phototropic and geotropic responses to unidirectional stimulation. In addition to phototropism and geotropism, parts of some plants make tropic responses to other stimuli, for example, to touch or rubbing. Parts of some plants also respond to certain stimuli by making nastic movements, in which the direction of the response is not related to the point of origin of the stimulus. Many nastic movements are brought about by turgidity changes rather than growth, but some involve growth. Flowering is induced by both internal and external factors. Many plants initiate flowering only when exposed to the right length of day. These are short day plants and long day plants. Others require exposure to a high enough, or low enough temperature before they will flower. Most plants are subjected to environmental stresses, including seasonal stresses. Many have mechanisms by which they avoid stress, lessen the effects of stress or tolerate stress without damage.

Questions

1 Discuss the relative advantages and disadvantages of using each of the following parameters to measure growth: (a) fresh weight; (b) dry weight; (c) height; (d) leaf area.

2 What is auxin? Describe how auxin is involved in (a) geotropism, (b) phototropism, and (c) apical dominance.

3 Distinguish between tropic responses and nastic responses. How do haptotropic responses differ from haptonastic responses? Give examples.

4 Select and name one savanna plant (herb, shrub or tree), one plant which grows in fresh water and one Mangrove. List the structural and other modifications shown by each. Show how each adaptation may help to avoid or ameliorate the particular stresses to which each plant is subjected in its natural environment.

5 With reference to the contents of this chapter and previous chapters, write an essay on 'Light and Plant Growth'.

14

Reproduction in flowering plants

Reproduction is the production of new individuals of a species. There are two types of reproduction, sexual reproduction and asexual reproduction. Sexual reproduction involves the fusion of two reproductive cells (**gametes**). Asexual reproduction does not involve the fusion of gametes.

Asexual reproduction

Many algae and fungi produce asexual spores. A **spore** is a unicellular reproductive body which can germinate and grow into a new plant. Such spores are formed by ordinary cell division (involving mitosis) from some part of the body of the plant. Flowering plants do not produce this type of spore. However, many flowering plants have the type of asexual reproduction known as vegetative reproduction. **Vegetative reproduction** is the production of new, independent individuals by growth, followed by the separation of newly formed parts of the plant. Structures responsible for vegetative reproduction in flowering plants are described in Chapter 4 (Modified Organs).

Vegetative reproduction is important because it enables a plant to reproduce quickly and to colonize ground that might otherwise become unavailable due to colonization by other species.

Vegetative reproduction results in the formation of **clones**, in which all the plants have exactly the same genetic constitution. It does not allow for the exchange of genetic material between plants and subsequent variation and adaptation to a changing environment.

Sexual reproduction

Sexual reproduction takes place in flowers.

A **flower** is a modified shoot. It develops from an apical meristem like that of an ordinary shoot, but the cells of the inner part of the meristem differentiate into permanent tissues at a very early stage. Longitudinal growth therefore ceases and further development is confined to the surface layers, from which appendages are formed. The appendages are all modified leaves.

The end of the stem, bearing the modified leaves, forms the **receptacle** of the flower.

A typical flower has four types of appendages (Fig. 14.1). The outermost appendages are the **sepals**, which together make up the **calyx**. Within the calyx are the **petals**, forming the **corolla**. Calyx and corolla together constitute the **perianth** of the flower. Within the petals are the **stamens**, which make up the **androecium**. At the centre of the flower, often at the end of the receptacle, are one or more **carpels**, forming the **gynaecium**.

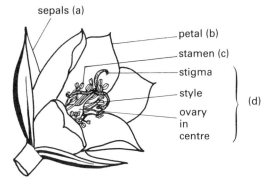

sepals (a)

petal (b)

stamen (c)

stigma

style

ovary in centre

(d)

14.1 A typical flower, *Talinum triangulare* (see also cover photograph): (a) calyx a whorl of two sepals; (b) corolla a whorl of five free petals; (c) androecium a whorl of many free stamens; (d) gynaecium of three fused carpels, with one style and three stigmas

In some primitive flowers, all the floral parts are attached to the receptacle in a continuous spiral (e.g. *Nymphaea* sp., Water Lily, Figs. 14.2 and 15.1). In more advanced flowers, some or all of the floral parts are in four or more separate **whorls** (e.g. *Talinum triangulare*).

Floral parts as modified leaves

Sepals, which have the function of protecting the inner parts of the flower, are quite leaf-like in structure. In some plants, one of the sepals of each flower

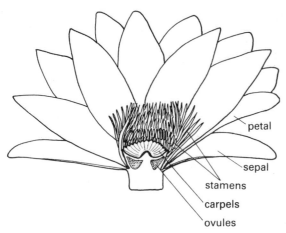

14.2 A primitive flower with many parts arranged in a spiral. Half flower of *Nymphaea* sp.

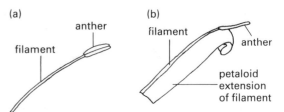

14.3 Stamens: (a) typical form; (b) petaloid stamen

regularly develops into an almost normal leaf (e.g. *Mussaenda* sp., Fig. 15.25). Sepals may also develop leaf blades at their tips, as abnormalities. This may happen, for instance, when a plant has been affected by a small dose of hormone weedkiller.

Petals are attractive in function. Their structure is less leaf-like than that of sepals but, in some primitive flowers, organs intermediate in structure between sepals and petals may occur.

A typical stamen is not leaf-like. It usually has a thin stalk (**filament**) with a lobed, pollen-producing **anther** at its tip. However, some flowers produce organs which are intermediate in structure between petals and stamens (Fig. 14.3) and, in some cultivated garden plants, some of the stamens are replaced by petal-like structure, to produce so-called 'double' flowers (e.g. *Hibiscus rosa-sinensis*, Hibiscus).

The gynaecium of a flower consists of one or more carpels. A **carpel** is a fertile leaf, bearing one or more ovules close to each margin, on the adaxial surface. Ovules are attached along **placentas**, which may or may not be swollen.

When the gynaecium is composed of only one carpel (e.g. *Caesalpinia pulcherrima*, Pride of Barbados, and other Leguminosae), or when it is composed of several separate carpels (e.g *Clematopsis* sp.), each carpel may be regarded as a modified leaf which is folded along the midrib with the edges fused together. The ovules are attached to almost fused placentas and are then attached in two rows, along one side of the carpel. The hypothetical origin

of this type of carpel is shown in Figure 14.4. This interpretation of the origin of this type of carpel is supported by the existence of primitive families of flowering plants in which the carpel is folded but fusion does not take place, so that the carpel is open between the placentas. Families with this type of primitive structure (e.g. Degeneriaceae) occur in countries around the margin of the Pacific Ocean but not in Africa.

The tip of the carpel is usually extended into a more or less elongated **style**. At the tip of the style is the **stigma**, on which pollen grains are received. The lower part of the carpel, containing ovules, is the **ovary**.

The ovary of the carpel described has only one cavity (**loculus**) and is described as **unilocular**. It is also said to have **marginal placentation**, because the placentas are at one side of the ovary.

In many plants with more than one carpel in the flower the carpels are fused together, at least in the region of the ovary. In such cases, the manner in which they have fused, and subsequent changes, affect the number of loculi present and the positions of the placentas within the ovary.

When there are two or more carpels, which may be supposed to have folded before they fused, the ovary has as many loculi as carpels (e.g. *Crinum* sp., Crinum Lily; *Solanum* spp.). In these cases, the ovules are attached at the centre of the ovary and the ovary has **axile placentation**. In some plants, the partitions between the loculi may break down leaving the placentas attached only at the bottom of the ovary. Such an ovary is unilocular and has **free-central placentation** (e.g. *Talinum triangulare*). Where the carpels have fused by their edges without folding, the resulting ovary is unilocular and has **parietal placentation** (e.g. *Passiflora* sp., Passion Fruit).

The term **pistil** is sometimes used to describe the

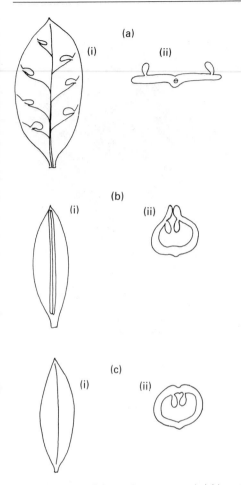

mentioned so far, it is attached above the level of the other parts of the flower. In these cases, the ovary is **superior**. In many plants, however, the receptacle grows up around the ovary during development, so that the ovary is to a greater or lesser extent embedded in and fused with the receptacle. The other floral parts are attached above the ovary. Here, the ovary is **inferior**.

Variations in ovary structure and placentation are illustrated in Fig. 14.5. Other variations in the structure and position of the ovary and of flower structure as a whole are dealt with in the next chapter.

(a) unilocular marginal   (b) unilocular free-central

(i) TS   (ii) LS   impression from ovule on other side of carpel   (i) TS   ovary wall of three carpels

ovule attached at margin

placenta

one loculus

central placenta

(ii) LS   ovules

one loculus

single folded carpel

placenta

ovules borne alternately on margin (sides) of the carpel

(c) unilocular parietal TS

ovary of five carpels

line of fusion between carpels

one loculus

ovules on placenta where carpels meet, later develop over whole inner surface

14.5 Ovary structure showing different types of placentation: (a) *Caesalpinia* sp. (b) *Talinum* sp. (c) *Carica* sp. (continued overleaf)

14.4 The origin of the angiosperm carpel: (a) hypothetical megasporophyll showing two rows of ovules close to the margins on the adaxial side, (i) from above, (ii) in section. (b) Intermediate condition in which the megasporophyll is folded, with its edges in contact but not fused, (i) whole, (ii) in section. This is similar to the condition which exists in some living, primitive angiosperms. (c) Margins fused so that the ovules are completely enclosed, (i) whole, (ii) in section. Apart from the omission of the style and stigma, this condition exists in legumes and many other angiosperms

unit from which a gynaecium is made up. The pistil comprises an ovary, style and stigma. Thus a flower with a single carpel, or with several carpels with the ovaries fused together, has a single pistil. A flower with several separate carpels (e.g. *Clematopsis* sp.) has several pistils.

As has been stated, the ovary of a flower is formed at the tip of the receptacle. In the examples

(d) trilocular axile
  TS

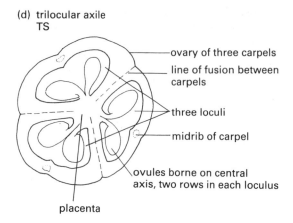

ovary of three carpels

line of fusion between carpels

three loculi

midrib of carpel

ovules borne on central axis, two rows in each loculus

placenta

(e) multilocular axile
  TS

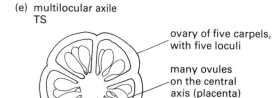

ovary of five carpels, with five loculi

many ovules on the central axis (placenta)

14.5 (continued) (d) *Gloriosa* sp.; (e) *Hibiscus esculentus*

(a)

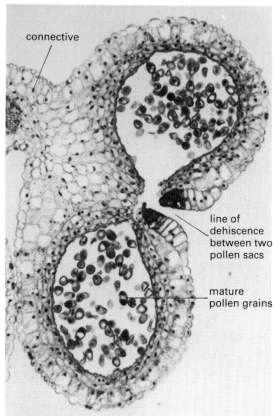

connective

line of dehiscence between two pollen sacs

mature pollen grains

Stamens and pollen

The stamen is the organ which, indirectly, supplies the male gametes of the plant. It is not itself an organ of sexual reproduction, but contains sporangia which produce microspores (pollen grains) which later produce sperms.

The anther of a typical stamen contains two pairs of microsporangia (**pollen sacs**), one pair each side of a central **connective** (Fig. 14.6). Each of the pollen sacs is lined with a specialized layer of cells forming the **tapetum**. The tapetal cells later break down and provide nutritive materials for the developing pollen grains. In the young anther, the tapetum surrounds a mass of **pollen mother cells**. Like the rest of the flowering plant, pollen mother cells are diploid. The nucleus of a diploid cell contains two identical sets of chromosomes. At a certain stage, each pollen mother cell undergoes the type of nuclear division known as **meiosis** (Chapter 25), in which the chromosome number is halved (becomes **haploid**) and four cells are produced. Each

pollen mother cell produces four haploid pollen grains. As the pollen grains continue to develop, they are nourished by materials from the tapetum, which has broken down.

Each pollen grain initially has a thin cellulose wall. Subsequently, a second wall is laid down. This outer wall is the **exine**. The inner wall is the **intine**. The exine contains a material called **sporopollenin**, which is extremely resistant to decay and protects the pollen grain from desiccation. The exine is often 'sculptured' – its surface is irregular in thickness and, in surface view, has a pattern that is characteristic of the species (Fig. 14.7).

As the pollen grain matures, its nucleus divides once (mitosis). One of the nuclei produced lies at one side of the cell, surrounded by cytoplasm and a thin cell wall. This forms the **generative cell**. The other nucleus lies free in the cytoplasm of the pollen grain. This is the **tube nucleus** (Fig. 14.7). In grasses, the generative cell divides again (mitosis) to form two

(b)

pollen mother cells ⎫ pollen
tapetum   ⎬ sac
                     ⎭

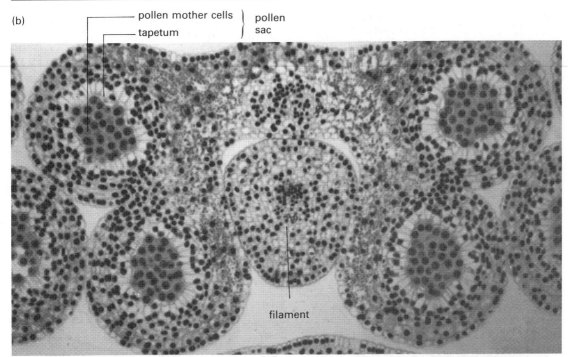

filament

14.6 Anther structure; (a) TS half anther at time of
dehiscence; (b) TS young stage showing four
microsporangia (pollen sacs) containing pollen mother cells

sperms. In other plants, formation of sperms is
delayed until after the pollen has been released.

Pollen is released by the dehiscence of the anther.
The anther usually splits down each side, by slits
formed between neighbouring pollen sacs.

Ovules

An ovule starts to develop as a small mound of tissue
on the placenta. This mound of tissue is the **nucellus**
(**megasporangium**). At the same time, one or two
rings of tissue bulge from the surface of the placenta,
encircling the nucellus. In the simplest case, the
nucellus increases in size and the two rings of tissue
keep pace with it, so that the ovule consists of the
nucellus, surrounded by two integuments, with a
small pore, the **micropyle**, at the end farthest from
the placenta (Fig. 14.8).

This type of ovule, the **orthotropous** ovule, is
uncommon but occurs, for instance in members of
the Polygonaceae and members of the Commelinaceae. More commonly, the ovule becomes inverted
during development to form an **anatropous** ovule,

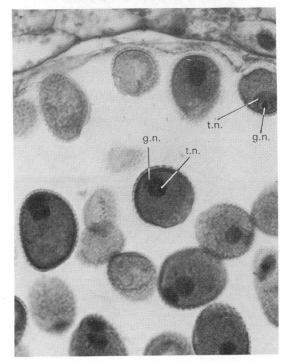

t.n.

g.n.

g.n.

t.n.

14.7 Mature pollen grains showing large, round tube
nucleus (t.n.) and smaller generative nucleus (g.n.) (HP)

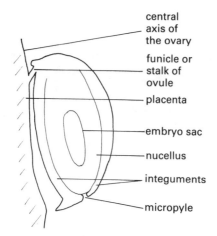

central axis of the ovary

funicle or stalk of ovule

placenta

embryo sac

nucellus

integuments

micropyle

14.8 Orthotropous ovule of *Oxalis* sp. (see also Fig. 14.18)

in which the micropyle faces downwards towards the placenta (Fig. 14.9).

While the ovule is still small, one cell of the nucellus becomes much larger than the rest. This is the **megaspore mother cell**. This cell divides by meiosis to form a row of four haploid cells. Development from this stage varies in different groups of plants, but in the commonest case three of the four cells disintegrate and one haploid **megaspore** remains (Fig. 14.10).

As the megaspore, nucellus and integuments increase in size, the nucleus of the megaspore divides three times (mitosis) and what was the megaspore becomes the **embryo sac**. At first, the embryo sac is a large, highly vacuolate cell containing eight nuclei, which are evenly distributed throughout the cell. The nuclei then migrate within the embryo sac

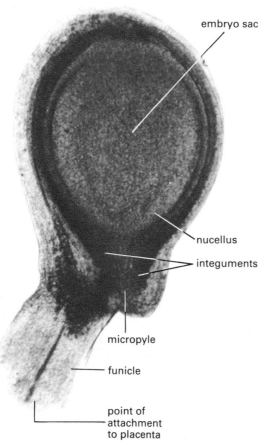

embryo sac

nucellus

integuments

micropyle

funicle

point of attachment to placenta

14.9 Anatropous ovule of *Carica papaya*, made semi-transparent by warming in lactophenol (LP)

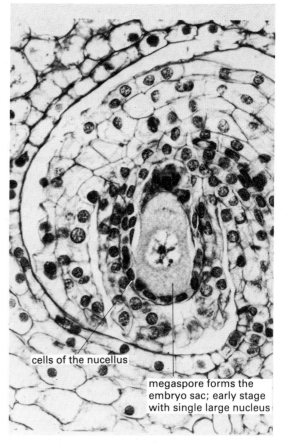

cells of the nucellus

megaspore forms the embryo sac; early stage with single large nucleus

14.10 Developing ovule in LS, from TS ovary (LP)

and take up characteristic positions. Three migrate to the micropylar end of the embryo sac and form separate cells, with thin cell walls. The middle one of these is the **ovum**. The other two are the **synergids**. Similarly, three nuclei migrate to the far end of the embryo sac, where they form the **antipodal cells**. The remaining two nuclei come to lie at the centre of the embryo sac and do not form cell walls. These are the **polar nuclei**. The polar nuclei sometimes fuse at this time or fusion may be delayed for some time. If fusion occurs at this time, the resulting diploid nucleus is the **central nucleus**.

The mature embryo sac (Fig. 14.11) is enclosed in a usually thin layer of nucellus, which is enclosed in the integuments, except where the nucellus is exposed at the micropyle. The ovule is attached to the placenta by a stalk, the **funicle**. There may be two integuments all round the nucellus or there may be only one at the side of the ovule next to the attachment of the funicle. The funicle of the anatropous ovule continues up the side of the ovule as the **raphe**. The vascular supply of the ovule travels up the funicle into the raphe and branches at the top (**chalazal** end of the ovule), at the base of the nucellus.

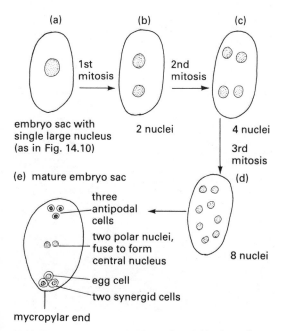

14.11 Diagrams to show divisions of nuclei in the embryo sac, up to the mature eight-nucleate condition

Pollination

Pollination is the transfer of pollen grains from the stamens of a plant to the stigma of, usually, a different plant.

Plants are pollinated by many different agencies, including wind, water and animals. The type of pollination mechanism is reflected in the structure and other characteristics of the flower.

In the artillery plant (*Pilea* sp.) and other members of the Urticaceae (e.g. *Fleurya* spp.), the filaments of staminate flowers are curved inwards at first, but finally spring outwards and discharge all their pollen at the same time. Pollen is thrown violently from the anthers and is then carried away by air currents.

Pollination by water occurs in some water plants. In *Ceratophyllum* spp., each plant has either flowers with stamens only (staminate or 'male' flowers) or flowers with pistils only (pistillate or 'female' flowers). The flowers are all formed under water. Pollen grains are released into the water and carried away by water currents. Pollination occurs if pollen grains come into contact with the long, sticky stigmas of pistillate flowers.

Valisneria spiralis is another water plant with separate staminate and pistillate individuals. Pistillate flowers have long stalks and float on the surface of the water. Staminate flowers are formed under water but break away and float to the surface when they are mature. Pollination occurs when the anthers of a staminate flower make contact with the long, protruding stigmas of a pistillate flower.

Pollination by animals is almost always associated with colourful, showy flowers. The flowers are often scented and produce nectar, which animals collect. **Nectar** is a sugary secretion formed by nectaries, which are usually located within the base of the flower. Flowers which do not have nectaries are visited by insects which collect pollen.

Some plants with relatively small and inconspicuous flowers are made more attractive by the development of large coloured bracts in the neighbourhood of the flowers (e.g. *Euphorbia pulcherrima*, Poinsettia, *Bougainvillaea* sp.) or by the enlargement and bright colouration of one or more sepals (e.g. *Mussaenda erythrophylla*).

Animals known to be involved in pollination include insects (bees, wasps, flies, butterflies, moths and beetles), gastropods (snails and slugs), birds (e.g. sun birds) and mammals (bats).

Bee pollinated flowers are usually brightly coloured and medium sized, or smaller and in showy inflorescences. The flowers are often scented and usually have nectaries. Especially when nectar is absent, they have much pollen. Both pollen and nectaries are easily accessible to the bee, either because the flower is wide open or because it is shaped in such a way that a bee has easy access (e.g. *Sesamum* sp., Sesame). Anthers and stigma are so placed that they will touch the bee's body or mouthparts when it visits the flower.

In some flowers which have no nectaries, some of the stamens (fodder stamens) are greatly enlarged and so placed that they attract the bee to the right part of the flower (e.g. some *Cassia* spp.).

Butterfly pollinated flowers have long, tubular corollas. They are usually brightly coloured, scented, and are open during the day (e.g. *Lantana camara*). Butterflies have very long mouthparts which are normally rolled up into a spiral but can be extended to a length of several cm and can thus reach nectar formed in the innermost part of a flower. Butterflies feed during the day.

Moth pollinated flowers are similar to butterfly pollinated flowers, except that those pollinated by night-flying moths are usually paler in colour, have a stronger scent and often open in the evening (Fig. 14.12b). Moths mostly feed at night. Some moth pollinated flowers open at dusk and shrivel soon after dawn the following day (e.g. *Ipomoea bona-nox*).

Flies are attracted to what, to us, is the unpleasant smell of rotting meat. *Aristolochia* spp. and their forest relative *Pararistolochia goldieana* are pollinated by flies and have such a smell. Flies are attracted and enter the flowers. Escape is hindered by backward-pointing hairs inside the base of the corolla. Flies may bring pollen from another flower, in which case, they effect pollination. After the anthers have dehisced, the flies escape as the flower decays. A similar mechanism exists in members of the Araceae (e.g. *Amorphophallus* sp; *Anchomanes* sp.). Some members of the Asclepiadaceae also have evil-smelling flowers (*Caralluma* sp.; *Stapelia* sp., Carrion Flower).

Wasp pollination occurs in *Ficus* spp. (Figs). Figs have many small flowers of separate sexes arranged inside a nearly spherical, hollow inflorescence (Fig. 14.13). There is a small opening at the end of the inflorescence. A species of wasp lays its eggs in the inflorescence and the larvae eat flowers and parts

(a)

(b)

14.12 Insect pollination: (a) butterfly on a flower of *Ipomoea involucrata*, midday; (b) moth feeding from flower of *Carica papaya*, early evening

of the fleshy wall of the inflorescence. Adult insects eat their way out covered with pollen and carry out pollination when they enter another inflorescence to lay eggs.

Bird pollinated flowers are large and showy and produce much nectar. Flowers of *Hibiscus rosa-sinensis* (Hibiscus) are visited for their nectar by humming birds in South America and by sun birds in Africa.

Bat pollinated flowers or inflorescences are generally showy, often pendulous and produce much nectar (e.g. *Adansonia digitata*, Baobab). Others have pendulous inflorescences made up of many small, nectar-containing flowers (e.g. *Parkia* sp.; *Chlorophora excelsa*, Iroko). The bats involved are fruit bats, which visit the trees at dusk and during the night. In their visits to the Iroko tree, bats bite pieces from the pendulous inflorescences of small, scented flowers but pollinate many of those which they do not destroy.

14.13 Insect pollination: inflorescences of *Ficus* sp., one cut open to show the fleshy receptacle and small flowers within. Insects (wasps) enter by the small opening at the apex of the receptacle (arrowed)

Wind pollinated flowers are usually small and inconspicuous. The corolla is absent or very much reduced and both styles and anthers hang freely outside the rest of the flower. Pollen grains are dry and small and are produced in very large quantities. Plants which are pollinated by wind often have unisexual flowers with male and female flowers in separate inflorescences (e.g. *Acalypha* spp., Fig. 14.15). Grasses are all wind pollinated.

Cross-pollination

The particular importance of sexual reproduction is that it allows for the exchange of genetic material between different individuals of a species. It thus leads to the formation of new combinations of genetic material. The resultant variability makes it possible for a species to adapt to new or changing conditions. However, these advantages depend largely on cross-fertilization. In flowering plants, cross-fertilization depends on cross-pollination. There are several mechanisms which ensure that cross-pollination is at least preferred to self-pollination.

Some plants are dioecious and have staminate and pistillate flowers on different individuals (e.g. short-fruited varieties of *Carica papaya*, Pawpaw). Here, cross-pollination is assured. Other plants are monoecious and have separate staminate and pistillate flowers, usually in separate inflorescences, on the same individual (e.g. long-fruited varieties of Pawpaw; *Acalypha* spp., Fig. 14.14; *Ricinus* sp.). In such cases, staminate and pistillate flowers on the same plant often mature at different times.

In most flowers which contain both stamens and pistils, stamens do not shed pollen at the same time as the stigma is ready to receive pollen. The flowers are either **protandrous**, with the stamens ripening first, or **protogynous**, with the stigma ready to receive pollen before the stamens are ripe. These mechanisms ensure that a flower cannot be self-pollinated, or give preference to pollen brought from another flower. Protandry and protogyny are found in flowers pollinated by all sorts of agencies.

Heterostyly is found in a number of plant genera. For example, the flowers of some individuals of *Bauhinia monandra* have long styles, roughly equalling in length the single stamen. Other individuals have flowers in which the style is very short, while the stamen is as long as in the other variety. Heterostyly is also found in the garden flower known as Californian Poppy (*Eschscholtzia californica*).

Another mechanism which ensures cross-fertilization, if not cross-pollination, is **self-incompatibility**. In *Mangifera indica* (Mango), for example, pollen produced by one tree will not develop, or develops very slowly, on the stigma of a flower of the same tree. Self-fertilization is therefore impossible.

(a)

(b)

14.14 Monoecious plants with separate staminate ('male') and pistillate ('female') flowers: (a) *Acalypha* sp., with long, catkin-like male inflorescence (right) and dark red stigmas of a female inflorescence (left) at the apex; (b) *Euphorbia* sp., with a single, large female flower at the centre of the flower-like inflorescence (cyathium) surrounded by many male flowers, each reduced to a single stamen

Self-pollination

Just as there is a genetical advantage in cross-pollination, there is also a general biological advantage in ensuring the production of seeds, after cross-pollination has failed.

In many plants, stigmatic surfaces are exposed for a time, during which cross-pollination may occur, after which they are brought into contact with pollen from the same flower. This happens, for example, in many members of the family Compositae (e.g *Tridax procumbens*, Fig. 14.15).

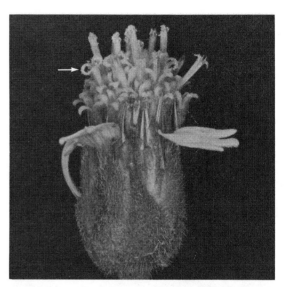

14.15 Inflorescence (capitulum) of *Tridax procumbens* showing the curling back of stigmatic surfaces in an old flower (arrowed), thus ensuring self-pollination if cross-pollination has not already occurred

Some or all of the flowers of most plants open fully and expose their anthers and stigmas (**chasmogamous** flowers). In some plants, a proportion of the flowers do not open (**cleistogamous** flowers) and self-pollination occurs within the unopened perianth. Examples of plants with cleistogamous flowers are *Commelina benghalensis*, Fig. 14.16, *Oxalis* spp. and *Arachis hypogaea*, Groundnut.

Fertilization

Pollen which has been transferred to the receptive surface of a stigma generally begins to germinate very soon. The exine opens at a predetermined point of weakness and the intine bulges out and forms the beginning of a pollen tube. The **pollen tube** grows into the surface of the stigma, pushing between the cells.

The style may be hollow or solid. A hollow style is lined with **transmitting tissue**, a tissue formed by the outgrowth of epidermal cells on the inner surface of the style. In a solid style the central cavity is completely filled with transmitting tissue, which forms a spongy core to the style. The pollen tube passes down the surface of the transmitting tissue, if the style is hollow. If the style is solid, the tube penetrates between the cells of the transmitting tissue. At this stage, the tube nucleus is usually close to the growing end of the pollen tube and the generative cell (or pair of sperms, in grasses) follows behind.

A style may be only a few millimetres long, or may be much longer. The styles ('silks') of Maize, for instance, may be as long as 30 cm. As the pollen tube grows in length, material is deposited in the tube behind the nuclei, closing the tube. This means that only a short length of pollen tube is alive and active at any one time.

The growing tip of the pollen tube enters the ovary, where it travels along the inside of the ovary wall over the surface of transmitting tissue like that in the style.

The pollen tube of a grass already contains two sperms. In other plants, the generative cell divides at some stage during the passage of the pollen tube down the style and into the ovary. Each of the two sperms formed consists of an elongated nucleus contained in a small amount of cytoplasm, limited by a plasma membrane. Each sperm is, therefore, a cell.

Finally, the tip of the pollen tube enters the micropyle and penetrates the nucellus. Modern work, including observations made with the electron microscope, shows that the tube then enters one of the synergid cells, into which the tube nucleus, sperms and some cytoplasm are discharged. One of the sperms travels from the synergid to the ovum, where the plasma membranes of the two cells merge and the sperm nucleus passes on and fuses with the nucleus of the ovum. This process of fertilization produces a diploid **zygote** nucleus.

The second sperm travels to the centre of the embryo sac and lies next to the central nucleus, or next to the two polar nuclei if they have not already

(a)

(b)

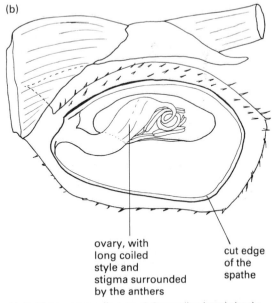

ovary, with long coiled style and stigma surrounded by the anthers

cut edge of the spathe

14.16 Cleistogamous flowers of *Commelina benghalensis*: (a) a plant dug up to show underground creeping stems with flowers enclosed in pale spathes (arrowed); (b) spathe cut open to show the cleistogamous flower, with part of the perianth removed to show ovary and stamens.

(a)

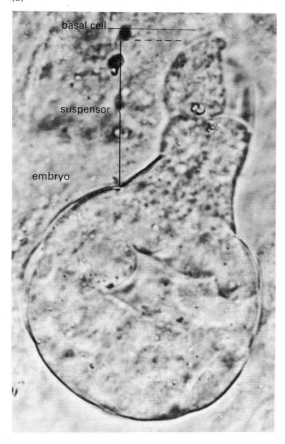

(b)

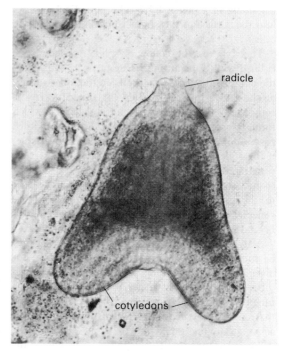

14.17 Embryo development in *Oxalis* sp. (dicot): (a) HP of embryo showing parts, with the basal cell and suspensor much less conspicuous than in many dicots; (b) later stage of embryo (LP)

fused to form a central nucleus. The sperm then fuses with the central nucleus, or with the two polar nuclei. The resulting triploid (three times haploid) nucleus is the **endosperm nucleus**.

Because both sperms fuse with nuclei in the embryo sac, the process is known as **double fertilization**. Double fertilization is unknown outside the flowering plants.

From ovule to seed
The ovule begins its development, to form the seed, soon after fertilization has taken place.

The embryo
The zygote usually starts its development, to form the embryo, soon after fertilization. There are several different patterns of development in different families but, in most, the first stage is repeated transverse division of the zygote to form a filament of cells. The **basal cell**, at the end of the filament closest to the micropyle, is often enlarged (Fig. 14.17). It is connected by a filamentous **suspensor** to the terminal cell which forms the embryo proper. This terminal cell divides successively in three planes to form a globular mass of eight cells. Each of the eight cells now divides periclinally, to form the outer protoderm and a core of eight cells. Further cell divisions, accompanied by enlargement of the whole of the embryo proper, lead to the formation of a lobed (dicots) or rod-shaped (monocots) embryo in which ground meristem and procambial meristem differentiate and the apical meristems of root and shoot become established.

The endosperm
The triploid endosperm nucleus starts to divide before, at the same time as, or after the zygote starts

to divide. The first divisions are usually **free-nuclear divisions**, in which the nucleus divides repeatedly without the formation of cell walls. As the embryo sac enlarges, nuclear divisions continue and cell walls are laid down between the nuclei, dividing the endosperm into cells. At this stage, the enlarged embryo sac is filled with **endosperm**, a nutritive tissue, in which the embryo is embedded.

The seed

The amount of development of the embryo during seed development varies a great deal. In orchids the embryo stops development at a very early stage and endosperm fails to develop, but in most plants there is a stage at which the embryo is small compared with the amount of endosperm. Outside the endosperm, the nucellus undergoes little change and forms no more than a thin, papery layer. In only a few seeds (e.g. *Beta vulgaris*, Beetroot), the nucellus develops into an additional nutritive layer, the **perisperm**. The integuments increase, keeping pace with the development of the inner structures and

form the **testa** of the seed. The original micropyle remains as a small hole in the testa.

In endospermous seeds, the development of the embryo ceases while there is still much endosperm present (e.g. *Ricinus* sp., Fig. 2.2). In non-endospermous seeds, the embryo enlarges until it completely absorbs and replaces the endosperm (e.g. *Arachis* sp., Groundnut). However, in some seeds with large embryos, a very thin layer of endosperm may remain up to the time the seed germinates (e.g. *Caesalpinia pulcherrima*, Pride of Barbados, Fig. 14.18).

When development of the seed is complete, it usually dries out and enters a period of dormancy. The seeds of some tropical plants have no period of dormancy and germinate immediately, even before they have been shed from the fruit (e.g. *Sechium edule*, Shushu).

From ovary to fruit

Just as the ovules are enclosed within and protected by the ovary of the flower, so the seeds are enclosed within the fruit. However, the fruit is important not only as protection for the developing seeds but also in their dispersal.

True fruits are the result of the development of the ovary wall only, or sometimes the ovary wall plus the style. A true fruit can be recognized because it is attached to the plant above the remains of the calyx and any other floral parts that remain (e.g. Tomato, Fig. 14.19).

The wall of a true fruit, developed from the ovary wall, is called the **pericarp**. In some fleshy fruits, the pericarp has three layers. The surface layer ('skin') is the **epicarp**, inside this is the **mesocarp** ('flesh') and the tough, inner-most layer is the **endocarp** (e.g. *Mangifera indica*, Mango, Fig. 14.19).

False fruits, from a single flower, are developed when a flower has an inferior ovary and the receptacle or other floral parts are included in the structure of the fruit. Such a false fruit can be recognized by the presence of the remains of the calyx or other floral parts at the end of the fruit (e.g. *Psidium guajava*, Guava, Fig. 14.20).

Compound or **multiple fruits** are false fruits in which several or many flowers contribute to the formation of a single fruit. The best example is *Ananas sativus* (Pineapple), in which the fruit is

(a) × 1

fruit wall, pod still green but hard

fully developed seed

thick green testa

cut edge of testa

thick rubbery layer of white endosperm surrounds the embryo

(b) LS seed × 2

thin endosperm layer at micropylar end

radicle

plumule

two cotyledons, green at this stage

as the seed matures the cotyledons become thicker and a creamy-yellowish colour, the testa hardens and becomes brown, and the endosperm remains as a very thin white layer

14.18 Fully-developed seed of *Caesalpinia pulcherrima*, (a) in the pod and (b) seen in LS

formed from the ovaries, sepals and the axis of the whole inflorescence.

The classification of fruits can be long and complicated since it appears that those responsible for naming different kinds of fruits have felt it necessary to invent a very great many terms to cope with rather superficial differences. The principal categories of dry and fleshy fruits are listed in Tables 14.1 and 14.2 respectively and illustrated in Figs. 14.19 to 14.28.

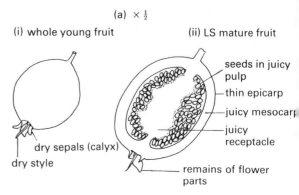

(a) × ½

(i) whole young fruit

(ii) LS mature fruit

- seeds in juicy pulp
- thin epicarp
- juicy mesocarp
- juicy receptacle
- remains of flower parts

dry sepals (calyx)
dry style

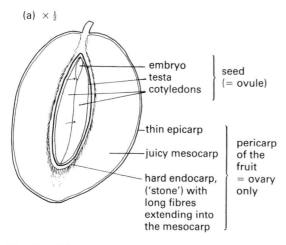

(a) × ½

- embryo ⎱
- testa ⎰ seed
- cotyledons ⎰ (= ovule)

- thin epicarp ⎱
- juicy mesocarp ⎰ pericarp of the fruit
- hard endocarp, ('stone') with long fibres extending into the mesocarp ⎰ = ovary only

(b) × ½ at right angles to (a)

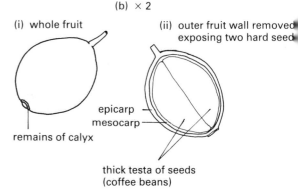

(b) × 2

(i) whole fruit

(ii) outer fruit wall removed exposing two hard seeds

remains of calyx

epicarp
mesocarp

thick testa of seeds (coffee beans)

14.20 False fruits, from an inferior ovary and receptacle: (a) *Psidium* sp. (Guava), a many-seeded inferior berry; (b) *Coffea* sp. (Coffee), an inferior berry with two seeds

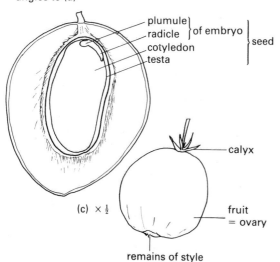

- plumule ⎱ of embryo
- radicle ⎰
- cotyledon ⎱ seed
- testa ⎰

(c) × ½

- calyx
- fruit = ovary

remains of style

14.19 True fruits, derived from the ovary only: (a) and (b) LS drupe of *Mangifera* sp.; (c) whole fruit (berry) of *Lycopersicum* sp. (Tomato)

14.21 A compound fruit, from an inflorescence: *Lantana camara*, fruit derived from a whole inflorescence with many flowers close together, each of which forms a drupe. See Fig. 15.10 for the inflorescence

Table 14.1 Dry fruits.

Type of fruit	Description	Examples
achenes	indehiscent fruits containing only one seed	
(a) simple achene	pericarp leathery, testa and pericarp separate, no appendages	*Cyperus* spp.
(b) caryopsis or grain	achene with pericarp and testa fused	*Zea mays* (Maize) and other grasses
(c) cypsella	achene from inferior ovary with a hairy pappus at the distal end	*Tridax procumbens* and many other Compositae
(d) samara	winged achene	*Pterocarpus* spp.; *Triplochiton scleroxylon* (Obeche); *Combretum* spp.
(e) nut	achene with a woody pericarp	*Anacardium occidentale* (Cashew nut)
follicle, legume and capsule	dehiscent, dry fruits, each containing several seeds	
(a) follicle	from one carpel of monocarpellary or apocarpous multicarpellary ovary, opening down one side only	*Cola* spp. (Cola Nut); *Plumeria* sp. (Frangipani)
(b) legume	from a monocarpellary ovary, opening down two sides	*Cassia* spp.; *Crotalaria* spp. many other Leguminosae
(c) capsule	from a multicarpellary, syncarpous ovary, opening at the top or down the sides	*Talinum triangulare* (Water Leaf); *Canna* sp. (Canna Lily)
schizocarpic fruits	fruit containing several seeds, breaking into one-seeded parts (mericarps)	
(a) lomentum	a legume-like fruit, constricted between the seeds, breaking into mericarps when ripe	*Entada* spp.; *Mimosa pudica* (Sensitive Plant); *Desmodium* spp.
(b) schizocarpic capsule	a capsule which breaks into mericarps, down each septum	*Ricinus communis* (Castor Oil);

Fruits and seed dispersal

The principal means of seed dispersal are explosive mechanisms, wind, animals and water.

Explosive mechanisms

Explosive mechanisms occur when a fruit dehisces violently and throws its seeds to some distance. The pods of *Caesalpinia pulcherrima* (Pride of Barbados, Fig. 14.23), and many other members of the Caesalpiniaceae and Papilionaceae dehisce in this way. On a dry day, especially during the harmattan, it is possible to hear the pods exploding and to see seeds being thrown for distances of many meters from the parent plant. The capsules of *Ricinus communis* (Castor Oil, Fig. 14.24) also open violently in very dry weather, often after they have fallen to the ground, and fling the three seeds from each capsule in three different directions.

Most explosive mechanisms depend on drying out, but some fruits explode due to increasing turgidity and the gradual weakening of their walls along predetermined lines. Thus, the fleshy capsules of *Impatiens* sp. (Balsam) and *Oxalis* spp. (Fig. 14.25) explode, usually early in the morning when the air is very humid.

Table 14.2 Fleshy fruits.

Type of fruit	Description	Examples
drupe	one-seeded fruit formed from a single flower with a superior ovary; pericarp of three layers, epicarp ('skin'), mesocarp (fleshy) and endocarp (woody)	*Mangifera indica* (Mango), with fleshy mesocarp; *Elaeis guineensis* (Oil Palm), with oily mesocarp; *Cocos nucifera* (Coconut Palm), with fibrous mesocarp, drying when mature
berry	two- to many-seeded fruit formed from a single flower, from a superior ovary (true fruit) or from an inferior ovary (false fruit); pericarp (etc.) without a woody layer	*Lycopersicum esculentum* (Tomato), superior ovary; *Citrus* spp. (Orange, etc.), superior ovary with fleshy endocarp; *Musa* spp. (Banana, etc.), inferior ovary; *Cucurbita* spp. (Pumpkin, etc.), inferior ovary
fleshy capsule	from a multilocular, superior ovary dehisces as a result of stresses induced by increasing turgidity	*Impatiens* spp. (Balsam)
aggregate fruit	from one flower with several ovaries united by the receptacle to form a single fruit	*Rubus pinnatus* (Orangeberry), fruit from ovaries only; *Annona* spp. (Custard Apple, etc.), flesh includes receptacle; *Fragaria* sp. (Strawberry), aggregate of achenes on fleshy receptacle
inflorescence fruits (also called compound, multiple or collective fruits)	from many flowers of an inflorescence, usually including the fleshy axis of the inflorescence	*Ananas sativus* (Pineapple), elongated axis with fleshy receptacles of inferior ovaries; *Ficus* spp. (Figs). fleshy, cup-shaped inflorescence axis encloses embedded drupes; *Artocarpus* spp. (Breadfruit, Jac Fruit), whole inflorescence axis fleshy, with embedded achenes

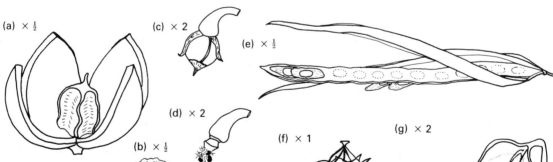

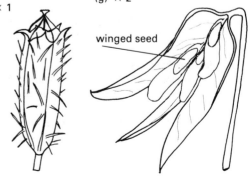

14.22 Seed dispersal from dry, dehiscent capsules:
(a) *Khaya* sp., of which the woody capsule contains four rows of winged seeds; (b) a winged seed of *Khaya* sp.; (c) and (d) *Talinum* sp. capsule before and after dehiscence, when the three valves of the capsule are shed explosively and scatter most of the tiny seeds; (e) *Tecoma* sp., fruit splits open to expose winged seeds on the central septum; (f) spiny capsule of *Argemone* sp., which opens at the top to release many small seeds; (g) *Agapanthus* sp., a three-valved capsule

all to same scale, × ½

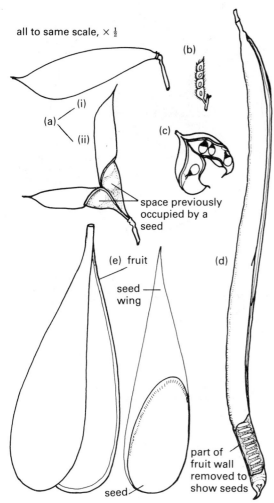

14.23 Dispersal of seeds from a variety of legumes:
(a) *Caesalpinia* sp., seeds are flung out when the pod
(i) splits and the two sides curl up (ii); (b) *Mimosa* sp., in
which the pod breaks into one-seeded mericarps; (c) *Abrus*
sp., pod splits to expose the red and black seeds, some of
which drop out and others are removed by birds;
(d) *Cassia* sp., the seeds of which are dispersed only when
the pod rots and falls to pieces; (e) *Schizolobium* sp., pod
opens to release a single, large winged seed which is
blown away, spinning as it slowly falls

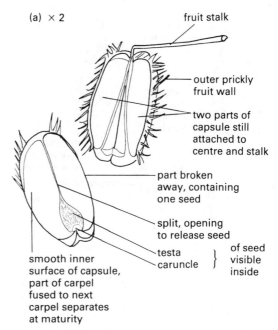

14.24 Schizocarpic capsule: *Ricinus* sp., capsule dried out
and splitting into three parts, each containing one seed

Wind dispersal

Wind dispersal is associated either with very small
seeds (e.g. orchids) or with structures which give the
seed buoyancy in the air. Such structures include
various forms of hairy or feathery outgrowths, or
wings, attached either to the fruit (Fig. 14.26) or to
the seed (Figs. 14.22, 14.23, 14.27).

Hairy seeds are found, for example, in *Ceiba
pentandra* (Silk Cotton Tree, Fig. 14.27) and the
related *Bombax malabaricum*. In both of these trees,
the fruits initially dehisce explosively in dry weather
and the seeds are afterwards carried by the wind for
long distances. Many members of the Apocynaceae
also have hairy seeds (e.g. *Funtumia elastica*, Lagos
Rubber; *Nerium oleander*, Oleander; *Strophanthus
hispidus*, Arrow Poison Plant).

Hairy fruits occur in plants in which the fruit is
the unit of dispersal. The cypsellas of many members
of the Compositae have a hairy pappus (a modified
calyx) at the upper end (e.g *Tridax procumbens*).

Winged seeds are found in some members of the
Apocynaceae (e.g. *Plumeria japonica*, Frangipani,
Fig. 15.24) and Bignoniaceae (e.g. *Tecoma stans*,
Golden Shower, Fig. 14.22; *Spathodea campanulata*,
West African Tulip Tree).

Some seeds, although themselves wingless, are
released attached to part of the fruit in which they
were formed. Thus, seeds from the inflated capsules
of *Cardiospermum halicacabum* are released at dehiscence, each attached to a folded structure which is

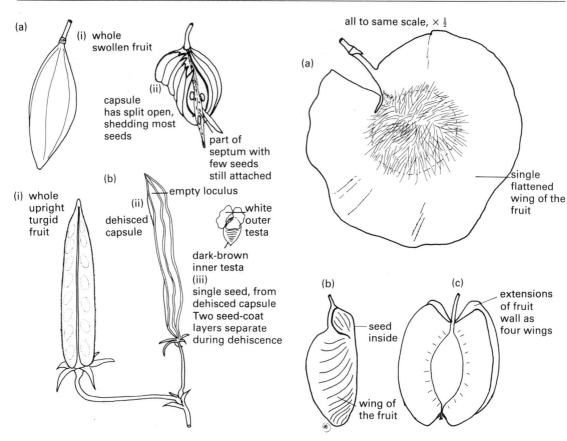

14.25 Seed dispersal from fleshy capsules: (a) *Impatiens* sp.; (b) *Oxalis* sp., both shown (i) before and (ii) after dehiscence

14.26 Wind dispersed fruits: single seeded fruits of (a) *Pterocarpus* sp., (b) *Tipuana* sp., and (c) *Combretum* sp.

formed from the septa between the loculi of the fruit.

Winged fruits are found in *Triplochiton scleroxylon* (Obeche) and *Dioscorea* spp. (Yams) as well as in many other plants (Fig. 14.26).

Animal dispersal

Seed dispersal by animals is achieved in one of two ways. Many fleshy fruits are attractive to animals and are thus eaten or collected by them. Other fruits have sticky hairs or hooks, which become attached to the fur of passing animals.

Almost any brightly coloured, fleshy fruit may be assumed to be animal dispersed. In some cases, the seeds are swallowed by the animal, pass through the gut and are deposited with the faeces (e.g. *Lycopersicum esculentum*, Tomato; *Capsicum* spp., Peppers).

Fruits with larger seeds must depend on the flesh being eaten and the seeds dropped (e.g. *Mangifera indica*, Mango). The dry or fleshy fruits of other plants open to expose colourful seeds, which are collected by animals (e.g. *Abrus precatorius*, Crab's Eyes, Fig. 14.23; *Momordica* spp., Fig. 14.28).

Boerhaavia sp. has fleshy fruits with sticky hairs so that the fruits adhere to animals. Examples of fruits with hooks include *Bidens pilosa* and *Acanthospermum hispidum*.

Animals involved in seed dispersal include all land animals and birds. Birds which feed in or near water have been shown to carry seeds and fruits of types not necessarily adapted to animal dispersal, in mud attached to their bodies. Of animals, man is responsible for much short and long distance dispersal,

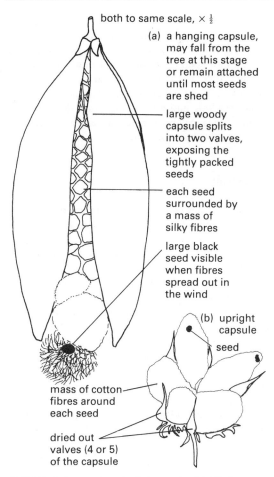

both to same scale, × ½

(a) a hanging capsule, may fall from the tree at this stage or remain attached until most seeds are shed

large woody capsule splits into two valves, exposing the tightly packed seeds

each seed surrounded by a mass of silky fibres

large black seed visible when fibres spread out in the wind

(b) upright capsule

seed

mass of cotton fibres around each seed

dried out valves (4 or 5) of the capsule

14.27 Wind dispersed seeds: hairy seeds of (a) *Ceiba* sp., (b) *Gossypium* sp.

14.28 Animal dispersal: *Momordica* sp., in which the prickly, orange-coloured fruit splits open and exposes the bright red, shiny seeds

both intentionally and otherwise. Evidence for this is the world-wide distribution of many tropical weeds. Weed seeds have travelled mixed with the seeds of crop plants and as seedlings in soil in which plants of economic importance have been introduced, especially before strict quarantine restrictions were imposed. It has also been shown that seeds may be carried long distances in the caked mud under motor car mudguards and on people's clothes. With air travel now so common, many weed seeds must be carried this way also.

Water dispersal

Water dispersal of fruits and seeds is met with in water plants and in plants that grow close to water.

Thus, the inflated capsules of some *Crinum* spp. (Crinum Lilies) will float in a stream or river for some time before releasing their seeds. The fruits of *Cocos nucifera* (Coconut Palm) have a mesocarp which is fleshy when young, but which dries out and contains a great deal of air when mature. The fruits are therefore buoyant, so that they float in sea water and may be carried for long distances by sea currents. The fringe of coconut palms to be found on many tropical beaches confirms the effectiveness of water dispersal.

Rhizophora spp. (Red Mangroves) are viviparous and seeds germinate while still attached to the parent tree. If the large, javelin-like seedlings which fall from trees do not lodge in the mud where they fall, they are afterwards water-dispersed.

Key words

Reproduction: sexual, gametes; asexual, spores; vegetative, clone. **Flower:** modified shoot; calyx, sepals; corolla, petals; androecium, stamens; gynaecium, carpels, ovary. **Carpels:** ovary, style, stigma; loculus; ovule, placenta. **Placentation:** axile, free-central, parietal; unilocular, multilocular. **Stamen:** filament; anther, pollen sac, microsporangium; pollen mother cells, meiosis; pollen grain, haploid. **Pollen:** exine, intine; generative cell, tube nucleus. **Ovule:** nucellus, megasporangium; megaspore mother cell, meiosis; megaspore, embryo sac; orthotropous, anatropous; funicle, raphe; chalazal end; integuments, micropyle. **Embryo sac:** ovum, synergids, antipodal cells, polar nuclei, central nucleus. **Pollination:** wind, water, animals; self-pollination, cross-pollination. **Flowers:** staminate, pistillate; dioecious, monoecious; protandrous, protogynous. **Fertilization:** pollen tube, sperms; double fertilization; zygote, endosperm. **Dispersal:** seed dispersal, fruit dispersal; explosive mechanisms, wind dispersal, water dispersal, animal dispersal.

Summary

Flowering plants reproduce sexually and asexually. Asexual (vegetative) reproduction is by the separation of plant parts. Flowers produce the organs of sexual reproduction. The male gametes are produced by the pollen grain, after it has been transferred from the stamen to a stigma. One female gamete is produced in the embryo sac of each ovule, inside the ovary. Two sperms are delivered into the ovule by a pollen tube and double fertilization results in the formation of an embryo and endosperm. As the ovule develops into a seed, the ovary develops into a fruit. Seeds may be dispersed together, in the fruit, or separately, by various means.

Questions

1 What is a flower? What parts are essential for the process of reproduction? Where are male and female gametes produced?

2 Distinguish between a pollen grain and a male gamete.

3 Define the following terms: (a) placentation; (b) pollination; (c) fertilization.

4 Briefly describe the characteristics you would expect to find in a flower that is pollinated by (a) bees, (b) moths, and (c) bats.

5 Where does meiosis take place in a flower? What structures are produced as a direct result of meiosis? What is double fertilization and where, exactly, does it take place?

6 What is the advantage of cross-pollination in flowering plants?

7 Briefly describe the characteristics of seeds or fruits that are dispersed by (a) wind, and (b) animals.

15

The classification of flowering plants

The classification of flowering plants is based very largely on floral structure. This is mainly because it is believed that reproductive structures, in this case flowers, change less quickly during the course of evolution than other plant characters.

A botanist must be able to identify plants, with the aid of a Flora. A **Flora** is a book in which all the plants which have been collected and recorded in a particular area are described. **Dichotomous keys** are usually included, as an aid to the identification of all those plants described in the Flora. Such keys are based largely, though not exclusively, on floral characteristics.

The use of a Flora requires a very extensive technical vocabulary. It is usual for a Flora to include a glossary of technical terms, usually at the beginning of the first volume (e.g. *Flora of West Tropical Africa*). However, it is necessary to have some prior knowledge of the ways in which flower structure varies and to have a basic knowledge of the principal technical terms. These will be provided in the first part of this chapter, together with an account of inflorescences.

Flower structure

The basic structure of the flower has been described in Chapter 14. There are many variations on this basic structure which must be understood in order to interpret the structure of any individual flower and to classify and identify it.

In biology, the term 'primitive' is used in its strict sense. A primitive structure is a structure which exists in its original form, before it was modified by subsequent evolution. When we find a structure which we believe has been modified during the passage of time, through evolution, we call it an 'advanced' structure.

The primitive structure of the flower is usually taken to be that in which there are many sepals, petals, stamens and carpels, all separate from each other and attached to the receptacle in a continuous

15.1 A primitive flower, *Nymphaea* sp. (see Fig. 14.2 for LS)

spiral (Fig. 15.1). In more advanced flowers, the different sorts of floral parts are arranged in a series of whorls, each whorl attached to the receptacle at a different level from the others (Fig. 15.2).

Further variations in floral structure result from the fusion of similar parts (**connation**), the fusion of dissimilar parts (**adnation**), the reduction or loss of parts and changes in the relative sizes of parts.

The receptacle

Variations in the shape of the receptacle (Fig. 15.3) affect the relative positions of other parts. If the receptacle is dome-shaped (convex), with the ovary or ovaries attached at the end, the ovary is **superior**. In this case, the other floral parts are attached to the receptacle below the ovary and are said to be **hypogynous**. In other plants, the receptacle is cup-shaped at the end, with the ovary attached at the bottom of the cup. In this case, the ovary is still superior, but the other parts, attached around the rim of the cup, are **perigynous**. Where the end of the receptacle completely encloses the ovary and is (usually) fused to it, the ovary is **inferior** and the other floral parts are **epigynous**.

Part of the receptacle, usually immediately below

15.2 An advanced flower, *Spathodea* sp., whole and half flower. The sepals and petals are connate, the calyx opening along the lower side only. The stamens are adnate to the corolla tube. Parts of the androecium and gynaecium are arranged in such a way that they facilitate cross-pollination. There is a swollen nectary (arrowed) formed from the disc at the base of the ovary

(a) hypogynous flower; superior ovary, above all other parts

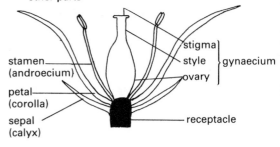

(b) perigynous flower; superior ovary, other parts around edge of receptacle

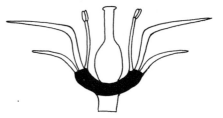

(c) epigynous flower; inferior ovary, other parts borne above and around the ovary

15.3 Hypogynous, perigynous and epigynous flowers. Diagrams to show the attachment of floral parts to the receptacle (shaded)

the ovary in flowers with a superior ovary, may be swollen to form a **disc**. In flowers with an inferior ovary, a disc may be formed at the top of the receptacle, around the base of the style. The epidermis of the disc secretes nectar, a sugary fluid, and forms a nectary. The nectar is attractive to insect pollinators.

Note that the nectaries of some flowers are formed instead from the inner surfaces of corolla parts, sometimes in a tubular outgrowth (**spur**) which projects from the outer surface of one or more petals (e.g. orchids). Nectaries formed within the flower are **floral nectaries**. Some plants (e.g. *Ricinus communis*) have **extra-floral nectaries**, outside the flowers (e.g. on petioles, in *R. communis*).

The gynaecium

In primitive flowers, the gynaecium (Fig. 15.4) consists of many separate (free) carpels (e.g. most Annonaceae). This type of gynaecium is described as **apocarpous**. When several or many carpels are present but have their ovaries fused into one structure (connate), the gynaecium is **syncarpous**. A syncarpous ovary may have a single loculus (**unilocular**) or may have two, three or more loculi (**bi-, tri-**, etc. to **multilocular** ovary).

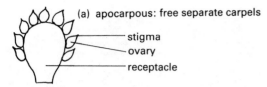

(a) apocarpous: free separate carpels

stigma
ovary
receptacle

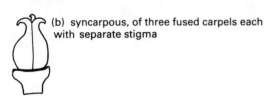

(b) syncarpous, of three fused carpels each with separate stigma

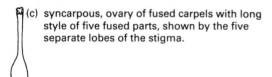

(c) syncarpous, ovary of fused carpels with long style of five fused parts, shown by the five separate lobes of the stigma.

15.4 Primitive and advanced gynaecium structure: (a) primitive; (b) and (c) more advanced

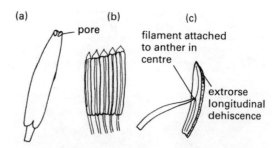

(a)   (b)   (c)

pore

filament attached to anther in centre

extrorse longitudinal dehiscence

15.5 Variation in the androecium: (a) stamen opening by apical pores (e.g. *Solanum* sp.); (b) syngenesious stamens, with free filaments and fused anthers (e.g. *Tridax* sp.); (c) stamen with versatile anther (e.g. *Gloriosa* sp.)

A syncarpous ovary may have as many styles and stigmas as there are carpels involved, or the styles, or styles and stigmas, may be fused together.

In plants where **staminate** ('male') and **pistillate** ('female') flowers are separate, the staminate flower contains only an aborted ovary (pistillode, e.g. *Carica papaya*).

The androecium
Stamens may be spirally arranged or in one or two whorls. Filaments are usually thin, but may be broad. The upper end of the filament may be attached along the length of the anther, merging with the connective (Fig. 14.3), or may be attached only at its extreme tip, to the centre of the anther (**versatile** anther, Fig. 15.5). There are usually four pollen sacs but there may be fewer or more. The pollen sacs may dehisce by longitudinal slits or by terminal pores. If they dehisce by slits, the slits may open towards the centre of the flower (**introrse** stamens) or towards the outside (**extrorse** stamens). Pollen may be dry or sticky, or may be united into a mass (**pollinium**) in each pollen sac, so that the entire contents of a pollen sac is picked up by a pollinating insect and transferred as such to the stigma of another flower (e.g. orchids; Asclepiadaceae).

The filaments of stamens may be fused together to form a tube around the style (**staminal tube**, e.g. *Hibiscus rosa-sinensis*, Hibiscus) or the filaments may be free and the anthers joined together (e.g. Compositae, Fig. 15.5).

Filaments may be attached directly to the receptacle or their lower ends may be adnate to the petals (**epipetalous** stamens) or to the sepals (**episepalous** stamens).

Branched stamens, with an anther at the end of each branch, are found in *Ricinus communis* and some of its relatives.

In plants with separate staminate and pistillate flowers, pistillate flowers may contain aborted stamens (**staminodes**). Petal-like (**petaloid**) staminodes are found in some flowers in which the number of functional stamens is reduced (e.g. *Canna indica*, Canna Lily, Fig. 14.3).

Perianth
The perianth usually consists of separate spirals or whorls of petals and sepals. Either petals or sepals may be absent or the two types of perianth parts may be indistinguishable from each other. Petals and sepals may be free from each other, attached separately to the receptacle. If connate, the petals form a **corolla tube** and the sepals form a **calyx tube**.

Floral symmetry
Many flowers are radially symmetrical. A radially symmetrical flower can be divided into two equal halves by an imaginary line drawn in any direction, through the centre of the flower. Radially symmetrical flowers are described as **actinomorphic** (**regular**, Fig. 15.6).

15.7 Floral symmetry: a zygomorphic flower, *Bauhinia* sp

15.6 Floral symmetry: an actinomorphic flower, *Talinum* sp. (see Fig. 14.1 for floral details, and front cover)

Zygomorphic (**irregular**) flowers are bilaterally symmetrical (Fig. 15.7). They can be divided into equal halves only by a line drawn in one direction, usually from the centre of the adaxial side of the flower to the centre of the abaxial side of the flower.

In actinomorphic flowers, all the whorls of floral parts are regular – all the component parts are of one size and shape. In zygomorphic flowers, the members of at least one whorl of floral parts are irregular – of different sizes or shapes.

The term **aestivation** describes the way in which the sepals and petals are folded together in the flower bud (Fig. 15.8). If the perianth parts do not touch, aestivation is **open**. If they meet at the edges but do not overlap, aestivation is **valvate**. If they overlap, aestivation is **imbricate**. In some families (e.g. Apocynaceae), there is a special type of imbricate aestivation in which the petals overlap in such a way

that one edge of a petal is underneath that of its neighbour, while the other edge lies on top of that of its other neighbour, and so on all round the flower. This is **contorted** aestivation.

Inflorescences

There are plants in which the flowers are formed singly (**solitary** flowers) and not in groups, but most plants have their flowers grouped into **inflorescences**. Inflorescences are usually formed on new growth, at the ends of young branches. Some tropical trees are **cauliflorous** and have their inflorescences growing from the trunk or woody branches (e.g. *Theobroma cacao*, Cocoa).

An inflorescence stalk is a **peduncle**. A flower stalk is a **pedicel**. Flowers are formed in the axils of bracts. A **bract** is a leaf which subtends a flower and may be as large as an ordinary leaf, but is usually scale-like. When a bract-like structure is formed on a pedicel, it is known as a **bracteole**.

There are two basic types of inflorescence, racemose and cymose. In a **racemose** inflorescence the first flower is formed in the axil of a bract, while the apex of the inflorescence continues to grow, producing a series of bracts and axillary flowers. Growth of a raceme is usually terminated when the apical bud forms a flower. In a **cymose** inflorescence, the first flower is formed from the apical bud. The next flower is formed from the apex of an axillary shoot behind the apex and successive flowers may be formed in the same way, each terminating the growth of an axillary shoot.

(a) (i) open aestivation:
parts do not meet, as in
calyx of Rubiaceae (Fig. 15.25)

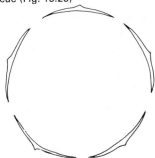

(ii) valvate aestivation:
parts meet at their edges,
as in corolla of
Asclepiadaceae
(see Fig. (b), *Stapelia* sp.),
and in calyx of *Mimosa* sp.,
Fig. 15.22

(iii) imbricate aestivation:
contorted corolla of
Apocynaceae (Fig. 15.24)
Each petal is overlapping
at one side (a)
and is underlapping at the
other side (b)
(see Fig. (c) *Plumeria* sp.)

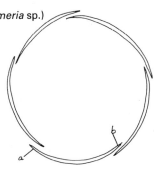

(iv) Imbricate, quincuncial.
Two parts (a and a) overlapping
at both edges, two parts (b and b)
underlapping at both edges,
and one part (c) overlapping
and underlapping, as in corolla
of *Momordica* sp. (Fig. 15.19)
and in calyx of *Allamanda* sp.
(Fig. 15.24)

(iv) to (vi) are further
types of imbricate

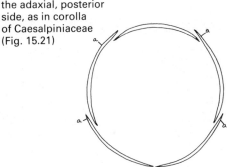

(v) ascending imbricate:
overlapping (a) towards
the adaxial, posterior
side, as in corolla
of Caesalpiniaceae
(Fig. 15.21)

(vi) descending imbricate:
overlapping (a) towards the
abaxial, anterior side,
as in corolla of
Papilionaceae (Fig. 15.23)

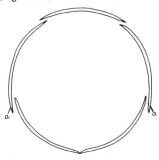

15.8 Aestivation: (a) diagrams of different types of
aestivation (continued overleaf)

(b)

(c)

15.8 (continued) (b) flower bud of *Stapelia* sp. showing valvate aestivation, with perianth parts in contact but not overlapping; (c) opening flowers of *Plumeria* sp. with imbricate, contorted aestivation

Racemose inflorescences (*Figs. 15.9, 15.10*)

The simplest racemose inflorescence is the **raceme**, in which each successive bract subtends a single flower with a pedicel (**pedicellate** flower). In a **spike**, the flowers are without pedicels (**sessile** flowers). A **panicle** is a compound raceme, in which each bract on the main inflorescence axis subtends a raceme of flowers, instead of a single flower.

A **corymb** is a raceme in which bracts are formed

successively higher and higher on the inflorescence axis, but the successive pedicels are shorter and shorter, so that all the flowers are at the same level and form a flat or slightly convex group. The formation of the corymb of *Lantana camara* results from differences in the lengths of the corolla tubes, not the pedicels (Fig. 15.10).

An **umbel** is a racemose inflorescence in which all the flowers are formed in the axils of a single whorl of bracts at the end of the peduncle and the flowers are arranged to form a single flat, convex or globose mass.

Corymbs and umbels can also be compound, with each branch of the inflorescence repeating the same pattern of branching.

A **capitulum** (**head**) is a racemose inflorescence in which the end of the peduncle is enlarged to form a flat, convex or globose surface to which are attached several to many small, sessile flowers, surrounded by one or more whorls of bracts. This type of inflorescence is especially characteristic of the

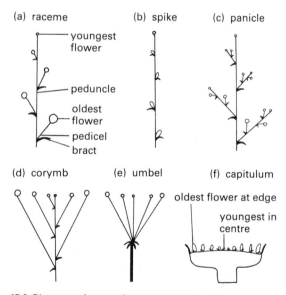

15.9 Diagrams of types of racemose inflorescences

family Compositae in which the flowers may be of two types. In some genera, the outer florets (**ray florets**) are zygomorphic, with strap-shaped corollas radiating out from the centre, while the inner florets are actinomorphic and inconspicuous. This arrangement gives the whole inflorescence the appearance

(a)

(b)

(c)

(d)

15.10 Examples of plants with racemose inflorescences:
(a) *Caesalpinia pulcherrima*, a raceme (sometimes
compound at the base); (b) *Lantana* sp., a corymb;
(c) *Haemanthus* sp., an umbel; (d) *Helianthus* sp., a
capitulum (see Fig. 15.26)

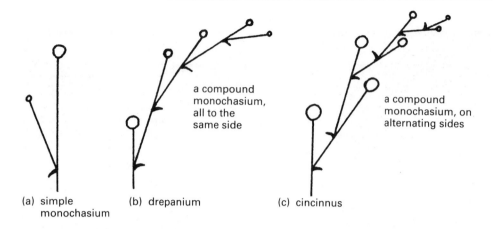

(a) simple
monochasium

a compound
monochasium,
all to the
same side

(b) drepanium

(c) cincinnus

a compound
monochasium, on
alternating sides

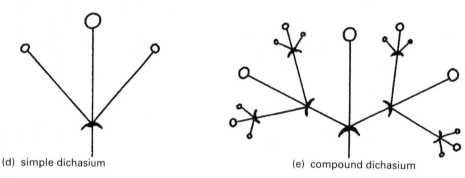

(d) simple dichasium

(e) compound dichasium

15.11 Diagrams of types of cymose inflorescences

of a single flower (e.g. *Tridax procumbens*; *Helianthus* sp., Fig. 15.10).

Corymbs, umbels and capitula can all be recognized as racemose inflorescences because the first flowers to open are those around the margin of the inflorescence.

Cymose inflorescences (Figures 15.11, 15.12)

The simplest form of cymose inflorescence is the **monochasial cyme** (**monochasium**). This type of inflorescence is formed when bracts are single at each node. Variations on this pattern are produced, according to whether successive bracts are on the same side of the inflorescence (**drepanium**) or alter-

nate (**cincinnus**).

If bracts are in opposite pairs, the result is a **dichasium**. Cymose forms of the umbel and capitulum also exist. The rain forest herb *Dorstenia* sp. has a wide cymose capitulum of small florets, which can be recognized as a cymose inflorescence because it is not the florets around the margin that open first. The hollow inflorescence of *Ficus* spp. (Fig. 14.13) is thought to be a development of the cymose capitulum.

There are also compound inflorescences which combine the basic pattern of the raceme with cymose branches (e.g. *Clerodendron paniculatum*, Pagoda Plant).

15.12 Example showing cymose inflorescence: *Carica* sp., inflorescence is a compound dichasium with the main axis ending in a female flower (here, young fruit) and side branches bearing male flowers

Describing a plant

In any flora (e.g. *Flora of West Tropical Africa*) you will find descriptions of plants belonging to various families, genera, species and even subspecies. These descriptions, taken together, include information about the type of plant (tree, shrub, climber, herb), the size of the plant, the habit of the plant (prostrate, erect, etc.), leaf shape and leaf arrangement, the type of inflorescence, the structure of the flower and the type of fruit it produces. There is also more detailed information, for example about the type of ovule, the shape and size of the embryo and the amount of endosperm the seed contains. The contents of some of the earlier chapters of this book should help you to understand the descriptions, but you will have to

refer to the glossary in the flora in order to understand some of the more detailed descriptions, e.g. of leaf shape, leaf margin and leaf base.

In making descriptions of floral structure, there are certain formal procedures which are always followed. These include making a drawing of the whole flower, making a drawing of the half flower (cut lengthways), drawing a floral diagram and writing the floral formula. Each of these items is a form of description of the flower and all are needed to present all possible information.

The drawings

The drawing of the **whole flower** (Fig. 15.13) is made from such an angle that as much as possible of the structure of the flower is shown.

The **half flower** drawing (Fig. 15.14) is made after cutting the flower into two parts. This is best done with a sharp razor blade, starting at the pedicel, dividing the flower into two equal parts along a line passing through the centre of the adaxial side of the flower and the centre of the abaxial side of the flower. The flower is then laid on the bench, cut side uppermost, and drawn. The drawing of the half flower will show details of the attachment of parts to the receptacle which could not be seen in the intact flower.

The floral diagram

The **floral diagram** is a diagram to show the numbers and relationships of the parts of the flower in a very formal way. It does not look at all like a flower.

The parts of the flower are represented according to accepted formalities. The symbols used are shown in Table 15.1. These include a symbol for the axis of the inflorescence to which the flower is attached and for the bract which subtends the flower. Between these are sketched four concentric circles to represent the positions of the four types of floral parts, calyx, corolla, androecium and gynaecium (Fig. 15.15). There will be more circles, one for each whorl, if there is more than one whorl of petals or stamens. When some of the floral parts are spirally arranged, a spiral is superimposed on the appropriate circle. The symbols for each part (Table 15.1) are then added (Fig. 15.15), allowing for the aestivation of calyx and corolla and, where possible, for the direction in which the stamens open (introrse, extrorse). Parts showing connation or adnation are joined by lines, as shown. At the centre, a diagra-

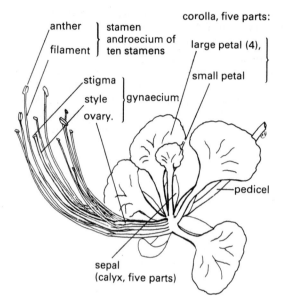

15.13 Drawing of a whole flower to show all the parts as clearly as possible; flower of *Caesalpinia pulcherrima*

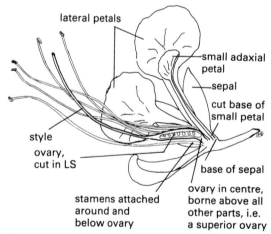

15.14 Drawing of a half flower to show attachment of parts to the receptacle and the internal structure of the ovary: *Caesalpinia pulcherrima*

mmatic TS of the ovary is used to show the number and positions of loculi and the placentation (Fig. 14.5).

When the floral diagram is completed (Fig. 15.16) it is possible to 'read off' most of the characteristics of the flower, with the exception of the level of attachment of the floral parts with respect to the ovary.

Table 15.1 Symbols used in floral diagrams.

Floral parts, etc.	Symbol	Example (Figure)
floral axis	○	Fig. 15.15
subtending bract		Fig. 15.15
sepal		Fig. 15.15
petal		Fig. 15.15
perianth parts (when all petaloid)	as for petals	Fig. 15.31
aestivation	overlapping of edges in diagram	Fig. 15.8
connation of petals, sepals	edges joined by loop or line	Fig. 15.16
stamens introrse		Fig. 15.15
extrorse		Fig. 15.30
connation of stamens		Fig. 15.26
anthers filaments (staminal tube)		Fig. 15.23
adnation of stamens (e.g. epipetalous)		Fig. 15.24
carpels gynaecium	drawn as TS, showing placentation	Figs. 15.15, 15.20

The floral formula

The floral formula (Fig. 15.17) is another very formal representation of the structure of the flower. The first symbols show whether the flower is actinomorphic or zygomorphic and whether it contains both stamens and pistil or only stamens or pistil. The symbols 'K' and 'C' refer to the numbers of parts in the calyx and corolla respectively. If the perianth is not clearly divided into calyx and corolla, the symbol 'P' is used, for perianth parts. Brackets around the numbers indicate connation of parts. After the number of stamens has been added ('A'), adnation of parts is indicated by adding a single bracket, above the letters representing the parts involved. After adding the number of carpels in the gynaecium, a line is drawn under or over the number to show whether the ovary is superior or inferior respectively.

e.g. *Caesalpinia pulcherrima*

floral axis on adaxial (posterior) side

(a) Draw four circles to represent the four whorls of parts, (i) calyx, (ii) corolla, (iii) androecium, (iv) gynaecium. Put in floral axis above, and subtending bract below. In a practical drawing book these lines, the four circles, should be put in lightly with pencil, as guide lines for putting in details of each whorl as in (b) to (e) below.

subtending bract on abaxial (anterior) side.

(b) Draw in the five sepals. Note that the abaxial sepal, in this flower, overlaps almost the whole bud in the early stage.

(c) Draw in the five petals. Note the petals alternate with the sepals, the petals are ascending imbricate aestivation, and there are four larger petals, and one smaller adaxial petal.

(d) Draw in the stamens. Note they are in two whorls of five, the anthers are introrse (opening inwards), and all stamens are free, i.e. there is no connation and they are not adnate to the calyx or corolla.

(e) Draw in the ovary, as seen in TS. Note it is a single folded carpel, with marginal placentation.
     After the ovary is added, the remaining parts of the guide lines can be erased. The floral diagram is then complete.

15.15 Construction of a floral diagram, using symbols shown in Table 15.1: *Caesalpinia pulcherrima*

(a) *Bauhinia* sp. (Fig. 15.21),
Family Caesalpiniaceae, same as
Caesalpinia sp. (Fig. 15.15)
Note: calyx different, joined parts;
corolla, same as *Caesalpinia* sp., with
ascending imbricate aestivation;
androecium, only five stamens;
gynaecium, same as *Caesalpinia* sp.

(b) *Crotalaria* sp. (Fig. 15.23),
Family Papilionaceae; this family
with the Caesalpiniaceae is in
the order Leguminosae.
Note: calyx all joined;
corolla descending imbricate
aestivation; large adaxial petal
forms the standard, and two
abaxial petals form the keel;
androecium, ten stamens with a
staminal tube;
gynaecium same as *Caesalpinia* sp.

(c) *Gloriosa* sp. (Fig. 15.30), a monocot,
in Family Liliaceae.
Note: parts in threes;
calyx and corolla the same, i.e.
has a perianth;
androecium, with extrorse
anthers;
gynaecium, of three fused
carpels.

15.16 Examples of completed floral diagrams

(a) *Caesalpinia* sp. (Figs. 15.13, 15.14, 15.15), Family Caesalpiniaceae, floral formula: ·|· ♂ K5 C5 A5 + 5 G1̲
(b) *Bauhinia* sp. (Fig. 15.21), Family Caesalpiniaceae, floral formula: ·|· ♂ K(5) C5 A5 G1̲
(c) *Crotalaria* sp. (Fig. 15.23), Family Papilionaceae, floral formula: ·|· ♂ K(5) C5 A(5+5) G1̲
(d) *Mimosa* sp. (Fig. 15.22), Family Mimosaceae, floral formula: ⊕ ♂ K4 C4 A4 G1̲
above examples are all in the order Leguminosae.
(e) *Gloriosa* sp. (Fig. 15.30), Family Liliaceae, a monocot, floral formula: ⊕ ♂ P3 + 3 A3 + 3 G(3̲)

15.17 Examples of floral formulae, using symbols shown in Table 15.2.

Table 15.2 Symbols used in floral formulae

Floral part or characteristic	Symbol	Example
symmetry		
actinomorphic	⊕	–
zygomorphic	·|·	–
sex of flower		
staminate (male)	♂	–
pistillate (female)	♀	–
hermaphrodite	♂♀	–
calyx	K	K5 = 5 sepals, free
corolla	C	C5 = 5 petals, free
perianth parts (if all the same)	P	P5 = 5 perianth parts, free
androecium	A	A5 = 5 stamens, free
gynaecium		
ovary superior	G_	G3̲ = 3 carpels, superior
ovary inferior	G⁻	G3̄ = 3 carpels, inferior
connation	brackets	K(5) = 5 connate sepals
adnation	horizontal bracket	C5̄A5 = 5 epipetalous stamens
indefinite number of parts	∞	A∞ = indefinite number of stamens

Selected families of flowering plants

The system of classification used here is that of Hutchinson, the same as that used in *Flora of West Tropical Africa*. As with other modern systems of classification, the authors have attempted to arrange orders and families according to presumed evolutionary relationships. However, it must be recognized that evolutionary relationships are complex and that the fact that one order or family is listed after another does not imply that the second is directly descended from the first. For instance, the monocots are listed after the dicots because it is believed that the dicotyledonous condition is primitive. However, the evolutionary connection between the two groups is obscure and certainly lies amongst plants which are extinct.

In considering the dicots, it is supposed that the primitive groups were probably woody plants, with actinomorphic flowers with many free sepals, petals, stamens and carpels arranged in a continuous spiral. Subsequent evolution led to the reduction of numbers of parts, with many examples of adnation and connation and with zygomorphy arising many times, independently.

Descriptions of families and genera
The descriptions given on succeeding pages are an attempt to summarize the characteristics of families and genera, as defined in Hutchinson's classification. Less common variations within families and genera are omitted and there are some compromises. For instance, the 'general floral formula' for each family is that which is commonest or most characteristic in each family and does not take account of variations which occur within each family. The same applies to the descriptions of family characters, to which exceptions can often be found.

The families described have been selected to illustrate typical variations in the dicot and monocot pattern and to give a reasonable cross-section of those families which are of importance in tropical countries. Under each family name, a general description of family characters is given, followed by brief descriptions of one or more common examples. This is followed by lists of some of the most important economic and cultivated species, weeds and ornamental plants. Ornamental plants, most of which are introduced and not indigenous, are included because most of them are grown throughout

tropical lands and most have large flowers, which are convenient for the study of family characters. Where any of these categories of plants (economic, weeds, ornamentals) is omitted, this is because examples are rare or do not exist.

Dicotyledoneae and Monocotyledoneae
The embryo of a dicot has two cotyledons. Leaves are usually broad and have reticulate venation. Floral parts are usually in fives or fours or multiples of these numbers but sometimes are in large, indefinite, numbers. Stem vascular bundles are usually open (with cambium) and arranged in a ring. Secondary thickening is common.

The embryo of a monocot has one cotyledon. Leaves are commonly narrow and have parallel venation. Floral parts are in whorls of three. Stem vascular bundles are closed and usually scattered throughout the thickness of the ground tissue. Secondary thickening is rare and, when present, of an anomalous type.

Dicotyledoneae

Family: Annonaceae
This mostly tropical family (Fig. 15.18) includes trees, shrubs and woody climbers. The leaves are entire, exstipulate (without stipules) and alternate. Flowers are solitary or in small or large terminal or axillary groups.

Flowers are actinomorphic. The perianth of the flower is in whorls of three. There are three sepals, free or connate at the base. Petals are in two whorls of three and free from each other. Stamens are usually many and hypogynous. The gynaecium is usually apocarpous, with numerous free carpels, but may be of a few, connate carpels. Species with apocarpous gynaecia usually form a fleshy fruit which is an aggregate of berries (Table 14.2), sometimes involving the receptacle and making a false fruit. Species with syncarpous gynaecia produce berries.

The general floral formula is:

$$\oplus \, \male \, \text{K3 C3 + 3 A}\infty \, \text{G}\infty$$

Annona spp. are trees and large shrubs. Their flowers have a reduced calyx and two whorls of fleshy petals. They form large, fleshy, false fruits which are edible.

(a)

(b)

(c) ⊕ ♂♀ K3 C3 + 3 A ∞ G ⚭

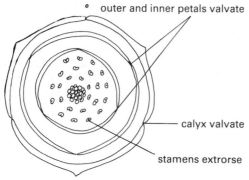

outer and inner petals valvate

calyx valvate

stamens extrorse

15.18 Family Annonaceae, *Annona* sp.: (a) young flower with the three outer, fleshy petals open and the inner whorl of more delicate petals still closed; (b) developing fruit; (c) floral diagram

Monodora spp. (Orchid Trees) are small forest trees and shrubs with very showy, multicoloured flowers. The ovary of the flower is syncarpous and the fruit is a berry.

The economic importance of the family is restricted to fruit production. Several introduced *Annona* spp. are grown commercially (e.g. *A. reticulata*, Custard Apple; *A. muricata*, Soursop).

Family: Cucurbitaceae

This is a mainly tropical and subtropical family (Fig. 15.19) of climbing or scrambling herbs with either simple or branched tendrils. Most are annuals but a few are perennials. The leaves are simple, often deeply lobed, exstipulate, alternate and frequently very hairy. Plants may be monoecious or dioecious. The flowers are actinomorphic and unisexual and may be large and showy. The fruit is usually a fleshy berry and many are edible.

The flowers have a regular, five-part calyx and corolla. The sepals may be united into a tube which, in the female flower, is adnate to the ovary. Petals are free, or joined with five lobes, with valvate or imbricate aestivation. The female flower usually has three fused carpels, with parietal placentation. The ovary is inferior. The styles may be united but the three stigma lobes are usually free. Staminodes are sometimes found in female flowers. The fruit is usually a many-seeded fleshy berry which may either remain fleshy or dry out at maturity.

Male flowers are very variable, basically with five stamens. Fusion of filaments or anthers often reduces this to three. The normal four-celled anther does not occur in this family. When there are three stamens, one always has a single-celled anther. The other two stamens have variously folded or coiled anther lobes, but each stamen has two-celled anthers (e.g. *Momordica* spp., male).

The general flora formula for the family is:

⊕♀ K5 C5 A0 G($\overline{3}$)
⊕♂ K5 C5 A3 G0

The family is sometimes known as the 'Gourd Family' because of the type of fruit. Hollowed fruits are often used as containers for food or drink and are commonly called calabashes. It is *Lagenaria siceraria* (previously known as *L. vulgaris*) which is best known, as a calabash, over the whole of tropical Africa.

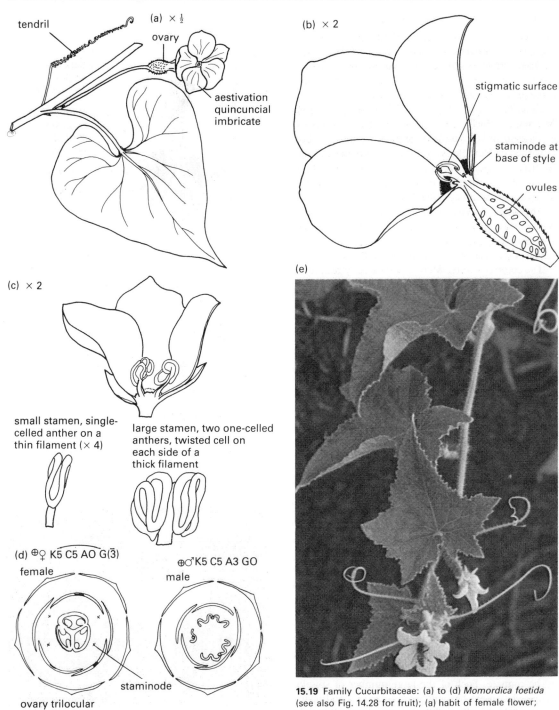

(a) × ½

tendril

ovary

aestivation
quincuncial
imbricate

(b) × 2

stigmatic surface

staminode at
base of style

ovules

(c) × 2

small stamen, single-
celled anther on a
thin filament (× 4)

large stamen, two one-celled
anthers, twisted cell on
each side of a
thick filament

(d) ⊕♀ K5 C5 A0 G(3̄)

female

⊕♂ K5 C5 A3 G0

male

staminode

ovary trilocular
at first, fruit
ripens to unilocular
in *Momordica* sp.

(e)

15.19 Family Cucurbitaceae: (a) to (d) *Momordica foetida*
(see also Fig. 14.28 for fruit); (a) habit of female flower;
(b) half flower (female); (c) half flower (male); (d) floral
diagrams; (e) *Cucumis sativa*, bee entering male flower,
female flower to right

There are many economically important plants in this family, of the genera *Citrullus*, *Cucurbita*, *Cucumis*, *Colocynthus*, *Lagenaria*, *Luffa*, *Telfairia* and *Sechium*. Most of these have edible fruits and/or seeds and some have edible leaves which are commonly used as 'spinach' or 'green'.

Cucurbita spp. include *C. pepo* (Vegetable Marrow), which has many different varieties. *C. maxima* and *C. moschata* are both commonly called Squashes or Pumpkins in different agricultural and horticultural publications from different parts of the world. Some varieties of *C. pepo* are also called Pumpkin or Summer Squash in America.

The genus *Cucumis* includes *C. sativa* (Cucumber), well-known world-wide and not given any other common (English) name. *C. melo* is a small, sweet melon, often called Honey Melon or Rock Melon.

The common Water Melon, with usually red, juicy flesh, is *Colocynthus citrullus*. *Citrullus lanatus* (formerly *Colocynthus vulgaris*) produces the edible seeds known as Egusi. The large seeds of *Telfairia occidentalis*, which has very large, ridged fruits, are also eaten.

Some *Luffa* spp. have fruits which are eaten in the young condition, but the commonest species over most of Africa, *L. aegyptica*, is grown for the fibrous skeleton of the dried fruit, which is used as a sponge, for cleaning.

Momordica charantia and *M. foetida* often grow as weeds of cultivated ground.

Family: Malvaceae

This is a family (Fig. 15.20) with world-wide distribution and includes mostly herbs and shrubs. The leaves are stipulate and either entire with usually digitately spreading veins or digitately compound. The flowers are solitary or in cymose inflorescences.

The flowers are actinomorphic, with perianth parts in fives and usually with an epicalyx of bracteoles, looking like a second calyx outside the true calyx. The sepals are free or connate at the base to form a short floral tube. The petals are free but, because the numerous stamens are united at the base, to form a staminal tube and the staminal tube is attached to the inside of the base of the corolla, they may appear to be connate. Each of the large number of spirally arranged stamens has only two pollen sacs and opens by a single slit. The ovary is superior and usually multilocular (five to many loculi). Placentation is axile. The fruit is a capsule which opens by a vertical

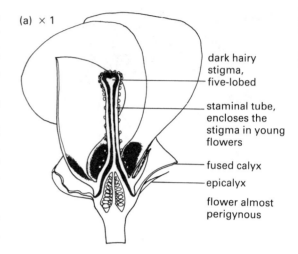

(a) × 1

dark hairy stigma, five-lobed

staminal tube, encloses the stigma in young flowers

fused calyx

epicalyx

flower almost perigynous

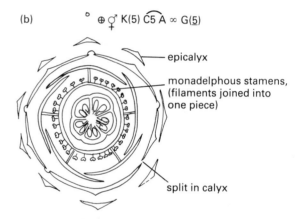

(b)   ⊕ ♂♀ K(5) C̑5 A ∞ G(5)

epicalyx

monadelphous stamens, (filaments joined into one piece)

split in calyx

(c)

15.20 Family Malvaceae, *Hibiscus esculentus*: (a) half flower; (b) floral diagram; (c) habit of plant showing open flower, bud to the right and fruit below

slit along each loculus or splits along the septa into several indehiscent, one seeded parts (mericarps).

The general floral formula for the family is:

$\oplus \, \male\female$ K(5) C5 A(∞) G$(\underline{1-5-\infty})$.

Hibiscus spp. are tropical herbs and shrubs. The leaves are entire. The flowers are usually solitary, yellow, pink or red. The ovary has five loculi and the fruit is a capsule which opens by slits. The genus includes *H. rosa-sinensis* ('Garden' Hibiscus), *H. esculentus* (*Abelmoschus · esculentus*, Okra) and *H. sabdariffa* (Roselle).

The Malvaceae are economically important as food plants and as producers of useful fibres. Food plants include Okra, the fruits and leaves of which are commonly eaten, and Roselle, which has edible, mucilaginous calyces. Fibre producers include *Gossypium* spp. (Cotton), which has hairy seeds (Fig. 14.27), *Sida rhombifolia* and *Urena lobata*, the last two of which provide fibres from the stem.

Several *Sida* spp. and *Urena* spp. are common as weeds of cultivated ground.

Ornamental plants include *H. rosa-sinensis*, *H. mutabilis* (Blushing Hibiscus) and *H. schizopetalus*.

Family: Caesalpiniaceae

The Caesalpiniaceae (Fig. 15.21) are the first family of the order Leguminosae. The family is almost entirely confined to tropical lands. It includes trees, shrubs and a few climbers and herbs. The leaves are stipulate and paripinnate or bipinnate or occasionally merely bilobed. Flowers are usually in racemose inflorescences.

The flowers are more or less zygomorphic with perianth parts in fives. The sepals and petals are free. The adaxial petal is enclosed within the others in the bud and is usually a slightly different size and shape from the others. There are usually two whorls of five free stamens but the number is sometimes reduced. The ovary is monocarpellary and superior, with marginal placentation. The fruit is a legume, opening explosively in dry weather.

The general floral formula is:

$\cdot|\cdot$ $\male$ K5 C5 A5 + 5 G$\underline{1}$.

Caesalpinia pulcherrima (Pride of Barbados) is an ornamental shrub, indigenous to Asia. The leaves are bipinnate. The yellow or red flowers are in terminal racemes which are usually compound at the base.

Bauhinia spp. are characterised by their bilobed

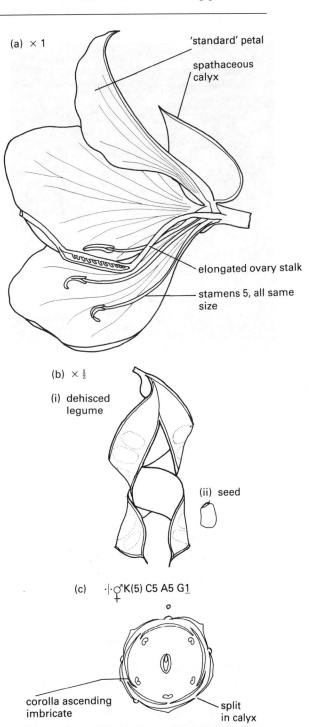

(a) × 1

'standard' petal

spathaceous calyx

elongated ovary stalk

stamens 5, all same size

(b) × ½

(i) dehisced legume

(ii) seed

(c) $\cdot|\cdot$ $\male$ K(5) C5 A5 G$\underline{1}$

corolla ascending imbricate

split in calyx

15.21 Family Caesalpiniaceae: (a) to (c), *Bauhinia purpurea*: (a) half flower; (b) fruit and seed; (c) floral diagram (see Fig. 15.7 for whole flower) (continued overleaf)

(d)

15.21 (continued) (d) *Cassia spectabilis*. See also Figs. 15.13 and 15.14 for *Caesalpinia* sp.

leaves. The genus includes several savanna trees and shrubs as well as a number of cultivated species. The flowers are showy, similar to those of *C. pulcherrima* above but larger.

Cassia spp. are mostly trees and shrubs with paripinnate leaves and yellow or purple flowers similar to those of *Caesalpinia pulcherrima* but less strongly zygomorphic. *Cassia rotundifolia* is a small herbaceous weed with yellow flowers and leaves with two leaflets only.

The economic importance of the family lies in the production of firewood and timber. Several *Cassia* spp. are planted to supply firewood. *Distemonanthus benthamianus* and *Gossweilerodendron balsamiferum* are rainforest trees which are exploited for timber. Forest species of *Brachystegia* and *Daniellia* also provide valuable timber.

Several small species of *Cassia* (e.g. *C. mimosoides*; *C. rotundifolia*) are important as weeds of cultivation.

In addition to *Caesalpinia pulcherrima*, ornamental species include *Delonix regia* (Flame Tree or Flamboyante), several *Cassia* spp. (e.g. *C. fistula*) and species of *Bauhinia* (e.g. *B. purpurea*).

Family: Mimosaceae

Like the Caesalpiniaceae, the Mimosaceae (Fig. 15.22) belong to the order Leguminosae and are mostly tropical. The family includes trees and shrubs and a few climbers and herbs. The leaves are almost always bipinnate, with conspicuous glands on the rachis. Inflorescences are racemose and consist of small flowers massed closely together on a globular or elongated axis.

The flowers are actinomorphic with perianth parts usually in fives. The sepals form a calyx tube, lobed

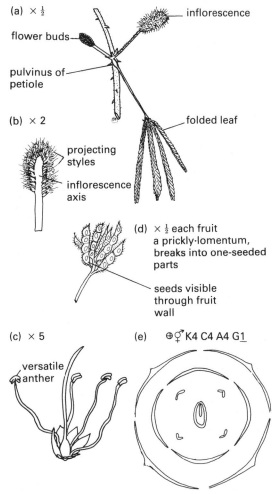

15.22 Family Mimosaceae: (a) to (e), *Mimosa pudica*; (a) flowering branch with leaf in 'sleep' position; (b) LS inflorescence; (c) single flower; (d) fruits, (e) floral diagram.

at the end. The petals are free or connate only at the base. There are usually two whorls of five stamens each. The ovary is monocarpellary and superior, with marginal placentation and the fruit is a legume.

The general floral formula is:

$$\oplus \male \; K(5) \; C5 \; A5 + 5 \; \underline{G1}.$$

Acacia spp. are mostly savanna shrubs and small trees, with elongated or spherical inflorescences of small, pale-coloured flowers and paired stipular spines at each node (Fig. 4.10).

Mimosa pudica (Sensitive Plant) is a common herb or small shrub with leaves which fold up when touched. The flowers are pink and differ from most members of the family in having sepals and petals in whorls of four and only one whorl of four stamens.

The economic importance of the group is mostly in timber production from forest trees. *Piptodeniastrum africanum* and *Cylicodiscus gabunensis* produce important timbers.

Albizia zygia, an introduced relative of several indigenous savanna trees, is commonly planted for ornament and shade in gardens and along roadsides. *Leucaena glauca*, an American tree, is also planted for ornament and sometimes as a shade tree in Cocoa and Coffee nurseries.

Family: Papilionaceae

This is the third family of the order Leguminosae. It is a cosmopolitan family (Fig. 15.23) of trees, shrubs, climbers and herbs. The leaves are characteristically imparipinnate and stipulate. Inflorescences are usually racemose.

The flowers are zygomorphic, with sepals, petals and stamens in whorls of five. The calyx usually forms a calyx tube. The petals are free. The adaxial petal (**standard**) enfolds the others in the bud and, later, almost always stands upright. The lateral petals (**wing** petals) lie outside the abaxial petals in the bud. The two abaxial petals are fused along their abaxial margins and form the **keel** of the flower. The structure of the corolla is very constant and makes members of the family very easy to recoginze. There are ten stamens, in two whorls of five, with the lower parts of the filaments connate into a staminal tube, or with nine stamens connate and the adaxial stamen free. The ovary is superior and monocarpellary, with marginal placentation. The fruit is a legume.

The general floral formula is:

$$\cdot | \cdot \; \male \; K(5) \; C5 \; A(5+5) \; \underline{G1}.$$

Crotalaria spp. are shrubby herbs, common and widely distributed in tropical countries. Their leaves may be 3- to 5-foliate or simple (Fig. 3.1), with or without stipules. Inflorescences are terminal and axillary racemes. In some species the fruit is strongly

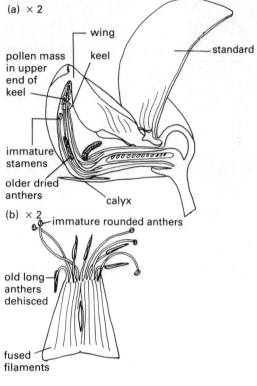

(a) × 2

wing

pollen mass in upper end of keel

keel

standard

immature stamens

older dried anthers

calyx

(b) × 2 — immature rounded anthers

old long anthers dehisced

fused filaments

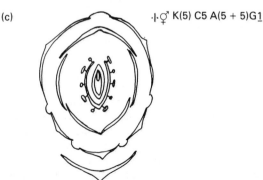

(c)

$\cdot | \cdot \male \; K(5) \; C5 \; A(5 + 5)\underline{G1}$

15.23 Family Papilionaceae, *Crotalaria spectabilis*: (a) half flower; (b) staminal tube cut open to show two sizes of stamens; (c) floral diagram (continued overleaf)

(d)

15.23 (continued) (d) inflorescence (raceme). See Fig. 3.1 for whole plant

inflated and the seeds become loose inside the legume when they are ripe. This gives such species the common name, Rattle Box.

Clitoria ternatea is a herbaceous climber with pinnate leaves and large, blue or white flowers. The flowers are inverted, with the standard petal pointing downwards and forwards.

The Papilionaceae are of very great economic importance, in several ways. In the first place, all have nitrogen fixing bacteria (*Rhizobium* spp.) in nodules on their roots (Chapter 24). Because of this, growing a crop of a plant belonging to this family increases the nitrogen content of the soil. Some members of the family are grown as 'green manure' and are ploughed into the soil after they have grown for a time (e.g. *Mucuna* sp.).

Forest species of *Pterocarpus* are exploited for timber.

Food crop plants include *Cajanus cajan* (Pigeon Pea), several types of beans (e.g. *Phaseolus lunatus*, Lima Bean), *Vigna unguiculata* (Cow Pea), *Arachis hypogaea* (Groundnut), *Voandzeia subterranea* (Bambara Groundnut) and *Cicer arietinum* (Chick Pea).

The commercial fibre known as Sunn Hemp (or San Hemp) is produced from the stem fibres of *Crotalaria juncea*.

The family also includes a number of weeds. *Mucuna pruriens* is a common climbing plant which seeds itself very rapidly on abandoned farm land. It is notable for the very unpleasant stinging hairs on its pods. Another climber, in similar situations, is *Abrus precatorius* (Crabs Eyes). The brilliant red and black seeds, which become visible when the pod opens (Fig. 14.23), are often used for ornament and also contain a deadly poison.

Among the ornamental species are *Gliricidia sepium* and *Erythrina indica*. *G. sepium* is often planted as a quick-rooting and quick-growing hedge or barrier plant or as support for climbing crop plants and is also grown as an attractive tree, for its purple or white flowers in hanging racemes. *E. indica* is prized for its racemes of scarlet flowers.

Family: Apocynaceae

This is a mostly tropical family (Fig. 15.24) and includes trees, shrubs, woody climbers and only a few herbs. All parts of the plants release milky juice (latex) when cut or broken. The leaves are simple, usually without stipules, and are usually in opposite pairs or in whorls. Flowers are solitary or in cymose inflorescences.

The flowers are actinomorphic, the calyx and corolla parts in fives. Sepals are free. Petals are connate, forming a long corolla tube and have contorted aestivation (Fig. 15.8). There are five stamens, epipetalous in the corolla tube. The ovary is bicarpellary and superior, with the carpels free from each other and styles and stigmas fused, or connate and partly free at the base. Where the carpels are free, placentation is marginal and each carpel ripens, usually, to form a follicle. Where the carpels are connate, placentation is axile or parietal and the fruit is often a berry. Seeds from follicles are often hairy or winged and are wind dispersed (Fig. 15.24).

The general floral formula is:

$$\oplus \male\female \; K5 \; \widehat{C(5)} \; A5 \; \underline{G2}.$$

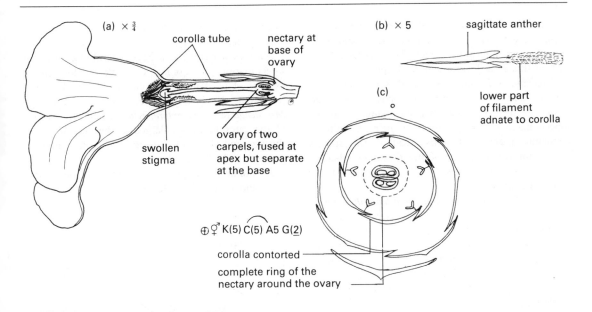

(a) × ¾

corolla tube

nectary at base of ovary

swollen stigma

ovary of two carpels, fused at apex but separate at the base

(b) × 5

sagittate anther

lower part of filament adnate to corolla

(c)

$\oplus \, \male\female \; K(5) \; C(5) \; A5 \; G(\underline{2})$

corolla contorted

complete ring of the nectary around the ovary

15.24 Family Apocynaceae: (a) to (d), *Allamanda* sp.; (a) half flower; (b) single stamen; (c) floral diagram; (d) habit showing yellow flowers and whorls of large, glossy leaves; (e) inflorescence of *Catharanthus roseus* showing the two follicles (arrowed) which develop from bicarpellary ovary

Rauwolfia vomitoria (Ordeal Tree) is a shrub or small tree with whorled, dark green leaves and small white flowers in terminal cymes. The carpels are free and, in this case, form a pair of drupes.

Strophanthus hispidus (Arrow Poison Plant) is a woody climber with leaves in opposite pairs and flowers with very long, twisted, hanging corolla lobes. Each flower produces a pair of follicles containing many hairy seeds. Seeds of this species are a commercial source of the poisonous alkaloid strophanthin, which is used in the treatment of some forms of heart disease.

In addition to *S. hispidus*, *Catharanthus roseus* (formerly *Vinca rosea*, etc., Madagascar Periwinkle) is grown to produce an important drug, extracted from the underground parts. *Funtumia elastica* (Lagos Rubber) is a source of rubber latex but its place as a rubber producer has been largely taken over by *Hevea braziliensis* (Para Rubber), a member of the Euphorbiaceae.

Ornamental plants include *Allamanda* spp., *Catharanthus roseus*, *Nerium oleander*, *Plumeria japonica* (Frangipani) and *Thevetia neriifolia* (Yellow Oleander).

Family: Rubiaceae

This is a family of world-wide distribution (Fig. 15.25) which is represented in tropical lands mostly by trees and shrubs, nearly all with stipulate simple leaves in opposite pairs. The stipules are very prominent and are often interpetiolar and fused in pairs between the leaf bases of opposite leaves. Flowers are solitary or in small axillary cymes.

The flowers are actinomorphic. The calyx is of five free sepals and epigynous. Petals are connate, forming a corolla tube. There are usually five parts to the corolla but the number is variable. Stamens are epipetalous and the same number as the petals. The ovary is inferior and most often bicarpellary, with axile or parietal placentation. Fruits include capsules, berries and drupes.

The general floral formula is:

$$\oplus \female^{\male} \; K5 \; \overparen{C(5)} \; A5 \; G(\overline{2}).$$

Macrosphyra longistyla is a scrambling shrub with opposite, stipulate leaves, which are covered with rather stiff hairs. The flowers are cream to yellow, in clusters at the ends of branches. The corolla tube is very long (up to 50 mm) and the style is much longer, with a much enlarged stigma at the tip. The

ovary has parietal placentation and ripens into a berry (inferior berry).

Mussaenda elegans is a large scrambling shrub of forest margins, with orange to red flowers in terminal cymes. The fruits are elongated, with long, persistent sepals. In several other species of *Mussaenda* (e.g. *M. erythrophylla*, Ashanti Blood) one of the calyx lobes is enlarged at the end to almost the size and shape of a foliage leaf, coloured pink, white or red (cf. Fig. 15.25).

Economically important species include timber trees and beverage producers. *Mitragyna ciliata* and

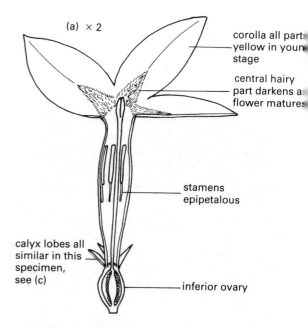

(a) × 2

corolla all parts yellow in young stage

central hairy part darkens as flower matures

stamens epipetalous

calyx lobes all similar in this specimen, see (c)

inferior ovary

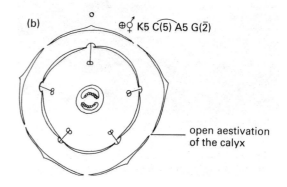

(b)

$\oplus \female^{\male} \; K5 \; \overparen{C(5)} \; A5 \; G(\overline{2})$

open aestivation of the calyx

(c)

(d)

15.25 Family Rubiaceae; (a) to (c) *Mussaenda arcuata*; (a) half flower; (b) floral diagram; (c) inflorescence, showing variation in size of calyx lobes in different flowers; (d) *Borreria* sp., a small weed with regular, 4-partite flowers

Nauclea diderrichii are important rain-forest timbers. *Coffea arabica* and other species of the same genus are grown to produce the Coffee beans of commerce (Fig. 14.20).

Common weeds of cultivation include species of *Borreria* and *Oldenlandia*.

The family includes many popular ornamental plants. These include exotic species of *Gardenia*, *Ixora* and *Rothmannia* (formerly *Randia*). The West African plant, *Mussaenda erythrophylla*, is also widely cultivated in gardens.

Family: Compositae

This is a very large family (Fig. 15.26) with world-wide distribution. It includes many herbs and a few shrubby plants, climbers and small trees. Leaf shape and attachment are variable and there are no stipules. The family is characterized by its inflorescences, which are capitular. The receptacle of the capitulum, to which the sessile florets are attached, is usually convex but may be flat or concave. It is surrounded by one or more series of involucral bracts, which subtend the outer florets. Inner florets may be subtended by more or less membranous receptacle bracts or receptacle bracts may be absent.

The calyx of each floret is epigynous and reduced to a whorl of hairs or bristles or absent. There are two sorts of florets with respect to the symmetry of the corolla.

Tubular (**disc**) florets are actinomorphic. The five petals are fused into a floral tube with very short lobes or teeth round the end. Ligulate (**ray**) florets are zygomorphic. There is a short corolla tube but the upper part of the corolla is drawn to one side and forms a strap-shaped structure, with three to five lobes or teeth at the end. In either case, the corolla represents five connate petals. Some genera have only tubular florets or only ligulate florets, while others have tubular florets at the centre of the receptacle and ligulate florets around the margin of the inflorescence (Fig. 15.26).

The five stamens are epipetalous in the corolla tube, with free filaments but with the five anthers partially fused to form a tube around the style (Fig. 15.26). Stamens may be lacking in the ligulate florets of some species which have both tubular and ligulate florets in the inflorescence.

The ovary is inferior, bicarpellary and unilocular and contains a single ovule, attached to the base (**basal** placentation).

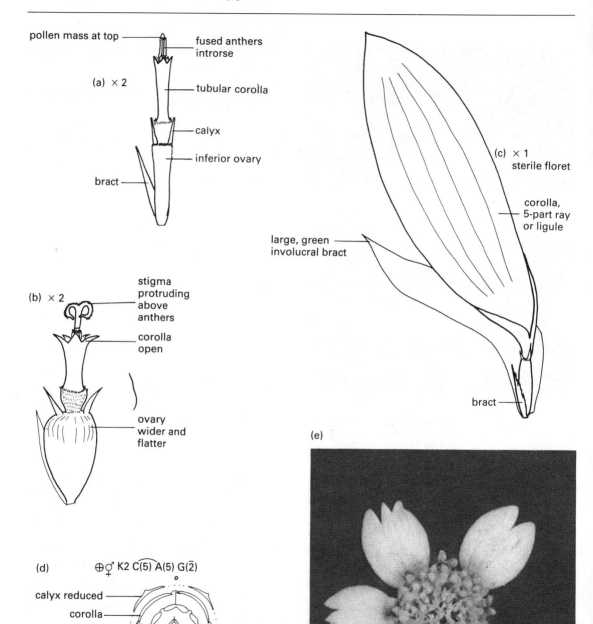

pollen mass at top

fused anthers
introrse

(a) × 2

tubular corolla

calyx

inferior ovary

bract

(c) × 1
sterile floret

corolla,
5-part ray
or ligule

large, green
involucral bract

bract

(b) × 2

stigma
protruding
above
anthers

corolla
open

ovary
wider and
flatter

(e)

(d)   ⊕♂ K2 C(5) A(5) G(2̄)

calyx reduced

corolla

anthers connate

bract

15.26 Family Compositae: (a) to (d) *Helianthus annuus*:
(a) young tubular floret from the centre of the capitulum;
(b) older tubular floret; (c) ligulate floret from margin of
capitulum; (d) floral diagram of tubular floret; (e) capitulum
of *Tridax procumbens* (see also Fig. 14.15)

The dry, one-seeded, indehiscent fruits (**achenes**) often have a pappus of hairs or bristles and are wind dispersed. Others have hooks or recurved hairs on the outside of the pericarp and are animal dispersed.

The general floral formulae are:

tubular florets $\oplus \male\female$ K0 $\overbrace{C(5)}$ A5 G$\overline{(2)}$;

ligulate florets $\cdot|\cdot$ $\male$ K0 $\overbrace{C(5)}$ A5 G$\overline{(2)}$.

Tridax procumbens and *Bidens pilosa* are among the many members of the family in which inflorescences contain both tubular and ligulate florets. Both are common, pantropical weeds. *T. procumbens* is annual or perennial with a procumbent stem and opposite leaves. Inflorescences are terminal and axillary, on long leafless peduncles. Fruits are wind-dispersed. *B. pilosa* is an upright annual with terminal and axillary inflorescences. The fruits have hooks and are dispersed by animals.

Inflorescences of *Ageratum conyzoides* and *Emilia sonchifolia*, both common weeds, contain only tubular florets. *A. conyzoides* (Goatweed) is a small erect herb with opposite leaves and pale blue flowers. Its leaves and stems give off a slightly unpleasant smell when broken and this gives the plant its common name. *E. sonchifolia* is an erect plant with alternate, rather elongated leaves of a greyish-green colour. The flowers are purple. Both have seeds which are dispersed by wind.

There are rather few examples of members of the family with ligulate florets only. Amongst these are *Lactuca* spp. *L. capensis* is an upright perennial herb which varies in height from about 10 cm to 1.5 m. It has opposite leaves and yellow inflorescences. *L. sativus* (Lettuce) has similar inflorescences.

The economic importance of the family is mostly as food. The leaves of *Lactuca sativa* and the rhizomes of the Jerusalem Artichoke (*Helianthus tuberosus*) are eaten as vegetables. The Sunflower (*Helianthus annuus*) (Fig. 15.26) is cultivated in tropical and subtropical countries for oil which is obtained from the seeds and is used, for instance, in the manufacture of margarine.

The list of weeds is very long indeed. Amongst the commonest are *Acanthospermum hispidum*, *Aspilia* spp., *Chrysanthellum americanum*, *Eupatorium odoratum* and *Synedrella nodiflora*.

Species of *Ageratum*, *Coreopsis*, *Cosmia*, *Gerbera*, *Helichrysum*, *Mikania* and *Zinnia* are grown as garden ornamentals. *Cosmia* sp. and *Zinnia* sp. often grow on waste land as garden escapes.

Family: Solanaceae

This is a world-wide family (Fig. 15.27) of herbs, with rather few shrubs, climbers and small trees. Leaves are alternate and exstipulate, usually simple. Flowers are solitary or in terminal or axillary cymes.

The flowers are actinomorphic. The calyx is of five sepals, connate at the base. The five petals are connate into a sometimes long floral tube. Stamens are epipetalous, alternating with the corolla lobes. The ovary is superior and bicarpellary, with axile placentation. The septum of the ovary is always at an angle of about 45° to the adaxial-abaxial axis of the flower. Fruits are capsules or berries.

The general floral formula is:

$\oplus \male\female$ K(5) $\overbrace{C(5)}$ A5 G$\underline{(2)}$.

Solanum torvum is a large herb or small shrub with entire leaves. The stem, petioles and major veins of leaves bear long prickles. The blue flowers are rather showy, with large yellow anthers grouped at the centre around the style. Cymose inflorescences are axillary in origin but displaced up the internode above the subtending leaf, as in other species of the genus and, for instance, in *Lycopersicum esculentum* (Tomato). The fruit is a berry.

Nicotiana tabacum is a large, upright herb with large, pale green leaves and white to pink flowers in terminal inflorescences.

Many members of the family are grown commercially, mostly for food. *Solanum melongena* (Garden Egg), *Lycopersicum esculentum*, *Capsicum annuum* (Sweet or Bell Pepper) and *C. frustescens* (Chilli Pepper) are grown for their fruits. Note that both nomenclature and species limits are unclear in some genera. For example, *Lycopersicum esculentum* is referred to by some writers as *Solanum lycopersicum* or *Capsicum lycopersicum*. Also, *C. frutescens* and *C. annuum* hybridise freely and varieties exist which share the characteristics of the two species.

Physalis peruviana (Cape Gooseberry) is also grown for its berries.

Solanum tuberosum (Irish Potato) is cultivated for its stem tubers at high altitudes in the tropics.

Nicotiana tabacum is grown for its leaves, which are dried and fermented and made into tobacco. Certain varieties with a very high nicotine content are grown as a source of nicotine, which is used to combat insect pests on crop plants.

Physalis angulata and *Datura stramonium* are weeds of cultivation.

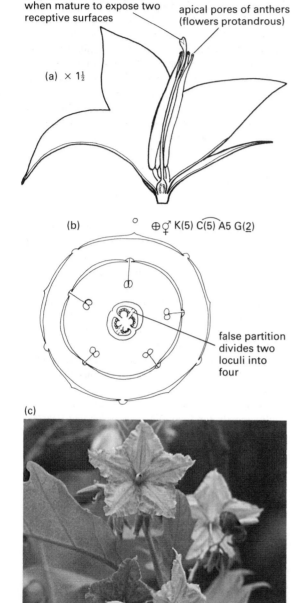

stigma, two lobed, opens when mature to expose two receptive surfaces

apical pores of anthers (flowers protandrous)

(a) × 1½

(b)   ○   ⊕♂♀ K(5) C(5̄) A5 G(2̲)

false partition divides two loculi into four

(c)

15.27 Family Solanaceae, *Solanum wrightii* (*S. macranthum*), (a) half flower; (b) floral diagram; (c) inflorescence, showing a newly-opened dark purple flower and older, paler flowers

Ornamental species include *Solanum wrightii* (Potato Tree), *S. seaforthianum* (Potato Creeper), *Brunfelsia calycina* (Yesterday, Today and Tomorrow), *Solandra grandiflora* (Golden Cup) and *Petunia* spp.

Family: Acanthaceae

Most of the plants in this tropical family (Fig. 15.28) are herbs or herbaceous climbers. Leaves are opposite and exstipulate. Inflorescences are usually compact, axillary, dichasial cymes.

Flowers are zygomorphic. The calyx, of four or five connate sepals, forms a calyx tube which may be very much reduced and its function taken over by a pair of bracteoles which enfold the bud and resemble sepals. The corolla is of five parts, connate into a tube which has an upper (adaxial) lip and a usually wider lower (abaxial) lip. There are four stamens in two pairs of unequal lengths, the filaments of which are free or connate in pairs. The ovary is superior, with axial placentation. The fruit is a capsule, dehiscing by slits from the top downwards.

The general floral formula is:

·|· ♂♀ K(5) C(5̄) A4 G(2̲).

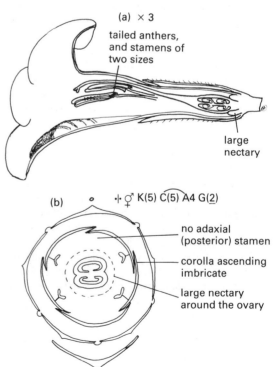

(a) × 3

tailed anthers, and stamens of two sizes

large nectary

(b)   ○   ·|·♂♀ K(5) C(5̄) A4 G(2̲)

no adaxial (posterior) stamen

corolla ascending imbricate

large nectary around the ovary

(c)

(d)

15.28 Family Acanthaceae: (a) to (c) *Asystasia gangetica*; (a) half flower; (b) floral diagram; (c) inflorescence, a one-sided cyme; (d) *Phaulopsis imbricata*, a common weed, with the white corolla divided into upper and lower lips

Asystasia gangetica is a spreading herb, common as a weed in forest areas. The flowers, about 2 cm long, are usually white, sometimes yellow, with purple markings in the tube.

Thunbergia erecta is a shrubby herb which may grow to as much as five metres high in the forest but is often much smaller and flowers in the open when only about thirty centimetres high. The flowers are purple, with a white corolla tube and about six centimetres long. The calyx is reduced and the base of the flower is enclosed between two leafy bracteoles.

The family includes several garden ornamentals. *T. erecta* (above) is sometimes grown in gardens. Its relative, *T. grandiflora*, a Malaysian species, is a large climber with white flowers. This species is naturalized in Africa and may be found growing wild in forest.

Monocotyledoneae

Family: Commelinaceae

This is a family (Fig. 15.29) of succulent annual or perennial herbs which are mostly confined to tropical parts of the world. Plants may be upright or often trailing. Leaves are alternate and have sheathing leaf bases. Terminal or axillary inflorescences are cymose and often enclosed within mucilage-filled sheathing bracts, so that the buds are not visible and flowers appear from between the bracts one at a time.

Flowers are usually actinomorphic, with three free sepals and three free petals. The stamens are free, sometimes in two whorls of three but often reduced to a single whorl. The ovary is superior, usually of three connate carpels. Placentation is axile and the fruit is usually a capsule.

The general floral formula is:

$$\oplus \, \male \; K3 \; C3 \; A3{+}3 \; G(\underline{3}).$$

Commelina diffusa is a very common weed, often in moist places. It has a straggling prostrate stem which roots at the nodes. Its bright blue or white flowers open early in the morning and shrivel up before noon. The two adaxial petals are well developed but the abaxial petal is reduced to such an extent that, at first sight, the flower seems to have only two petals. There are several other common species belonging to the same genus, including

(a) × 4

lateral pollen
sacs

staminal
hairs (see
Fig. 5.1)

stamens
epipetalous

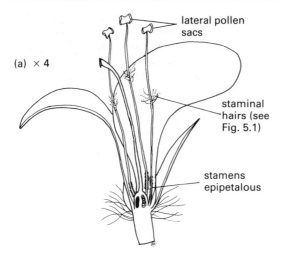

(b)   ⊕♂̦ K3 C͡3 A3 + 3 G(3)

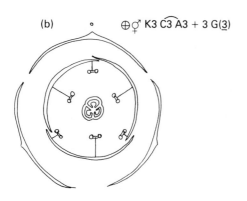

(c)

15.29 Family Commelinaceae: (a) *Setcreasea purpurea*, half flower; (b) *S. purpurea*, floral diagram; (c) *Commelina benghalensis*, inflorescence, mostly enclosed in a hairy spathe (see Fig. 14.16 for cleistogamous flowers)

C. africana which has yellow flowers. The other species have all three petals equally developed.

Palisosta hirsuta is a large, broad-leaved, upright herb which may reach a height of over three metres. Its inflorescences lack a sheathing bract and the flowers are white or purple and open in the evening.

There seem to be no species of this family which are of economic importance.

Several exotic genera are grown as garden plants, or in houses. Most are grown for the colour and ornamentation of their leaves, rather than for their flowers. These include *Rhoeo spathacea*, formerly *R. discolor*, an upright plant with purple and grey leaves, *Setcreasea purpurea*, with purple stems and leaves and *Zebrina pendula*, a trailing plant rather like *Commelina* spp. but with purple and grey leaves.

Family: Liliaceae

Members of this family (Fig. 15.30) are herbs and climbers which perennate by corms, bulbs or rhizomes or which have perennial leafy stems. Their leaves are alternate and their inflorescences are terminal or axillary racemes.

The flowers are actinomorphic, with two whorls of three, usually petaloid, perianth parts. Perianth parts may be free or connate into a perianth tube. Stamens are in two whorls of three, free or with their filaments adnate to perianth parts at the base. The ovary is superior, tricarpellary and usually trilocular with axile placentation. Fruits are capsules or berries.

The general floral formula is:

⊕♂̦ P3+3 A3+3 G(3).

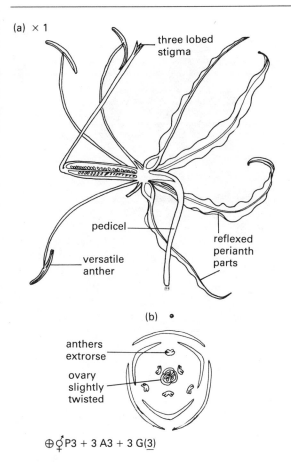

(a) × 1

three lobed stigma

pedicel

reflexed perianth parts

versatile anther

(b)

anthers extrorse

ovary slightly twisted

$\oplus \female \, P3 + 3 \, A3 + 3 \, G(\underline{3})$

15.30 Family Liliaceae: (a) *Gloriosa* sp., half flower (see Fig. 1.20 for whole flower); (b) floral diagram of *Gloriosa* sp.; (c) inflorescence of *Urginea* sp., a common plant of rocky outcrops in savanna

Gloriosa superba is a climber with leaf-tip tendrils (Fig. 4.11), found in forest margins, extending into savanna. The perianth parts are sharply reflexed when the flower is fully open. The flowers are yellow, changing to orange or pale red with age. *G. simplex*, with smaller flowers, is commoner in the savanna.

Aloe spp. are succulent plants with stout, perennial above-ground stems. Stems bear many fleshy leaves up to about 60 cm in length. Tall racemose inflorescences appear, usually, in the dry season. The tubular flowers are red, orange or yellow. *Aloe* spp. are characteristic of rocky areas in savanna.

There are no members of the family, as defined by Hutchinson, which are of great economic importance.

Indigenous and introduced species of *Aloe* are often planted for ornament. Other cultivated garden plants include introduced species of *Asparagus*, *Kniphofia*, *Lilium*, *Chlorophytum* and *Sansevieria*.

Family: Amaryllidaceae

Members of this family (Fig. 15.31) are bulbous perennials, usually with long, narrow leaves. Inflorescences are carried on long, bare peduncles (**scapes**), with a solitary flower, or an umbel of flowers, initially enclosed in one, two or more bracts, at the top. Perianth parts are in two whorls of three, all petaloid and either free from each other or forming a perianth tube at the base. There are two whorls of three stamens, free or adnate with the perianth tube. The ovary is tricarpellary, inferior or superior, with axile placentation. The fruit is a capsule or berry.

Note that this family is here defined as set out by Hutchinson, who takes the scapose, umbellate inflorescence as the most important character. Because of this, several genera with superior ovaries are included here rather than in the Liliaceae. Plants with superior ovaries are more usually included in the Liliaceae (e.g. *Allium* spp.). Note also that the common names of several members of the Amaryllidaceae include the word 'Lily', which is botanically misleading.

The general floral formula is:

$\oplus \male\female$ P(3+3) A3+3 G$\overline{(3)}$.

Haemanthus spp. (Blood Lilies) occur in forest and savanna but may be found also on the beaches next to lagoons and next to the sea. The almost spherical, bright red inflorescences appear before the leafy stems, which bear broad leaves.

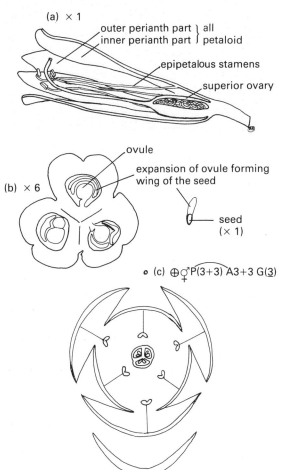

(a) × 1

outer perianth part ⎱ all
inner perianth part ⎰ petaloid

epipetalous stamens

superior ovary

(b) × 6

ovule

expansion of ovule forming wing of the seed

seed (× 1)

(c) $\oplus \male\female$ P$\overline{(3+3)}$ A3+3 G$\underline{(3)}$

(d)

(e)

15.31 Family Amaryllidaceae: (a) to (c) *Agapanthus* sp.; (a) half flower; (b) TS developing fruit and seed (see (Fig. 14.22 for capsule); (c) floral diagram; (d) *Hippeastrum* sp.; (e) *Haemanthus* sp. (see also Fig. 15.10)

Crinum spp. (Crinum Lilies) usually grow close to water in forest or savanna regions. The large white, or white and pink flowers are few, at the top of the scape.

As defined by Hutchinson, this family includes the genus *Allium*, which includes *A. cepa* (Onion), *A. ascalonicum* (Shallot), *A. porrum* (Leek) and *A. sativum* (Garlic). These are all widely cultivated food plants.

Garden ornamentals include *Agapanthus umbellatus* (Blue African Lily), *Hippeastrum equestre* (Harmattan Lily), *Hymenocaulis littoralis* (Spider Lily), *Pancratium* spp. and *Zephyranthes* spp.

Family: Gramineae

This family (grasses) (Fig. 15.32) is one of the largest and most important families of flowering plants. The family includes annuals and perennials, ranging from very small grasses to bamboos of tree-like size. The family is characterised by alternate leaves with sheathing leaf bases, with a small, usually membranous outgrowth (ligule) standing up at the junction of leaf base and leaf blade.

The grass inflorescence is often complicated. The ultimate unit of the inflorescence is the **spikelet**, which consists of one or a few small flowers enclosed between scale-like reduced leaves (**glumes**). Spikelets may be sessile on a single axis, forming a compound spike (usually called a spike), or may be part of a raceme or panicle.

Each flower is enclosed between two bracts, the **palea** (adaxial) and the **lemma** (abaxial). There is no perianth, but two small fleshy **lodicles** at the abaxial side may represent two reduced petals. There are three free stamens with versatile anthers. The monocarpellary ovary is superior and contains only one ovule with basal placentation. There is usually a divided, feathery style and stigma at the top of the ovary.

After fertilization, the testa of the developing seed becomes fused to the inside of the pericarp and the fruit is a caryopsis (grain) (Fig. 2.4), characteristic of grasses.

The general floral formula is:

$$\oplus \male \; P0 \; A3 \; \underline{G1}.$$

Andropogon spp. are large, perennial savanna grasses. The two-flowered spikelets are in pairs, forming parts of racemes, each of which is in the axil of a bract. The bract consists of a leaf base with a

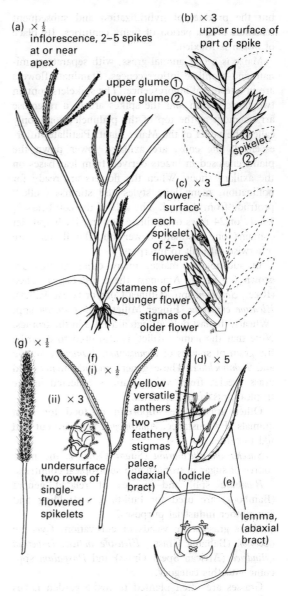

15.32 Family Gramineae: (a) to (e), *Eleusine indica*; (a) whole plant; (b) upper surface of spike: (c) lower surface of spike showing spikelets; (d) single flower; (e) floral diagram; (f) *Paspalum* sp.; (g) *Setaria* sp.

small leaf blade. Each lemma bears a stiff terminal bristle (**awn**). Spikelets drop from the plant intact and hygroscopic bending and unbending of the awns propels the spikelet along the surface of the soil.

Zea mays (Maize) is probably not a natural species,

but the product of hybridization and subsequent selection over a period of many centuries. Its place of origin is Mexico.

Maize is a tall, annual grass, with separate staminate and pistillate inflorescences. Staminate flowers are in a terminal inflorescence. The spikelets contain two flowers each and the spikes are in a racemose arrangement at the top of the peduncle, commonly called the tassel of the Maize plant. Pistillate inflorescences, the cobs, are axillary, lower down the plant, enclosed in bracts derived from leaf bases on the axillary shoot. When the flowers are ready for pollination, many long styles and stigmas ('silks') protrude from the end. The axis of the cob bears 8, 10, 12 or 24 rows of pistillate flowers. Each spikelet contains one pistillate flower but the flowers are otherwise difficult to interpret.

In addition to maize, many other grasses are grown for grain. Amongst these are *Oryza sativa* (Rice), *Sorghum* spp. (Guinea Corn or Great Millet), *Eleusine coracana* (Finger Millet) and *Triticum* spp. (Wheat), the last only at high altitudes in the tropics. Note that the name 'Millet' is also used to describe the grain of species of *Pennisetum*, *Panicum*, *Setaria* and *Echinochloa*. These genera include many wild grass species from which grain is collected in the tropical grasslands of Africa.

Other grasses are important as food for farm animals. For instance, *Andropogon* spp. are cut and fed to cattle.

Saccharum officinarum (Sugar Cane) is the main source of sugar in tropical and subtropical countries.

Bambusa spp. and species of related genera (Bamboos) are used for building, scaffolding and some other industrial purposes.

Many grasses are weeds of cultivation. *Cynodon dactylon* (Bahama Grass), *Eleusine indica*, *Imperata cylindrica* (African Spear Grass) and *Paspalum* spp. come into this category.

Grasses are often planted to make garden lawns and to cover playing fields, airfields and so on. Amongst these are *Cynodon dactylon*, *Axonopus compressus* and *Oplismenus* spp.

Key words

Flora. Flower arrangement: solitary; inflorescence. **Flower structure:** primitive, advanced; fusion, connation, adnation. **Receptacle:** hypogynous (superior ovary), perigynous (superior ovary), epigynous (inferior ovary); disc, nectary. **Gynaecium:** apocarpous, syncarpous; ovary, style, stigma; pistillate flower. **Androecium:** episepalous, epipetalous; stamen, filament, anther; extrorse, introrse; staminate flower. **Symmetry:** actinomorphic (regular), zygomorphic (irregular). **Inflorescence:** racemose, cymose; peduncle, bract, pedicel, bracteole. **Racemose:** raceme, pedicellate; spike, sessile; corymb, umbel; capitulum. **Cymose:** monochasium, dichasium. **Description:** drawing, whole flower, half flower; floral diagram; floral formula.

Summary

Flower structure is the basis of the classification of flowering plants. A primitive flower has many floral parts in spirals. An advanced flower has whorls of parts, variously joined and reduced in number. Symmetry of the flower may be radial or bilateral, giving actinomorphic or zygomorphic flowers respectively. Attachment of parts to the receptacle varies greatly, as does the degree of fusion of carpels with each other and with the receptacle. This variation also leads to differences in fruit structure.

Most flowers are grouped in inflorescences, mostly strictly racemose or cymose but some of mixed, compound structure.

Dicots and monocots differ in flower structure, general morphology and anatomy. A typical dicot flower has 5-partite flowers, sometimes reduced to 4-partite. A typical monocot flower has parts in threes and multiples of three. Most families of flowering plants have world-wide distribution, but some are mainly tropical or mainly temperate.

Questions

1 What reproductive and vegetative structures are most used in classifying plants as dicots and monocots?

2 What do we mean by evolutionary changes in flower structure? What is a 'primitive flower'? What is an 'advanced flower'?

3 Describe a named dicot zygomorphic flower with which you are familiar. In what ways could it be described as an advanced flower?

4 Distinguish between the following terms by writing a definition of each: (a) a hypogynous flower; (b) an inferior ovary; (c) a syncarpous ovary; (d) a protandrous flower; (e) a staminate flower.

16

Special nutrition in flowering plants

Most flowering plants are autotrophic and require only water, carbon dioxide, sunlight and certain mineral ions to provide all their nutritional needs. There are a few flowering plants which have adopted other types of nutrition which either supplement or totally replace the autotrophic habit. These include parasitic, saprophytic and carnivorous plants.

Parasites

Parasites obtain their nutritional requirements from other living plants by means of haustoria. A **haustorium** (Fig. 16.1) is an absorbing organ which penetrates the tissues of the host plant and makes close connections, usually with the xylem and phloem of the host. These connections are so close that it has been found that, in at least one case, cells of the host and cells of the parasite form plasmodesmata between them where they come into contact.

Thonningea sanguinea (dicot, Balanophoraceae) is a root parasite. It is a rain forest plant and parasitises the roots of many species of trees. The stem of *T. sanguinea* is a naked, horizontal rhizome, one end of which is attached to the root of a tree. The rhizome puts up vertical branches, covered with closely overlapping scale leaves, which form inflorescences at ground level. Each inflorescence consists of a mass of small, tubular flowers, surrounded by many bright red bracts.

Striga spp. (dicot, Scrophulariaceae) are of great importance in tropical countries, as pests of Guinea Corn and Maize (Fig. 16.2). They are root parasites, but since they have leafy stems and are not totally dependent on their hosts they are classified as **semiparasites** (Fig. 16.2).

The most perfect examples of stem parasites are members of the dicot family Cytinaceae (formerly classified in the Rafflesiaceae). These plants are reduced to strands of tissue within the stems of their woody hosts and, in some species, only the flowers appear at the surface, when they burst through the

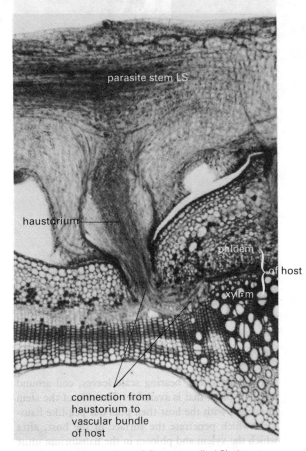

16.1 Stem and haustorium of *Cuscuta* sp. (in LS), the haustorium connecting to vascular tissues of stem of *Coleus* sp. (in TS)

bark. In other species, they appear as small moss-like plants, with green scale leaves and small flowers, appearing on the branches of the host. Members of this family are known from tropical South America and from East and Central Africa.

Cuscuta australis (Dodder, dicot, Convolvulaceae) is a stem parasite which usually attacks plants growing on wet ground. Its seeds germinate on the

16.2 *Striga* sp., a semiparasite on grass roots: parasitic on *Zea mays*

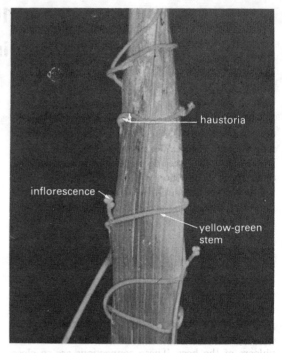

16.3 *Cassytha* sp., a twining semiparasite with yellow–green stems

soil and its stems, bearing scale leaves, coil around any host plant that is available. The side of the stem in contact with the host then puts out peg-like haustoria which penetrate the surface of the host, after which the xylem and phloem in the haustorium unite with the xylem and phloem of the host. Once it is established on a host, the parasite loses contact with the ground and its stems continue to grow around the host and form more and more haustoria. *Cuscuta* sp. lacks chloroplasts and its stems are yellow or red. It is a total parasite.

Cassytha filiformis (dicot, Lauraceae) is very like *Cuscuta* sp. and is much more widespread in tropical vegetation (Fig. 16.3). It has yellow or pale green stems in which chloroplasts are present. It is a semiparasite.

Tapinanthus spp., *Viscum* spp. and other members of the dicot family Loranthaceae (Mistletoes) are common as semiparasites on the branches of many sorts of trees. They are shrubby or hanging plants with deep green cladophylls or leafy stems. The large, woody haustorium connects to the xylem of the host only.

Saprophytes

Saprophytes lack chloroplasts and it is usually only the leafless inflorescences which appear above ground. The underground parts contain a symbiotic fungus (**mycorrhiza,** Chapter 23) which spreads out into the surrounding soil and leaf litter. The fungus obtains organic and other nutrients from the leaf litter, some of which pass back to the flowering plant.

Sciaphylla ledermannii (monocot, Triuridaceae) appears as small inflorescences with pink stems and flowers, on the floor of rain forest. There are also several species of saprophytic orchids (e.g. *Eulophia* spp.) and genera in the family Burmanniaceae (monocot).

16.4 *Drosera* sp.: rosette plant. Note the large glandular hairs on the leaves which close slowly and press the prey against the leaf surface, where it is digested

animal, such as a water flea, touches the trigger hairs, the lid opens inwards and the animal is sucked inside. The lid closes again when pressures are equalised and the animal cannot escape. After the animal has died, products of decomposition are absorbed by way of the hairs lining the bladder.

Drosera spp. (Sundews, dicot, Droseraceae) are small rosette plants which grow on waterlogged ground (Fig. 16.4). Their leaves have long petioles and spoon-shaped or elongated leaf blades. The leaf blade bears many long, multicellular hairs, each tipped with a gland which secretes sticky mucilage. There are also digestive glands on the leaf surface. When an insect or other small animal is trapped in the mucilage, the long hairs bend slowly inwards and press the prey against the leaf surface, where it is acted on by enzymes secreted by the digestive glands. Products of digestion are absorbed by the leaf and the remains of the prey are washed away by rain after the long hairs have straightened again.

Other carnivorous plants include *Dionaea muscipula* (Venus Fly Trap, Droseraceae) and Pitcher Plants (*Nepenthes* spp., dicot, Nepenthaceae; *Sarracenia* spp., dicot, Sarraceniaceae). *Dionaea* sp. traps its prey in its leaves, which fold inwards along the midrib and are edged with long, soft spines which form a cage when the trap closes. Like *Drosera* spp., they produce enzymes which digest the prey. The leaves of pitcher plants form large upright containers which catch rain water, in which small animals are trapped, drown and decay.

Carnivorous plants

All carnivorous plants are either water plants or plants which grow on boggy ground. Both habitats are poor in available nitrogen compounds and the carnivorous habit supplements the plants' supply of this part of their mineral requirements.

Utricularia spp. (Bladderworts, dicot, Lentibulariaceae) are mostly free-floating water plants. Their much branched·stems bear leaves which are divided into long, narrow segments with bladder-like traps attached. Each bladder has a door-like lid, branched sensitive hairs which form a 'trigger' mechanism outside the opening and water-absorbing hairs lining the inside. The lid is normally closed and the pressure inside the bladder is reduced because of the absorbing action of the hairs inside. When a small

Key words

Flowering plants: autotrophic, heterotrophic. **Heterotrophic**: parasitic, saprophytic, carnivorous. **Parasite**: haustorium; root parasite; stem parasite. **Saprophyte**: mycorrhiza, symbiotic fungus. **Carnivore**: water plants, nitrogen shortage; hairs, digestive glands, enzymes.

Summary

Most flowering plants are autotrophic and make their own food. Some flowering plants supplement their ordinary intake by trapping and digesting small animals, or depend partially or entirely on other plants, or on dead organic material, for their needs.

These heterotrophic plants, with special nutrition, are carnivores, parasites or saprophytes.

Questions

1 What different types of heterotrophic nutrition are found in flowering plants?
2 Distinguish between plants which are semiparasites and plants which are full parasites.
3 List three different ways in which carnivorous plants trap their prey.

Further reading for Section B

Baron, W. M. M. 1979. *Organization in Plants*, 3rd edn, Edward Arnold, London.
Coult, D. A. 1973. *The Working Plant*, Longman, London.
Cutler, D. F. 1978. *Applied Plant Anatomy*, Longman, London.
Cutter, E. G. 1978. *Plant Anatomy, Part 1, Cells and Tissues*, 2nd edn, Edward Arnold: London

Cutter, E. G. 1971. *Plant Anatomy, Part 2, Organs*, Edward Arnold: London.
Esau, K. 1977. *Anatomy of Seed Plants*, 2nd edition, Wiley: New York.
Exell, A. W., Wild, H. and Fernandes, A. 1960. *Flora Zambeziaca*, Crown Agents: London.
Hall, J. L., Flowers, T. J. and Roberts, R. M. 1982. *Plant Cell Structure and Metabolism*, 2nd edn, Longman: London.
Hubbard, C. E. and Milne-Redhead, E. 1952. *Flora of Tropical East Africa*, Crown Agents: London.
Hutchinson, J. and Dalziel, J. M., *Flora of West Tropical Africa*, (1927–36), 2nd edition (revised by R. W. J. Keay et al.) 1954, Crown Agents: London.
McLean, R. C. and Ivimey-Cook, W. R. 1967. *Textbook of Theoretical Botany*, vol. 1, new edition, Longman: London.
McLean, R. C. and Ivimey-Cook, W. R. 1956. *Textbook of Theoretical Botany*, vol. 2 Longman: London
Nielsen, M. S. 1965. *Introduction to the Flowering Plants of West Africa*, University of London Press: London.
Olorode, O. 1984. *Taxonomy of West African Flowering Plants*, Longman: London.
Sporne, W. R. 1975. *The Morphology of Angiosperms*, Hutchinson: London.
Thomas, M., Ranson, S. L. and Richardson, J. A. 1973. *Plant Physiology*, 5th edition, Longman: London.

Section C
FLOWERLESS GREEN PLANTS

In studying flowerless green plants, we are studying plants of which some must resemble the evolutionary ancestors of flowering plants. We are, in effect, making an evolutionary survey of the plant kingdom.

There was a time in the history of the earth when each of the divisions we are to study, with the exception of the bryophytes, formed a major part of the vegetation of the earth. Algae were the first green plants and lived almost exclusively in water. Pteridophytes were the first true land plants and also the first plants with true xylem and phloem. Gymnosperms were the first seed plants and largely displaced the pteridophytes as dominant vegetation before they were themselves displaced, over most of the surface of the earth, by flowering plants.

17
Algae

Algae are mostly water plants. They are of simple structure and their reproductive organs (sporangia, gametangia), unlike those of land plants, are naked and not protected by a jacket of sterile cells. Because of their common simplicity of structure and their similar ecology, algae were once regarded as a natural group and were classified as such. For the same reasons, it is convenient to treat them together, before passing on to land plants.

The principal divisions of algae are listed in Table 1.3. The Cyanobacteria are separated from other divisions because they are procaryotic and must be placed in the kingdom Monera. All other algal divisions are eucaryotic and are placed in the kingdom Plantae. The eucaryotic divisions are separated from each other on the basis of fundamental differences in biochemistry and ultrastructure. These differences include differences in chloroplast pigments, cell wall substances, storage substances and in the type and arrangement of flagella. All of these differences are referred to below, in the introduction to each division.

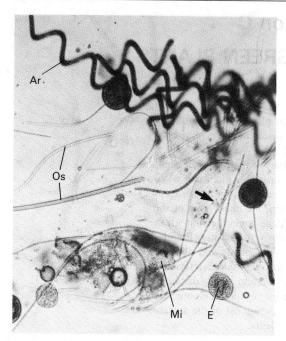

17.1 A mixture of mainly blue–green algae in a sample of water from a sewage pond, including unicellular (arrowed), cell aggregates (Mi = *Microcyctis* sp.) and filamentous types (Ar = *Arthrospira* sp.; Os = *Oscillatoria* sp.). The large, round cells (E) are encysted *Euglena* sp. (Euglenophyta) (LP)

Procaryotic algae – Cyanobacteria

Most members of the Cyanobacteria (blue-green algae) live in fresh water, but some are subaerial (grow on soil or rock surfaces) and some are marine (live in the sea).

Morphology

Some blue-green algae are unicellular, others form more or less irregular cell aggregates and others are filamentous (Fig. 17.1). Filamentous species may be unbranched or may have false or true branching.

Cell structure

The cell wall is composed mostly of mucopeptides. These are substances in which amino acids are bonded to sugar molecules. There is always a mucilage sheath outside the cell wall.

The protoplast is bounded by a plasmalemma,

similar to that of eucaryotic cells. However, there is no nuclear envelope or endoplasmic reticulum and there are no Golgi bodies, no mitochondria and no plastids.

DNA lies free as a thread, usually at the centre of the cell. It is not associated with protein to form DNP, as it is in eucaryotic organisms. There are no structures comparable to the chromosomes of higher organisms. **Chlorophyll lamellae** also lie free, in the outer part of the protoplast (**chromoplasm**). The chlorophyll lamellae are similar to those of eucaryotic plant cells, but there is no chloroplast envelope. Pigments in the chlorophyll lamellae include chlorophyll *a* (green), xanthophylls (yellow), carotene (yellow-orange), phycocyanin (blue) and phycoerythrin (red). The colour of the chromoplasm, usually blue-green, is the result of the combination of the colours of all these pigments.

Ribosomes, smaller than those in the cytoplasm of eucaryotic cells, are distributed throughout the protoplast. There are also granules of various storage substances, including cyanophycin (proteinaceous) and other substances which have not been identified. The protoplasts of many blue-green algae also contain **gas vacuoles**. Changes in the size of gas vacuoles alter the specific gravity of the cells as a whole and cause them to float or sink in the water at different times.

Under the light microscope, it is possible to see the chromoplasm as a coloured region. Storage granules and gas vacuoles also can be distinguished but all other features of the protoplast can be resolved only with the electron microscope.

The cells of blue-green algae do not have flagella at any stage in their life history.

Cell division

Blue-green algae do not have chromosomes and there is no process similar to mitosis or meiosis. There is presumably replication of the DNA and a separation of daughter threads towards opposite ends of the cell prior to cell division, but nothing is known of the details of the process.

Division of the cell is usually by the gradual ingrowth of a septum from the sides of the cell, until the cell is divided into two daughter cells.

Reproduction

Unicellular species reproduce by **fission**, the division of a cell to form two daughter cells. In filaments, cell division results in the growth in length of the filament. Breakage of filaments (**fragmentation**) results in vegetative reproduction. Most blue-green algae produce asexual spores, some of which are able to survive desiccation.

There is no known form of sexual reproduction.

Special features of blue-green algae

Some blue-green algae are able to fix atmospheric nitrogen, when their environment is deficient in nitrates. In this process, atmospheric nitrogen is incorporated into compounds which can be used by the cell. Nitrogen fixation by blue-green algae is of great ecological importance, especially in savanna soils, where blue-green algae may be an important source of nitrogen compounds for the vegetation as a whole.

In some filamentous blue-green algae, fixation of nitrogen is known to be a function of specialized cells known as **heterocysts**, which occur at intervals along the filament. Heterocysts are usually larger than ordinary cells of the filament, have thicker walls and colourless contents (Fig. 17.2). Their function is, however, not fully understood. Although they are a site of nitrogen fixation in some species, they occur widely in filamentous species which are not known to fix nitrogen. Also, some unicellular blue-green algae, in which there can be no equivalent of a heterocyst, are known to fix nitrogen.

Some blue-green algae are capable of **chromatic adaptation**. If they are exposed to light of a colour other than that of ordinary sunlight, the balance of the pigments in their chromoplasm alters in such a way that they take up a colour complementary to that of the incident light. Chlorophyll *a* absorbs and uses mostly red light, its complementary colour. Other (accessory) pigments absorb light of other wavelengths and are able to pass on the absorbed energy to the chlorophyll, which is the only pigment directly involved in photosynthesis.

Classification and examples of blue-green algae

The division Cyanobacteria contains a single class, which is divided into five orders. Separation into the five orders is based on morphology (unicellular, filamentous, etc.) and on the types of spores and other asexual reproductive structures formed.

Chroococcus spp. are common in fresh water. The cells are more or less spherical. After cell division, the daughter cells form new cell walls, inside the wall

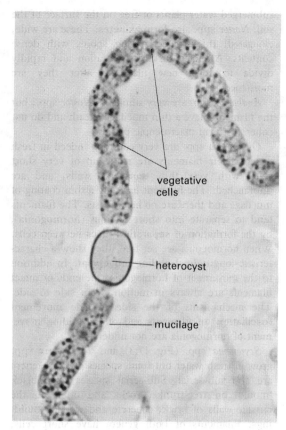

17.2 Part of a filament of *Nostoc* sp., showing a heterocyst and the wide, rounded vegetative cells (HP). The filaments of *Nostoc* sp. are embedded in a mass of mucilage

of the parent cell, after which the parent cell wall persists for some time. Further divisions may occur before the first cell wall breaks down. Aggregates of four to eight cells are commonly seen.

Microcystis spp. have very small, spherical cells which form loose, irregularly-shaped aggregates at the surface of nutrient-rich bodies of water, for instance, in sewage ponds.

Nostoc spp. (Fig. 17.2) are filamentous and have the ability to fix nitrogen. They are found in fresh water and on wet soil. Some species are symbiotic (Chapter 24) and live within the tissues of land plants (e.g. in the thallus of *Anthoceros* spp.), or are the algal partner in some lichens.

The filaments of *Nostoc* spp. secrete a great deal of mucilage and cohere to form macroscopic colonies, often attached to the stems and leaves of submerged water plants or free on the surface of the soil. *Nostoc* spp. also form **akinetes**. These are wide, elongated, thick-walled resting spores with dense contents. Akinetes survive desiccation and rapidly divide to form new filaments after they are moistened.

Anabaena spp. are very similar to *Nostoc* spp., but the filaments have a thin mucilage sheath and do not cohere to form macroscopic bodies.

Oscillatoria spp. are very common indeed in fresh water. Their filaments are made up of very short cells, with very thin transverse walls, and are unbranched. The filament has only a thin coating of mucilage and there are no heterocysts. The filaments tend to separate into short sections (**hormogonia**), by the formation of **separation discs** between cells. When hormogonia are set free, they show a characteristic longitudinal **gliding movement**. In addition to the movement of hormogonia, the ends of intact filaments are always in motion, from side to side. The mechanisms of the side to side movement (**oscillation**) of filaments and of the gliding movement of hormogonia are not understood.

Scytonema spp. (Fig. 17.3) and *Tolypothrix* spp. grow in fresh water but some species of both genera are also sub-aerial. Sub-aerial species form black 'mould' on tree-trunks, rocks and often on the outside walls of brick, concrete and wooden buildings. Filaments of both genera have short cells, inside a distinct mucilage sheath. Both also have false branching. In *Scytonema* spp., false branching occurs in sections of filaments between heterocysts and there may be one or two branches from the same point. In *Tolypothrix* spp. a single false branch is formed, always immediately next to a heterocyst.

Eucaryotic algae

Of the divisions of eucaryotic algae listed in Table 1.3, we shall concentrate attention on the Chlorophyta (green algae), Euglenophyta (euglenoids), Bacillariophyta (diatoms) and Phaeophyta (brown algae) and refer briefly to the Rhodophyta (red algae). Before dealing with these divisions, we shall first look at common features of all the divisions.

Morphology

Most divisions of eucaryotic algae include both unicellular and multicellular forms. There are also

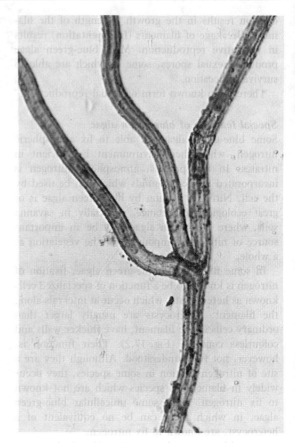

17.3 *Scytonema* sp.: filament showing false branching (LP)

examples of siphonaceous forms, in which the whole plant body consists of a single large cell which contains many nuclei.

Unicellular algae may be motile (**flagellate**) or non-motile (**non-flagellate**).

Cell aggregates are formed when cells fail to separate immediately after cell division and remain together in a more or less irregular mass. A cell aggregate may be any size or shape and may contain any number of cells. Cell division results in increase in size of the aggregate, which may also break up to form smaller aggregates.

A coenobium is a colony of fixed maximum size, shape and cell number. Once a coenobium is formed, increase in size is by the enlargement of cells. Cell division occurs only in asexual reproduction and in the formation of gametes. Coenobia may be motile or non-motile.

A filament is made up of cells arranged end to end and cell divisions are all, or almost all, in the transverse plane. If cell divisions are strictly in the transverse plane, the filament is unbranched. Branched filaments are the result of occasional oblique divisions, which produce branches.

A parenchymatous thallus is formed when cells divide in two or more planes and all the products remain closely attached to each other. If cell divisions occur in two planes, an expanded thallus, one cell in thickness, is produced. If divisions are in three or more planes the thallus is two or more cells thick and either flattened or rod-shaped. In a transverse section, the tissue resembles a section through parenchyma of a higher plant.

A pseudoparenchymatous thallus is formed by a filamentous alga, when the filaments cohere closely and form a usually macroscopic plant body. The appearance of a transverse section may resemble a section through parenchyma of a higher plant.

In most **siphonaceous** algae, the cell (**coenocyte**) is a much-branched filament with a continuous vacuole and a lining of cytoplasm which contains many nuclei. Septa are formed only at the bases of reproductive structures. In some siphonaceous green algae, filaments cohere to form a pseudoparenchymatous thallus.

Cell structure

Cells of eucaryotic algae are broadly similar in organelle content to those of flowering plants. Variations are in the composition of cell walls and other biochemical features, in the type and number of chloroplasts and in the type and arrangement of flagella.

Biochemical variations will be dealt with in the introduction to each division.

Chloroplasts are similar to those of flowering plants, but the chlorophyll lamellae do not form grana and each chloroplast may contain one or more pyrenoids. **A pyrenoid** is a protein body which acts as a centre for the formation of starch or another storage product.

The basic structure of flagella produced by eucaryotic organisms is extremely constant, whether in plants, animals or fungi. Each flagellum (Fig. 17.4) is bounded by an extension of the plasmalemma. Within the plasmalemma there is a ring of nine pairs of fused tubules and a pair of unfused tubules at the centre. The tubules are proteinaceous. At the base

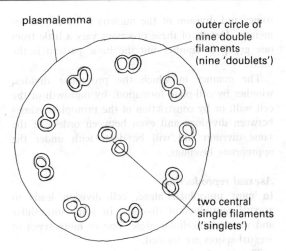

17.4 Diagram of TS flagellum of a eucaryotic organism, as seen with EM

of the flagellum, within the main part of the cell, the tubules are joined to a **basal body**, which is also composed of protein tubules.

In eucaryotic algae, there are two types of flagella (Fig. 17.5), both of which have the basic structure described above. First, there is the 'whiplash' (**acronematic**) type of flagellum, in which the outline of the flagellum is smooth. Secondly, there is the 'tinsel' (**pantonematic** or **pleuronematic**) type of flagellum, in which the flagellum bears many fine outgrowths on all sides. The type of flagellum, the number of flagella present and the point of attachment of flagella (anterior, lateral, posterior) is characteristic of each division of the eucaryotic algae.

Cell division

Like higher plants, eucaryotic algae have chromo-

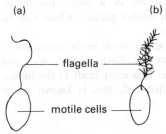

17.5 Types of flagella in algae: (a) acronematic (whiplash type), with a smooth outline; (b) pantonematic (tinsel type), with fine outgrowths from the axis

somes and division of the nucleus is by mitosis or meiosis. Details of these processes vary a little from one group to another but the basic pattern is the same.

The manner in which the protoplast divides, whether by cell-plate formation, by ingrowth of the cell wall, or by constriction of the protoplast, varies between divisions and even between orders of the same division and will be dealt with under the appropriate headings.

Asexual reproduction

In most unicellular algae, cell division leads to asexual reproduction (fission). In some unicellular and most multicellular algae, one or more types of asexual spores are formed.

The two basic types of spores are **zoospores**, which are motile, and **aplanospores**, which are non-motile. Other names are given to different types of aplanospores.

Filamentous algae often increase the number of filaments by fragmentation (breaking into two or more pieces). This is a type of vegetative reproduction.

Sexual reproduction

There are three basic types of sexual fusion (**syngamy**), isogamy, anisogamy (heterogamy) and oogamy.

Isogamy is the fusion of two gametes of the same size, structure and behaviour. There may be slight variation in the sizes of gametes which fuse, but these are not regular and depend on the stage of development of the cell from which gametes are formed.

Anisogamy is the fusion of gametes of which one is always larger than the other, both with the same structure and behaving in the same way.

Oogamy is the fusion of a large, non-motile gamete with a much smaller gamete, which may be motile or non-motile.

Isogamy and anisogamy result in the formation of a zygote which may lay down a thick wall and become a **zygospore**. If oogamy leads to the formation of a thick-walled cell, this is known as an **oospore**.

Life histories

There are three types of life cycles in eucaryotic organisms, all of which are found within the algae.

In the **haplontic** life cycle (Fig. 17.6), the vegetative stage of the organism is haploid (with a single set of chromosomes). Cell division by mitosis results in asexual reproduction of unicellular algae, or growth of multicellular algae. Gametes are also produced by mitosis. Syngamy results in the formation of a diploid zygote, which may form a thick wall and become a zygospore or oospore. This is the only diploid cell in the life cycle. Sooner or later, this diploid cell undergoes meiosis and produces four haploid **meiospores**. There are no diploid vegetative (non-reproductive) cells in a haplontic life cycle.

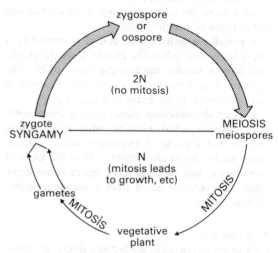

17.6 Diagram of a haplontic life cycle (in this and other life cycle diagrams, N = haploid phase, 2N = diploid phase)

In the **haplodiplontic** life cycle (Fig. 17.7), vegetative cells are produced, by mitosis, in both the haploid phase and the diploid phase. The haploid plant grows and, sooner or later, produces gametes. After syngamy, the zygote undergoes mitosis and a diploid plant body is produced. Some of the cells of the diploid plant undergo meiosis and produce meiospores. Meiospores germinate to form haploid plants. The haploid plant (haploid generation) is the **gametophyte** generation. The diploid plant is the **sporophyte** generation. The full life cycle involves an alternation of a haploid gametophyte generation with a diploid sporophyte generation. Plants with a haplodiplontic life cycle are said to show **alternation of generations**.

In the **diplontic** life cycle (Fig. 17.8), only the diploid phase produces vegetative cells. Some of the

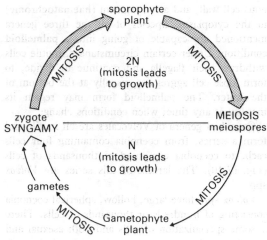

17.7 Diagram of a haplodiplontic life cycle

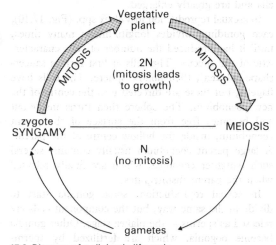

17.8 Diagram of a diplontic life cycle.

diploid cells undergo meiosis and produce gametes. After syngamy, the zygote divides by mitosis and produces diploid vegetative cells. The diplontic life cycle is uncommon in plants, but it is the life cycle found in all multicellular animals.

Chlorophyta – green algae

Green algae are mostly fresh-water organisms, but some are sub-aerial and a few are marine.

The division includes examples of all morphological types, from motile and non-motile unicellular forms to parenchymatous and pseudoparenchymatous forms. The structure and biochemistry of their cells are very similar to those of land plants. The main cell wall material is cellulose, the chloroplast pigments are chlorophyll *a*, chlorophyll *b*, xanthophylls and carotene, and the main storage product is starch.

Motile cells almost always have two flagella attached at the anterior end of the cell. Flagella are always of the whiplash type.

The division Chlorophyta is divided into two classes, Chlorophyceae and Charophyceae. The Chlorophyceae are divided into twelve or more orders, according to which classification is used. The Charophyceae include only one order. We are going to look at members of seven orders of Chlorophyceae (Volvocales, Ulotrichales, Ulvales, Oedogoniales, Zygnematales, Chlorococcales and Caulerpales) and the one order of Charophyceae (Charales). Examples are chosen for their botanical importance and for their availability in tropical Africa. The deepest treatment is given to the most important examples.

Volvocales
Members of the Volvocales are mostly motile and include unicellular and coenobial forms. The life cycle is always haplontic.

Chlamydomonas spp. (Fig. 17.9) are unicellular and motile, with two whiplash flagella at the anterior end of the cell. There is a thin cellulose cell wall. The protoplast contains one cup-shaped chloroplast with a single pyrenoid. There is a single red eye-spot (**stigma**) embedded in the surface of the chloroplast,

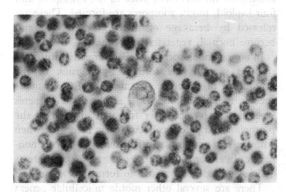

17.9 Volvocales, a unicellular motile form: many motile cells of *Chlamydomonas* sp., each of which is about 10 μm in diameter. The large cell is an encysted *Euglena* sp. (Euglenophyta). (LP)

usually near the anterior end of the cell. The eyespot is part of a light-sensitive system which enables the organism to make directional movement responses to differences in light intensity (**phototaxis**). There are also two contractile vacuoles close to the front end. These are responsible for osmoregulation, the control of the amount of water within the protoplast.

Asexual reproduction in *Chlamydomonas* spp. is by longitudinal fission. Flagella are first withdrawn, after which the nucleus divides by mitosis and the protoplast becomes constricted between the daughter nuclei and divides one to four times, to form two, four, eight or sixteen cells, according to species. Each daughter cell then secretes a new cell wall and flagella are formed. Daughter cells are released by the breakdown of the parent cell wall.

There is some variation between different *Chlamydomonas* spp. in the details of sexual reproduction. Some species are homothallic and others are heterothallic. **In homothallic** species, any pair of gametes can fuse together, even if they come from a single clone and are therefore genetically identical. In **heterothallic** species, only gametes of different strains will fuse. The genus also includes species which are isogamous, anisogamous and oogamous. In *C. eugametos*, an isogamous species, ordinary individuals act as gametes when two compatible strains meet. They come together by their anterior ends and start to fuse. The gametes then cast off their cell walls and the protoplasts merge to form a naked, spherical cell. After nuclear fusion, the zygote forms a thick cell wall and becomes a zygospore. After a resting period and under suitable conditions, meiosis takes place in the zygospore and four haploid, motile zoospores are formed. These are released by breakage of the zygospore wall and become normal, vegetative individuals.

In anisogamous species (e.g. *C. braunii*), different sizes of gametes are formed by division of the protoplasts of different parent cells. In oogamous species (e.g. *C. coccifera*), small male gametes (sperms) are formed after repeated division of the protoplast of a parent cell. An **ovum** is formed when a normal cell withdraws its flagella and becomes non-motile. In either case, the zygospore or oospore rests and later divides by meiosis to form four zoospores.

There are several other motile unicellular genera which are similar to *Chlamydomonas*. *Carteria* spp. are very similar indeed, except that each cell has four flagella. *Haematococcus* spp. have a very thick, gelati-nous cell wall, and a red pigment (**haematochrome**) in the cytoplasm. Species of all the three genera mentioned are capable of going into a **palmelloid** condition. Under certain circumstances, motile cells withdraw their flagella and continue to divide, to form loose cell aggregates, usually at the bottom of the water. The palmelloid form may regain its motility at any time, when conditions change.

Coenobial genera of Volvocales are all motile and form a series, from coenobia containing four cells each, to coenobia composed of thousands of cells (Fig. 17.10). The largest of this series are *Volvox* spp.

Volvox spp. have large, hollow, spherical coenobia consisting of hundreds or thousands of cells. There is some specialization of cells and both asexual and sexual reproduction involve only a few of the cells of the coenobium. These cells (**gonidia**) have no flagella and are greatly enlarged.

In asexual reproduction in *Volvox* spp. (Fig. 17.10), each gonidium divides longitudinally many times, until it has produced the number of cells characteristic of the species. The cells at first form a saucer-shape (**plakea**), then hollow sphere. The cells have flagella, but these are directed into the centre of the new coenobium. The sphere then turns inside-out and becomes free from the surface of the parent coenobium, inside the hollow centre of the parent. A large parent coenobium usually contains several such daughter coenobia, which are finally released when the parent disintegrates.

In sexual reproduction, some gonidia start to divide in the same way, but the component cells are released as sperms at the plakea stage. Other gonidia become oogonia, which are fertilized by sperms while they are still in position as part of the parent coenobium. Oospores form thick walls and are eventually set free. After a resting period, meiosis occurs but only one of the four nuclei which are produced survives. The contents of the oospore are then liberated as a single zoospore, which then divides to form a small coenobium. Repeated asexual reproduction leads to the formation of larger and larger coenobia, until the full size characteristic of the species is attained.

Chlamydomonas spp. and species of various coenobial genera are common in the tropics. They occur wherever there is clear, fresh water. They are especially common in temporary pools which form in hollows in rocky outcrops in the wet season.

(a)

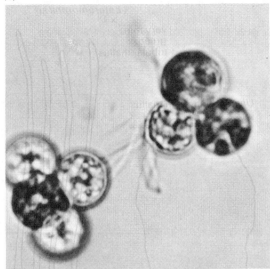

(b)

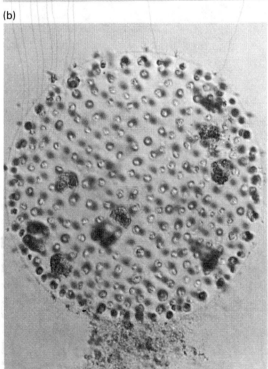

17.10 Volvocales, coenobial forms: (a) two coenobia of *Gonium* sp., each consisting of four flagellate cells similar to *Chlamydomonas* sp. (HP); (b) a coenobium of *Volvox* sp., made up of hundreds of small flagellate cells and a few large cells (gonidia) which will subsequently be involved in asexual or sexual reproduction (LP)

Ulotrichales

Members of the Ulotrichales are filamentous, unbranched or branched, with a single parietal chloroplast, containing several pyrenoids, in each cell. Life cycles are haplontic or haplodiplontic.

The life cycle of *Ulothrix* spp. may be taken to start when a biflagellate or tetraflagellate zoospore settles down on a firm surface by its anterior end. The anterior end widens slightly and becomes firmly attached to the substrate to form a **holdfast**. A series of transverse divisions produces an unbranched filament, which continues to increase in length as more transverse cell divisions take place. All the cells are green, except for the holdfast. Each green cell has a single parietal chloroplast containing several pyrenoids (Fig. 17.11).

(a)

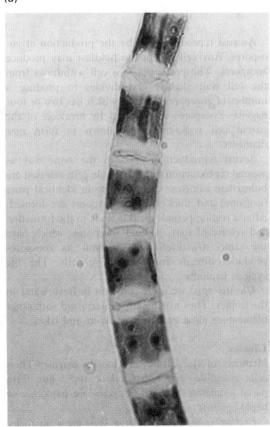

17.11 Ulotrichales, *Ulothrix* sp.: (a) part of filament (continued overleaf)

(b)

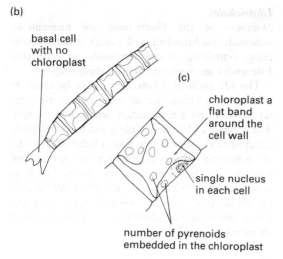

basal cell
with no
chloroplast

(c)

chloroplast a
flat band
around the
cell wall

single nucleus
in each cell

number of pyrenoids
embedded in the chloroplast

17.11 (continued) (b) basal part of filament; (c) a single cell

(a)

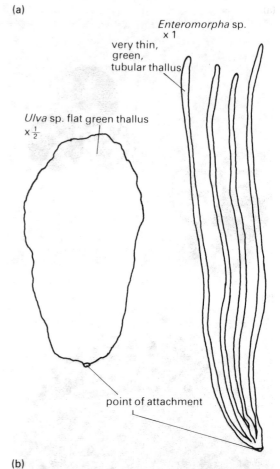

Enteromorpha sp.
x 1

very thin,
green,
tubular thallus

Ulva sp. flat green thallus
x $\frac{1}{2}$

point of attachment

Asexual reproduction is by the production of zo-ospores. Any cell except the holdfast may produce zoospores. The protoplast of a cell withdraws from the cell wall slightly and divides to produce a number of zoospores, each of which has two or four flagella. Zoospores are released by breakage of the parent cell wall and settle down to form new filaments.

Sexual reproduction starts in the same way as asexual reproduction but the motile cells released are biflagellate gametes. Gametes fuse in identical pairs (isogamy) and thick-walled zygospores are formed. After a resting period, meiosis leads to the formation and release of four, haploid zoospores, which have the same structure and behaviour as zoospores produced directly from vegetative cells. The life cycle is haplontic.

Ulothrix spp. are quite common in fresh water in the tropics. They usually grow intermixed with other filamentous algae on stones in rivers and lakes.

Ulvales

Members of the Ulvales are mostly marine. Their cells resemble those of *Ulothrix* spp., but form parenchymatous thalli. Life cycles are haplontic or haplodiplontic.

Enteromorpha spp. and *Ulva* spp. grow on rocky seashores throughout the world. *Enteromorpha* spp. usually grow high up on the seashore, where they

(b)

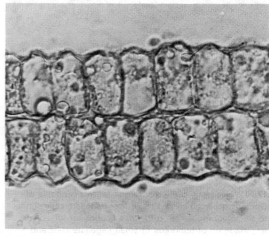

17.12 Ulvales: (a) habit of *Ulva* sp. and *Enteromorpha* sp.;
(b) TS *Ulva* sp., two cells in thickness (HP)

may be covered by sea water for only part of each day and often where there is some fresh water running into the sea. *Ulva* spp. grow in rock pools, or in shallow coastal water, where they are permanently immersed in sea water.

In either genus, we may take the life cycle to begin when a haploid zoospore becomes attached to a rocky surface, in the same way as that of *Ulothrix* sp. The first cell divisions are transverse and a short filament is formed. Further divisions are in three planes, and a parenchymatous thallus, two cells thick, is produced. In *Ulva* spp., the thallus remains two cells thick (Fig. 17.12). In *Enteromorpha* spp., the two layers separate, except at the edges, so that the thallus becomes a branched tube, the wall of which is one cell thick. The thallus is attached by the original holdfast, formed by the attachment of the zoospore, supplemented by rhizoid-like outgrowths of green cells which grow down between the layers of cells close to the point of attachment.

What has been described is the haploid, gametophyte generation of the plant. Sooner or later, the protoplasts of cells around the margin of the thallus divide and subsequently release biflagellate gametes. Syngamy occurs between gametes of two sizes (anisogamy) and the diploid, now tetraflagellate zygotes soon settle down, exactly like haploid zoospores, and divide to form thalli identical with the gametophyte thalli, but made up of diploid cells.

When reproduction starts in the diploid, sporophyte thallus, meiosis occurs in the cells involved. The four cells produced from each cell by meiosis divide twice more, by mitosis, and sixteen tetraflagellate zoospores are released. Zoospores settle down to form new gametophytes.

The life cycles of *Enteromorpha* spp. and *Ulva* spp. are haplodiplontic. They have alternation of generations. Because the gametophyte and sporophyte plants are morphologically identical, we say that they have **isomorphic** alternation of generations.

Enteromorpha spp. usually branch and proliferate at the base and form tufts of tubular thalli which may reach a length of 12 cm or more. *E. clathrata*, a West African species, is shorter than this and the tubular branches are only about 0.5 mm wide. Some tropical and subtropical species of *Ulva* form tough, bright green thalli which may be many centimetres long and wide. In *U. fasciata*, an African species, the longitudinally divided thallus may reach a length of 50–60 cm.

Oedogoniales

The order Oedogoniales comprises three genera, all of which are filamentous. They are characterised by a unique type of cell division and by the production of unique, multiflagellate motile cells. Their life cycle is haplontic.

Oedogonium spp. (Fig. 17.13) are very common in the tropics, in fresh waters of all types. They grow mixed with other filamentous algae, or form masses of coarse, yellowish-green filaments floating close to the surface of the water.

The life cycle of *Oedogonium* sp. may be taken to start when a multiflagellate zoospore settles down by its anterior end on a solid surface. It then divides transversely and transverse divisions continue, to form a long filament, attached to the substrate by a holdfast cell. Filaments often fragment and unattached filaments are common.

The cell of *Oedogonium* sp. has a cellulose wall, without any coating of mucilage. Because of the absence of mucilage, the outer surface of the filament offers firm attachment to small epiphytic algae and protozoans. Each cell has a single, parietal, reticulate chloroplast with many pyrenoids situated at junctions in the network.

Because of the unique type of cell division (Fig. 17.13), the walls of the daughter cells produced by the division of a parent cell differ from each other. The proximal daughter cell, closer to the holdfast, has a cell wall composed almost entirely of the wall of the parent cell. The distal daughter cell has a cell wall composed mostly of new material, with a distal cap of wall material from the parent cell. This cap of old cell wall is easily visible and cells bearing one or more caps are known as **cap cells**. Cap cells often divide repeatedly and a cap cell may bear a whole series of caps at its distal end (Fig. 17.13).

In asexual reproduction, the contents of a single cap cell withdraw from the wall and form a single, large zoospore. The zoospore has a clear region at its anterior end and has a whorl of many flagella and an eye-spot. The zoospore is released when the end of the cap cell breaks open.

The first sign of sexual reproduction is when some of the cap cells swell until they are almost spherical and become oogonia. Each oogonium contains a single ovum and has a pore near the distal end, close to the cell cap. On the same filament, or on different filaments, cap cells divide repeatedly to form rows of

(a)

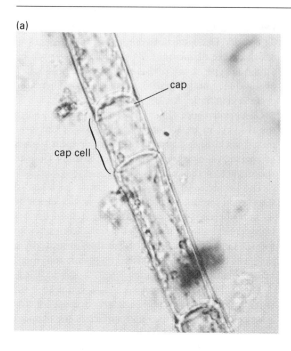

(b)

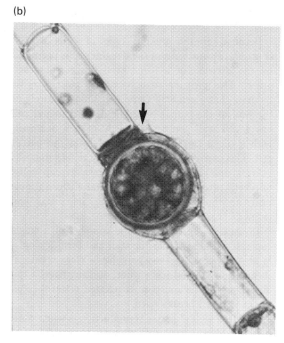

17.13 Oedogoniales, *Oedogonium* sp.; (a) filament showing the formation of a cell cap after cell division; (b) after fertilization, an oospore within the oogonium, showing the pore through which a sperm entered (HP)

short cells. The details of what happens next are different according to the species concerned.

In **macrandrous** ('large-male') species, each of the small cells is an antheridium and its protoplast divides to form two sperms, exactly like small zoospores. Like zoospores, they are released by breakage of the filament. In **nannandrous** ('dwarf-male') species, the small cells are **androsporangia** and each cell releases a single androspore, which also resembles a small zoospore. Each androspore settles down on another filament, close to an oogonium and often on the cell immediately proximal to an oogonium. The androspore then divides to form a dwarf male filament. A dwarf male filament is only two or three cells long and consists of a holdfast and one or two antheridia. Each antheridium produces two sperms, as in macrandrous species.

Fertilization takes place after a sperm enters through the pore of an oogonium and the nuclei of sperm and ovum fuse. The zygote secretes a thick wall and becomes an oospore, which is released by the decay of the wall of the oogonium. Later, the zygote nucleus divides by meiosis and the oospore opens to release four zoospores, which settle down to form new filaments as described above.

Zygnematales

Members of the Zygnematales differ from other green algae in having no flagella and no motile stages in any part of the life cycle. They all grow in fresh water. Their life cycle is haplontic.

The cells of members of the Zygnematales have cellulose walls with an outer coating of mucilage. Each cell contains one or more chloroplasts, which may be strap-shaped or stellate and either parietal or axile (Fig. 17.14). Each chloroplast contains one or more pyrenoids.

Sexual reproduction is by the fusion of naked gametes which have a limited degree of **amoeboid** (*Amoeba*-like) movement. The gametes which fuse are the entire protoplasts of otherwise normal vegetative cells. Fusion is isogamous, as far as the size and structure of the gametes are concerned, but the two gametes which fuse may differ slightly in behaviour.

There are three families in the order, Zygnemataceae, Mesotaeniaceae and Desmidiaceae.

Members of the Zygnemataceae are all filamentous. There are three basic types, represented by the genera *Mougeotia*, *Zygnema* and *Spirogyra*, which

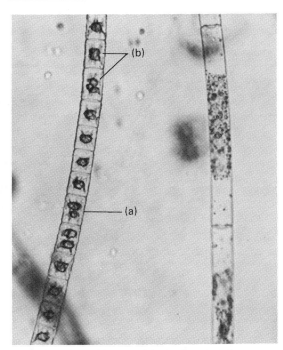

17.14 Zygnematales, Zygnemataceae: filaments of *Zygnema* sp. (left), with stellate chloroplasts, two in each mature cell (a) and a single chloroplast dividing in some recently-divided cells (b); cells of *Mougeotia* sp. (right) are larger and have a single strap-shaped chloroplast in each cell (LP)

differ in the shapes and position of their chloroplasts. *Mougeotia* spp. have a single, axile, strap-shaped chloroplast containing several pyrenoids in each cell. *Zygnema* spp. have two axile, stellate chloroplasts, each with one pyrenoid, in each cell. Cells of *Spirogyra* spp. have one to three parietal, spiral chloroplasts and each chloroplast contains several pyrenoids. Cell division is by the ingrowth of a septum from the sides of the cell. There are no asexual spores in any of the genera and the only asexual reproduction is by fragmentation.

Spirogyra spp. (Fig. 17.15) are very common in all tropical fresh waters. They can usually be recognized by their bright green colour and by the slimy nature of their filaments. Because of the coating of mucilage on the filaments, they usually bear no epiphytes.

Sexual reproduction is by a process known as **conjugation**. In heterothallic species, there must first be close side to side contact between filaments belonging to different strains of the same species. In homothallic species, conjugating filaments may be of the same strain. The first sign that conjugation is beginning is the formation of short, tubular outgrowths (**conjugation tubes**) from cells of both filaments, with their tips in close contact (Fig. 17.15(b)). Extension of these conjugation tubes first pushes the filaments apart. The tips of the tubes then fuse and a **conjugation canal**, open to both cells, is formed. The contents of the cells of one of the conjugating filaments then contract slightly from the cell wall and the same contraction occurs in the cells of the other filament shortly afterwards. When the two plastids have contracted, the plastids from the filament in which they were first to contract start to move into the conjugation canal and, by amoeboid movement, move into the cells of the other filament (Fig. 17.15(c)). Fusion of the protoplasts (**amoeboid gametes**) is followed by nuclear fusion and a zygote is formed. The zygote secretes a thick wall and the oval zygospore remains within the cell wall of the receiving filament (Fig. 17.15(d)) until the latter decays. The zygospore rests for a time. Under suitable conditions, the zygote nucleus undergoes meiosis, three of the four haploid nuclei degenerate, and a new haploid filament grows out, following rupture of the zygospore wall.

The process described above is **scalariform** (ladder-like) conjugation. In some homothallic species, **lateral** conjugation may take place between neighbouring cells of a single filament. In this case, conjugation tubes form side by side at opposite sides of a septum between two cells. Subsequently, a conjugation canal is formed and the contents of one cell pass into the other. The process is otherwise the same as scalariform conjugation.

Scalariform conjugation in *Spirogyra* spp. is common at the beginning of the dry season, or at any time when the water containing the alga is drying up. Masses of *Spirogyra* sp. in which conjugation is taking place can be recognized with the naked eye, because of a greyish discolouration due to the presence of many dark-coloured zygospores. In the laboratory, *Spirogyra* spp. can be stimulated to conjugate by placing them in a dilute solution of sugar and allowing the liquid to concentrate by evaporation.

Scalariform and lateral conjugation take place in the same way in some species of *Zygnema*, but in other species both gametes migrate into the conjugation tube and the zygospore is formed in the

(a)

(b)

(c)

(d)

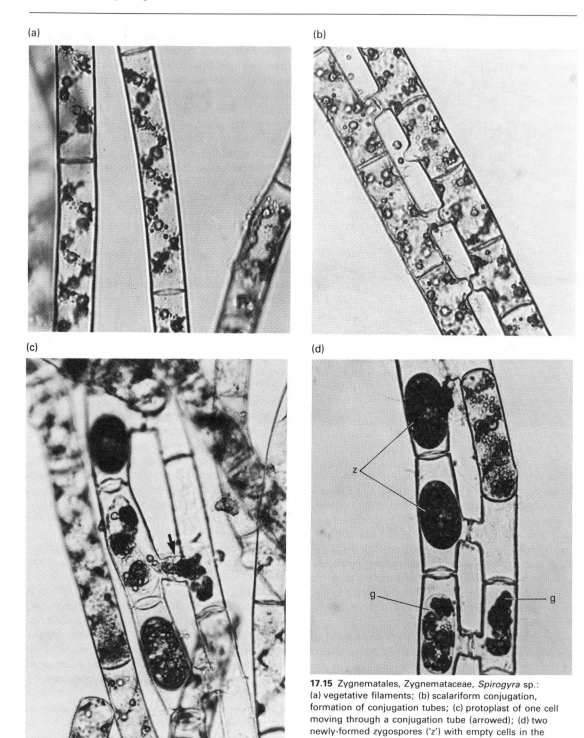

17.15 Zygnematales, Zygnemataceae, *Spirogyra* sp.:
(a) vegetative filaments; (b) scalariform conjugation,
formation of conjugation tubes; (c) protoplast of one cell
moving through a conjugation tube (arrowed); (d) two
newly-formed zygospores ('z') with empty cells in the
adjacent filament and an earlier stage below, with gametes
('g') just forming in each filament

conjugation tube, between the filaments. In almost all *Mougeotia* spp., the zygospore forms in the conjugation tube.

Members of the Mesotaeniaceae (**saccoderm desmids**) are either unicellular or have cells in loose filaments. The same three types of chloroplasts are found in different genera and similar forms of conjugation occur.

In members of the Desmidiaceae (**placoderm desmids**), chloroplasts are stellate and axile and the family is characterized by the fact that their cells are all made up of two identical parts (**semicells**), usually with a narrow connection (**isthmus**) between (Fig. 17.16). In this family, cell division is followed by **semicell regeneration**. After the nucleus has divided, one daughter nucleus remains in each semicell and the two semicells separate. Each semicell then regenerates a second semicell. Sexual reproduction occurs when naked protoplasts (amoeboid gametes) emerge from their cell walls and fuse in pairs.

Saccoderm and placoderm desmids are common in all types of fresh water and are often found on moist soil.

Chlorococcales
Members of the order Chlorococcales differ from members of all other orders of green algae except the Volvocales in the total absence of vegetative cell divisions. Cell division always results in the formation of spores or gametes. The order includes non-motile unicellular types, which may form cell aggregates, and coenobial types. Most of the genera are common in all sorts of fresh water. Most genera are thought to have haplontic life cycles.

Chlorococcum spp. are unicellular and are found in water or on wet soil. The cells are spherical and, in content, very like cells of *Chlamydomonas* spp. but have no flagella and no eye-spots. In asexual reproduction, the protoplast of each cell divides to form a large number of zoospores. After they are liberated, each zoospore loses its flagella and settles down as a vegetative cell. Sexual reproduction is isogamous and the life cycle is probably haplontic.

Chlorella spp. are similar to *Chlorococcum* spp., but produce aplanospores. Since the aplanospores exactly resemble the parent cell in which they are produced, they may be called **autospores**.

Ankistrodesmus spp. are very common indeed in all fresh waters. They are essentially unicellular but their elongated, sickle-shaped cells often adhere together in aggregates (Fig. 17.17). Each cell has a parietal chloroplast, with or without pyrenoids. In asexual reproduction, each protoplast·divides to form

(a)

(b)

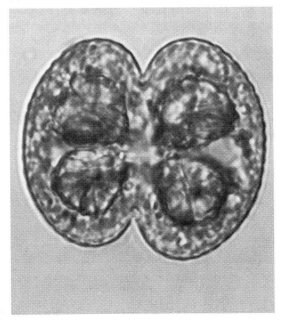

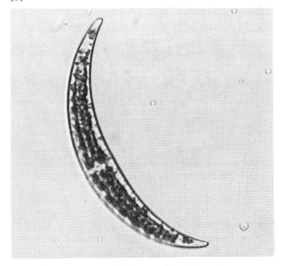

17.16 Zygnematales, Desmidiaceae: (a) *Cosmarium* sp., showing two semicells joined by an isthmus (HP); (b) *Closterium* sp., a crescent-shaped desmid showing a chloroplast in each semicell but no isthmus (LP)

(a)

(b)

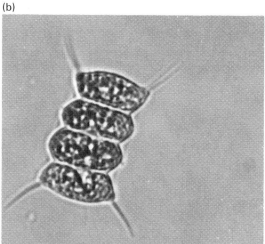

17.17 Chlorococcales: (a) cell aggregate of *Ankistrodesmus* sp. (HP); (b) *Scenedesmus* sp., one coenobium of four cells (HP)

four autospores. Sexual reproduction is not known.

Scenedesmus spp. (Fig. 17.17) are also extremely common. The elongated cells here form coenobia which consist of two to sixteen elongated cells arranged side by side. The cells at the ends of a coenobium often have long, pointed processes at each end. Any cell may divide to form an 'autos-

pore', which is a complete, miniature coenobium. Sexual reproduction is not known.

In addition to the genera described, there are also some with larger coenobia. *Pediastrum* spp. have disc-shaped coenobia of four to one hundred and twenty-eight cells. *Hydrodictyon* spp. have large coenobia, several to many centimetres in length and width. The coenobium is in the form of a network (Fig. 17.18) in which each mesh of the net is bordered by six cylindrical cells. Cells of *Hydrodictyon* spp. and larger species of *Pediastrum* become multinucleate as they become larger. Both genera form isogametes. It is thought that the life cycle of *Hydrodictyon* spp. may be diplontic.

Caulerpales

Members of the order Caulerpales have thalli which are made up of branched, coenocytic filaments. All genera are marine and representatives of the order

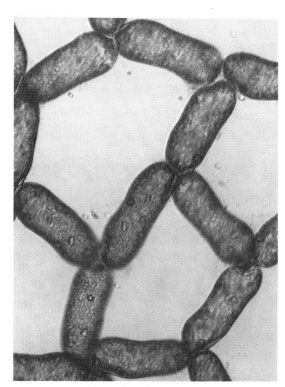

17.18 Chlorococcales: part of the network of cells forming the coenobium of *Hydrodictyon* sp. (LP). Each cell is multinucleate and has a single reticulate chloroplast with many pyrenoids

are common in shallow water, attached to rocks or sand. It is probable that all have diplontic life cycles.

In *Bryopsis* spp., the coenocyte is several millimetres wide and the outline of the cell forms the macroscopic outline of the plant. There is a prostrate stolon-like part, attached to the substrate, with an upright system of pinnate branches which reach a height of several centimetres.

Caulerpa spp. are very common along the coast of West Africa and in warmer waters elsewhere. The thallus is generally similar to that of *Bryopsis* spp. but the plants are mostly larger, in some species up to a height of twenty cm. The pale green upright branches are variously lobed or divided. The greater size and usually more complex shape of the thallus are maintained by a thick cell wall. Stiff transverse and longitudinal rods of pectic material within the cell also probably help to maintain the shape and give mechanical support.

In *Codium* spp., the dark green spongy thallus is pseudoparenchymatous, composed of fine coenocytic filaments which branch and intertwine. Each plant is attached at the base by fine, rhizoid-like threads. The upright part of the thallus branches frequently and usually dichotomously. In West African species, each branch is from a few millimetres to about two centimetres in width, sometimes flattened, and a whole plant may be anything from a few centimetres high to about one metre high, according to species.

The only type of asexual reproduction in the above genera is by fragmentation. Sexual reproduction is anisogamous.

Charales

The structure and biochemisry of the cells of members of the order Charales are essentially the same as those of other green algae. However, their morphology is strikingly different and they have uniquely complex structures associated with their oogonia and antheridia. For these reasons, they are placed in a different class, the Charophyceae.

Of the eight genera, *Chara* and *Nitella* are common in tropical lakes and rivers. Plants are attached to the solid substrate at the bottom of the water by rhizoid-like outgrowths and have upright branches which may be from 5 to 50 cm high (Fig. 17.19). Their life cycles are haplontic.

Each upright branch grows as a result of repeated transverse divisions of an apical cell. The axis consists of alternating internodes and nodes. The

whole internode, or the central part of the internode, consists of a single multinucleate cell, which may be as long as 7 cm and as wide as 2 mm. Each node bears a whorl of branches of limited growth. These branches themselves branch, but soon stop growing and remain as the main photosynthetic part of the plant. Whorls of branches of limited growth sometimes include one or more branches of unlimited growth, which continue to grow and repeat the structure of the main axis.

Oogonia are borne at the tips of short side branches, on branches of limited growth. As it develops, each oogonium becomes enclosed in filaments which grow up, in a spiral pattern, from around the base. The whole structure is called a **nucule**. Antheridial filaments are made up of rows of unicellular antheridia, each of which produces one biflagellate sperm. The antheridial filaments are formed within a spherical structure known as a

(a)

17.19 Charophyceae, Charales, *Nitella* sp.: (a) part of a plant showing long internodal cells and whorls of branches at each node (× ½) (continued overleaf)

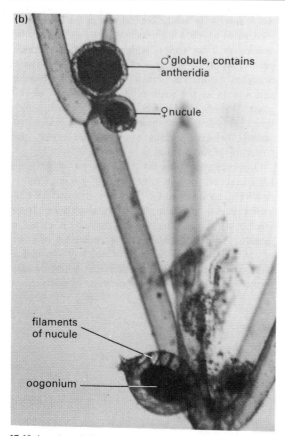

(b)

♂globule, contains antheridia

♀nucule

filaments of nucule

oogonium

17.19 (continued) (b) reproductive structures (LP)

glòbule. Each globule is formed at the tip of a short branch, usually close to the base of a nucule in monoecious species.

After fertilization, the oospore forms a thick wall, with the spiral pattern of the enclosing filaments impressed upon its surface. After a resting period, the nucleus of the oospore undergoes meiosis but only one of the four nuclei takes part in further developments. Germination leads to the formation of one branched rhizoidal filament, which attaches the new plant, and one or several green filaments. The green filaments grow on for a time as **protonema**, but apical cells soon differentiate and establish typical branches with nodes and internodes.

Euglenophyta – euglenoids

Euglenoids are almost all motile and unicellular. A few genera form aggregates in which the cells are held together by mucilage. Most are green and autotrophic but some are capable of living heterotrophically, when their environment contains suitable organic material. Others lack chlorophyll and are totally heterotrophic.

Cell structure

The protoplast of most euglenoids is limited by a **pellicle**, composed of plates of proteinaceous material, inside the cell membrane, which may be rigidly connected or able to slide over each other. There is no cell wall, but, in some genera, the cell is partially enclosed in a gelatinous **lorica**, which is open at the anterior end and is cast off and reformed as the cell grows.

Each motile cell has one, two or three flagella. Flagella are of the tinsel type and emerge through a gullet. The **gullet** is a flask-shaped invagination at the anterior end of the cell. Autotrophic euglenoids do not ingest solid food, but some of their colourless relatives may do so.

Close to the gullet is a single contractile vacuole, which discharges through the gullet. Nearby is a red eye-spot. The nucleus lies at the centre of the cell. There are several to many chloroplasts, which are usually discoid. Chloroplast pigments are chlorophyll *a*, chlorophyll *b*, xanthophylls and carotene, the same as in green algae. Granules of **paramylum**, a substance similar to starch, are present in the cytoplasm and stain brown with iodine solution.

Reproduction

Asexual reproduction is by longitudinal fission. There is no firm evidence of the existence of any form of sexual reproduction.

Classification and examples

The Euglenophyta are divided into two classes, one of which includes all motile forms and the other includes the relatively few non-motile forms.

Euglena spp. are common in all types of fresh water, especially where the water has a high organic content. Their cells are elongated, rounded at the anterior end and pointed at the posterior end. There is a single flagellum at the anterior end (Fig. 17.20). The pellicle is not rigid and, when at rest, the cell may become wider in the middle or bend. This type of change in shape is called **euglenoid movement**. Under adverse conditions, cells may become spherical, secrete a thick wall, and accumulate a red

(a)

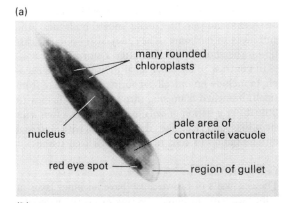

many rounded
chloroplasts

nucleus

pale area of
contractile vacuole

red eye spot

region of gullet

(b)

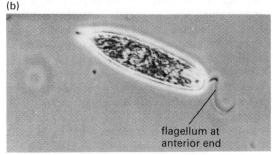

flagellum at
anterior end

17.20 Euglenophyta, *Euglena* sp.: (a) motile cell (HP);
(b) view of one cell using phase contrast illumination, to
show the flagellum (HP)

pigment in the cytoplasm.

Phacus spp. are also common. In this genus, the
pellicle is rigid and gives the cell a constant shape,
flattened from side to side.

Trachelomonas spp. are the commonest of the
euglenoids which form a lorica. A lorica is an incomplete covering of gelatinous material, open at the
anterior end. Here, the lorica is colourless when
newly formed but turns brown with age, due to the
accumulation of iron compounds.

Bacillariophyta – diatoms

Diatoms include unicellular forms, cell aggregates
and loosely filamentous forms. The plants are found
in fresh water, on soil, amongst the leaves of mosses
and other small plants and in the sea.

Cell structure
The cells of diatoms have walls composed of silica,
associated with pectic substances. Each cell wall is
in two overlapping halves which are arranged exactly
like the two halves of a petri dish. The half which

overlaps the other, like the lid of a petri dish, is the
epitheca. The other half is the **hypotheca**.

The protoplast has one central nucleus and one or
more discoid or elongated chloroplasts, with pyrenoids. Chloroplast pigments are chlorophyll *a*, chlorophyll *c*, xanthophylls and carotene and the storage
substances do not include starch.

Vegetative cells are non-motile. Motile male
gametes are known in a few species and these each
have a single, anterior, tinsel-type flagellum.

The cells of diatoms (Fig. 17.21) are either radially
symmetrical (circular, triangular, etc.) or bilaterally
symmetrical (elongated). Radially symmetrical
(**centric**) diatoms have irregular, concentric or radial
patterns of thin spots (**areolae**) in their walls. Bilaterally symmetrical (**pennate**) diatoms often have a
longitudinal groove running along the centre-line of
the cell wall, with radiating lines spreading outwards
from this. In some, the central groove is an actual
slit in the wall (**raphe**) and these diatoms are capable
of longitudinal gliding movements. It is thought that
this movement results from a circulation of cytoplasm outside the cell, so that the cell moves along
rather like a bulldozer or other tracked vehicle. In
other pennate diatoms, the groove does not penetrate
the wall and is called a **pseudoraphe**.

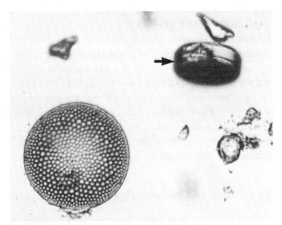

17.21 Bacillariophyta, diatoms: a centric diatom,
Coscinodiscus sp., from the Warri River (Nigeria). Note
oblique view of a smaller cell (arrowed) showing the petri
dish-like form of the cell wall. See Figure 1.11 for pennate
diatoms

Asexual reproduction
Asexual reproduction is by fission. After mitosis, the
protoplast separates into two parts, each part having

one half of the original cell wall. Each daughter cell then secretes a new second part to its cell wall, within the sides of the old cell wall. Thus, the epitheca of the parent cell becomes the epitheca of one daughter cell, while the (narrower) hypotheca of the parent cell becomes the epitheca of the second daughter cell. In this way, the average diameter of cells in a population decreases each time asexual reproduction takes place. This is remedied by the occasional formation of **auxospores**. An auxospore is formed when the entire protoplast emerges from a cell. It then enlarges slightly and forms a completely new epitheca and hypotheca, of the maximum size characteristic of the species.

Sexual reproduction

Sexual reproduction is known in only a few species of diatoms. In pennate diatoms, each protoplast forms one, two or four amoeboid gametes which emerge from the cell and fuse isogamously with gametes from other cells. In some centric diatoms, the protoplasts of some cells divide repeatedly to form many uniflagellate sperms, while others release their protoplast, which functions as an ovum. In both isogamous and oogamous forms, there is evidence that meiosis precedes gamete formation, so that the diatom life cycle is probably diplontic.

Classification and examples

Diatoms are divided into two orders, Centrales and Pennales, according to the symmetry of their cells.

Diatoms are numerous both in fresh water and in sea water and are not very easy to identify. However, *Coscinodiscus* sp. is a common West African representative of the Centrales. *Navicula* spp. and similar, boat-shaped pennate diatoms are especially common in fresh water, often on the surfaces of the leaves of submerged water plants.

Phaeophyta – brown algae

Brown algae are almost all marine and grow attached to rocks, or as epiphytes on larger algae. They are usually abundant on rocky coasts. Each species occupies a characteristic position between the highest point reached by the sea, down into deep water. Since they favour a rocky substrate, they are not common along those parts of the West African coast where there is a lagoon system protected by a sand bar. In such areas, they are found only where the

lagoons connect with the open sea, where they grow on rocks and on the rocky moles built near the entrances to harbours. They may also be found cast up on the shore, especially after a storm.

In contrast to all other algal divisions, the Phaeophyta include no unicellular forms. They range from small, branched filaments, through simple thalli with little differentiation of cells, to large thalli, often with stem-like and leaf-like regions and with internal differentiation of cells to form specialized tissues.

Cell structure

The cell walls of brown algae have an inner layer of cellulose, with an outer covering of pectic materials. Cells are uninucleate and contain many discoid chloroplasts. Chloroplast pigments are chlorophyll *a*, chlorophyll *c*, xanthophylls and carotene. There is an abundance of xanthophylls, often referred to collectively as **fucoxanthin**, which give the chloroplasts and the plants as a whole their characteristic brown colour. The cells store the carbohydrate **laminarin**.

Motile stages are biflagellate, with the flagella attached at one side. One of the flagella is of the whiplash-type, the other is tinsel-type.

Reproduction

Some of the smaller forms of brown algae produce asexual zoospores and many have vegetative reproduction by fragmentation.

Sexual reproduction may be isogamous, anisogamous or oogamous.

Life cycles and classification

Brown algae are separated into three classes on the basis of the type of life cycle.

Members of the class Isogeneratae have a haplodiplontic life cycle in which the gametophyte and sporophyte are of the same size and structure, or of similar size and structure (isomorphic alternation of generations). Members of the class Heterogeneratae have heteromorphic alternation of generations, in which the gametophyte is small and filamentous and the sporophyte is large and thalloid. Members of the third class, Cyclosporae, have a diplontic life cycle.

Isogeneratae

This class includes branched filamentous forms and simple thalloid forms. In filamentous forms, growth occurs as a result of cell divisions which occur at various points along the filament, though cell divisions may be more frequent close to the ends of branches. Simple thallose forms usually have an apical cell at the end of each branch, cutting off initials which subsequently divide and add to the length of the thallus (**apical growth**). There are also genera in which the growing end of the thallus ends in a tuft of hairs and it is divisions of the cells at the bases of the hairs which add to the length of the simple parenchymatous thallus (**trichothallic growth**).

The class Isogeneratae includes five orders, of which we shall study members of two, Ectocarpales and Dictyotales.

Ectocarpales

Members of this order are branched, filamentous plants which grow on rocks, or as epiphytes on larger algae. Species of *Ectocarpus* are mostly temperate but some occur in tropical waters. *E. breviarticularis* occurs on rocky parts of the coast of West Africa.

The gametophyte of *Ectocarpus* sp. forms matted, woolly-looking brown tufts up to about 3 cm long, attached to rocks. The gametophyte bears **plurilocular** gametangia (Fig. 17.22). A plurilocular gametangium is formed by repeated cell division in three planes and consists of many small, almost cubical cells. Each cell forms a single, biflagellate isogamete. Gametes are released into the sea and fuse in pairs. The zygote is motile for a time, after which it settles on a suitable surface and becomes attached. Repeated cell division produces a branched, filamentous plant body exactly like that of the haploid gametophyte.

(b)

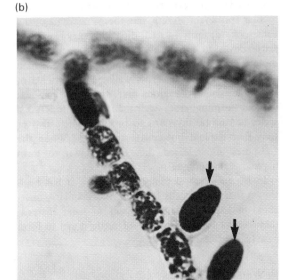

(a)

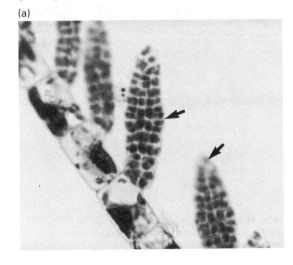

(c)

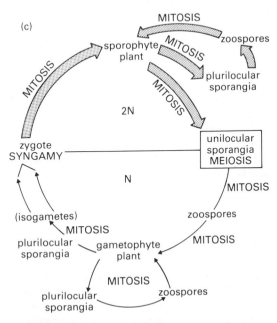

17.22 Phaeophyta, Ectocarpales, *Ectocarpus* sp.: (a) part of plant with plurilocular sporangia; (b) plant with unilocular sporangia (arrowed); (c) life cycle, an example of isomorphic alternation of generations

Diploid plants form two sorts of sporangia. **Pluri-locular sporangia**, exactly like the gametangia of the gametophyte, form asexual, diploid zoospores which settle down to form new sporophyte plants. The diploid plant also bears **unilocular sporangium**, which are not divided up by cross walls. The contents of the unilocular sporangium first divide by meiosis, without forming cross walls, and then undergo a series of mitotic divisions to produce thirty-two or sixty-four haploid biflagellate zoospores. When the zoospores are released they settle down to form haploid, gametophyte plants.

In *E. siliculosus*, a temperate species, it has been shown that motile cells from the plurilocular gamet-angia of haploid plants may fail to pair. Instead, they behave like zoospores and settle down to form new haploid plants.

Dictyotales

Species of *Dictyota* are common on rocky shores in all parts of the world.

The thallus of *Dictyota* spp. is up to about one centimetre wide and about one millimetre thick. It branches dichotomously (into two equal parts) at regular intervals. Growth of the thallus is by the activity of a single apical cell which can be easily observed under the microscope. The apical cell divides transversely and each derivative divides many times, in three planes. This produces a thallus with a central layer consisting of one row of large cells, with a single layer of small, pigmented cells at each surface (Fig. 17.23). Dichotomy occurs when the apical cell divides lengthways into two equal parts and each daughter cell becomes the apical cell of a new branch.

The gametophyte thallus bears groups (**sori**) of gametangia on its surface. There are separate male and female plants. In male plants, cells of the surface layer divide to produce plurilocular gametangia. In female plants a group of oogonia, each containing a single ovum, is formed in the same way. After the gametes have been liberated, fertilization takes place in the sea.

If the zygote settles on a suitable surface, it forms a diploid plant, exactly like the gametophyte except that it forms scattered sporangia, instead of sori of gametangia. Each sporangium (**tetrasporangium**) forms when a superficial cell divides periclinically and the outer daughter cell becomes greatly enlarged. The nucleus of the enlarged cell divides by

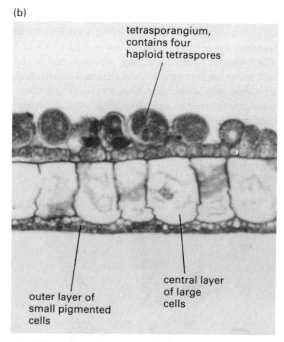

(a) × 1

flattened brown thallus with regular dichotomous branching

plants may be male, female or tetrasporic, but all of same appearance with same vegetative features

(b)

tetrasporangium, contains four haploid tetraspores

outer layer of small pigmented cells

central layer of large cells

17.23 Phaeophyta, Dictyotales, *Dictyota* sp.: (a) habit; (b) TS of a thallus of the diploid generation, with tetraspores

meiosis and the contents divide to form four tetraspores (Fig. 17.23), which become spherical when released. Tetraspores produce gametophyte plants.

Heterogeneratae

The Heterogeneratae are well represented in all temperate waters but are apparently absent from tropical seas. They include all the genera of brown algae in which the sporophyte is macroscopic and the gametophyte is very small and filamentous.

The **kelps** are among the larger members of the class. Species of *Laminaria* are characteristic of the North temperate zone, while *Ecklonia* spp. occur in the South, reaching as far north, in Africa, as the coast of southern Namibia.

In each of these genera, the plant is attached by a dichotomously branching system which forms the holdfast. Above this is a cylindrical stipe, with a flattened lamina at the end. Growth is **intercalary** at the base of the lamina and, in some species, a new lamina is formed each year as the old lamina is cast off.

Both sides of the lamina bear large sori of unilocular sporangia. In each sporangium, meiosis is followed by mitosis and many haploid, biflagellate zoospores are released.

In *Ecklonia* spp. and *Laminaria* spp., zoospores become attached to a rock surface and form very short branched filaments.

The gametophyte filaments are of two sexes. In the male filaments, terminal cells enlarge into antheridia and each antheridium forms a single, biflagellate sperm. Terminal oogonia on female filaments form one ovum each. The ovum emerges from the oogonium but remains at the mouth of the oogonium while it is fertilized. The zygote remains in the same position and the developing sporophyte becomes attached in the same place.

Kelps are notable for the amount of cell differentiation in the sporophyte thallus, which is greater than that seen in any other algae. There is a layer of small, almost cubical cells containing many chloroplasts forming an **epidermis** or **meristoderm**. Cells of this layer divide periclinally and add to the thickness of the cortex within. The **cortex** is composed of cells that are very like the parenchyma of higher plants. At the centre of the stipe and lamina is a **medulla** formed of greatly elongated cells embedded in mucilage and joined end to end. These cells (**trumpet hyphae**) are narrow in the middle and wide

at the ends, where they are in contact, and the cross walls between adjacent cells are perforated in a manner very similar to that of sieve tubes in angiosperms.

The class Heterogeneratae is divided into two subclasses and into a total of five orders, all of which occur in temperate seas.

Cyclosporae

The thallus of members of the class Cyclosporae is formed on a basically dichotomous plan, though the basic plan is often modified by unequal development of branches. Growth is apical, by division of a single apical cell.

As seen in TS, there are one to several layers of cells with chloroplasts, grading from the small cubical cells of the epidermis (**meristoderm**) to larger, parenchyma-like cells of the outer part of the cortex. At the centre of the thallus is a medulla made up of longitudinally elongated cells embedded in mucilage. The overall pattern resembles that of the kelps, but the cells of the medulla lack the wide ends and sieve-like perforations of the trumpet hyphae of the latter.

The class Cyclosporae includes only one order, the Fucales. This includes the genus *Fucus*, species of which are used as the principal example of the order in text books designed for use in the North temperate zone. *Fucus* spp., with their rather simple dichotomous thallus, are strictly North temperate in distribution. *Sargassum* spp. are probably the commonest representative of the Fucales in warmer seas.

Species of *Sargassum* (Figs. 1.9, 17.24) are common in large rock pools and below low tide mark from the Mediterranean sea southwards as far as Australia. They are especially common on tropical coasts.

The plant of *Sargassum* sp. is attached by a small, more or less circular **holdfast**. Above this is a cylindrical **stipe**. Branching is dichotomous, but different development of the two branches produced by each dichotomy gives the overall impression of a plant with stems, leaves and axillary branches. At each dichotomy in the vegetative part of the plant, one of the two branches grows on as a stem-like organ, circular in TS. The other branch dichotomizes again almost immediately. The branch farthest from the axis develops into a flattened, leaf-like organ which ceases to grow in length. The other branch, in the angle between the leaf-like branch and the axis,

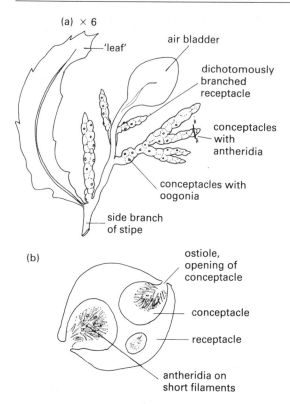

(a) × 6

'leaf'

air bladder

dichotomously branched receptacle

conceptacles with antheridia

conceptacles with oogonia

side branch of stipe

(b)

ostiole, opening of conceptacle

conceptacle

receptacle

antheridia on short filaments

17.24 Phaeophyta, Fucales, *Sargassum* sp.: (a) details of a part bearing receptacles; (b) TS receptacle with antheridia (see also Fig. 1.9.)

grows on and repeats the structure of the main axis, or may produce fertile branches (**receptacles**).

Some of the leaf-like branches become inflated and gas-filled and, as air bladders, help to keep the plant in an upright position under the water.

Apparently axillary branches which are to produce receptacles branch again repeatedly, this time by equal dichotomy, and form bunches of cylindrical organs which become slightly swollen at the ends. These are the receptacles. Each receptacle has many flask-shaped invaginations of the surface, the mouths of which appear as pores on the surface of the receptacle. These are **conceptacles**.

In the West African species of *Sargassum*, male and female reproductive organs are in separate conceptacles on the same receptacle. Conceptacles near the tip of the receptacle are usually male and those at the base are usually female.

The conceptacle has many long, slightly branched, multicellular hairs (**paraphyses**) which originate in the epidermal lining and extend outwards through the pore of the conceptacle. In a male conceptacle, there are also shorter and more profusely branched hairs. The end cell of each branch becomes enlarged and forms an antheridium. Meiosis takes place in the antheridium and this is followed by several mitotic divisions, so that each antheridium produces, usually, sixty-four biflagellate sperms. The fertile hairs in the female conceptacle are much shorter and usually do not branch. The terminal cell of each hair enlarges until it is nearly spherical and forms an oogonium. Meiosis in the oogonium is followed by mitotic division of the four haploid nuclei and eight nuclei are produced, without division of the cytoplasm. In *Sargassum* spp., seven of the eight nuclei degenerate and the oogonium contains only a single ovum.

Sperms are liberated from the conceptacle into the sea. Each ovum at first remains attached by the middle part of its wall, protruding from the mouth (**ostiole**) of the conceptacle, where it is fertilized. The zygote is then released into the sea, where it quickly settles down on a solid surface. It first becomes attached by a single rhizoid, after which mitotic cell divisions, growth and development lead to the establishment of a new diploid plant.

Rhodophyta – red algae

Red algae are almost all marine. They occur on rocky sea coasts throughout the world but make up a greater proportion of the algal vegetation of coastlines in the tropics than in the temperate zones.

The division includes unicellular, filamentous and thalloid forms. Some of the thalloid forms are parenchymatous (*Porphyra* spp.) but most are pseudoparenchymatous.

Cell structure

The cell wall of red algae contain cellulose and pectic substances. Cells are almost always uninucleate but there are cases in which large cells contain many nuclei. In some of the simpler members of the division, each cell contains one axile chloroplast with a pyrenoid. In the majority, each cell contains many discoid parietal chloroplasts and there are no pyrenoids. Chloroplast pigments are chlorophyll *a*, chlorophyll *d*, xanthophylls, carotene, phycoerythrin and phycocyanin. The phycoerythrin (red) and phycocyanin (blue) pigments are similar to, but not identical to, the pigments which are given the same names in blue-green algae. The red pigment usually predominates but thalli of various species vary from green, through red and brown to almost black.

The cells store carbohydrate in the form of granules of **floridean starch**, which has a smaller molecule than true starch and does not stain blue with iodine.

There are no flagellate cells in red algae.

Asexual reproduction

Some of the simpler red algae produce monospores. A **monospore** is the protoplast of a vegetative cell which escapes from its cell wall, forms a new wall of its own and may then settle down to form a new plant. Otherwise, asexual reproduction is uncommon and results from fragmentation.

Sexual reproduction

Sexual reproduction in red algae is different from that in any other algae. Non-motile male gametes (**spermatia**) are formed, each in a unicellular **spermatangium** (antheridium). The female reproductive organ is the carpogonium. A **carpogonium** is an oogonium with a long, tapering extension called a **trichogyne**. Spermatia adhere to the trichogyne and the contents of a spermatium enter the trichogyne, after which the nucleus travels down into the body of the carpogonium, where the nuclei fuse.

Life cycles

Life cycles in red algae are haplontic or haplodiplontic. The haplodiplontic life cycle is essentially of the isomorphic type, but is sometimes complicated by the fact that the zygote germinates while attached to the parent gametophyte and forms a diploid, dependent, filamentous plant which subsequently releases diploid spores (**carpospores**), which grow into the free-living diploid generation.

Classsification and examples

The division Rhodophyta contains a single class, which is divided into two subclasses and a total of seven orders. Red algae which occur on the coast of West Africa include filamentous and thalloid genera. Thalloid genera include *Rhodymenia* sp. *Polysiphonia* sp. is illustrated in Fig. 1.10.

Key words

Algae: procaryotic, eucaryotic. **Procaryotic**: Cyanobacteria (blue-green algae). **Eucaryotic**: Chlorophyta (green algae), Euglenophyta (euglenoids), Bacillariophyta (diatoms), Phaeophyta (brown algae), Rhodophyta (red algae). **Cyanobacteria**: unicellular, cell aggregate, filament; mucilage sheath, DNA, chromoplasm, chlorophyll lamella, pigment; fission; hormogonium, akinete, heterocyst; true branching, false branching. **Eucaryotic divisions. Morphology**: unicellular, cell aggregate, multicellular. **Multicellular**: coenobial, filamentous, thalloid; parenchymatous, pseudoparenchymatous; siphonaceous (coenocytic). **Cells**: chloroplast, pyrenoid; flagella, acronematic (whiplash), pantonematic (tinsel). **Reproduction**: zoospore, aplanospore; isogamy, anisogamy (heterogamy), oogamy; zygospore, oospore. **Life cycle**: haplontic, haplodiplontic, diplontic. **Alternation of generations**: gametophyte, sporophyte; isomorphic, heteromorphic.

Summary

Algae are all fairly simple in structure. Their reproductive organs do not cut off a surface layer of sterile cells during development, as do those of land plants. Blue-green algae are placed in the kingdom Monera, because their cells are procaryotic. Other divisions are eucaryotic and are placed in the kingdom Plantae. Asexual reproduction is usually by the formation of zoospores or aplanospores. Sexual reproduction is unknown in blue-green algae and euglenoids. Other divisions have isogamy, anisogamy or oogamy and their life cycles, where completely known, are haplontic, haplodiplontic or diplontic. In haplontic genera, zygotes form zygospores or oospores, which usually rest for a time before undergoing meiosis to produce haploid spores. In haplodiplontic genera, the zygote divides by mitosis and forms a diploid generation, in which meiosis finally occurs and haploid spores are produced. Algae are of ecological importance as pioneers and as producers of oxygen.

Questions

1 Make a table in which you compare and contrast the characteristics of different algae divisions (structural and biochemical aspects of the cell, morphology variations, sexual reproduction, life cycles, etc.).

2 In some systems of classification, the division Chlorophyta is taken to include the green algae and all land plants, including flowering plants. How might this be justified?

18

Bryophytes – non-vascular land plants

Bryophytes are green plants in which the reproductive organs have a surface layer of sterile cells outside the fertile cells and which lack vascular tissue. Like members of all the groups which follow in this Section, their cells have cellulose walls and store starch. Their chloroplast pigments are the same as those in flowering plants. Motile cells (sperms) have two whiplash flagella attached at the anterior end.

Reproductive organs

Sexual reproductive organs (Fig. 18.1) are **antheridia** (male) and **archegonia** (female). Both are carried on short stalks. The antheridium is spherical or elongated and has a wall of sterile cells (jacket), one cell thick, enclosing a large number of fertile cells (**spermatocytes**). Each spermatocyte metamorphoses into an elongated, biflagellate sperm. Sperms are released by rupture of the jacket.

The archegonium consists of two parts, a lower, swollen **venter** and a long, narrow **neck**. The wall of the venter is one or more cells thick and contains a basal **ovum** and a **ventral canal cell**. The neck has a wall which is one cell thick, surrounding an axial row of **neck canal cells**. At maturity, the ventral

(a)

sperms
mature antheridium open
younger antheridium

paraphyses, sterile hairs

18.1 Bryophyte reproductive organs: (a) antheridia with associated paraphyses (sterile hairs) dissected from the antheridial cup of a moss (LP); (b) LS archegonia of a liverwort, young (right), and an older one (left) in which the neck is incomplete (HP)

(b)

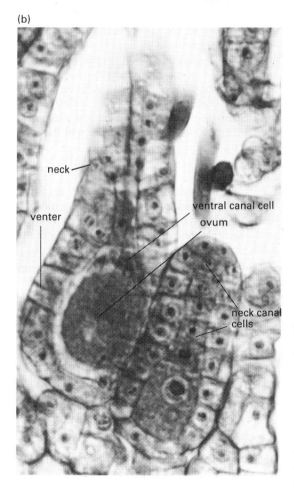

neck

venter

ventral canal cell

ovum

neck canal cells

canal cell and the neck canal cells degenerate and form a strand of mucilage and the end of the neck opens. Sperms are attracted and swim down the neck. Fertilization takes place when one sperm enters the ovum and the nuclei of sperm and ovum fuse to form a zygote.

Life cycle

The bryophyte life cycle is haplodiplontic (Fig. 18.2). The haploid gametophyte is a free-living green plant, which is often perennial and has one of a number of types of asexual reproduction. Gametophytes may have male and female organs on the same individual (**monoecious** plants), or there may be separate male and female gametophytes (**dioecious** plants). Fertilization requires the presence of water. Sperms are attracted to the necks of archegonia and fertilization takes place. The zygote develops into an embryo within the wall of the venter. The lower part of the embryo penetrates further into the tissues of the gametophyte and forms a haustorial organ, the **foot**, through which materials are absorbed from the gametophyte. The distal end of the sporophyte develops into a capsule. The capsule contains sporocytes which undergo meiosis to produce meiospores. The spore-producing region becomes raised above the gametophyte on a stalk (**seta**), or by increase in length of the capsule itself, and spores are shed when the capsule opens. Spores develop into new gametophyte plants.

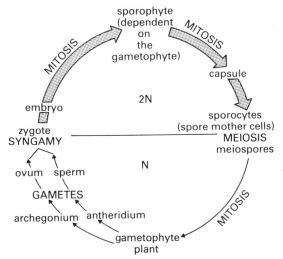

18.2 Bryophyte life cycle. This is a haplodiplontic life cycle in which the gametophyte (haploid) generation is dominant

The sporophyte of a bryophyte is always less conspicuous than the gametophyte, is largely dependent on the gametophyte and is short lived. Because of this, we say that bryophytes have alternation of generations in which the gametophyte is **dominant**.

Classification

The division Bryophyta is divided into three classes, Anthocerotae (hornworts), Hepaticae (liverworts) and Musci (mosses).

Anthocerotae

Cells of members of the Anthocerotae resemble algal cells in that they contain one, occasionally two to four, large chloroplasts, each of which has a pyrenoid (Fig. 18.3).

The gametophyte is a deep green thallus attached to the substrate by unicellular rhizoids. The thallus branches dichotomously. In some genera (e.g. *Anthoceros*), the lower part of the thallus contains mucilage-filled cavities which open to the underside by a slit between two cells. The opening resembles a stoma. The mucilage usually contains filaments of the blue-green alga, *Nostoc* sp. The cavities are then visible as dark green spots within the thallus and are called *Nostoc* cavities.

Plants may be monoecious or dioecious. Antheridia differ from those of other bryophytes in that they are produced, usually in groups, in a mucilage-filled cavity in the upper part of the thallus, enclosed under a single layer of surface cells. Archegonia also differ from those of other bryophytes. They are completely sunken in the thallus, with the tissues of the venter and neck walls continuous with those of the thallus and with the tip of the neck opening at the upper surface of the thallus.

After fertilization, the embryo develops into a sporophyte with a bulbous foot, separated from an elongated capsule by a short meristematic region. Growth of the capsule is from this intercalary meristem and, in most species, continues over a period of months. The epidermis of the capsule is covered with cuticle and contains stomata almost exactly like those of flowering plants. Inside the epidermis there is a layer of chlorenchyma and a fertile layer (**archesporium**), surrounding a central strand of sterile cells (**columella**). The columella

(a)

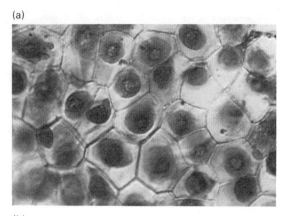

(b)

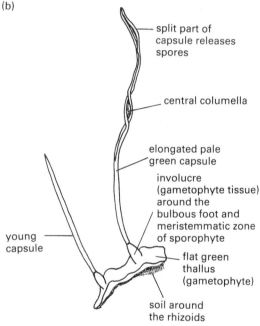

split part of
capsule releases
spores

central columella

elongated pale
green capsule

involucre
(gametophyte tissue)
around the
bulbous foot and
meristemmatic zone
of sporophyte

young
capsule

flat green
thallus
(gametophyte)

soil around
the rhizoids

18.3 Anthocerotae, *Phaeoceros* sp.: (a) part of a thallus showing a single large chloroplast, containing a pyrenoid, in each cell (LP); (b) drawing of a single plant with sporophytes (see also Fig. 1.12)

reaches nearly to the tip of the capsule, where it is spearated from the epidermis by a dome of archesporium and chlorenchyma. The archesporium at first contains spherical diploid spore mother cells and pseudo-elaters. The **pseudo-elaters** are branched, elongated cells with thin cellulose walls, sometimes with irregular or spiral thickenings and are connected together to form a three-dimensional network.

The spore mother cells divide by meiosis and each forms four haploid meiospores. The spores separate from each other and form thick outer walls.

Dehiscence of the capsule is by two longitudinal slits, which may or may not meet at the tip. At first, the slits are short, but they gradually extend downwards at about the same rate as new tissues are added from the intercalary meristem. As the slits form, the inside tissues dry out and the pseudo-elaters separate and twist with sudden hygroscopic movements, thus throwing out the spores. Spores are wind dispersed. A single capsule may continue to produce and release spores for the whole of the growing season.

Phaeoceros spp. (e.g. *P. laevis*, with world-wide distribution) are common in rain forest and on shaded, muddy soil beside streams and rivers in other areas. They have short, much branched thalli which form almost circular rosettes. They have no *Nostoc* cavities. Capsules are 3 to 4 cm high and have yellow-brown spores. *Anthoceros* spp. are similar but have *Nostoc* cavities and the spores are black.

Notothylas sp. is very common in all parts of Africa. It forms small, dark green rosettes on level soil which is wet and sometimes water-logged during the wet season. It differs from other hornworts in that capsules stop growing when the tip has just appeared, as a short yellow point, through the surface of the thallus. The structure of the sporophyte is also simple and there is no columella.

Hornworts can be identified as such very easily. If a piece of thallus is held up, with the light behind it, a hand lens makes it possible to see that there is only one large chloroplast in each cell (Fig. 18.3). Apart from that, the elongated capsules are unique to the Anthocerotae.

Hornworts are hygrophytes and always grown on moist substrates, including soil and rotting logs. Most grow in shade, but *Notothylas* sp. tolerates full sunlight.

Hepaticae

The cells of liverworts contain many discoid, parietal chloroplasts without pyrenoids.

Gametophytes (Figs. 18.4, 18.5) are either thalloid or consist of a stem-like axis with two or three rows of leaf-like organs. They are attached to the substrate by unicellular rhizoids. Plants are monoecious or dioecious. Antheridia are formed either on

18.4 Hepaticae, Jungermanniales, *Plagiochila* sp., a rainforest epiphyte (× ½)

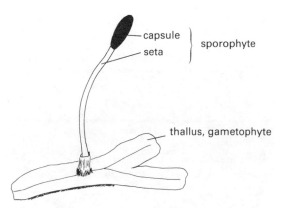

18.5 Hepaticae, Metzgeriales, *Symphyogyna* sp.: thallus of a female plant bearing a mature sporophyte, the seta of which has elongated and lifted the capsule away from the thallus (× 2)

the upper surface of the thallus, or in the axils of leaves. Archegonia are formed either on the upper surfaces of thalli or, in many leafy forms, terminally in groups, protected by modified leaves (**perianth**).

When the **sporophyte** is first formed, the capsule develops fully while it is still enclosed in the enlarged venter of the archegonium (**calyptra**). At this stage, the seta is short and composed of longitudinal rows of very short cells. At maturity, the cells elongate very rapidly by vacuolation. The calyptra is broken and the capsule is carried up above the gametophyte in a period of only a few hours (Fig. 18.5). The capsule contains many meiospores, with thick, dark-coloured walls, mixed with many separate, usually spirally thickened elaters. When the capsule wall splits, the elaters twist and turn with changes in humidity and thus scatter the spores into the air. There are no stomata in liverwort capsules.

Classification

Hepaticae are classified in six orders, of which only the Marchantiales, Metzgeriales and Jungermanniales occur in tropical Africa.

Marchantiales

Members of the Marchantiales are all thalloid and almost all have dichotomous thalli containing air chambers in the upper part. The underside of the thallus bears two or more rows of ventral scales, as well as rhizoids. The rhizoids are of two types. **Smooth rhizoids** are like the rhizoids of other liverworts, with smooth walls. **Peg rhizoids**, which are found only in this order, have peg-like ingrowths on their walls. The thallus contains scattered oil body cells. These cells contain many small droplets of oil.

Antheridia are sunk in pits in the upper surface of the thallus. In one genus (*Marchantia*) they are confined to specialized branches (**antheridiophores**) raised up on stalks above the level of the thallus (Fig. 18.6). Archegonia are also usually sunk in pits in the upper surface of the thallus. In several genera (e.g. *Marchantia*, *Plagiochasma*) they are formed on specialised, raised branches (**archegoniophores**).

Sporophytes have a rather short seta and a spherical capsule, with many unattached elaters. The capsule usually opens by slits, which meet at the apex.

Riccia spp. (Fig. 18.7) are very common and often grow on level soil which is wet during the wet season but totally dry during the dry season. The dichotomous thallus is simpler in structure than most of

18.6 Hepaticae, Marchantiales, *Marchantia*: male and female thalli with antheridiophores (♂) and archegoniophores (♀), in side view, showing the long stalks which raise the fertile branches of the gametophyte above the general level (× 1)

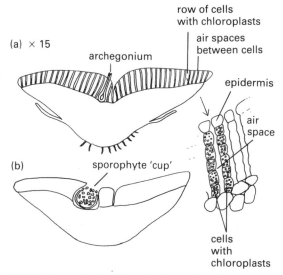

18.7 Hepaticae, Marchantiales, *Riccia* sp., TS thallus: (a) diagram showing regions of the thallus and embedded archegonium (× 15) and greater detail of photosynthetic tissue; (b) showing embedded sporophyte releasing spores as calyptra and thallus tissue start to disintegrate. (See Fig. 1.13 for whole plants.)

the other members of the Marchantiales. There is a lower region of more or less colourless parenchyma and an upper region composed of vertical plates of photosynthetic cells, with narrow air slits between. Antheridia and archegonia are in rows, either side of the centre line of the thallus. Antheridia are in pits, the openings of which are marked by small papillae. Archegonia are also sunk in pits in the thallus. Their necks emerge just above the surface and look like short hairs.

After the archegonium has been fertilized, the sporophyte develops within the calyptra. It has no foot or seta and consists only of a wall of cells containing spores. There are no elaters. This is the simplest sporophyte known in bryophytes. By the time the spores are mature, the capsule wall has disintegrated and spores lie free in the calyptra. Spores are released by the decay of gametophyte tissue. Decay may be localized, above the capsule, so that the spores come to lie in small depressions in the thallus, from which they are dispersed by rain. This occurs in the common West African species, *R. nigrosquamata*. In other species, release of spores is delayed until the total decay of the part of the thallus in which they are contained.

Several West African species of *Riccia*, including *R. nigrosquamata*, have thalli which fold up, from each side, under dry conditions. They survive the dry season in a folded condition and open up again, within a few minutes, as soon as they are moistened.

The spores of *Riccia* spp. are large, up to 0.125 mm in diameter, and are dispersed by water.

Marchantia spp. have large, dichotomous thalli and may be found growing on moist soil in shady places or, on wet soil, exposed to the full light of the sun. The thallus of the commonest species, *M. plani-loba*, is 0.8 cm to over 1.0 cm wide and has two rows of ventral scales, one each side of the thick centre portion (**midrib**). The lower part of the thallus is of parenchyma and there is a single row of air chambers beneath the upper epidermis (Fig. 18.8). The outlines of the air chambers can be seen as four to five sided areas on the upper side of the thallus (Fig. 18.9). Each air chamber has a conspicuous central pore, which opens on the upper surface. The pore is limited by an almost cylindrical structure, which stands above the surface of the thallus and projects a short distance into the chamber. There are many photosynthetic filaments, some branched, standing up from the floor of each chamber.

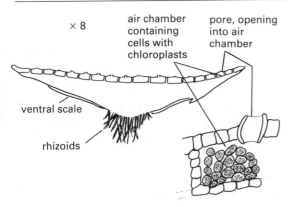

18.8 Hepaticae, Marchantiales, *Marchantia* sp., TS thallus to show the complex, chambered structure (× 8)

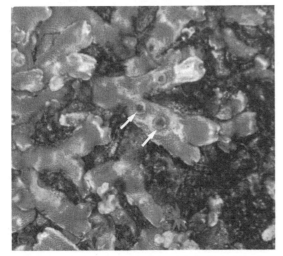

18.9 Hepaticae, Marchantiales, *Marchantia* sp., surface view of thalli with gemma cups (arrowed). The pale spots on the thallus are pores leading into air chambers (× ¾)

Most thalli of *M. planiloba* have **gemma cups** on the upper surface (Fig. 18.9). These are raised, cup-like areas which contain many more or less discoid gemmae, which are formed at the base of the cup and gradually discharged. Each gemma is attached by a short filament of cells and has an apex at each side. After they are set free, gemmae become attached by rhizoids and a new thallus grows from each apex. *Marchantia* spp. propagate themselves very rapidly by means of these gemmae.

All *Marchantia* spp. are dioecious. They produce sexual branches most profusely when growing under fairly bright light. In deep shade, they produce very wide thalli with gemma cups only.

The female plant produces umbrella-shaped archegoniophores on the dorsal surface of the thallus, close to the apex. The head of the archegoniophore is a set of eight laterally fused branches and is attached to the thallus by a stalk, which is at first very short, so that the head lies almost in contact with the upper surface of the thallus. The stalk of the archegoniophore is interpreted as two branches of the thallus fused side by side. Later, the stalk extends and carries the head to a level which varies between about 1 cm and 5 cm above the thallus. The head of the archegoniophore has chambers on the upper surface and the apex of each of the eight branches is turned down. Archegonia are formed on the morphologically upper surface of the archegoniophore, close to each apex, but because of the angle of the apex they face downwards (Fig. 18.10). Each group of archegonia is enclosed between two hanging, rather curtain-like sheets of cells which together form the **involucre**.

The antheridiophore has a similar origin to the archegonia, but the apices do not turn downwards and antheridia are formed in two rows, embedded in the upper surface of each of four or eight radiating branches.

Fertilization takes place while the stalk of the archegoniophore is still short. As the zygote begins its development to form the sporophyte, a cyclindrical **sheath** grows down around each archegonium. The sporophyte has an arrow-shaped foot and the capsule is spherical and contains many spores and unattached, spirally thickened elaters. The seta elongates only sufficiently to bring the capsule just beyond the involucre and sheath, and the capsule opens by irregular slits. Because of the inversion of the apices of the archegoniophore, the capsules point downwards.

Marchantia spp. are the most complex of liverworts. Other members of the Marchantiales have similar thallus structure, but archegoniophores occur in only a few other genera (e.g. *Asterella*; *Plagiochasma*) and antheridiophores are found in only one other genus, which is confined to New Zealand. Gemma cups of the *Marchantia* type are found only in *Marchantia* spp. and *Lunularia* spp. The latter genus does not occur in tropical Africa.

(a)

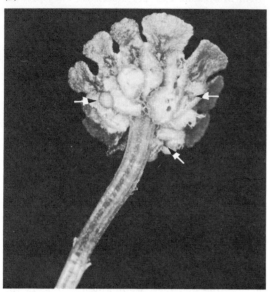

(b)

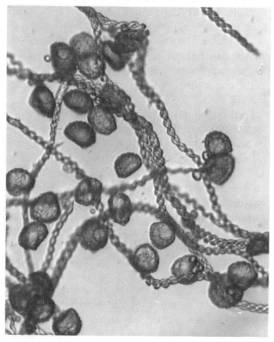

18.10 Hepaticae, Marchantiales, *Marchantia* sp.: (a) the under surface of the head of an archegoniophore showing a capsule projecting from between the sides of the involucre (arrow, left) while other capsules (arrowed) are still enclosed (× 3); (b) spores and elaters with double spiral thickening (HP)

Metzgeriales

Most members of the Metzgeriales are thalloid but some genera have a strong midrib, with the sides (**wings**) of the thallus divided into two rows of leaf-like lobes (e.g. *Fossombronia* spp.). Attachment is by smooth rhizoids. In most genera, many of the cells contain oil bodies, which may be small and many in each cell (simple oil bodies), or few and larger, each composed of several droplets of oil (compound oil bodies).

Riccardia spp. grow on moist soil and on rotting wood, often in rain forest. The thallus is dark green and without a midrib. Plants are dioecious or monoecious. Archegonia are formed on the upper surfaces of short side-branches mixed with mucilage-secreting hairs. Antheridia are embedded in the upper surface of the thallus, also usually on short side-branches. The sporophyte has a long seta and the capsule is elongated. Spores are mixed with elaters, some of which are free and some attached to the inside of the end of the capsule, forming a brush-like **elaterophore**. The capsule opens by four longitudinal slits, which meet at the end and divide the elaterophore into four tufts.

Symphyogyna spp. are fairly common, growing in shade on muddy soil, often on the banks of streams and rivers. The thallus has a strong midrib, with wings which are one cell thick towards the edges. All are dioecious. Male plants have antheridia in rows along the upper surface of the midrib, each in the axil of a small scale. Female plants have dorsal groups of archegonia, surrounded by scales, along the mid-line. After fertilization, the tissue of the receptacle (the thallus tissue where the archegonia are attached) grows up and protects the developing capsule, so that the calyptra protects only the upper part. The seta is long and the capsule is elongated. Spores are mixed with free elaters (Fig. 18.5).

Fossombronia spp. grow on soil, often in exposed places. Each thallus has a strong midrib and the wings are divided to form two rows of leaves. Plants are dioecious or monoecious. Antheridia are spherical and are in groups on the upper surface of the midrib. Archegonia are similarly placed. After fertilization, each capsule is protected by a well-developed calyptra and by a cylindrical pseudoperianth, which grows up from the surface of the axis. By this time, the sporophyte is attached to the gametophyte some distance behind the apex. Spores are mixed with free elaters, which have annular or spiral thickenings.

Jungermanniales

Jungermanniales (Fig. 18.11) the true leafy liverworts, are very common indeed in tropical countries, where most grow as epiphytes on the trunks, branches and even on the leaves of trees. In most genera, almost all the cells of leaves and stems contain several large oil bodies each. The stem usually bears three rows of leaves, one at each side of the stem (**lateral leaves**) and one under the stem (**underleaves**). The underleaves are usually of a different shape from and smaller than the lateral leaves, or may be absent. Attachment is by groups of smooth rhizoids, usually in groups close to the bases of underleaves.

Some leafy liverworts are dioecious, others are monoecious. Antheridia are in the axils of leaves (**male bracts**). Male bracts may be identical to ordinary leaves or may be much reduced. Archegonia are formed in groups at the tips of branches. The apical cell of a female branch is always used up in the formation of one of the archegonia. Each group of archegonia is enclosed in a **perianth**, composed of fused, modified leaves. The perianth is small at first, so that the necks of archegonia protrude until after fertilization. Subsequently, the perianth enlarges and encloses the developing capsule until it is mature. Only one archegonium in a group produces a capsule.

Plagiochila spp. mostly grow as epiphytes, on the bark of forest trees. They have two rows of lateral leaves which are obliquely inserted on the stem with the lower margin of each leaf sloped forwards, lying underneath the upper margin of the leaf in front (**succubous** leaves). Underleaves are very small or absent.

All *Plagiochila* spp. are dioecious. Male plants have groups of reduced male bracts which alternate with groups of normal leaves along the length of the stem. There are several archegonia at the tip of each fertile female stem. The mature perianth is flattened and trumpet-shaped. The capsule is spherical, on a fairly long seta, and contains spores and free elaters.

Members of several families (e.g. Frullaniaceae; Lejeuneaceae) have their lateral leaves folded,

(a)

(b)

18.11 Hepaticae, Jungermanniales, *Plagiochila* sp.
(a) ventral view of a male plant with groups of male bracts (arrowed); (b) ventral view of a female plant with perianths (P) from some of which project capsules (C), one with the white seta (S) visible. (× 2)

usually forming a large, upper lobe and a smaller, lower lobe. The upper lobes are inserted so that their upper edges lie over the lower edges of the leaves in front (**folded insertion, incubous** upper lobes). *Frullania* spp. are common as epiphytes on the bark of trees, often exposed to full sunlight. *F. squarrosa* is brown or black in colour, owing to the development of pigment in the cell walls. *Frullania* spp. have the lower lobe of each lateral leaf modified into a cup-shape, with the opening of the cup facing backwards along the stem. Underleaves are bilobed and symmetrical.

All *Frullania* spp. are dioecious. Male bracts are much reduced and strongly overlapping, on short side branches. Archegonia are in groups of two at the ends of fertile female stems. The perianth is inflated and has only a small terminal opening. The seta grows just far enough to expose the capsule above the perianth. The capsule is spherical and contains spores and attached elaters. Each elater runs for the length of the capsule and is attached at the top and the bottom of the capsule. When the capsule starts to open, by four vertical slits meeting at the apex, the elaters are first stretched and then become detached at the lower end, so that the spores are all flung out in a single explosive movement.

Members of the Lejeuneaceae are very common, usually on bark. They are similar to *Frullania* spp., except that the lower lobe of each lateral leaf is flat and there is only one archegonium at the end of each fertile female stem.

Some members of the Lejeuneaceae and other leafy liverworts grow on the surfaces of the leaves of rain forest shrubs and trees (**epiphyllous** liverworts). One of the commonest epiphyllous liverworts is *Radula flaccida*, which belongs to the Radulaceae. *R. flaccida* is similar in structure to members of the Lejeuneaceae but lacks underleaves. It has been shown that *R. flaccida* is more than an epiphyte. Its rhizoids penetrate the cuticle of the host's leaves and draw water from the host. Most epiphytic leafy liverworts (Fig. 18.4) are able to survive drying out and continue to grow as soon as they are moistened after a dry period.

Musci

The cells of mosses contain many discoid, parietal chloroplasts. The chloroplasts lack pyrenoids and there are no oil bodies in the cells (Fig. 18.12).

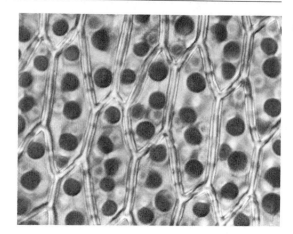

18.12 Musci, cells of a moss leaf with many discoid chloroplasts (HP)

All mosses have leafy gametophytes, with the leaves almost always arranged in a spiral on the stem (Fig. 18.15). Moss stems mostly have a central strand of elongated and thickened fibre-like cells, surrounded by a parenchymatous cortex and an epidermis. Moss leaves (Fig. 18.13) may be one cell thick throughout their width, but often have one or more nerves. The **nerve** is more than one cell thick and is composed of elongated cells like those of the central strand of the stem. The base of the nerve ends in the outer cortex of the stem. The central strand of the stem and the nerve of the leaf give mechanical support to the stem and leaf respectively but do not have any conducting function.

Mosses are attached by branched, multicellular, filamentous rhizoids. Some mosses grow in upright tufts of stems (**orthotropic** mosses), others have branched, prostrate stems (**plagiotropic** mosses) (Fig. 18.14).

Plants are monoecious or dioecious. Antheridia, which are usually elongated, are axillary to leaves or in terminal cup-like structures at the end of a stem (**antheridial cups**). The antheridial cup is formed of a broad stem tip, surrounded by many rather broad leaves which form the walls of the cup (Fig. 18.15). Antheridia are mixed with sterile hairs (Fig. 18.1) (**paraphyses**) which secrete mucilage. The formation of an antheridial cup does not end the growth of the stem, which grows on after the antheridia have shed their sperms. Archegonia are usually in small groups at the tips of leafy stems, mixed with paraphyses and

(a)

(b)

18.13 Musci, examples of moss leaves: (a) *Ectropothecium* sp., wide, concave leaf with no nerve; (b) *Bryum* sp., leaf with a nerve running out beyond the end of the leaf, margin toothed

(a)

(b)

18.14 Musci, habits: (a) *Polytrichum* sp., an orthotropic moss with the capsule (in calyptra) at the end of each main stem; (b) *Ectropothecium* sp., a plagiotropic moss

18.15 Musci, a male stem of *Polytrichum* sp. with an antheridial cup. In this plant, the leafy stem has continued to grow from the base of the antheridial cup. (× 2)

(a)

(b)

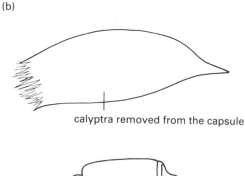

calyptra removed from the capsule

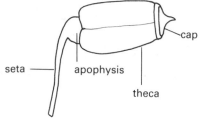

18.16 Musci, *Polytrichum* sp.: (a) female gametophyte with a capsule enclosed in a calyptra; (b) drawing of the calyptra removed, with capsule below

enclosed within **perichaetial leaves**, in the apical bud. If a sporophyte starts to develop, growth of the stem apex ceases.

The moss sporophyte usually has a pointed foot, a long wiry seta and a capsule which opens by a cap. During its development, the embryo forms an elongated structure, pointed at both ends. As it increases in length, the upper part of the wall of the archegonium forms a pointed calyptra which breaks free at the base and is carried up on the end of the sporophyte (Fig. 18.16). The calyptra increases in size and encloses the developing capsule, at the end of the seta.

The moss capsule (Fig. 18.16) has two regions. The lower region, attached to the top of the seta, is the **apophysis**. The epidermis of the apophysis contains stomata and, like the rest of the capsule, has a cuticle. The apophysis contains a central strand of elongated cells, continuous with the inner tissue of the seta, surrounded by chlorenchyma. Above the apophysis is the **theca**, with its cap. The central strand of the apophysis runs all the way to the top of the capsule and forms the **columella**. The columella is surrounded by a layer of sporocytes which divide by meiosis to form many tetrads of spores. There are no elaters. Outside the archesporium there is a cylindrical air space and a few layers of cells between the air space and the wall of the theca.

When the capsule is ripe, first the calyptra drops away. Then, the lid separates along a line of weakness. This exposes one or two rows of hygroscopic teeth, which form the **peristome**, closing the top of the theca (Fig. 18.17). The peristome teeth are very sensitive to changes in humidity. In humid weather, they close the top of the theca. In dry weather, they open and either pull spores from the capsule (upright capsule) or allow spores to fall through the opening (hanging capsule). Spores are dispersed by wind.

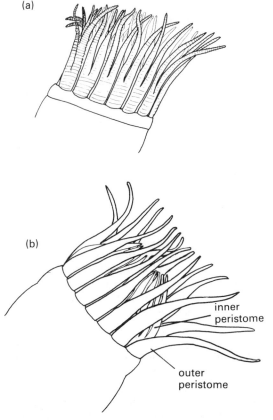

(a)

(b)

inner peristome

outer peristome

18.17 Musci, peristomes: (a) capsule with a single row of deeply divided peristome teeth; (b) capsule with a double peristome

Spores develop to form a green, filamentous branching **protonema**. The protonema has rhizoidal branches, with elongated cells and oblique cross walls, which penetrate the substrate. After a short time, side branches of the protonema divide to form buds, each of which grows by an apical cell and forms a leafy shoot. The protonema that grows from the spore is the **primary protonema** and may form many leafy shoots. Rhizoidal protonema also grows from the base of the stem and from leaves. This is **secondary protonema**. Rhizoidal protonema produces green branches when it rises above the surface of the soil and may produce more leafy branches. This is one type of vegetative reproduction. Other types of vegetative reproduction include regeneration from detached parts (stems, leaves), and gemma formation. Gemmae of different types are formed by many genera, on leaves or stems.

Classification

Mosses are divided into five subclasses containing a total of seventeen orders. Of these, we shall look at members of two subclasses, Bryidae (Dicranales, Fissidentales, Eubryales) and Polytrichidae (Polytrichales). Members of two other subclasses (Sphagnidae, Andreaeidae) have protonema and capsule structure quite different from that described above.

Dicranales

Members of this order are orthotropic and have capsules with a single row of deeply forked peristome teeth.

Calymperes spp. are very common epiphytes in forest areas. They form yellowish-green tufts, 2 to 3 cm high, with tapering leaves about 3.5 mm long and 1 mm wide. Each leaf has a wide nerve which extends beyond the end of the lamina, which is one cell thick and is made up of small cells, more or less square in surface view. The end of the nerve initially bears a cluster of elongated gemmae, tapered at both ends, each consisting of a single row of cells. These are lost as the leaf ages.

The capsule is upright, short and wide, and has only a rudimentary peristome consisting of very short teeth.

Octoblepharum albidum (Fig. 1.14) is a tufted, epiphytic moss, at once recognized by its yellowish-white colour when dry, pale green when wet. The leaves are stiff and pointed. In TS, there is a single irregular row of green cells which runs from one edge of the leaf to the other, with two rows of air-filled cells above and below. The empty cells have pores in their walls and take up and retain water when they have been moistened. The capsule is upright, with a pointed cap. The peristome is made up of sixteen short, deeply forked teeth.

Leucobryum spp. are very similar to *Octoblepharum albidum*, to which they are closely related, but their leaves are softer and tend to curl when dry.

Fissidentales

Members of the Fissidentales are easily recognized because their leaves are in two rows and flattened in the plane of the stem. Each leaf has a flap-like outgrowth along the nerve near the base and this flap encloses the stem at its base. The leafy plant looks rather like a small, pinnate, fern leaf.

The capsules of most members of this order are upright and cylindrical, pointed at both ends. The peristome is of sixteen deeply forked teeth.

There are three genera in this order. *Fissidens* spp. are extremely common on moist earth in shady places and are amongst the few mosses which grow on banks of soil on the floor of the rain forest, as well as on river banks in savanna country. *Nanobryum dummeri* is an especially interesting species. It is known as a **protonema moss**, because the protonema forms the main part of the plant. The protonema makes a green coating on worm casts and the bases of termitaria in the forest. The leafy stage consists of small, separate, male and female branches. Each male branch has only two or three small leaves and a small group of antheridia. Female plants are taller and bear about six leaves each, similar to those of *Fissidens* spp. After its development, the capsule points obliquely downwards from the top of a pale seta about 4 mm high.

Eubryales

Gametophytes of members of the Eubryales (Fig. 18.18) include orthotropic and plagiotropic types.

The capsule of the sporophyte has a double peristome. There is an outer row of stiff, hygroscopic teeth and an inner row of membranous teeth. The inner peristome teeth do not move with changes in humidity. In humid weather, the outer teeth turn into the mouth of the capsule, between the inner teeth. As they dry, they bend outwards, bringing spores out of the capsule with them.

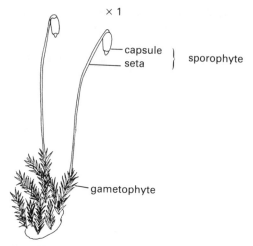

18.18 Musci, *Bryum* sp. A common moss, often found growing on soil around tree bases and on concrete blocks and similar substrates in Africa (× 1)

Bryum spp. are mostly small mosses. *B. argenteum* is very common growing in crevices in rocky outcrops and similar places. It forms tufts which have a silvery sheen when dry. *B. coronatum* forms larger tufts and is very common growing at the bases of buildings and on concrete blocks and bricks in built-up areas. The stems are 2 to 3 cm high and the leaves about 5 mm long, tapering and with a nerve that runs into the pointed apex. The capsule has a seta 1 to 3 cm long and hangs downwards (is **pendant**).

Ectropothecium spp. (Fig. 18.14) are creeping mosses, often on fallen logs and tree bases, close to the soil. They branch profusely and all the branches are prostrate. Leaves are short and overlapping, with elongated cells and are either nerveless or have two short nerves at the base. Antheridia are in the axils of ordinary leaves and archegonia are at the tips of very short side branches, where capsules are later formed. The capsule is oval, held horizontally at the end of a long, upright seta.

There are many genera of Eubryales, including both orthotropic and plagiotropic types. They vary in their manner of branching, shape of leaves, shape of the cells in the leaf and number, length and positions of leaf nerves. Capsules vary in shape and the angle at which they are held. Identification of these mosses is made difficult by the fact that descriptions of genera and species are scattered throughout many separate scientific papers.

Polytrichales

Members of the Polytrichales are fairly common in the tropics, but are most common at higher altitudes. All grow on soil and are most often seen on roadside banks, though they may form extensive ground cover in forest clearings at altitudes above about 1000 m.

Members of the order (e.g. *Polytrichum* spp., Fig. 18.14) have several structural features which are different from those of other groups of mosses. The central strand of the stem is more highly differentiated and contains two sorts of cells, some thick-walled and some thin-walled, and has been shown to conduct water and dissolved substances. The leaf, as seen in TS, has a lower, parenchymatous part, several cells thick, with separate vertical sheets of green cells running the length of the leaf. The capsule is like that of other mosses, except for the dispersal mechanism. There are many, very short peristome teeth with their tips attached to a circular membrane (**epiphragm**). The epiphragm is drawn down and closes the mouth of the capsule in humid weather. When the air is dry, the teeth raise the membrane and spores are shaken out between the teeth (pepperpot mechanism of spore dispersal).

When dry, the capsule of *Polytrichum* spp. is roughly square in TS. In the related and sometimes commoner *Pogonatum* spp., the capsule is always cylindrical.

Key words

Bryophytes: non-vascular. **Sexual reproduction**: antheridia, archegonia; jacket, sterile cells. **Antheridium**: spermatocytes, sperms. **Archegonium**: venter, ovum, ventral canal cell; neck canal cells. **Alternation of generations**: gametophyte, sporophyte; haplodiplontic life cycle, dominant gametophyte generation. **Gametophyte**: monoecious, dioecious. **Sporophyte**: foot, seta, capsule, calyptra. **Classification**: Anthocerotae (hornworts), Hepaticae (liverworts), Musci (mosses). **Anthocerotae**: thallus, pyrenoid; sporophyte, intercalary meristem, columella, elaters. **Hepaticae**: thalloid, leafy, rhizoids, oil bodies; leafy liverworts, stem, lateral leaves, underleaves; capsule, spores, elaters, elaterophore. **Musci**: protonema, stem, spirally arranged leaves, nerve; orthotropic, plagiotropic; protonema, rhizoids; sporophyte, capsule, apophysis, theca, columella, peristome, peristome teeth, epiphragm.

Summary

Bryophytes are simple land plants with no true vascular tissues (i.e. xylem, phloem). Their organs of sexual reproduction are antheridia and archegonia, in which the fertile cells are protected within a jacket of sterile cells. The life cycle is haplodiplontic, with heteromorphic alternation of generations in which the gametophyte generation is dominant. Gametophytes are free-living and either thalloid or leafy, attached to the substrate by rhizoids. Sporophytes develop attached to the gametophyte and remain largely dependent on the gametophyte. Meiospores are released and grow to form a new generation of gametophytes.

Questions

1 Suppose you were given a small piece of thallus without reproductive organs of any kind. How would you quickly determine whether it was a piece of a hornwort or a piece of a liverwort?

2 Suppose you were given a piece of a leafy bryophyte bearing a fully developed capsule. How would you quickly determine whether it was a leafy liverwort or a moss (a) on gametophyte characters, and (b) on sporophyte characters?

3 Construct a table to show similarities and differences between hornworts, liverworts and mosses, using both gametophyte and sporophyte characters.

4 What are the male and female reproductive organs of bryophytes? How do they differ from the corresponding organs of algae?

5 Describe how spores are released and dispersed in named examples of hornworts, liverworts and mosses.

19

Pteridophytes – spore-dispersed vascular plants

Pteridophytes were once classified together, as one complete division of the plant kingdom. Although we are following a different classification, in which they are divided amongst several subdivisions of the division Tracheophyta (vascular plants), it is convenient to look first at some important features which they have in common. Their common features are in their morphology, anatomy and life cycles.

Morphology and anatomy

All pteridophytes have well-defined stems and leaves and almost all have roots. Root, stem and leaf all have vascular systems which include xylem and phloem similar to that found in flowering plants.

In the most primitive form, the xylem of pteridophytes is made up of **tracheids** alone. These have lignified secondary walls with the same patterns of thickening as are found in flowering plants (annular, spiral, scalariform, reticulate and with bordered pits). In more advanced pteridophytes, some of the tracheids are replaced by parenchyma and each strand of xylem consists of a mixture of tracheids and parenchyma. There are a few of the most advanced forms in which there are vessels as well as tracheids.

The basic phloem cell is the sieve cell. A **sieve cell** is elongated and tapered at both ends. Its cellulose walls include many **sieve areas**, similar to the sieve plates of flowering plants. Sieve cells have nuclei and a full complement of cytoplasmic organelles. There are no companion cells.

The vascular tissues in the stems of primitive pteridophytes are at the centre of the stem and consist of a central strand of xylem surrounded by a layer of phloem. There is usually also an endodermis, immediately outside the phloem. This type of vascular cylinder (**stele**) is a **protostele** (Fig. 19.1). In more advanced forms, the stele consists of a ring of xylem with phloem and endodermis on the outside only, or outside and inside the xylem, with a medulla of parenchyma. This is a **solenostele**. Where part of a solenostele turns outwards through the cortex to form a leaf trace, this leaves a gap (**leaf gap**) in

the solenostele which is now C-shaped. Such a leaf gap persists for a little way above the insertion of the leaf but subsequently closes and the solenostele then remains as a closed cylinder until the departure of a further leaf trace.

Where leaves are attached close together on a stem, leaf gaps overlap so that the stele is interrupted by several leaf gaps at any one level. In TS, this appears as a circle of **meristeles**, and the whole stele is then a **dictyostele**.

The vascular system of the root is always a protostele, often similar to that of flowering plants.

Vascular tissues occur only in the sporophyte.

Reproduction and life cycle

The sporophyte is the dominant generation in all pteridophytes (Fig. 19.2). The sporophyte bears a few or many sporangia, from which meiospores with thick, resistant walls are dispersed by wind. Each spore germinates to form a small, inconspicuous and usually short-lived gametophyte.

Gametophytes are usually monoecious and produce antheridia and archegonia which have the same basic structure as those of bryophytes. However, antheridia are usually small and the venters of archegonia are embedded in the gametophyte, with only their necks standing above the surface (Fig. 19.15).

Fertilization is followed by the development of the young sporophyte within the venter of the archegonium. The early **embryo** soon differentiates to form a swollen, haustorial foot, a primary shoot apex and a primary root apex. The shoot and root then grow out through the tissues of the gametophyte and the new sporophyte becomes an independent plant (Fig. 19.15).

Classification

Pteridophytes include members of four subdivisions of the division Tracheophyta. These are the Psilopsida, Lycopsida and Sphenopsida, also the class Filicinae of the subdivision Pteropsida. Sphenopsida (Horsetails) are rare in tropical Africa and grow only

(a)

(i) simplest form of protostele

solid central xylem surrounded by phloem

protoxylem points at outside of xylem cylinder (exarch xylem)

phloem

pericycle

endodermis

cortex

epidermis

leaf trace, does not cause a break in the central stele

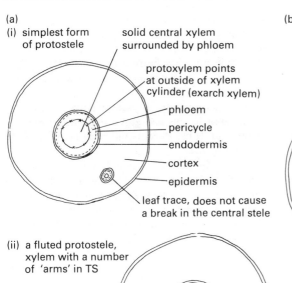

(ii) a fluted protostele, xylem with a number of 'arms' in TS

protoxylem at ends of the arms

phloem continuous around xylem

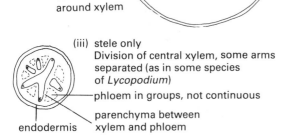

(iii) stele only
Division of central xylem, some arms separated (as in some species of *Lycopodium*)

phloem in groups, not continuous

parenchyma between xylem and phloem

endodermis

centre of xylem may be replaced by parenchyma to give a medulla (in a medullated protostele)

(iv) stele only

endodermis

protostele divided into 'bands' as in TS

protoxylem at outer edge of xylem

phloem penetrates between the xylem bands

(as in *Lycopodium clavatum*)

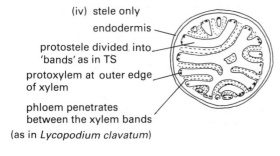

(b) (i) solenostele, cylinder of xylem with a central medulla of parenchyma

when continuous only on outer edge of xylem gives an ectophloic solenostele; if present, a leaf trace breaks the cylinder with a leaf gap,

endodermis

pericycle

phloem

xylem

medulla

epidermis

cortex

(ii) solenostele with continuous inside and outside the xylem gives an amphiphloic solenostele

endodermis

pericycle

phloem

xylem

leaf gap

leaf trace

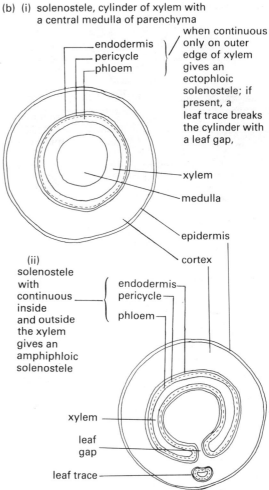

(c) a dictyostele, with a number of separate meristeles

epidermis

cortex

meristele

perforation, a gap in the stele not associated with a leaf trace

leaf gap

leaf trace

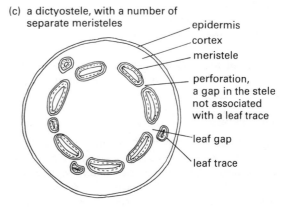

19.1 Vascular tissues in pteridophytes: (a) protosteles; (b) solenosteles; (c) a dictyostele

(a) homosporous life cycle

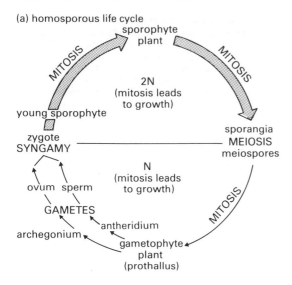

(b) heterosporous life cycle

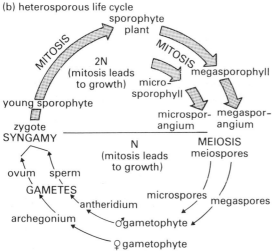

19.2 Pteridophyte life cycles: (a) in homosporous forms, as in most Filicinae and in Lycopodiales; (b) in heterosporous forms, as in water ferns, Selaginellales and Isoetales

at very high elevations (above 1000 m). Our attention will be concentrated on the other groups.

Psilopsida

Sporophytes of members of the subdivision Psilopsida have a rhizome and aerial shoots which are both dichotomous. The rhizome has unicellular rhizoids but no roots. The aerial branches bear small leaves, with or without vascular connections with the vascular system of the stem, which is a protostele. Sporangia are large and are borne in groups of two or three in the axils of structures which are probably short side branches bearing two small leaves. The pair of leaves is sometimes interpreted as a single, forked leaf.

The gametophyte is saprophytic, contains a mycorrhizal fungus and grows under the surface of the ground. It has rhizoids and bears both antheridia and archegonia.

The subdivision includes only two living genera, *Psilotum* and *Tmesipteris*. The latter genus occurs only in Australia, New Zealand and neighbouring islands. *Psilotum* spp. are much more widely distributed in tropical and subtropical areas and *P. nudum* occurs in Africa (Fig. 19.3).

P. nudum is not a common plant but is often abundant in those areas in which it occurs. It grows as an epiphyte, or in soil-filled cracks between rocks.

The sporophyte has a horizontal, branching rhizome with upright, dichotomously branched green stems which grow to a height of 25 cm or more. The stem bears small leaves at rather widely spaced intervals. On the upper part of the plant, groups of three large sporangia are carried on very short side branches, in the axils of what appear to be forked leaves.

The stem has a protostele in which the xylem is star-shaped in TS, with protoxylem at each point of the star. There is a wide parenchymatous cortex, with an outer zone of chlorenchyma. The epidermis contains stomata and has a thick cuticle. None of the leaves has any vascular tissue and there are no leaf traces. The leaf epidermis contains no stomata and the mesophyll is chlorenchyma like that in the outer cortex of the stem.

Spores are released when each sporangium opens by a single slit.

The gametophyte is creamy-white in colour, more or less cylindrical and grows to a length of 1 cm or more. It has unicellular rhizoids and bears antheridia and archegonia. Antheridia are more or less hemispherical and attached to the surface of the gametophyte. Archegonia have sunken venters, with the necks protruding above the surface.

Lycopsida

Sporophytes of plants belonging to this subdivision have stems, leaves and roots. Branching of stems and

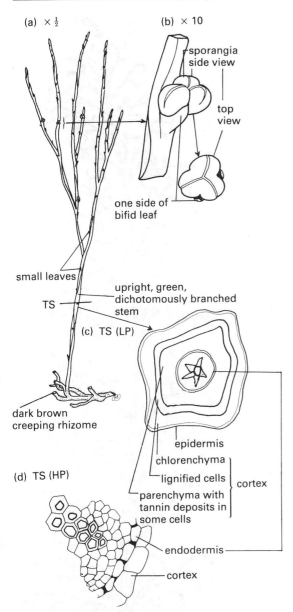

(a) × ½

(b) × 10

sporangia
side view

top
view

one side of
bifid leaf

small leaves

TS

upright, green,
dichotomously branched
stem

(c) TS (LP)

dark brown
creeping rhizome

(d) TS (HP)

epidermis
chlorenchyma
lignified cells
parenchyma with
tannin deposits in
some cells

cortex

endodermis

cortex

19.3 *Psilotum nudum*, a primitive pteridophyte: (a) whole plant; (b) three fused sporangia (a synangium) on a short side branch; (c) TS aerial stem; (d) cell detail of the stele

roots is dichotomous but, in the stems of some species, subsequent inequality in development between the two branches of a dichotomy leads to an apparently monopodial pattern of branching.

In all living genera, the vascular system of the stem is some variety of protostele, with varying amounts of parenchyma included in the xylem. Roots also have protosteles, often diarch and similar to the roots of dicots.

The leaves are small and entire and usually taper to a point. The epidermis of the leaf contains stomata and the mesophyll is like the spongy mesophyll of flowering plants. There is always a single, unbranched vein at the centre of the leaf. The vein is connected to the stele of the stem by a small leaf trace. There is no reduction in the diameter of the stele above the point of attachment of the leaf trace. Leaves of this description are **microphylls**, as distinct from the megaphylls of ferns and seed plants. A **megaphyll** has a branched vascular system, or several veins running the length of the leaf, and its leaf trace diverts a substantial amount of vascular tissue from the stele. In the case of a solenostele, the leaf trace of a megaphyll is associated with a leaf gap.

Leaves of members of the class Eligulatae are without appendages. In the class Ligulatae, there is a small, pointed, membranous appendage on the upper surface of the leaf, close to the point of attachment to the stem and directed towards the tip of the leaf. This is a **ligule**. The function of the ligule is not known.

Sporangia are borne singly on the upper surfaces of fertile leaves (**sporophylls**), close to the attachment to the stem. Sporophylls may be mixed with vegetative leaves along the stem, or may be gathered together into terminal cone-like **strobili** on some of the branches.

Sporangia of members of the Eligulatae are all of one type. They produce spores which are all the same size and germinate to form monoecious gametophytes. These are **homosporous** plants (life cycle, Fig. 19.2). In the Ligulatae, there are two sorts of sporophylls, **microsporophylls** and **megasporophylls**. A microsporophyll has a microsporangium which contains many small spores (**microspores**), which produce very small male gametophytes. Each male gametophyte is made up of little more than a single antheridium. A megasporophyll bears a **megasporangium** which contains very few, usually four, very large **megaspores**. Each megaspore germinates to form a female gametophyte which bears archegonia. Plants which produce two types of spores, in this way, are **heterosporous** (life cycle, Fig. 19.2).

Classification

Some of the orders of the classes Eligulatae and Ligulatae are known only as fossils. Living genera are classified in the order Lycopodiales in the Eligulatae and the orders Selaginellales and Isoetales in the Ligulatae. Members of all three orders are common in the tropics.

Eligulatae

All genera of the class Eligulatae are homosporous. Their leaves lack ligules, and adventitious roots arise directly from leafy stems.

Lycopodiales

The Lycopodiales include only two living genera, *Lycopodium* and *Phylloglossum*. *Phylloglossum* has only one species and is known only from Australasia, but *Lycopodium* has a world-wide distribution and is especially common in the tropics.

Lycopodium spp. have creeping, hanging or upright sporophytes, attached to the substrate by adventitious roots. Leaves are spirally attached and sporophylls are usually in terminal strobili. Figure 19.4 shows an upright and a pendant sporophyte, both tropical African species. Hanging, epiphytic forms are attached to host trees by adventitious roots at the base of the stem. Terrestrial forms (e.g. *L. cernuum*) produce adventitious roots wherever the prostrate part of the stem is in contact with the soil.

The vascular system of the stem is a protostele, with strands of parenchyma alternating with strands of tracheids. The xylem is surrounded by a narrow region of phloem and a pericycle, several cells thick. An endodermis is present but is often very difficult to recognise, except in the youngest part of the stem.

Spores are released when a sporangium (Fig. 19.5) opens by a single slit. The gametophytes of some species grow on the surface of the substrate, attached by unicellular rhizoids, and are autotrophic. Other species have underground, saprophytic gametophytes which contain a mycorrhizal fungus. Archegonia are of the usual pteridophyte type, with sunken venters.

(a)

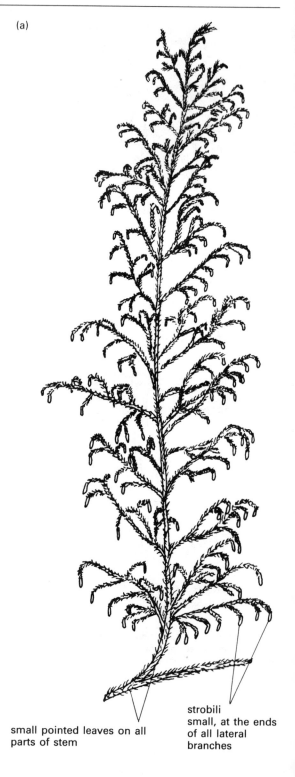

small pointed leaves on all parts of stem

strobili small, at the ends of all lateral branches

(b)

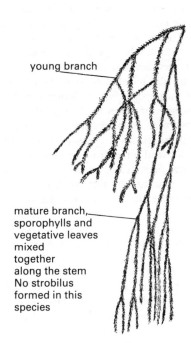

young branch

mature branch,
sporophylls and
vegetative leaves
mixed
together
along the stem
No strobilus
formed in this
species

19.4 Species of *Lycopodium* in tropical Africa: (a) terrestrial species, *L. cernuum*; (b) epiphytic species, *L. verticillatum* (× ⅓)

Antheridia are stalked, spherical and embedded in hollows in the thallus. Both are borne on the upper surface of the thallus. Sperms are biflagellate.

Ligulatae

Members of the class Ligulatae are heterosporous. Their leaves have ligules and adventitious roots are produced only from specialised, leafless branches (**rhizophores**).

Selaginellales
The Selaginellales include only one living genus, *Selaginella*. *Selaginella* spp. have a world-wide distribution, but are commonest in the tropics.

Most species of *Selaginella* (Fig. 19.6) have creeping or scrambling stems, with leaves in four rows. The leaves of some species are all alike and

(a)

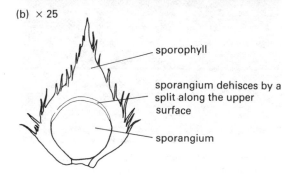

(b) × 25

sporophyll

sporangium dehisces by a
split along the upper
surface

sporangium

19.5 *Lycopodium cernuum*: (a) mature cone shedding spores; (b) a single sporophyll, with sporangium split open

spirally arranged (e.g. *S. njam-njamensis*) but the branches of most species are distinctly dorsiventral and bear two rows of larger leaves, attached to the underside of the stem, and two rows of much smaller leaves, attached along the upper side of the stem.

The vascular system of the stem is a protostele. The phloem is surrounded by a pericycle, usually one cell thick. The cells of the endodermis are radially elongated and suspend the protostele at the centre of a cavity. The endodermis is described as **trabecular**. In some larger species, the stem is **polystelic** and has two separate protosteles suspended in the cavity.

(a)

(b)

rhizophore

19.6 Species of *Selaginella* in tropical Africa: (a) *S. abyssinica*, small and erect or partly pendant, never creeping; (b) *S. njam-njamensis*, end of branch showing spirally-arranged leaves and short rhizophore with two roots emerging from the tip

Rhizophores grow from meristematic areas immediately below points at which the leafy stems branch. Their vascular system is a protostele, with no cavity around the stele. Adventitious roots arise in pairs from the tip of the rhizophore, as soon as it touches the ground.

Strobili are almost always terminal on short side branches (Fig. 19.7). Each strobilus is made up of four rows of sporophylls, strongly overlapping and usually equal in size. Both microsporophylls and megasporophylls occur in a single strobilus, but there are usually fewer megasporophylls than micro-

sporophylls. Microsporangia have a smooth outline and contain many very small spores. In most species, only one megaspore mother cell remains in the sporangium and, following meiosis, produces a single tetrad of very large megaspores. The megasporangium is usually lobed, following the outline of the tetrad of megaspores inside. Spores are shed when the sporangium dehisces by a single slit.

Both microspores and megaspores have thick outer walls with short spines. Also, each spore has a rounded side and three flat, triangular faces (**facets**), where it was previously compressed against the three other spores making up the tetrad. Cell division has started inside each type of spore before the spores are shed.

In the microspore, the first cell division cuts off a single prothallial cell, more or less towards the rounded side of the spore. Further divisions cut off a layer of cells at the surface, which form an antheridial jacket. The larger, central cell then divides to form many small cells which metamorphose into biflagellate sperms. About this time, the prothallial cell and the jacket cells disintegrate. The spore wall opens along the three edges between the facets and the sperms are released.

After the megaspore has been shed, the three facets of the spore coat separate along their edges and turn outwards. Gametophyte tissue grows out of the open end of the spore wall and forms three groups of unicellular rhizoids, between the open edges of the spore coat. There is no further growth and archegonia, of the usual pteridophyte pattern, are formed on the upper surface. Subsequently, the primary root and shoot grow out through the tissues of the gametophyte and a new sporophyte plant becomes established. The remains of the gametophyte, still partly enclosed within the spore coat, remain attached to the young plant for some time.

S. njam-njamensis, with spirally arranged leaves with hair-like extensions at their tips, occurs widely on shallow soil on rocky outcrops. It is able to dry out during the dry season and recover when moistened. *S. kraussiana* is a creeping plant of rain forest and river and stream banks. *S. myosurus* is a scrambling plant which may reach a height of 2 m and can be quite a serious agricultural weed in wetter areas. *S. abyssinica* is a small upright plant of wet areas, in which the sporophylls in the strobilus are of two sizes, like the leaves on vegetative stems.

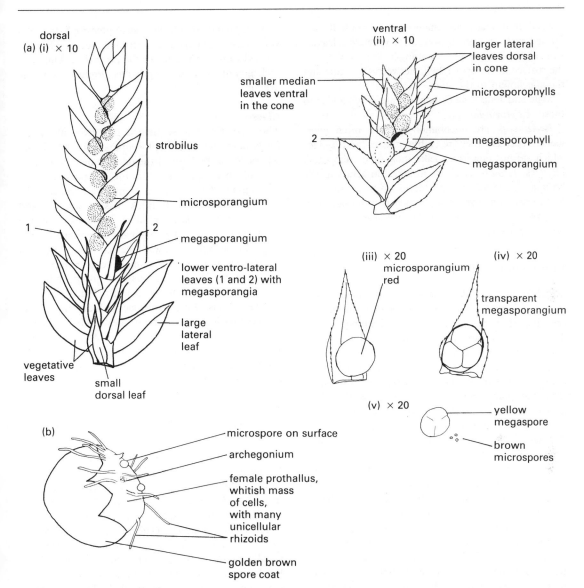

19.7 Reproduction in *Selaginella* spp.: (a) *S. abyssinica*, structure of the cone (strobilus); (b) *S. njam-njamensis*, megagametophyte (note relative size of adhering microspores)

Isoetales

Isoetes is the only living genus of this order, which has a world-wide distribution. *Stylites* is a recently discovered S. American genus.

Isoetes spp. have a lobed, corm-like underground stem with elongated, grass-like leaves arranged in a spiral. The lobes at the base of the stem have dark-coloured adventitious roots arising between them and are interpreted by some as rhizophores. The vascular system of the stem is a protostele, widened and lobed at the base. In older stems, a cambium differentiates outside the phloem of the protostele. *Isoetes* is the only living genus of pteridophytes that has secondary thickening and the type of secondary thickening closely resembles that found in extinct, fossil relatives.

The leaves are ligulate. The leaf blade is internally

divided into many air chambers. Sporophylls are formed each growing season, after vegetative leaves, which they resemble. Sporangia are sunken in the surface of the sporophyll, between the ligule and the attachment to the stem.

Male and female gametophytes are very similar to those of *Selaginella* spp.

Isoetes spp. are uncommon but are often overlooked because they resemble grasses and sedges which are not flowering. Like most of its temperate relatives, *I. biafrana* grows in lakes, attached to the bottom. *I. nigritiana*, with leaves up to about 7 cm high, grows on shallow soil which is waterlogged in the wet season but dry in the dry season, e.g. on rocky outcrops.

Pteropsida – class Filicinae

Members of the class Filicinae (ferns) have stems, leaves and roots. Their stems include upright or creeping rhizomes, stolons and climbing stems. Their leaves are **megaphylls**, often pinnate or bipinnate. Their roots are adventitious.

The vascular system of the stem of a very young plant is a protostele. The protostele may persist in the adult plant, but usually changes as the plant grows older and the stem becomes thicker, first to a solenostele and then, often, to a dictyostele. There are more complex steles in ferns with very thick stems (e.g. tree ferns).

Except in ferns with very simple leaves (filmy ferns), the internal structure of the leaf is like that of a flowering plant. The leaf blade often has a midrib and reticulate venation. There is no axillary bud where the leaf is attached to the stem. Large, compound fern leaves are often called **fronds**.

Ferns are mostly homosporous, but a few (water ferns) are heterosporous. Sporangia are usually numerous, gathered together into groups (**sori**) on the abaxial side or margin of each sporophyll.

Spores of homosporous ferns germinate to form small, usually unbranched, green thalloid gametophytes (**prothalli**), attached to the soil by unicellular rhizoids. Archegonia and antheridia are borne on the underside of the prothallus.

Antheridia are formed before archegonia, on the older part of the prothallus. They are more or less hemispherical and project from the surface. Each antheridium produces many spirally coiled multiflagellate sperms.

Archegonia are of the usual pteridophyte type, with sunken venters and protruding necks. They are formed close to the apex of the prothallus. Only one archegonium on each prothallus produces a viable zygote.

Classification

The class Filicinae is divided into two subclasses, Eusporangiatae and Leptosporangiatae, which are separated on the basis of the development and structure of their sporangia. In the Eusporangiatae a sporangium develops from a group of initial cells, has a wall several cells thick, has no annulus and contains a very large number of spores. In the Leptosporangiatae, the sporangium develops from a single initial cell, has a wall one cell thick, opens by an annulus and contains a small, regular number of spores (e.g. 32). An **annulus** is a group or line of cells with specially thickened walls (Fig. 19.14).

The subclass Eusporangiatae includes the orders Ophioglossales and Marattiales. The subclass Leptosporangiatae includes three orders. The order Filicales is very large and contains typical, homosporous ferns. The orders Marsileales and Salviniales are heterosporous water ferns.

Ophioglossales

This order includes only three genera, of which *Ophioglossum* has a worldwide distribution.

Most *Ophioglossum* spp. are terrestrial and erect, from 2 to 30 cm high. A few species are epiphytic and hang from the branches of their host.

The stem is a short, vertical rhizome. Its vascular system is a dictyostele. Leaves are alternate and usually produced singly from the stem apex. Each leaf has two parts. There is a broad, entire vegetative part (lamina) and a narrow **fertile spike**. The lamina has reticulate venation. Leaves are not spirally coiled during development. The fertile spike bears two rows of sporangia, which are partly embedded in the other tissues of the spike.

Gametophytes are all to some extent saprophytic, but some have green lobes above the surface of the ground.

Ophioglossum spp. (e.g. *O. costatum*) are most frequently found on level, often shallow, soil, growing among grasses and other short herbs (Fig. 19.8).

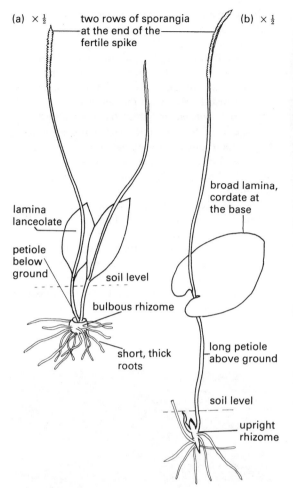

(a) × ½

two rows of sporangia at the end of the fertile spike

(b) × ½

lamina lanceolate

petiole below ground

soil level

bulbous rhizome

short, thick roots

broad lamina, cordate at the base

long petiole above ground

soil level

upright rhizome

(c)

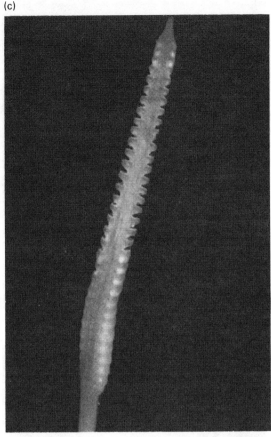

19.8 Species of *Ophioglossum* in tropical Africa: (a) *O. costatum*; (b) *O. reticulatum*; (c) fertile spike of *O. reticulatum*, sporangia in middle region dehisced, those at the tip and base still undehisced

Marattiales

The Marattiales include six genera, of which *Marattia* is the only tropical African representative (Fig. 19.9).

Marattia spp. are large ferns. The stem of *M. fraxinea* is an upright, very short, thick trunk, rather like that of a small palm tree. Leaves are spirally arranged at the end of the stem and are bipinnate and up to 2 m long. Like most terrestrial ferns except *Ophioglossum* spp., young leaves are spirally coiled at the ends and gradually uncoil as their tissues mature. This type of uncoiling of the young parts of a leaf is called **circinnate vernation** and is characteristic of ferns.

Fertile leaflets of *Marattia* spp. have a row of elon-gated, brown synangia at each side of the midrib, on the abaxial surface. Each **synangium** is made up of two rows of large, thick-walled sporangia which are fused together into a single structure.

M. fraxinea is quite common in forests and in gullies in more open country, from altitudes of 600 to 2500 m.

Filicales

The Filicales (Figs 19.10 to 19.15) include epiphytic, climbing, terrestrial and aquatic ferns with entire, pinnate or bipinnate leaves (Figs. 19.10, 19.11, 19.12, 29.8).

(a)

(b)

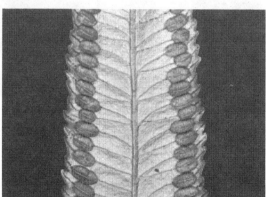

(c)

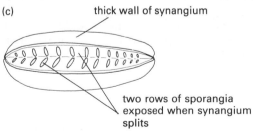

thick wall of synangium

two rows of sporangia
exposed when synangium
splits

19.9 *Marattia fraxinea*: (a) habit of whole plant, growing on
a river bank in forest; (b) lower surface of pinnule with a
row of synangia on each side, each synangium over a vein
(× 3); (c) an open synangium

In some members of the order, sporophylls are
different in size and shape from vegetative leaves
(**dimorphic** leaves). For instance, the epiphytic
ferns, *Platycerium* spp., have dimorphic leaves. At
the beginning of the growing season the plant
produces two upright leaves which become pressed
back against the surface of the host and two rounded
or divided sporophylls, which hang downwards.

Some other ferns have the fertile area confined to
two or more pinnae of a pinnate leaf (e.g. *Osmunda*
sp.). In *Anemia* spp., the two lowest pinnae have
much reduced laminas and are covered with
sporangia.

In most ferns, the first leaves produced each
season are purely vegetative and later leaves, of identical form, are fertile.

Sporangia may be scattered over the abaxial
surfaces of sporophylls, usually mixed with hairs
(e.g. *Platycerium* spp; *Pityrogramma calomelanos*), or
they may be grouped together into sori. A **sorus** may

(a)

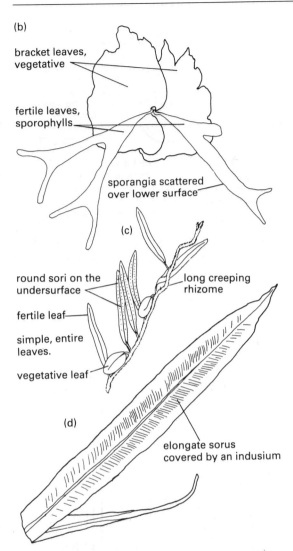

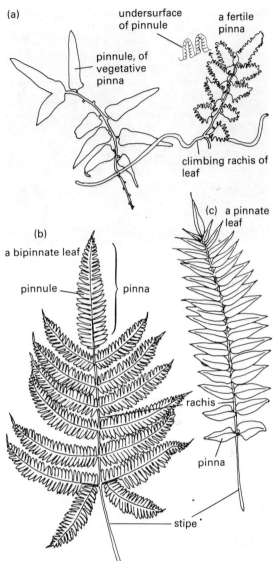

19.10 Order Filicales, some epiphytic ferns: (a) on an Oil Palm near Benin, Nigeria, (i) *Nephrolepsis biserrata*, (ii) *N. undulata*, both of which also grow on the ground in drier areas (see Fig. 19.15), and (iii) *Arthropteris* sp.; (b) *Platycerium stemaria*; (c) *Microgramma owariensis*; (d) *Asplenium currori*. Both (b) and (c) have dimorphic leaves

19.11 Order Filicales, some terrestrial ferns: (a) *Lygodium microphyllum*, a small part of a leaf; (b) *Pteris quadriaurita*, common at forest edges and often found with *Arthropteris* sp. (Fig. 1.18); (c) *Pellaea doniana*, common on rocky ground especially in savanna

be small and rounded, or elongated, over a vein or at the ends of veins when they are near the leaf margin.

Each sorus has a slightly swollen **receptacle**, to which sporangia are attached. The receptacle may also have a membranous outgrowth, the **indusium**, which shelters the sporangia until they are mature. In marginal sori, an indusium may be produced from the margin of the leaf. When sporangia are mature, the indusium dries and curls back and does not obstruct the dispersal of spores. In filmy ferns (*Hymenophyllum* spp; *Trichomanes* spp.) sori are

19.12 *Osmunda regalis*, a species which grows in moist places (e.g. river banks) in Africa, Asia, America and Europe. Close-up of a fertile leaf, with the upper pinnae replaced by groups of sporangia

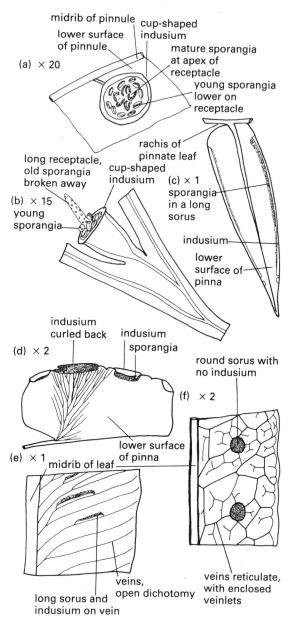

19.13 Sorus structure in the Filicales: (a) round sorus with a cup-shaped indusium (*Cyathea* sp., Fig. 29.8);
(b) marginal sorus of *Trichomanes* sp., a filmy fern;
(c) marginal coenosorus of *Pellaea* sp. (Fig. 19.11);
(d) reflexed leaf margin forming an indusium, in *Adiantum* sp.; (e) elongate sorus in *Asplenium* sp. (Fig. 19.10);
(f) round sorus in *Pleopeltis* sp. (similar to that in *Microgramma* sp., Fig. 19.10)

marginal at the tips of veins, with a long receptacle protected by a two-lipped or cup-shaped indusium. Different types of sori are illustrated in Figure 19.13.

Sporangia are stalked and open as a result of the action of an annulus. An **annulus** is a group or single row of cells which are strongly thickened on their inner cell walls and along the side walls between adjacent cells. Their thin outer cell walls are exposed to the atmosphere. An annulus may be (a) a group of cells at the tip of the sporangium, (b) a row of cells around the tip of the sporangium, (c) an oblique band of cells, or (d) a vertical band of cells running around the edge of a lens-shaped sporangium, from the stalk at one side, for about two-thirds of the circumference (Fig. 19.14).

The commonest type of annulus ((d) above) is regarded as the most advanced and gives a very efficient spore dispersal mechanism. When the sporangium is mature, cells of the annulus start to lose water by evaporation. Loss of water produces tension in the cells and the thin outer wall is drawn inwards, into each cell. The stretching of these walls draws the radial walls together, thus tending to straighten the annulus. The sporangium then breaks open at a point (**stomium**) just below the end of the annulus, where the cells have thin walls. The annulus slowly straightens and, as more water evaporates, curves back on itself. It carries with it the

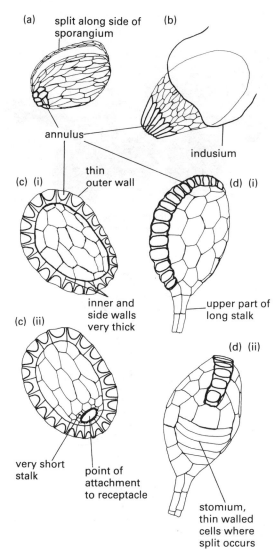

(a) split along side of sporangium

(b)

annulus

indusium

thin outer wall

(c) (i)

(d) (i)

inner and side walls very thick

upper part of long stalk

(c) (ii)

(d) (ii)

very short stalk

point of attachment to receptacle

stomium, thin walled cells where split occurs

19.14 Types of sporangia in the Filicales: (a) annulus a group of thick-walled cells, in *Osmunda regalis*; (b) apical annulus, at the tip of the sporangium, in *Lygodium* sp.; (c) continuous, oblique annulus in *Trichomanes* sp., (i) from above and (ii) from below, showing attachment to the receptacle; (d) vertical annulus extending from the top of the stalk to the stomium, in *Adiantum* sp

spores and the broken sides of the sporangium. Eventually, a point is reached when gas bubbles form in the cells. This instantly releases the tension and the annulus springs back to its original position. As it does so, the spores are violently ejected.

Spores germinate to form prothalli, usually of the heart shape that is characteristic of the Filicales (Fig. 19.15).

Most members of the Filicales are mesophytes which grow in moist and shaded places. *Ceratopteris cornuta* is a true hydrophyte and has dimorphic leaves. Its vegetative leaves are usually submerged in shallow water and its fertile leaves rise out of the water. *Pellaea* spp. are xerophytes (Fig. 19.11). They often grow from cracks in exposed rocks and their leaves are capable of drying out and later recovering their original form when it rains.

(a)

(b)

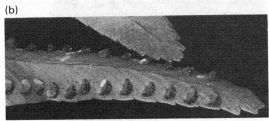

19.15 Stages in the life cycle of *Nephrolepis undulata*, as an example of the Filicales: (a) habit of plants growing with grasses at forest margin; (b) close-up of pinnae showing curved indusia with sporangia visible round the edges ($\times$ 1½) (see also front cover) (continued overleaf)

(c) undersurface of
     prothallus

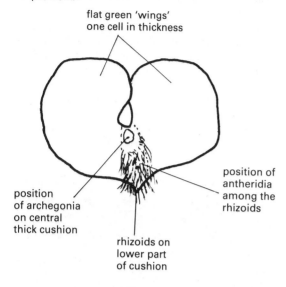

flat green 'wings'
one cell in thickness

position of
antheridia
among the
rhizoids

position
of archegonia
on central
thick cushion

rhizoids on
lower part
of cushion

(e)

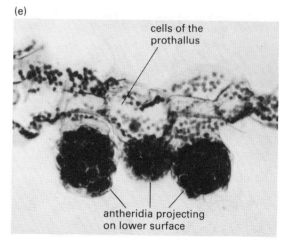

cells of the
prothallus

antheridia projecting
on lower surface

(f)   first leaf

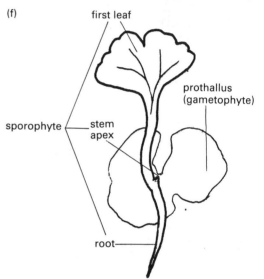

prothallus
(gametophyte)

sporophyte   stem
              apex

root

(d)

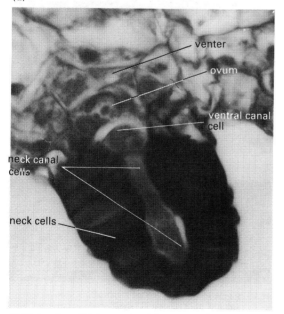

venter

ovum

ventral canal
cell

neck canal
cells

neck cells

19.15 (continued) (c) a gametophyte, a small (1.0 cm),
green prothallus; (d) VS archegonium (HP); (e) VS
antheridia (LP); (f) prothallus with young sporophyte

Marsileales

Of the three genera of Marsileales, only *Marsilea*
spp. occur in tropical Africa, where they grow in
shallow fresh water or on waterlogged ground
(Fig. 19.16).

The stem is creeping, with adventitious roots at
nodes and has a solenostele. Leaves are upright, with
long petioles, with four wedge-shaped leaflets at the
end. Some leaves have an additional, fertile leaflet
close to the base of the petiole. This leaflet is folded
along the midline and the margins of the leaflet are
fused together. Sori are formed within this structure, which is the **sporocarp**. Each sorus contains
both microsporangia and megasporangia.

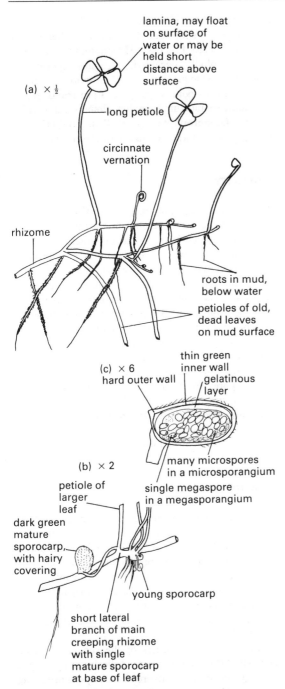

(a) × ½

lamina, may float on surface of water or may be held short distance above surface

long petiole

circinnate vernation

rhizome

roots in mud, below water

petioles of old, dead leaves on mud surface

(c) × 6

thin green inner wall

hard outer wall

gelatinous layer

many microspores in a microsporangium

single megaspore in a megasporangium

(b) × 2

petiole of larger leaf

dark green mature sporocarp, with hairy covering

young sporocarp

short lateral branch of main creeping rhizome with single mature sporocarp at base of leaf

19.16 Order Marsileales, *Marsilea* sp., a water fern: (a) young leaves at the end of a long rhizome; (b) older part of a plant with a sporocarp at the base of the petiole; (c) VS sporocarp, showing megasporangia and microsporangia

(a)

(i) × ½

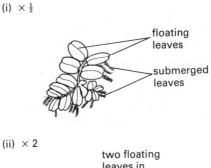

floating leaves

submerged leaves

(ii) × 2

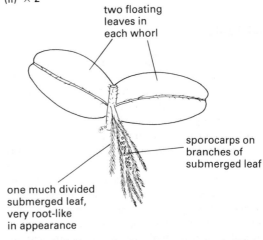

two floating leaves in each whorl

sporocarps on branches of submerged leaf

one much divided submerged leaf, very root-like in appearance

(b)

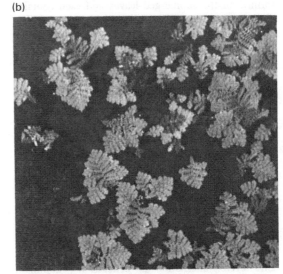

19.17 Order Salviniales, small floating water ferns: (a) *Salvinia* sp., (i) dorsal view of a whole plant, (ii) view of a whorl of leaves from underneath; (b) *Azolla africana*, small plants floating on a lake surface

Sporocarps are hard and tough. Spores are not set free in the usual way. No development takes place until the sporocarp decays, or is broken, and is submerged in water. The sori are then extruded in a worm-like string of mucilage and spores germinate *in situ* to form small, male and female gametophytes.

Sporocarps which have been dried and stored have been shown to contain viable spores for over forty years. When cracked and soaked in water, sori are extruded, spores germinate and a new sporophyte generation is established, all within a few days.

Salviniales

There are only two living genera of Salviniales. These are *Salvinia* and *Azolla* (Fig. 19.17), both of which are quite common in tropical Africa. Both are genera of free-floating water plants.

Salvinia spp. have a branched, horizontal stem, with leaves in whorls of three. There are two oval, entire leaves up to about 1 cm long in each whorl and these float upon the water. The third leaf in each whorl is submerged and divided into many narrow parts, resembling roots. There are no true roots. The floating leaves are folded slightly upwards at each side of the midrib. Their upper surfaces are densely covered with hairs and repel water. Sporocarps are formed on the submerged leaves and each contains either microsporangia or megasporangia.

Azolla spp. are much smaller plants, with two rows of closely overlapping leaves. Roots are present and have root hairs. Each leaf has two parts, the photosynthetic part which is visible from above and a colourless lower lobe which has cavities in which filaments of the blue-green alga *Anabaena* sp. are usually present. The lower lobes also produce sporocarps, each of which contains either microsporangia or megasporangia.

Key words

Spore-dispersed vascular plants. Vascular tissue: xylem, tracheids; phloem, sieve cells, sieve areas. **Stele**: protostele, solenostele, dictyostele, meristeles. **Life cycle**: alternation of generations, dominant sporophyte. **Classification**: division Tracheophyta; subdivisions Psilopsida, Lycopsida, Sphenopsida, Pteropsida; class Filicinae. **Morphology**: root, stem, leaf; microphyll, megaphyll; strobilus, sporophyll; sporangia, synangia, spores; homosporous; heterosporous, microspore, megaspore. **Fern frond**: sorus, indusium, sporangium, annulus, stomium; dimorphic leaves. **Gametophyte**: prothallus, antheridium, archegonium; motile sperms, fertilization, embryo.

Summary

Pteridophytes are spore-dispersed plants with true vascular tissues (xylem, phloem) in the sporophyte generation, which is the dominant generation in the life cycle. Unicellular spores germinate to produce the gametophyte generation, which bears archegonia and/or antheridia. Fertilization requires the presence of water. The zygote forms an embryo, which is dependent on the gametophyte for a short time but soon becomes independent. Some pteridophytes have microphylls, others have megaphylls. Sporangia are borne on sporophylls. Spores are released when sporangia dry out and open. In homosporous plants, all the spores are of one kind and the gametophytes produced are monoecious. In heterosporous plants, the spores are of two types, microspores and megaspores. Microspores germinate to form male gametophytes. Megaspores form female gametophytes.

Questions

1 In what ways are pteridophytes similar to flowering plants? How does their reproduction differ from that of flowering plants? Give reasons to support the view that pteridophytes are less well adapted to life on land than flowering plants.
2 What are (a) microphylls, and (b) megaphylls?
3 What is heterospory? Illustrate your answer by referring to two named genera of the subdivision Lycopsida.
4 How do the two orders of the subclass Eusporangiatae differ from the order Filicales of the Leptosporangiatae in the structure of their spore-producing organs?

20

Gymnosperms – flowerless seed plants

Gymnosperms are seed-dispersed plants in which ovules and seeds are carried naked on sporophylls or similar structures and are not enclosed in an ovary. In our classification, the most important living groups belong to the order Cycadales (cycads) and Coniferales (conifers) of the class Gymnospermae in the subdivision Pteropsida.

The seed habit

The seed habit is based on **heterospory**. It involves the retention of the megaspore and female gametophyte within the ovule of the parent sporophyte until the embryo is well established. The **seed**, developed from the fertilized ovule, is the unit of dispersal (Fig. 20.1). The seed habit also involves reduction of the microgametophyte and the development of the pollen tube, which plays a part in the delivery of male gametes to the archegonia of the female gametophyte.

The ovule
A mature gymnosperm **ovule** (Fig. 20.2) has a large female gametophyte, with archegonia, enclosed within tissue of the megasporangium (**nucellus**). Both are enclosed within an integument with a micropyle at the distal end. Just within the micropyle and outside the nucellus is a chamber (**pollen chamber**) in which microspores (pollen grains) are received.

The pollen grain
The pollen grain is a microspore in which one or more nuclear divisions have taken place before dispersal. In gymnosperms, a **pollen tube** develops after the pollen grain has entered the pollen chamber of an ovule. Further nuclear and cell divisions take place as the pollen tube grows. Finally, two **sperms** are formed.

At its maximum development, the male gametophyte consists of pollen grain and pollen tube. There are two or more cells which appear to have no function. These are thought to represent the vegetative

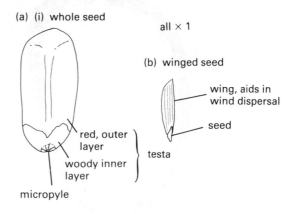

(a) (i) whole seed

all × 1

(b) winged seed

wing, aids in wind dispersal

seed

testa

red, outer layer

woody inner layer

micropyle

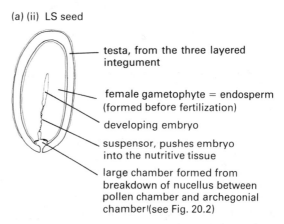

(a) (ii) LS seed

testa, from the three layered integument

female gametophyte = endosperm (formed before fertilization)

developing embryo

suspensor, pushes embryo into the nutritive tissue

large chamber formed from breakdown of nucellus between pollen chamber and archegonial chamber (see Fig. 20.2)

20.1 Gymnosperm seeds: (a) *Encephalartos* sp., a cycad; (b) *Pinus* sp., a conifer

tissue of the gametophyte. One nucleus (**tube nucleus**) appears to be concerned with the growth of the pollen tube, and another cell divides to form two cells, one of which divides again to form two sperms.

The seed
After fertilization, the zygote gives rise to an **embryo**, which is deeply embedded in the tissue of the female gametophyte. The haploid tissue of the

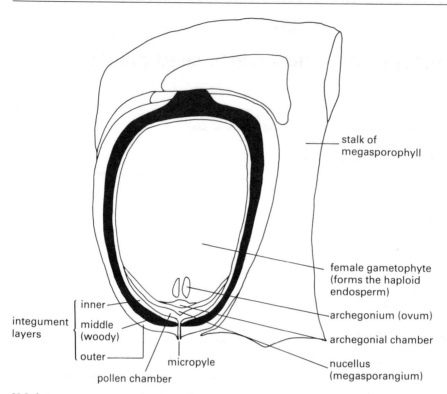

- stalk of megasporophyll

- female gametophyte (forms the haploid endosperm)

- archegonium (ovum)

- archegonial chamber

- nucellus (megasporangium)

integument layers
- inner
- middle (woody)
- outer

micropyle

pollen chamber

20.2 A gymnosperm ovule: LS ovule of *Encephalartos* sp., *in situ* on the megasporophyll

female gametophyte itself is the **endosperm** of the seed. The nucellus remains as a dry layer around part of the endosperm. The integument usually differentiates into two or more layers of tissue and remains as the **testa** of the seed. The seed is dispersed and a new sporophyte becomes established as the seed germinates.

Life cycle

The gymnosperm life cycle (Fig. 20.3) is haplo-diplontic with a dominant sporophyte generation.

Classification

In addition to five living orders (Cycadales, Conifer-ales, Taxales, Gingkoales and Gnetales), there are several orders of gymnosperms which are known only as fossils. The fossil orders include seed ferns (Cycadofilicales). **Seed ferns** were very like living ferns, except that they possessed ovules of a pattern very similar to that of present day cycads.

Here, we shall confine our attention to the cycads and conifers.

Cycadales

Cycads are tropical and subtropical plants and are mostly confined to the Southern Hemisphere. Different genera occur in different continents. There are two genera in southern Africa of which one, *Encephalartos*, has species which extend north as far as West and East Africa (Fig. 20.4).

In addition to indigenous species, the East Asian *Cycas circinalis* and *C. revoluta* are widely planted in tropical countries, for ornament (Fig. 20.5).

Most cycads are palm-like plants. The main axis (**caudex**) is a stem covered with scale leaves and leaf bases. The caudex is usually unbranched but may have adventitious branches. It is attached at the base by stout roots and has a crown of pinnate leaves at

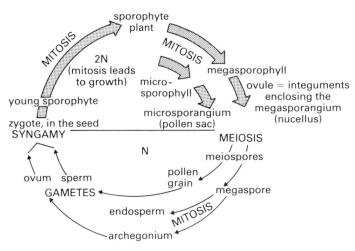

20.3 Life cycle diagram of a gymnosperm

the top. The caudex of the principal West African cycad, *Encephalartos barteri*, grows no higher than about 30 cm and is often shorter. It appears to be held at this level by the action of **contractile roots**. Other *Encephalartos* spp. reach a height of 2 to 3 m. The tallest cycad recorded is a specimen of *Macrozamia hopei* in Australia, which reached a height of over 18 m.

The stem has a wide medulla and a wide cortex,

20.4 Cycadales: *E. barteri* in savanna near Ilorin, Nigeria. Note that the caudex (stem with old leaf bases) of *E. barteri* rarely grows much above ground level

20.5 Cycadales: *C. revoluta* (female) with a tall trunk bearing alternating series of old leaf bases and sporophyll bases and scale leaves (arrowed)

with rather little vascular tissue (Fig. 20.6). The primary vascular tissues form a ring of vascular bundles with narrow rays. Secondary thickening starts at an early stage and is similar to that in dicots. However, *Cycas* spp. and *Encephalartos* spp. have a type of anomalous secondary thickening in which the first cambium ceases to act after a time and a new cambium is formed outside the phloem. The process may be repeated several times.

Xylem is of tracheids only. Phloem is of sieve cells, similar to those of pteridophytes.

The cycad leaf is similar in structure to the dorsiventral leaf of a flowering plant. The leaf is strongly xeromorphic. There is a thick cuticle and the abaxial epidermis contains sunken stomata. There is also a hypodermal layer of lignified cells next to each epidermis.

Cycads have a strong tap root system and also form adventitious roots. Most lateral roots grow normally, but some grow to the surface of the ground where they branch repeatedly to form groups of **coralloid roots** (Fig. 24.6). Coralloid roots have a zone containing the nitrogen-fixing blue-green alga *Anabaena* sp. in the cortex.

All organs of cycads secrete a sticky mucilage, which is exuded from ducts in the soft tissues when they are damaged.

Reproduction

Cycads are all dioecious. Some plants are **microsporangiate** and others are **megasporangiate**. It is convenient, if not strictly correct, to call them male and female plants respectively.

Male plants produce cones in which the axis is covered with a spirally-attached series of fleshy, scale-like microsporophylls. The abaxial surfaces of the microsporophylls bear microsporangia (Fig. 20.7). In *Encephalartos* spp. there are up to 200 microsporangia on a microsporophyll. The microsporangia are up to 2 mm in diameter and each produces many pollen grains. These are released when the axis of the cone lengthens, thus separating the scales slightly. Each microsporangium opens by a slit. The pollen is probably dispersed by wind.

After meiosis has taken place in the microspore mother cells, each pollen grain has a single, haploid nucleus. Before the pollen is shed, the nucleus divides and one of the daughter nuclei becomes the nucleus of a **prothallial cell**, at one side of the pollen grain. The other nucleus divides again and forms a

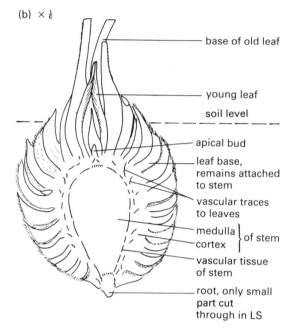

(b) × ⅙

base of old leaf

young leaf

soil level

apical bud

leaf base, remains attached to stem

vascular traces to leaves

medulla ⎱ of stem
cortex ⎰

vascular tissue of stem

root, only small part cut through in LS

20.6 The caudex of *E. barteri*, diagram of LS showing fleshy leaf bases round the stem

generative cell, which lies close to the prothallial cell, and a **tube nucleus**, which occupies the centre of the pollen grain. It is at this stage that the pollen is dispersed.

In *Encephalartos* and all other genera of cycads except *Cycas*, female plants produce cones similar to those of the male plant, but larger (Fig. 20.8). The megasporophylls are stalked and each has a wide, rhomboid head. There are two large ovules attached to the inner surface of the head, parallel to the stalk and with their micropyles facing the axis of the cone. The ovules undergo full development before pollination and each has a large female gametophyte contained within a three-layered integument (Fig. 20.2). The outer layer of the integument is fleshy and brightly coloured, red in *E. barteri*. Within the fleshy layer is a hard, stony, middle layer and an inner layer which is at first fleshy but later shrinks to form a dry, brown, papery layer.

The micropyle opens at the tip of a small projection. At the inner end of the micropyle is a small chamber, the **pollen chamber**, above a cap-like remnant of the nucellus, which separates the pollen chamber from another space, the **archegonial chamber**. The archegonial chamber is formed by a

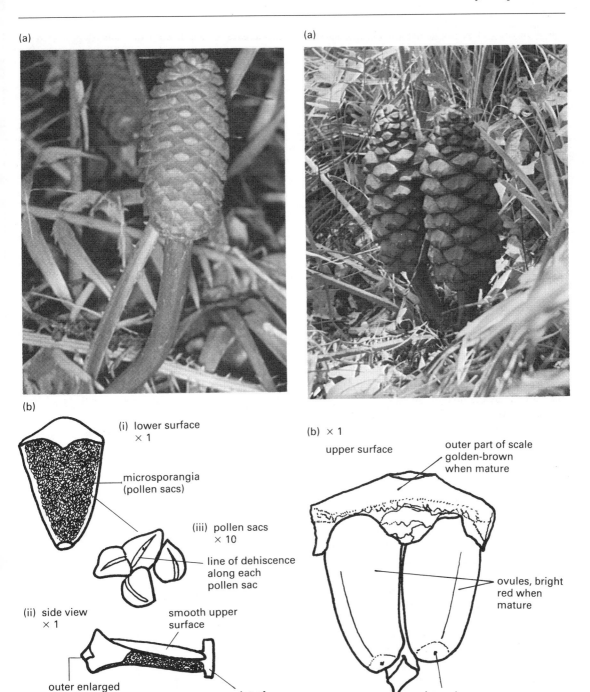

(a)

(a)

(b)

(i) lower surface
× 1

microsporangia
(pollen sacs)

(iii) pollen sacs
× 10

line of dehiscence
along each
pollen sac

(ii) side view
× 1

smooth upper
surface

outer enlarged
end of scale

point of
attachment
to cone

(b) × 1

upper surface

outer part of scale
golden-brown
when mature

ovules, bright
red when
mature

micropyle

20.7 *Encephalartos barteri*: (a) young male cone, not yet expanded to release pollen; (b) single microsporophyll from a mature cone, with open microsporangia

20.8 *Encephalartos barteri*: (a) female plant with cones at the time of pollination, with the megasporophylls just starting to separate; (b) single megasporophyll with two ovules

hollow in the end of the gametophyte. The surface of the gametophyte in the archegonial chamber bears several archegonia. The venters of the archegonia are sunken and each contains a large ovum, about 1.5 mm long and about 1 mm wide. The neck of each archegonium is formed by two inflated cells, which protrude above the surface of the gametophyte.

When the ovule is mature, a drop of secreted liquid, the **pollination drop**, is exuded through the micropyle. Pollen grains become trapped in the pollination drop, which is subsequently reabsorbed into the pollen chamber, taking pollen grains with it.

The three-celled pollen grain now grows a pollen tube, which penetrates into the nucellus. The pollen tube, unlike that of a flowering plant, is mainly a haustorial organ. In cycads for which full data are available, the generative cell of the pollen grain divides to form a stalk cell, next to the prothallial cell, and a body cell which then divides to form two sperms.

The sperms of cycads are so large that they are visible to the naked eye. Each is about 0.25 mm in diameter, rounded at the base and with many flagella attached in a continuous spiral around the conical anterior end. By the time the sperms are formed, the nucellus between the pollen chamber and the archegonial chamber has partly broken down and the two chambers are continuous. The sperms lie at the base of the pollen tube and are shot out violently, when the tube bursts, evidently by increasing turgidity in the pollen grain and pollen tube. One archegonium is fertilized and an embryo starts to develop.

The mature embryo is attached at the micropylar end of the endosperm by a long, convoluted **suspensor**. The embryo itself is nearly as long as the endosperm, with a stout radicle at the micropylar end and two cotyledons, pressed together into a cylindrical shape, with a small plumule between. There is a tough **coleorhiza** around the end of the radicle.

It is at this stage that the seeds are released, often by decay of the cone. They are probably dispersed by animals. Animals are attracted by the bright colour of the seeds and in some cycads the outer layer of the testa has a fruity smell and taste. The red outer layer of the seed of *E. barteri*, however, tastes extremely bitter.

The seeds of some cycads have no resting period. If conditions are not adverse, the embryo continues to increase in length until the radicle, protected by the coleorhiza, emerges through the testa in the region of the micropyle (Fig. 20.9). There are radiating lines of weakness in the woody layer of the testa around the micropyle, so that the testa opens into triangular valves as the radicle emerges. The radicle then penetrates the coleorhiza and grows downwards into the soil. The cotyledons then lengthen further and the plumule grows upwards from between the bases of the cotyledons, as soon as it is outside the seed. Germination of the seed is

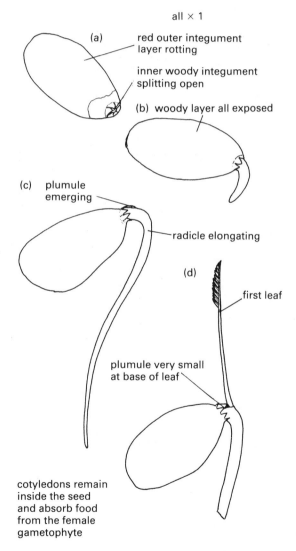

all × 1

(a) red outer integument layer rotting

inner woody integument splitting open

(b) woody layer all exposed

(c) plumule emerging

radicle elongating

(d)

first leaf

plumule very small at base of leaf

cotyledons remain inside the seed and absorb food from the female gametophyte

(e)

20.9 *Encephalartos barteri*: (a) to (d) stages in germination of seeds; (e) group of germinating seeds close to the parent plant. One seedling (arrowed) has reached the first leaf stage

therefore typical hypogeal germination and very similar to that of palm seeds.

In *Encephalartos* spp. and in some other genera, the ovule grows to its full size before pollination and there is no way of telling if a 'seed' contains an embryo without cutting it open. The situation is different in *Cycas* spp.

Male plants of *Cycas* spp. have male cones like those described above (Fig. 1.19). The female plant does not have a cone, but sporophylls are carried on the main stem. At the start of each growing season, the plant first unfolds a series of scale leaves which have been protecting the apex, then produces a crown of megasporophylls and, finally, produces a new crown of leaves. Each megasporophyll has a long stalk and a flat, pointed end which is pinnately divided or toothed. Ovules are carried in two rows, one row each side of the stalk. There are eight ovules in *C. circinalis*, usually four in *C. revoluta* (Fig. 20.10). This condition, in which the sporophylls are much more leaf-like than in other cycads, is thought to be primitive. It is thought that the T-

shaped megasporophylls of other genera have arisen by reduction from the more leaf-like form of those in *Cycas* spp.

(a)

(b) × ½

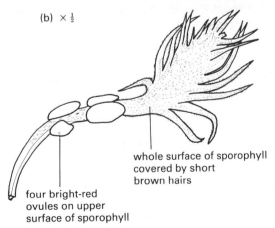

four bright-red ovules on upper surface of sporophyll

whole surface of sporophyll covered by short brown hairs

20.10 Female plant of *C. revoluta*: (a) close-up of the crown of megasporophylls, not yet fully expanded; (b) a single detached megasporophyll showing the leafy, pinnate apex

Ovules of *C. circinalis* produce their pollination drops when they are about 1 cm long and the female gametophyte is at a very early stage of development. If pollination does not take place, the ovules stop developing and most of them shrivel up completely. If pollinated, the ovules continue to enlarge until they are about 5 to 6 cm long. Development of the female gametophyte is completed after about a year and fertilization takes place at this time. It takes a further year for the embryo to develop, after which the seed is shed. There is thus a period of two years between pollination and the formation of mature seeds.

Coniferales

Conifers are mostly trees of temperate countries. Conifer forests dominate the vegetation in far northern and far southern latitudes, with different genera in the north and south. Some conifers occur at high altitudes in the tropics. For instance, *Podocarpus* spp. (Fig. 20.11) occur naturally in tropical Africa at altitudes above 1200 m.

Certain conifers are a very important source of softwoods. The trees grow to a useful size within thirty years of planting and the wood is useful for building and for making furniture, boxes and matches. For this reason, non-indigenous conifers are widely planted at high altitude in the tropics. Such trees include *Pinus patula* (Fig. 20.11), which grows naturally on mountains in Mexico, and *Araucaria excelsa*, from the South Pacific area.

All conifers are woody plants. Some are shrubs but most are trees. The pattern of their vascular tissues is very similar to that of woody dicots, but the details differ. Xylem is made up of tracheids only. Tracheids of the secondary xylem of conifers are notable for their **bordered pits** (Fig. 20.12). Bordered pits occur on the radial walls of tracheids only and are large, with a wide border. The membrane across the pit also has a thickened area (**torus**) at the centre. When the membrane across the pit is pushed to one side by pressure differences, perhaps following injury, the torus seals the pit opening.

Phloem consists of sieve cells only.

The xylem and other tissues of many conifers contain many conspicuous **resin canals**, which exude resin when they are damaged.

Most conifers are evergreen and have rather small, xeromorphic leaves. Leaves of *Podocarpus* spp. are

(a)

(b)

20.11 Conifers in Africa: (a) *Podocarpus milanjianus*, an indigenous species, growing in montane forest at an altitude of 1800 m; (b) *Pinus patula*, an introduced species, in a plantation in Central Africa, at an altitude of about 1000 m

lanceolate and up to 15 cm long (Fig. 20.11). In *Araucaria* spp., leaves are short and triangular and strongly overlapping. In both of these genera, leaves are attached directly to growing stems. The situation in *Pinus* spp. is different. The young seedling has elongated, dorsiventral leaves attached directly to growing stems (**long shoots**). Older stems bear only brown scale leaves and the foliage leaves are produced in groups, at the tips of very short shoots which grow in the axils of the scale leaves. The leaf-

(a)

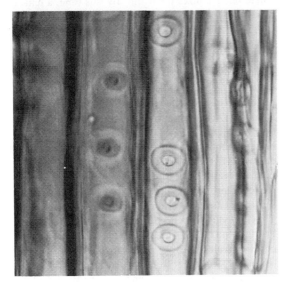

20.12 Tracheids of *Pinus* sp., showing bordered pits (HP)

bearing shoots (**dwarf shoots**) are only a few milli-metres long, have scale leaves on the lower part and bear one or more long, pointed, needle leaves close to the tip. They stop growing when they produce the needle leaves. The number of needle leaves on a dwarf shoot, which varies from one to five, is charac-teristic of the species.

Pinus patula has three needle leaves on each dwarf shoot and may be called a 'three-needle pine.' Each long shoot produces a succession of dwarf shoots with needle leaves and an occasional whorl of long shoots. Like most conifers, *P. patula* is monoecious and bears male and female cones (Figs. 20.13 and 20.14).

Male cones replace dwarf shoots at the very beginning of each season's growth. Each cone consists of a short axis up to about 1 cm long, with many spirally arranged microsporophylls (Fig. 20.13).

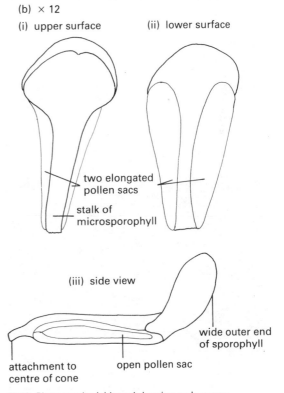

(b) × 12

(i) upper surface   (ii) lower surface

two elongated pollen sacs

stalk of microsporophyll

(iii) side view

wide outer end of sporophyll

attachment to centre of cone   open pollen sac

20.13 *Pinus patula*: (a) branch bearing male cones; (b) single microsporophyll

Each sporophyll has two pollen sacs on the abaxial surface. Each mature pollen grain contains three **prothallial cells**, a **generative cell** and a **tube nucleus**. Pollen grains of *Pinus* spp. and many other conifers are winged. The 'wings' are inflated areas, where there are air spaces between intine and exine, and are thought to give greater buoyancy when the pollen is dispersed by wind.

Female cones replace long shoots. Each has a main axis, to which many scales are attached in a continuous spiral. Each scale bears a pair of ovules on its adaxial surface, with the micropyle directed towards the axis of the cone. At the time of pollination, the female cone is upright, green and about 1.5 cm long (Fig. 20.14). Pollen is drawn into the pollen chamber of an ovule by a pollination drop, as in cycads. At the time of pollination, the female gametophyte is at a rudimentary stage of development. Full development and fertilization are not achieved until about a year following pollination. The structure of the mature female gametophyte is similar to that in cycads, with several archegonia at the micropylar end.

20.14 *Pinus patula*, branch bearing female cones. The small, upright cone (a) is ready for pollination. The larger cones (b) are a year older, at the stage when fertilization takes place

During the year between pollination and fertilization, the pollen grain grows a pollen tube, which penetrates into the nucellus. Meanwhile, the generative cell divides into a **stalk cell** and a **body cell**. When the pollen tube comes close to an archegonium, the body cell divides to form two **sperms**. Each sperm has a large nucleus enclosed in cytoplasm and has no cell wall. Conifer sperms are not motile, but are delivered by the pollen tube directly between the cells of the neck of an archegonium, where one fertilizes the ovum. By this time, a female cone of *P. patula* is about 6 cm long and is in a hanging position.

Development of the embryo takes a further twelve months. The mature embryo is similar to that of a cycad, except that it has a large number of cotyledons (e.g. nine). By the time the seeds have matured, the female cone, still in a hanging position, is about 10 cm long, with the scales closely pressed together. Opening of the scales of the cone is brought about gradually by desiccation, but is almost explosive if the cone is exposed to a high temperature. A forest fire, for instance, causes almost explosive opening of mature seed-bearing cones. Each seed consists of an **embryo, endosperm** (the female gametophyte), a very thin trace of **nucellus** and a brittle **testa**. Each seed separates from the surface of the scale of the cone attached to a **wing** (Fig. 20.1), formed from the surface of the scale. The seed spins as it falls and may be dispersed for some distance by wind. Germination is epigeal.

The life cycle of a conifer is like that of a cycad, except that the sperms are non-motile and that the pollen tube delivers the sperms all the way into the neck of the archegonium. Also, the scales of the female cone are not sporophylls. They are usually interpreted as stem-like structures with the vegetative parts of the megasporophylls reduced to the extent that they are not detectable. In some genera of conifers, the scales of the female cone are formed in the axils of scale leaves (**bracts**). In *Pinus* spp., the bracts are much reduced and attached to the scales.

The male and female cones of other genera of conifers vary in the number of pollen sacs (microsporangia) and number of ovules carried by each scale of the cone and in the shape and position of the cone. *Cupressus* spp., also widely planted in tropical countries, have very small male cones with about eight pollen sacs on each scale. The pollen is wingless.

Female cones have few scales and are nearly spherical. Each scale bears about twelve ovules.

Key words

Flowerless seed plants: ovule, pollen grain, seed. **Ovule**: megasporangium (nucellus), female gametophyte, archegonia; pollen chamber, integument, micropyle. **Pollen grain**: male gametophyte; tube nucleus, generative cell, sperms, pollen tube. **Seed**: embryo, endosperm (female gametophyte), testa. **Life cycle**: haplodiplontic, dominant sporophyte. **Classification**: division Tracheophyta; subdivision Pteropsida; class Gymnospermae; orders Cycadales (cycads), Coniferales (conifers). **Cycads**: caudex, xeromorphic leaves; mucilage. **Conifers**: evergreen, softwood; resin; long shoots, dwarf shoots; needle leaves.

Summary

Gymnosperms are seed-dispersed plants. Their seeds are not enclosed but are attached to scales, in cones. The gymnosperm seed differs from an angiosperm seed in having haploid endosperm, which is formed before fertilization. Cycads are palm-like and grow in the tropics, mostly in the Southern Hemisphere. Conifers are evergreen trees of temperate latitudes and are often planted at high altitudes in the tropics, for their softwood timber.

Questions

1 Describe the structure of a mature gymnosperm ovule. How does it differ from a typical angiosperm ovule?
2 Outline the vegetative features of cycads and conifers. Why are these two groups of plants included in the same class (Gymnospermae)?
3 How does a seed of *Encephalartos* sp. differ from a seed of *Ricinus* sp., in structure and germination?
4 Compare the structure of the male gametophyte of a cycad with that of a conifer and that of a flowering plant. How do the processes of fertilization differ in these three types of plant?

Further reading for section C

Bold, H. C. 1977. *The Plant Kingdom*, 4th edn, Prentice-Hall: New Jersey.

Bold, H. C. and Wynne, M. J. 1980. *Introduction to the Algae*, Prentice Hall: New Jersey.

Doyle, W. T. 1970. *The Biology of the Higher Cryptogams*, Macmillan: London.

Lawson, G. W. and John, D. M. 1982. *The Marine Algae and Coastal Environment of Tropical West Africa*, Cramer: Vaduz.

Morris, I. 1967. *An Introduction to the Algae*, 2nd edn, Hutchinson: London.

Prescott, G. W. 1978. *How to Know the Freshwater Algae*, 3rd edn, Wm C. Brown: Dubuque, Iowa.

Round, F. E. 1973. *The Biology of the Algae*, 2nd edn, Edward Arnold: London.

Smith, G. M. 1955. *Cryptogamic Botany, vol. 1, Algae and Fungi*, 2nd edn, McGraw-Hill: New York.

Smith, G. M. 1955. *Cryptogamic Botany, vol. 2, Bryophytes and Pteridophytes*, 2nd edn, McGraw-Hill: New York.

Sporne, K. R. 1974. *The Morphology of Gymnosperms*, 2nd edn, Hutchinson: London.

Sporne, K. R. 1975. *The Morphology of Pteridophytes*, 4th edn, Hutchinson: London.

Watson, E. V. 1971. *The Structure and Life of Bryophytes*, 3rd edn, Hutchinson: London.

Section D

PLANTS WITHOUT CHLOROPHYLL

In this Section, we are to study those groups of plants which are all, or almost all, heterotrophic. Apart from a few bacteria which are autotrophic, becteria and fungi depend on other organisms for their supplies of energy and organic material. We shall also study symbiotic relationships between autotrophic and heterotrophic organisms, in which each partner benefits from its association with the other.

21

Nutrition in bacteria and fungi

Autotrophic organisms obtain all their nutritional requirements from inorganic sources. The photosynthetic (**photoautotrophic**) organisms we have studied obtain their carbon from carbon dioxide, use water as the hydrogen donor and obtain their energy from light. Otherwise, they require only mineral ions from their environment.

Heterotrophic organisms require more complex carbon compounds as their carbon source and obtain their energy by the oxidation of organic compounds obtained from other organisms.

There are both autotrophic and heterotrophic bacteria. All fungi are heterotrophic.

Autotrophic bacteria

Autotrophic bacteria include those which are **photoautotrophic** and those which are **chemoautotrophic**.

Photoautotrophic bacteria possess **bacteriochlorophyll**, which replaces the chlorophyll a of green plants. They also have carotenoid pigments which differ from those of green plants. There are different bacteriochlorophylls in different groups of bacteria. The molecules of those which have been fully analysed are very similar to the molecule of chlorophyll a, but most absorb light of longer wavelengths, even into the infrared band. Their combinations of pigments make them appear green, red or brown.

All photosynthetic bacteria use carbon dioxide as the carbon source but their hydrogen donor may be hydrogen sulphide, organic compounds or molecular hydrogen, never water. Those which use hydrogen sulphide produce free sulphur, in the same way as green plants produce free oxygen.

Chemoautotrophic bacteria use carbon dioxide as their carbon source but obtain their energy by the oxidation of inorganic substances (e.g. ammonium ion, nitrite ion, hydrogen sulphide, sulphur).

Heterotrophic bacteria and fungi

The naked, amoeboid cells of some slime fungi are **phagotrophic** and ingest and subsequently digest bacteria in the same way as protozoa (e.g. *Amoeba* spp.) feed on smaller organisms. Other fungi and all heterotrophic bacteria are saprophytic, parasitic or symbiotic.

Saprophytes
Saprophytes are organisms which obtain their raw materials and their energy-containing compounds from dead organic matter. They secrete hydrolytic enzymes into their substrate and absorb some of the products of digestion (**external digestion**) into their cells.

Parasites
Parasitic bacteria and fungi obtain their raw materials and energy from compounds which they absorb from other living organisms. **Facultative parasites** can live as parasites or as saprophytes and can continue to obtain nourishment from host tissues even after the host is dead. **Obligate parasites** are more specialised and can exist only on a living host.

Symbionts
Some bacteria (e.g. root nodule bacteria, *Rhizobium* spp.) and some fungi (e.g. mycorrhizal fungi) have an apparently mutually beneficial relationship with green plants. Such a relationship, usually called **symbiosis**, may have evolved from a previously parasitic relationship.

Symbiotic relationships, including those involving bacteria and fungi, are dealt with in Chapter 24.

A note on viruses

Viruses are all parasites, but their mode of existence does not include any form of nutrition in the ordinary sense of the word. They do not absorb energy from outside sources. They do not ingest, digest or absorb food substances and they do not grow in size. Instead, their presence within a host cell causes the host cell to produce new copies of the virus, using its own resources of energy and chemicals and its own cell machinery. Viruses are dealt with in greater detail in the next chapter.

Key words

Nutrition: autotrophic, inorganic sources; heterotrophic, organic sources. **Bacteria**: autotrophic, heterotrophic. **Fungi**: heterotrophic. **Autotrophic nutrition**: photoautotrophic (photosynthetic), bacteriochlorophyll; chemoautotrophic. **Heterotrophic nutrition**: saprophytes; parasites, facultative, obligate; symbionts, symbiosis.

Summary

With few exceptions mentioned in Chapter 16, the plants we have studied so far in this book are photoautotrophic and possess chlorophyll *a*. Some bacteria are photoautotrophic and possess bacteriochlorophyll. Others are chemoautotrophic and use chemical energy to synthesize their own carbohydrates. Most bacteria and all fungi are heterotrophic and have a saprophytic or parasitic mode of life.

Questions

1 How does the photoautotrophic nutrition of photosynthetic bacteria differ from that of green plants?
2 How do saprophytic and parasitic fungi differ in their mode of nutrition? Why can we classify both as heterotrophs?

22

Viruses and bacteria

Bacteria, like blue-green algae, are procaryotic and are placed in the kingdom Monera. Viruses, which consist only of nucleic acid enclosed in a protein coat, are sometimes placed in a division, Viri, of the kingdom Monera, but are better regarded as belonging to a separate kingdom, Viri.

Viruses

Viruses are all intracellular parasites and can multiply only inside living host cells. There are plant viruses which attack plants, animal viruses which attack animals, and bacteriophages which attack bacteria. There are also bacteriophage-like viruses which attack blue-green algae.

A virus is made up of a coiled thread of nucleic acid enclosed in a protein coat (**capsid**). The virus particle is not regarded as a cell, but is called a **virion**. In animal viruses and bacteriophages, the nucleic acid is DNA. In plant viruses the nucleic acid is RNA. Viruses vary in shape and size, from isodiametric particles 25 nm in diameter through threadlike forms 10 nm by 500 nm, to more or less cuboid forms with a diameter of over 250 nm. The largest (e.g. vaccinia virus, the virus of smallpox vaccine) are, theoretically, just large enough to be seen with the light microscope.

Viruses were first discovered when fluids extracted from diseased plants and animals were found to have the power to transmit the disease to other individuals, even after being passed through a filter fine enough to remove all bacteria. They were first known as **filterable viruses**, which means poisons which will pass through a bacterial filter. All are effectively invisible with the light microscope and, at first, could be studied only by observing the symptoms of disease which they produced (Fig. 22.1) and by transmitting them from one host to another. It was only after the invention of the electron microscope that they could be studied directly.

Since they possess only nucleic acid and a protective capsid and have no other organelles, viruses are totally dependent on the organelles and chemical

(a)

(b)

22.1 Symptoms of virus diseases: (a) leaf of *Zea mays* with Maize streak virus; (b) healthy leaves of *Manihot* sp. (bottom) and leaves infected with Cassava mosaic virus

substances present in the host cell for their replication. When a virus enters a cell, the protein coat is lost. In the case of a plant virus, the virus RNA acts as messenger RNA and, attached to a host ribosome, acts as a template for the production of capsid protein and enzymes which are necessary for RNA replication. More virus RNA is then formed, using nucleotides, pentose sugar and phosphate present in the host cell. In the case of DNA viruses, the virus DNA first acts as a template for the formation of virus messenger RNA, after which the sequence of events is the same as for RNA viruses. In both cases, virions are later reformed from nucleic acid and capsid protein.

It is sometimes argued that viruses are not living things. Such arguments depend, of course, on how one defines a 'living thing', but viruses differ from procaryotic and eucaryotic organisms in a number of important ways. In the first place, viruses are incapable of reproducing themselves and depend for reproduction on the cells of their hosts, which they induce to make copies of themselves as described above. Some viruses can be extracted from cells, concentrated, and crystalized like chemical substances. Some, again, can be split into separate protein and nucleic acid components and will reform normal virions when the components are put together again, *in vitro*. It is also possible to infect host cells with the nucleic acid, alone, of some viruses, after which replication and virion formation continue in the normal way.

Bacteriophages (**phages**) are more complex than other viruses both in structure and behaviour (Fig. 22.2). The virion consists of a head, which contains a single thread of DNA, and a tail, with attaching fibres at the end. When a phage virion attacks a bacterium, it attaches itself to the outside of the bacterial cell wall by its attaching fibres and injects its DNA (only) into the bacterium through its tail. Usually, the bacterial chromosome is immediately broken down and the phage DNA takes over the machinery of the cell and causes it to produce many phage capsids and threads of phage DNA. Capsids and DNA are produced separately within the host cell but the DNA threads subsequently enter the capsids and new virions are reconstituted. About 30 to 40 minutes after the bacterium is infected, the cell wall ruptures (lysis occurs) and the new virions are released to attack other bacterial cells.

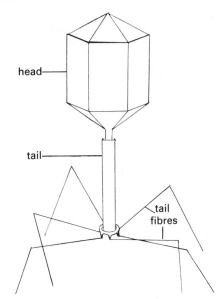

22.2 Bacteriophage: a diagram to show the principal parts. The visible parts are all protein. DNA is contained in the head of the virus

Virus diseases

Viruses are the causative agents of many plant and animal diseases. It is believed that the symptoms of disease result from the introduction of protein foreign to the host cells. These proteins are incompatible with the host cells. Also, viruses deplete their host cells of the nitrogen compounds which are necessary for their own metabolic activities.

There are several ways in which viruses are transferred from one host to another. Animal viruses are transmitted through bodily secretions. This may happen through direct contact, or, for instance, by inhaling droplets of saliva, etc., after an animal sneezes. Some diseases are transmitted in saliva injected by the bite of an animal (e.g. yellow fever, by the bite of an infected mosquito; rabies, by the bite of a rabid dog or other rabid animal).

Plant viruses are usually transmitted by way of aphids and other insects which suck the juices of plants, but can also be transmitted by contact if the organs which make contact have slight surface damage.

A list of some important animal diseases caused by viruses is given in Table 22.1. Plants showing the symptoms of virus disease are illustrated in Fig. 22.1.

Table 22.1 Some virus diseases of animals.

Animal affected	Disease	Vector
man	common cold	—
	dengue fever	mosquito
	influenza	—
	Lassa fever	rat or other rodent
	measles	—
	rabies	dog, hyena
	yellow fever	mosquito
cattle	foot and mouth disease	—
domestic fowl	Newcastle disease	—

Table 22.2 Descriptive names of different shapes, etc., of bacteria. Some of these names (with asterisk below) are also the names of genera, if spelled with a capital letter.

Shape of cell	Other features	Descriptive name
spherical	any	coccus
	cells in pairs	*diplococcus
	cells in chains	*streptococcus
	cells in clumps	*staphylococcus
elongated	straight	*bacillus
	spiral, rigid	*spirillum
	spiral, flexible	spirochaete

Bacteria

Bacteria are distinguished from blue-green algae by the fact that they do not possess chlorophyll *a*, which is the primary photosynthetic pigment of all green plants. As has been shown, a few bacteria are auto-trophic (photoautotrophic, chemoautotrophic), but the vast majority are heterotrophic.

There are about 1600 named species of bacteria. Classification within the group is complicated and is based on characters such as the mode of nutrition, whether aerobic or anaerobic, the presence or absence of a capsule enclosing the cell, the presence or absence of flagella and the staining properties of the cell.

The bacterial cell

The names given to different shapes and arrange-ments of bacterial cells are given in Table 22.2 and Fig. 22.3. The cells range in size from 0.3 μm by 1.0 μm to 1.0 μm by 20 μm.

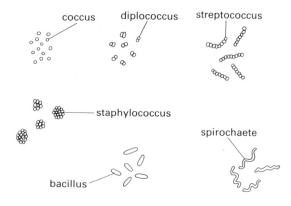

22.3 The shapes of bacteria

Little internal structure can be seen with the light microscope. It is sometimes possible to distinguish bodies of storage substances and one or more nucleoids, which stain positively for DNA. A capsule may be seen, enclosing the cell, and special staining techniques can be used to demonstrate flagella, when present.

Much more can be seen when ultra-thin sections are examined with an electron microscope (Fig. 22.4). Also, modern biochemical techniques tell us much more about the chemical components of the parts of the cell.

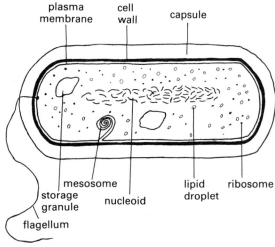

22.4 The structure of a generalized bacterial cell, diagram of LS as seen with EM

The **cell wall** is of several layers, with a basic framework of mucopeptide. Other cell wall components are variable but may include protein or phospholipids. Organisms known as Mycoplasma, usually classified as Bacteria, have no cell wall. Their naked cells are variable in shape and size.

Many bacteria secrete a **capsule** outside the cell wall. This may be very thin, or up to 10 μm thick. It may contain fibrils of cellulose but is otherwise composed of various polymers. The polymers vary from species to species and include polypeptides and polymers of nitrogen-containing sugars.

Within the cell wall, the plasma membrane is similar in structure to that of other organisms. It has sac-like invaginations (**mesosomes**) which are possibly a substitute for the endoplasmic reticulum of eucaryotic cells, which is otherwise absent.

Within the plasma membrane, the protoplast contains many ribosomes, smaller than but otherwise similar to the ribosomes of eucaryotic plants and animals. There are several types of non-living inclusions, including lipid droplets and granules of polysaccharides and other storage substances.

The relatively few photosynthetic bacteria have chromatophores of various types but always lacking the organised lamellar structure of the chromatophores of blue-green algae.

In those bacteria which possess flagella, these may be one or a few attached at one end, or at both ends (**polar** flagella) or may be attached on all sides. Bacterial flagella are much simpler than those of eucaryotic organisms. There is no membrane and each consists of a single protein thread similar to the tubules inside a eucaryotic flagellum. The flagellum is 10 nm to 20 nm thick and up to 70 nm long. The thickness is below the resolving power of a light microscope, but flagellar stains form a deposit on the surface so that they are thick enough to be seen, once they have been stained (Fig. 22.5). Each flagellum is attached within the plasma membrane of the cell by a hook-like basal body. In addition to flagella, many bacteria have **pili**. These are fine, hair-like protein threads, 3 nm to 25 nm thick and 5 nm to 20 μm long. Pili are probably organs of attachment.

Within the cell, the **nucleoid** consists of a long, circular strand of DNA, which is folded into loops. The circular DNA strand is sometimes called the 'bacterial chromosome', but unlike the eucaryote chromosome (Chapter 25), it has little associated protein and has no centromere.

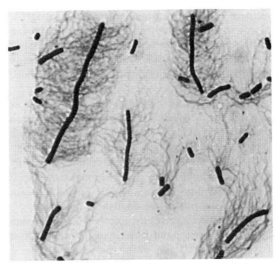

22.5 Bacterial flagella. Typhoid bacilli stained to show flagella (VHP)

Some bacteria form spores. These are formed singly within the bacterial cell. Each spore has a thick, resistant coat and may remain within the cell in which it was formed or may be released. The spores of some bacteria are very long-lived and have been shown to germinate after over fifty years of dormancy. Some are extremely heat resistant and can germinate after exposure to temperatures over 100°C.

Reproduction

Most bacteria reproduce asexually by **fission**. After replication of the DNA of the nucleoid, the nucleoid separates into two parts and one part migrates to each end of the cell. The formation of a septum, which grows inwards around the middle of the cell, divides the cell into two. A few bacteria reproduce by **budding**, in which a new cell is formed as an outgrowth from the parent cell and, after it has received a nucleoid, separates from the parent.

Bacteria do not form gametes and there is no meiosis or true sexual reproduction. There are, however, processes by which different strains achieve a limited degree of recombination of their genetical material. These mechanisms are transformation, conjugation and transduction.

In **transformation**, small lengths of DNA released from a **donor** cell enter another cell (**recipient**) through the cell wall. Part of the donor DNA then becomes incorporated in the chromosome of the recipient cell, replacing some of the original DNA. Most research on transformation has involved the use of DNA from donor cells which have first been killed by heating, but there is evidence that it can also occur naturally in a mixed culture of closely related strains of bacteria.

In **conjugation**, two bacterial cells come into close contact and a **conjugation tube** is formed between them. In one of the cells, the ring of DNA which is the bacterial chromosome breaks open at one point and starts to pass through the conjugation tube. Only part of the chromosome passes through and this is incorporated into the chromosome of the recipient cell, replacing part of the original chromosome.

Transduction involves the infection of a bacterium by a bacteriophage and may occur in either of two ways. In the first case, when new virions are being reconstituted before lysis of the bacterium, phage capsids may take in some of the original bacterial chromosome instead of phage DNA. Capsids containing bacterial DNA function normally when released and may inject (donor) bacterial DNA into another (recipient) bacterium. When this occurs, part of the injected DNA becomes incorporated in the chromosome of the recipient bacterium in the same way as in transformation or conjugation. The second type of transduction involves only certain strains of phage. Phage which causes lysis soon after attacking a host is known as **virulent** phage. There are also **temperate** phages, which do not immediately cause lysis. When the DNA of a temperate phage enters a bacterium, it becomes incorporated as part of the bacterial chromosome. It may then remain dormant through several bacterial cell generations replicating when the chromosome replicates. Under certain conditions, this phage DNA (**prophage**) separates from the bacterial chromosome and behaves like a virulent phage. Many new virions are formed and released by lysis of the bacterial cell. Sometimes, separation of virus DNA from the bacterial chromosome is inaccurate and some bacterial DNA is incorporated in new phage virions attached to virus DNA. When injected into another bacterium, the bacterial DNA carried by the virus may be inserted into the newly infected bacterial chromosome along with the virus DNA.

Bacteria and the soil

Bacteria are very important in the recycling of organic and other materials in the soil. Bacteria and fungi together are responsible for the breakdown of dead plant and animal remains and animal faeces. Their action results in the release of carbon dioxide, ammonia and other substances into the soil.

In the soil, ammonia is rapidly oxidised to nitrite by nitrifying bacteria (e.g. *Nitrosomonas* spp.). These bacteria are aerobic and chemoautotrophic. They obtain the energy they require from the reaction:

$$2NH_3 + 3O_2 \rightarrow 2HNO_2 + 2H_2O + 79\,000\ cal.$$

The nitrite produced then becomes the substrate and energy source for another group of chemoautotrophic nitrifying bacteria (e.g. *Nitrobacter* spp.). The reaction carried out is:

$$2HNO_2 + O_2 \rightarrow 2HNO_3 + 43\,200\ cal.$$

It is in the form of nitrate that nitrogen is taken up by most green plants.

Another source of nitrate in the soil is the action of organisms which are able to fix atmospheric nitrogen. These include some bacteria and some blue-green algae. Some of these bacteria (e.g. *Clostridium* spp.) and some of the blue-green algae (e.g. *Anabaena* spp.) are free-living in the soil. Other bacteria (e.g. *Rhizobium* spp.) and blue-green algae (e.g. *Anabaena* spp.) live within the tissues of higher plants in symbiotic associations (Chapter 24).

Bacteria and man

Apart from general ecological importance, as in the soil, bacteria affect human life. Some are used by man to suit his purposes. Others are agents of human, animal and plant disease.

Bacteria have been and are used for the benefit of man in many ways.

Sewage disposal schemes make use of the action of either aerobic bacteria or anaerobic bacteria. Most sewage schemes involve and encourage the action of aerobic bacteria. The commonest methods commence with the settling out of solid material, some of which is subsequently dispersed by bacterial action or used as manure (**sludge**). The liquid containing suspended and dissolved organic matter is then sprayed on to porous materials, through which it sinks while having a great surface area in contact with the air. An efficient system produces an effluent liquid without harmful organic compounds, which can safely be discharged into the sea, or into a river. In tropical countries, **sewage ponds** are now often used. In this case, the sewage passes through a series of shallow ponds, in which a luxuriant growth of algae, by producing oxygen, encourages the action of aerobic bacteria.

Anaerobic methods of sewage treatment have been in use for a long time and are now being encouraged, in rural areas, for the production of 'biogas' (methane) for local use. If air is excluded from closed containers of sewage, fermentation by anaerobic bacteria (e.g. *Methanobacterium* spp.) breaks down carbon compounds and releases methane gas. The gas produced can be used for cooking and other purposes which require a source of heat. Using the same method, animal dung is also used to make biogas.

Bacteria are also used in the production of human and animal food. In a cattle-dependent culture, milk may be preserved by making it into yogurt. **Yogurt** is made by innoculating fresh milk with a mixture of bacteria (e.g. *Lactobacillus* spp.) and fungi (*Saccharomyces* spp.). The fermenting agents and their products are harmless and protect the fermented milk from the actions of harmful bacteria, which might otherwise thrive.

In producing **silage**, for animal feed, the leaves and stems of plants are enclosed so that only anaerobic organisms can grow. Bacteria and fungi break down the cellulose and other plant materials and make a large proportion available as easily digestible food for cattle and other farm animals, when fresh food is not available.

There are many industrial uses for bacteria. Bacteria are used to produce many organic chemicals, including acetic acid, acetone and several vitamins.

Key words

Kingdom Monera: procaryotic; bacteria, viruses. **Viruses**: animal viruses, plant viruses, bacteriophages (phages). **Structure**: virion; capsid, nucleic acid. **Virus diseases**: plant diseases, animal diseases. **Bacteria**: cell structure, capsule, cell wall, nucleoid, flagella, pili; reproduction, fission, transformation, conjugation, transduction. **Soil bacteria**: nitrifying bacteria, nitrogen cycle. **Sewage treatment**: aerobic bacteria; anaerobic bacteria, biogas. **Food**: yogurt; silage, animal feed.

Summary

All viruses are parasites and use the synthetic machinery of plant, animal or bacterial cells to reproduce themselves. Viruses are very small and can only be observed with the EM. Viruses cause many diseases in plants, animals and bacteria. Bacteria are very small procaryotic organisms. They are visible with the light microscope but EM is required to observe most features of internal cell structure. They may be free-living as saprophytes, parasitic, or symbiotic. Many are of great ecological importance. Some are useful to man, others cause diseases of animals and plants or destroy materials that are useful to man.

Questions

1 Why is it that some scientists do not regard viruses as living organisms? Give as many reasons as you can.

2 Both blue-green algae and photosynthetic bacteria carry out photosynthesis. How do they differ?

3 Why are bacteria in the soil important to green plants? Give an outline of the nitrogen cycle, with emphasis on the role played by bacteria.

4 How do bacteria reproduce? To what extent may bacteria be said to have sexual reproduction?

23

Fungi

Like algae, fungi were once treated as a single taxon but have now been separated into two or more divisions. In our classification, they are placed in two divisions, Myxomycophyta (slime moulds) and Eumycophyta (true fungi). The Myxomycophyta are summarized in Chapter 1 and will not be treated further in this chapter.

It was once thought that fungi evolved from algae, by the loss of plastids and the adoption of a heterotrophic mode of nutrition. Modern work (electron microscopy, biochemistry) shows that this cannot be so. Some fungi have more in common with members of the animal kingdom (Protozoa) than with algae. All that is certain is that fungi, as a whole, are an unnatural group and that members of the group have more than one evolutionary ancestor.

Fungi are **eucaryotic** organisms. In the Eumycophyta, the fungal body is almost always a **mycelium**, made up of many branched filaments. In discussing fungi, a filament is always described as a **hypha** (pl. **hyphae**). Hyphae may be **septate** (multicellular, with cross walls) or **coenocytic** (without cross walls). Coenocytic hyphae have a continuous cell wall, a lining of cytoplasm containing many nuclei and a central vacuole. Cross walls are formed only where they separate off reproductive organs, injured parts of a hypha or, sometimes, in old hyphae.

Hyphae are microscopic, but some Eumycophyta have fruiting bodies or other structures which are made of many compacted hyphae and are macroscopic. The few unicellular fungi (e.g. *Saccharomyces* spp., yeasts) probably evolved from filamentous forms.

The cell of a true fungus is similar to that of a plant, but has no plastids and has a **centriole** associated with each nucleus. The commonest cell wall material is **chitin**. Chitin is a polysaccharide containing amino (NH_2—) groups and is more familiar as an important component of the exoskeletons of insects and crustaceans. Cellulose replaces chitin in the cell walls of members of one subclass of fungi (Oomycetidae). Storage products are glycogen and oil.

The life cycle of true fungi range from haplontic,

through haplodiplontic to diplontic. Details of the life cycles of some are not well understood. Sexual reproduction is isogamous or oogamous and is followed by the production of spores which are characteristic of each group.

Many fungi produce asexual spores. Some water fungi have flagellate, motile spores (**zoospores**). Other fungi have non-motile spores (**aplanospores**) which are dispersed by water or air. **Sporangiospores**, usually referred to simply as spores, are produced inside sporangia and are set free when the sporangium opens. **Conidia** are spore-like but are cut off from the tips of specialized hyphae (**conidiophores**). Some fungi form **chlamydospores**, also known as gemmae, which are parts of hyphae which have formed a thick wall, inside the original cell wall, and are capable of growing into new hyphae.

In this chapter, we shall deal only with fungi which belong to the three principal classes, Phycomycetes, Ascomycetes and Basidiomycetes (Table 23.1). A fourth class, Deuteromycetes (imperfect fungi), includes all multicellular fungi of which the sexual stages are unknown and which cannot, therefore, be placed with certainty in any other class. When sexual stages are discovered, members of the Deuteromycetes are transferred to the Ascomycetes or Basidiomycetes, as appropriate.

Phycomycetes

In some of the simplest members of the Phycomycetes, the fungal body is a simple multinucleate sac which increases in size and eventually releases uniflagellate zoospores or isogametes. In the majority, the fungal body is a mycelium which consists of a branching system of coenocytic hyphae, which produces sporangia and gametangia.

The Phycomycetes are divided into six subclasses, of which the Oomycetidae and Zygomycetidae are the most important.

Oomycetidae
The Oomycetidae are characterized by the possession

Table 23.1 Summary classification of Eumycophyta and examples given in this chapter. Genera marked with an asterisk may otherwise be placed in the Deuteromycetes.

Class Subclass	Orders and examples of genera	Common names
I. Phycomycetes		
1. Oomycetidae	Saprolegniales *Saprolegnia* *Achlya*	water fungi
	Peronosporales *Pythium* *Phytophthora*	downy mildews
2. Zygomycetidae	Mucorales *Mucor* *Rhizopus* *Choanephora*	black moulds
II. Ascomycetes		
1. Hemiascomycetidae	Saccharomycetales *Saccharomyces*	yeasts
2. Plectomycetidae	Erysiphales *Erysiphe* **Ovulariopsis*	powdery mildews
	Eurotiales **Aspergillus* **Penicillium*	
3. Pyrenomycetidae	Sphaeriales *Sordaria* *Neurospora* *Xylaria* *Daldinia*	fruit moulds
4. Discomycetidae	Pezizales *Pyronema* *Cookeina*	cup fungi
III. Basidiomycetes		
1. Hemibasidiomycetidae	Uredinales *Puccinia*	rusts
	Ustilaginales *Ustilago*	smuts
2. Hymenomycetidae	Agaricales *Lepiota* *Agaricus* *Coprinus* *Amanita*	mushrooms and toadstools
	Aphyllophorales *Polyporus* *Polystictus*	bracket fungi
	Clavaria	club fungi
	Tulasnellales *Tremella*	jelly fungi
3. Gasteromycetidae	Lycoperdales *Lycoperdon*	puff balls
	Geastrum	earth stars
	Nidulariales *Cyathus*	bird's nest fungi

of biflagellate zoospores and by their cell wall material. The flagella of zoospores are of two types, one whiplash-type and one tinsel-type. Cell walls contain no chitin but are made up of a small amount of cellulose, plus other polymers of glucose.

The life cycles of many Oomycetidae are not fully known. In those that have been studied most intensively, meiosis takes place immediately before gamete formation and the life cycle is diplontic. All members of the subclass are oogamous.

We shall concentrate our attention on members of the Saprolegniales and Peronosporales.

Members of the Saprolegniales are almost all found in water. Most are saprophytes but a few are parasites and cause diseases of fish and other water animals. Samples are most easily obtained by placing dead plant material (e.g. boiled seeds) or dead animal material (e.g. a dead fly) in a sample of pond water. In three to four days, the organic matter becomes the centre of a spreading circle of radiating, whitish hyphae.

Two of the commonest genera are *Saprolegnia* and *Achlya*.

In both genera, the mycelium consists of two types of hyphae. Fine, rhizoidal hyphae penetrate the substrate, from which they absorb nutrients. Wider hyphae spread out from the substrate into the water and produce sexual and asexual reproductive organs (Figs. 23.1, 23.2)

Asexual reproduction is by zoospores. Zoosporangia begin to form when the tips of hyphae are separated off by the formation of septa. The tip of the hypha distal to the septum becomes slightly swollen. The cytoplasm of the developing zoosporangium is denser than that of other hyphae, but is still vacuolate and contains many nuclei. The cytoplasm then divides, so that each nucleus becomes the nucleus of a zoospore. Each zoospore (**primary zoospore**) has two flagella, attached at the anterior end. The primary zoospores are released by the breakdown of the tip of the zoosporangium.

In *Saprolegnia* spp., the primary zoospores swim away in all directions and, very soon, encyst. After a short time, each cyst opens and releases a **secondary zoospore**. A secondary zoospore is kidney-shaped and has the two flagella attached at one side. Encystment may be repeated, until a cyst puts out a hypha (**germ tube**) which penetrates a suitable substrate.

In *Achlya* spp., the primary zoospores gather

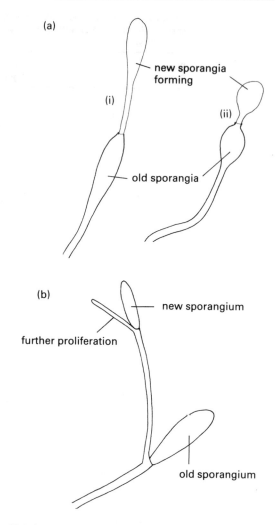

23.1 Asexual reproduction in Saprolegniales: proliferation of sporangia in (a) *Saprolegnia* sp., and (b) *Achylya* sp.

outside the mouth of the zoosporangium and encyst immediately. Kidney-shaped secondary zoospores are then released and subsequent events are the same as in *Saprolegnia* spp.

Empty zoosporangia are quickly replaced by new ones. This is called **proliferation** (Fig. 23.1). In *Saprolegnia* spp., proliferation takes place by the upgrowth of the septum in the base of the old zoosporangium, so that a succession of zoosporangia are formed, each inside the wall of the previous one. In

Achlya spp., proliferation is by the outgrowth of a new zoosporangium from the same hypha, immediately below the septum of the old one.

Some species of *Saprolegnia* and *Achlya* are monoecious (homothallic), others are dioecious (heterothallic).

An oogonium is formed when the tip of a hypha is separated off by a septum and swells until it is roughly spherical. The cytoplasm contains many nuclei and nuclear division continues for a time. Finally, most of the nuclei disintegrate and the cytoplasm divides so that each remaining nucleus becomes the nucleus of an ovum (**oosphere**) (Fig. 23.2). There may be only one ovum in each oogonium but, in most species, there are several. All of the cytoplasm is used up in the formation of ova.

In monoecious species, one or more antheridia are formed as branches of a hypha which bears an oogonium, or of neighbouring hyphae. In dioecious species, they form as branches of nearby hyphae of a different mycelium. Branches which are to form antheridia are attracted by a substance secreted by an oogonium and grow towards it (**chemotropism**).

The first stage in the development of an antheridium is similar to that of a zoosporangium. One or more antheridia become closely attached to the surface of each oogonium. Nuclear divisions occur in each antheridium, after the formation of a septum, and each puts out one or more fine fertilization tubes, which penetrate the wall of the oogonium. One fertilization tube grows towards each ovum and penetrates the wall of the ovum. A single male nucleus passes through the fertilization tube into the ovum. Nuclear fusion follows, and each ovum develops a thick, smooth wall and becomes an oospore. Oospores remain within the oogonium until it breaks up. After a resting period, each oospore puts out a single hypha, which immediately forms a zoosporangium.

Some members of the Peronosporales are water moulds, but many are parasites, especially of higher plants. *Pythium* spp. are often responsible for the 'damping off' disease of seedlings. If seeds are sown and kept under conditions of high humidity, the fungus attacks the hypocotyls of the seedlings at and just above the level of the soil and the seedlings subsequently fall over and die. This is 'damping off'. *Phytophthora* spp. cause diseases in several important crop plants. *P. palmivora* causes diseases in *Elaeis guineensis* (Oil Palm), *Cocos nucifera* (Coconut Palm),

(a)

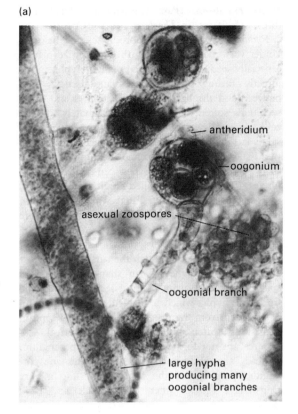

(b)

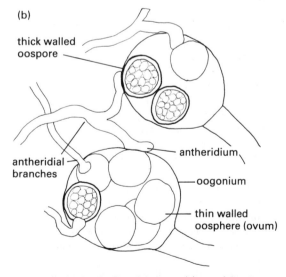

23.2 Reproduction in *Saprolegnia* sp.: (a) sexual structures and zoospores just released from a zoosporangium (HP); (b) drawing of antheridia and oogonia

Hevea braziliensis (Para Rubber) and black pod disease (fruit rot) in *Theobroma cacao* (Cocoa).

Pythium spp. grow as saprophytes in the soil. Under very moist conditions they parasitize plants, especially seedlings or damaged plants. Hyphae grow directly through the higher plant tissues, both between and through the cells. Enzymes and other secretions soften the tissues and ultimately kill the cells. Asexual reproduction takes place at an early stage. Sexual reproduction takes place later, when the tissues have been killed and the fungus is living as a saprophyte. Because they can live as saprophytes, *Pythium* spp. are called **facultative parasites**, to distinguish them from fungi which can obtain nourishment only from living host cells and are **obligate parasites**.

Pythium spp. form zoosporangia at the tips of hyphae which project through the surface of the host tissues. Zoosporangia vary in shape, according to species, but are usually more or less spherical. The tip of the hypha which is to form a zoosporangium is separated from the rest of the hypha by a septum and swells, often until it is nearly spherical. Zoospores are not formed within the zoosporangium. Instead, the undivided contents of the zoosporangium pass out of the sporangium into a very thin-walled vesicle, where they divide and form zoospores. Zoospores are released when the vesicle bursts.

Zoospores are kidney-shaped, with flagella attached at the side. They swim for a time and then encyst. An encysted zoospore puts out a germ tube which penetrates the tissues of a new host, or a suitable non-living substrate in the soil.

In some *Pythium* spp., zoosporangia may break away whole, without producing zoospores, and be dispersed by wind. A zoosporangium which acts in this way can produce a germ tube directly and may be called a conidium.

Most *Pythium* spp. are monoecious. Oogonia are produced inside the host tissue, after the host tissue has died. Their development is similar to that in *Saprolegnia* spp., except that each forms a single ovum and that formation of the ovum does not use up all of the cytoplasm in the oogonium. Some of the cytoplasm remains, as **periplasm**, outside the ovum.

Antheridia, also, are like those of *Saprolegnia* spp., and are formed close to oogonia. Each becomes closely attached, by its tip, to the surface of an oogonium. A fertilization tube is formed, which penetrates the oogonial wall, the periplasm and the wall of the ovum. A single male nucleus passes through the fertilization tube and a zygote is formed. The zygote forms an oospore. After a resting period, the oospore germinates to produce a single hypha, which forms a zoosporangium.

Phytophthora spp. are mostly parasites but some species may survive in the soil, as saprophytes, in the absence of a suitable host. Those which are obligate parasites survive in the soil as oospores or chlamydospores, both of which are originally formed inside host tissues.

The hyphae of *Phytophthora* spp. penetrate between the cells of host tissues, but put out fine branches (**haustoria**) into the cells of the host. It is the haustoria which absorb nutrients from the living cells of the host.

Zoosporangia are formed at the tips of stout hyphae (**sporangiophores**), which emerge through the surface of the host plant, sometimes through stomata. Sporangia are usually slightly elongated or pear-shaped (e.g. *P. palmivora*). In some species (e.g. *P. palmivora*), zoosporangia may become detached before they form zoospores. If so, they may subsequently form zoospores, after they have landed on a suitable substrate. Zoospores are kidney-shaped with two flagella at one side. Each zoospore swims for a time, encysts and then forms a germ-tube, which usually enters a new host through a stoma. Alternatively, a detached 'zoosporangium' may act as a conidium and form a germ-tube directly, without forming zoospores.

In most species in which sexual reproduction is known, the process is similar to that in *Pythium* spp. In other species, the antheridium develops first and the hypha which is to form an oogonium actually grows through the centre of the antheridium, before the tip enlarges to form the oogonium. In such cases, the antheridium forms a collar around the base of the oogonium.

Zygomycetidae

Members of the Zygomycetidae differ from other members of the Phycomycetes in their asexual and sexual reproduction. Asexual reproduction is by non-motile spores (aplanospores) or conidia, which are usually wind-dispersed. There are no structurally distinct gametes, but fusion takes place between pairs of identical coenocytic gametangia. Nuclei from the gametangia then fuse in pairs and a coenocytic zygospore is formed.

Cell walls contain chitin, together with other polymers, including polymers of glucose. Cellulose is absent.

The details of life cycles are known for only a few species, in which meiosis takes place soon after syngamy. These life cycles are therefore haplontic.

There are two orders, Mucorales and Ento-mophthorales. We shall concentrate on the Mucorales.

Most members of the Mucorales are saprophytes and grow on substrates rich in carbohydrates. Many grow in soil and on dead plant material such as rotting fruits. The easiest way to obtain samples of 'wild' species is to add a little soil or dust to the surface of a piece of moist bread or boiled vegetable and to keep the substrate covered, to maintain high humidity. After two or three days, the substrate is covered with a mat of white or greyish hyphae, with small, dark-coloured sporangia standing out from the surface. This method may also produce colonies of moulds belonging to other subdivisions, but members of the Mucorales form thick masses of pale hyphae, while colonies of other moulds are usually small, circular, low growing and coloured yellow, orange or green and can easily be distinguished.

Mucor spp. and *Rhizopus* spp. (Fig. 23.3) are among the commonest of the Mucorales. Species of the two genera are very similar and differ primarily in the arrangement of their sporangia.

Spores germinate to form a rather coarse mycelium, with finer hyphae penetrating the substrate. In *Mucor* spp., stouter hyphae (sporangiophores) stand up from the substrate, singly, and form sporangia at their tips. In *Rhizopus* spp. (Fig. 23.4), the sporangiophores are stouter and often slightly coloured and arise in groups from hyphae at the surface of the substrate. The bases of groups of sporangiophores are anchored in the substrate by narrow, much-branched hyphae (**rhizoidal hyphae**), in which septa sometimes occur.

As the tip of the sporangiophore starts to swell, a septum is formed at the base of the swelling. From the time of its formation, the septum bulges upwards into the developing sporangium, and it bulges further into the sporangium as development continues, and forms the **columella** of the sporangium. Nuclear division continues for a time, after which the contents of the sporangium divide up to

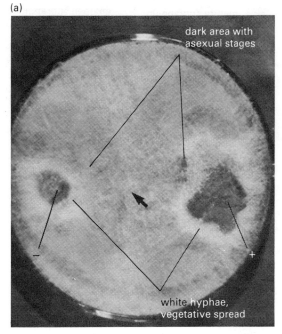

(a)

dark area with asexual stages

white hyphae, vegetative spread

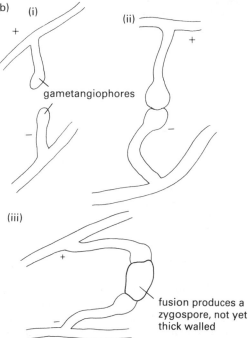

(b) (i)   (ii)

gametangiophores

(iii)

fusion produces a zygospore, not yet thick walled

23.3 Mucorales, *Rhizopus* sp.: (a) plus and minus strains growing together in a petridish, meeting down the centre of the dish; (b) drawings of sexual stages taken from the arrowed region in (a) (HP)

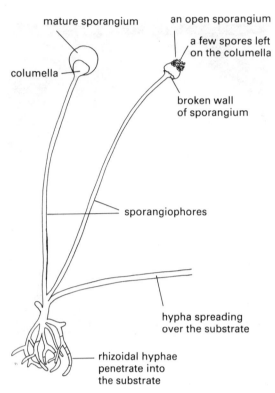

mature sporangium

an open sporangium

a few spores left
on the columella

columella

broken wall
of sporangium

sporangiophores

hypha spreading
over the substrate

rhizoidal hyphae
penetrate into
the substrate

23.4 Mucorales, *Rhizopus* sp., asexual reproduction, drawing of sporangia

form a number of spores, each of which may contain one or several nuclei. The ripe sporangium is dark in colour.

In some *Mucor* spp., the wall of the sporangium liquifies and the spores are left, in a drop of liquid, stuck to the columella until they disperse. In other *Mucor* spp. and in *Rhizopus* spp., the wall of the sporangium breaks and liberates the spores.

In some species of both genera, developing sporangiophores always grow towards a light source (are **positively phototropic**).

Some Mucorales are homothallic, others are heterothallic. Homothallic species reproduce sexually within a mycelium grown from a single spore. Heterothallic species produce gametangia only when hyphae of two different strains are present and close together. There is no distinction between male and female gametangia, so it is impossible to describe strains as 'male' and 'female'. Instead, sexually compatible strains of heterothallic species are

described as 'plus' ('+') and 'minus' ('−') strains respectively.

When sexually compatible hyphae grow close together, each hypha puts out short side branches (**gametangiophores**) which grow towards each other, in response to a chemical attraction (**chemotropism**). After the tips of gametangia make contact with each other, septa are formed which separate gametangia from the base of each gametangiophore. The tips of gametangia then dissolve and the gametangia fuse to form a single zygospore, which remains attached to the two gametangiophores. Within a few days, most of the nuclei fuse in pairs. Meanwhile, the zygospore forms a thick wall. After a resting period, the zygospore germinates to form a single sporangiophore, which forms a sporangium, Meiosis may take place before the zygospore germinates, or soon after it has germinated.

There are several other genera in which asexual and sexual reproduction take place in almost exactly the same way as in species of *Mucor* and *Rhizopus*. In other genera, there are variations in the pattern. *Choanephora cucurbitarum* shows an interesting variation and is readily available in tropical Africa.

C. cucurbitarum grows as a saprophyte on rotting fruits of members of the Cucurbitaceae, but can most readily be obtained from fallen flowers (corolla, androecium) of *Hibiscus rosa-sinensis* (Hibiscus). When the flowers have been on the ground in humid weather for a day or two, they are usually covered with mould, which is usually *C. cucurbitarum*.

C. cucurbitarum produces some sporangia similar to those of *Mucor* spp., except that the sporangiophore is curved so that the sporangia hang downwards. It also produces sporangiophore-like structures which are inflated at the end, and themselves bear several inflated outgrowths. Each of these outgrowths bears, on the outside, a number of small sporangia (**sporangioles**). Each sporangiole contains a single elongated spore, similar to the spores formed in a normal sporangium. Sporangioles are dispersed entire, but later open to release the spores. The sporangioles are sometimes referred to as conidia, which they resemble in all but the fact that they finally open and release the contained spore.

Ascomycetes

The fungal body of most members of the class

Ascomycetes is a multicellular mycelium. Chitin is the main constituent of cell walls. When a cell divides, a septum grows in from the periphery, but always leaves a small pore at the centre. This pore is much larger than the plasmodesmata between plant cells. There is protoplasmic continuity between adjacent hyphal cells and, therefore, throughout the whole mycelium. The multicellular nature of the mycelium is less clear cut than in plants. Cytoplasmic organelles and even nuclei can, and do, pass from one cell to another.

Cells may be uninucleate, binucleate or multinucleate. Uninucleate and binucleate cells may characterize different stages in the life cycle of a single species.

There are no motile stages in Ascomycetes.

Yeasts are unicellular forms with uninucleate cells. They reproduce asexually by budding or by fission.

Many multicellular forms reproduce asexually by conidia. Conidiophores are usually branched and often form conidia in series, so that each conidiophore bears a succession of conidia, with mature conidia at the free end and immature conidia close to the conidiophore (Fig. 23.5).

The most characteristic feature of the life cycle of Ascomycetes is the ascus (pl. asci). **Asci** are formed following the completion of sexual reproduction and are sac-like, usually elongated organs in which meiosis takes place and **ascospores** are formed. Meiosis is usually followed immediately by mitosis in each of the four nuclei and, usually, an ascus contains eight ascospores (Fig. 23.6), though the number can vary from one to many. Ascospores are discharged violently through the tip of the ascus and afterwards dispersed by wind.

Mycelia may be homothallic or heterothallic. In the simplest cases, mycelia which grow from ascospores are uninucleate (**monokaryotic**) and remain so until hyphae of sexually compatible strains meet and intermingle. At this stage, the structure of a fruiting body (**ascocarp**) starts to form. Female reproductive organs (**ascogonia**, sing. **ascogonium**) are formed by the enlargement of a single cell, which puts out an elongated **trichogyne** (Fig. 23.7). A neighbouring hypha, of a different strain if the species is heterothallic, forms a male reproductive organ (antheridium), and the contents of the antheridium pass into the ascogonium, through the trichogyne. This process is known as **plasmogamy**. It is a fusion of the cytoplasm. The two nuclei do not

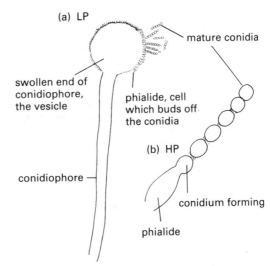

23.5 Ascomycetes, *Aspergillus* sp., production of conidia (see also Figs. 23.11, 23.12)

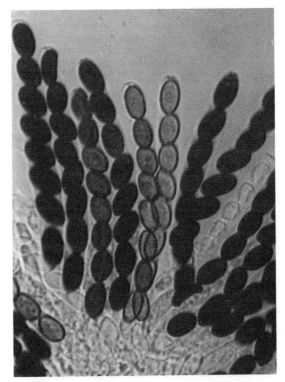

23.6 Ascomycetes, *Sordaria* sp., group of asci from a perithecium. Each ascus contains a row of eight spores. Mature spores are black, immature spores are white (HP)

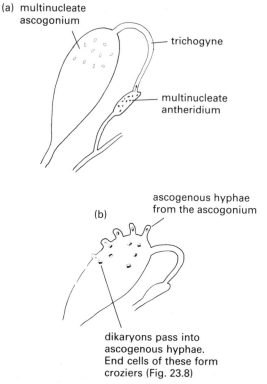

(a) multinucleate ascogonium

trichogyne

multinucleate antheridium

(b)

ascogenous hyphae from the ascogonium

dikaryons pass into ascogenous hyphae. End cells of these form croziers (Fig. 23.8)

23.7 Diagram of stages of sexual reproduction in an ascomycete: (a) plasmogamy, cytoplasm and nuclei pass from the antheridium into the ascogonium through the trichogyne; (b) the dikaryotic ascogonium produces dikaryotic ascogenous hyphae

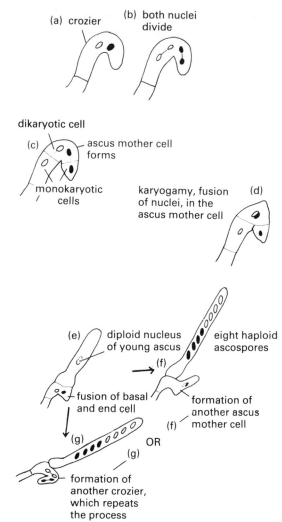

(a) crozier   (b) both nuclei divide

dikaryotic cell

(c) — ascus mother cell forms

monokaryotic cells

karyogamy, fusion of nuclei, in the ascus mother cell   (d)

(e) diploid nucleus of young ascus   eight haploid ascospores

(f)

fusion of basal and end cell   formation of another ascus mother cell   (f)

(g)   OR   (g)

formation of another crozier, which repeats the process

23.8 Production of an ascus from the hooked end-cell (crozier) of an ascogenous hypha

fuse, but the ascogonium becomes binucleate (**dikaryotic**). The two nuclei are a **dikaryon**. In species in which the ascogonium and antheridium are multinucleate, several male nuclei pass into the ascogonium, which then contains several dikaryons (pairs of sexually compatible nuclei). The nuclei divide several times and dikaryotic **ascogenous hyphae** grow from the surface of the ascogonium. In species with macroscopic fruiting bodies, they mingle with monokaryotic hyphae which form part of the fruiting body.

Ascogenous hyphae branch and the terminal cell of each branch finally folds upon itself to form a hook-like **crozier** (Fig. 23.8). There is next a simultaneous division of both nuclei, orientated in such a way that a central, dikaryotic cell (ascus mother cell) is cut off in the upper part of the crozier. At the same time, monokaryotic cells are cut off at the tip of the crozier and at the base of the crozier and these two cells may later unite to form a dikaryotic cell. The two haploid nuclei in the ascus mother cell now fuse (karyogamy). As the ascus mother cell elongates to form the ascus, the diploid nucleus undergoes meiosis and each of the four haploid nuclei divides again by mitosis. The eight nuclei become the nuclei of eight ascospores.

Ascus formation is the same in most Ascomycetes,

but the manner in which plasmogamy is achieved varies quite a lot. In the type described above, there are distinct male and female gametangia. In some Ascomycetes, the two gametangia are of the same size and shape and no sexual distinction can be made. In yet others, plasmogamy occurs between apparently normal vegetative cells, or between vegetative cells of one strain and uninucleate conidia (**spermatia**) of another strain.

Some of the simpler Ascomycetes form their asci naked and have nothing that can be described as a fruiting body. In most, an ascocarp is formed from hyphae which become so closely pressed together that the tissue formed can be called pseudoparenchyma. A mixture of asci and sterile hairs (**paraphyses**) forms a fertile layer (**hymenium**) as part of the ascocarp. If the hymenium is more or less flat and fully exposed, the ascocarp is an **apothecium** (Fig. 23.9). If the fruiting body is flask-shaped with a pore and with the hymenium on the inside surface, it is a **perithecium** (Fig. 23.13). If it is completely closed, with the hymenium on the inside surface, it is a **cleistothecium** (Fig. 23.11).

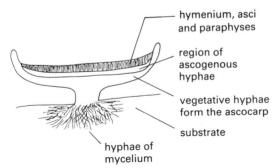

23.9 Diagrammatic VS of an open, cup-shaped ascocarp (apothecium) to show the hymenium (fertile layer) and other layers

The Ascomycetes are divided into five subclasses, of which we shall look at representatives of the Hemiascomycetidae, Plectomycetidae, Pyrenomycetidae and Discomycetidae.

Hemiascomycetidae
Hemiascomycetidae have no ascocarp and fusion takes place between cells which subsequently form asci directly. *Saccharomyces* spp. (yeasts) belong to this group.

Saccharomyces spp. are uninucleate, unicellular fungi. Most are saprophytic and are noted for their ability to ferment carbohydrates to produce alcohol, when they are growing anaerobically. They are also capable of carrying out aerobic respiration, in which case they do not produce alcohol.

The cells reproduce asexually by budding (Fig. 23.10). During this process, a small outgrowth develops on the surface of the otherwise spherical cell and enlarges until it approaches the size of the parent cell. The nucleus then divides by mitosis and one of the daughter nuclei becomes the nucleus of the daughter cell. The daughter cell may separate immediately, or remain attached until both the original cell and daughter cell have produced new buds.

Saccharomyces spp. may be homothallic or heterothallic. When cells of different strains of a haploid heterothallic species meet under suitable conditions, they fuse in pairs and the cell walls become continuous. Karyogamy occurs immediately and meiosis soon follows. The cytoplasm then divides around the four nuclei and four ascospores are formed. Ascospores have thick walls and germinate to form ordinary, haploid vegetative cells.

Some strains of *S. cerevisiae* are diplontic. Diplontic strains can be caused to form asci directly

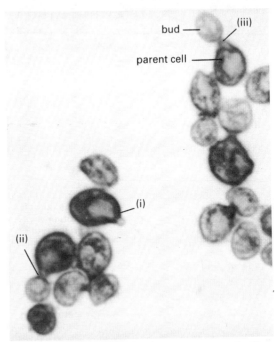

23.10 Hemiascomycetidae, *Saccharomyces* sp., yeast cells, including rounded, vegetative cells and stages of budding, (i) to (iii) (HP)

(a)

(c) *Erysiphe* sp.
  (i) whole ascocarp   (ii) ascocarp split open
                                  after squashing

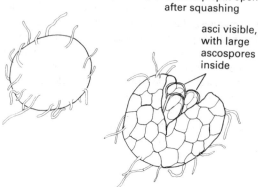

asci visible,
with large
ascospores
inside

23.11 Plectomycetidae, powdery mildews: (a) on undersurface of leaves of *Carica* sp. (Pawpaw); (b) conidia from Pawpaw leaf (LP); (c) cleistothecia from leaf of *Zinnia* sp. (LP)

by being placed in a rich nutrient medium. The diploid condition is re-established by fusion of vegetative cells or fusion of ascospores soon after they are set free.

Plectomycetidae

Members of this subclass have asci in simple asco-carps (cleistothecia), of loosely entangled hyphae. The commonest representatives are *Erysiphe* spp.

Erysiphe spp. are powdery mildews, which are obligate parasites (Fig. 23.11). They are commonly found on the upper surfaces of the leaves of members of the Cucurbitaceae (e.g. *Cucumis sativus*, Cucumber), and on *Carica papaya* (Pawpaw) leaves, especially in the wet season.

Spores of *Erysiphe* spp. are uninucleate and germinate on the surface of a host leaf to form a short germ tube. This becomes closely attached to the surface of the leaf and spreads out to form a disc-like **appressorium**. A narrow, peg-like haustorium is formed at the centre at the lower surface of the appressorium. The haustorium penetrates the cuticle and cell wall of an epidermal cell and inflates or branches within the host cell. A branching mycelium then gradually extends over the surface of the leaf, many of the hyphae attached by haustoria.

The attached hyphae produce upright, aerial branches, each made up of a single, inflated conidial mother cell. Each conidial mother cell produces a chain of elyptical conidia, which are dispersed as they mature. It is the conidia that give the surface

(b) *Ovulariopsis* sp.

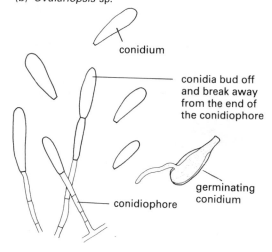

conidium

conidia bud off
and break away
from the end of
the conidiophore

germinating
conidium

conidiophore

of the leaf a powdery appearance, and give *Erysiphe* spp., and species of related genera, the name 'powdery mildews'.

Sexual reproduction takes place in the manner already described. Wide asci are formed inside a cleistothecium, which has appendages on the outside.

Aspergillus spp. and *Penicillium* spp. are common moulds, most often seen on substrates rich in carbo-hydrates (e.g. rotting fruits), on which they form brown, yellow or green, circular colonies (Fig. 23.12). Their cells are multinucleate.

Species of both genera form many conidia. Coni-diophores of *Aspergillus* spp. are swollen into a vesicle at the end (Fig. 23.5). Conidia are budded off from the tips of cells (**phialides**) which arise from the vesicle, or from phialides which are branches of cells which arise from the vesicle. The conidiophores of *Penicillium* spp. are superficially similar (Fig. 23.12) but there is no vesicle and phialides are the ultimate branches of a simple branching system.

Sexual reproduction and ascus formation are known in only a few species of *Aspergillus* and *Penicillium*. In fact, *Penicillium* was included in the Deutero-mycetes (imperfect fungi) until fairly recently. Both have multinucleate ascogonia and antheridia, but the sexual process is the same as that already described. Ascocarps are small cleistothecia.

Pyrenomycetidae

Members of the Pyrenomycetidae are characterized by the production of perithecia. The best known genera are *Sordaria* (Fig. 23.13) and *Neurospora*. Species of these genera are commonly grown in laboratories, where they are used in research on fungal nutrition and fungal genetics.

Sordaria spp. are saprophytes and grow mostly on herbivore dung. They are, therefore, **coprophilous** fungi. They do not produce conidia. Their cells are multinucleate and the exact mechanism of sexual fusion is not known, but it is probably through the fusion of cells of normal hyphae. Eight-spored, elon-gated asci (Fig. 23.13) are formed in perithecia. *S. fimicola*, the species most commonly used for experi-mental purposes, is homothallic.

Neurospora spp. are saprophytes, some of which occur naturally on substrates which have been subjected to heat, for example, after burning. Like *S. fimicola*, *N. crassa* has been widely used in nu-tritional and genetical studies. This species is hetero-

(a)

(i) green area of conidia

(ii) white mycelium of vegetative hyphae

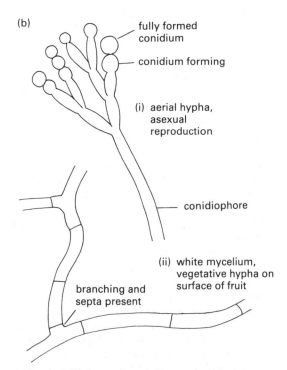

(b)

fully formed conidium

conidium forming

(i) aerial hypha, asexual reproduction

conidiophore

(ii) white mycelium, vegetative hypha on surface of fruit

branching and septa present

23.12 Plectomycetidae, *Penicillium* sp.: (a) growth of *Penicillium* sp. on a rotting orange; (b) structure of the fungus from the two regions indicated in (a) (HP)

(a) × 25

mature black perithecium, lower part embedded in mass of hyphae

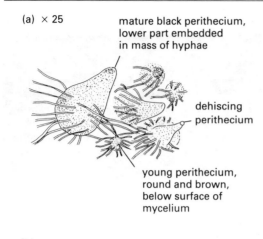

dehiscing perithecium

young perithecium, round and brown, below surface of mycelium

(b)

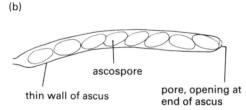

ascospore

thin wall of ascus

pore, opening at end of ascus

23.13 Pyrenomycetidae, *Sordaria* sp.: (a) a group of perithecia on agar (VLP); (b) structure of a single ascus (see also Fig. 23.6)

thallic. Hyphae have multinucleate cells and produce two types of conidia. Large, multinucleate **macroconidia** serve to propagate the species asexually. Smaller, uninucleate **microconidia** may act in the same way, but also function as non-motile male gametes (**spermatia**). A spermatium adheres to the trichogyne of an ascogonium produced by a mycelium of a different strain, after which plasmogamy occurs, by the movement of the contents of the spermatium into the trichoygne of the ascogonium. Perithecia are similar to those of *S. fimicola*, only just large enough to be called macroscopic.

Much larger fruiting bodies are formed by some Pyrenomycetidae. *Xylaria* spp. and *Daldinia* spp. occur as saprophytes on dead wood, but some species of *Daldinia* are parasites. The fruiting bodies of *Xylaria* spp. are grey or black, horn-like structures several centimetres high. *Daldinia* spp. form hemispherical or kidney-shaped structures several centimetres in diameter. They have concentric markings when cut or broken across. In both genera, young fruiting bodies (**stromata**, sing. **stroma**) are covered with conidiophores with

powdery conidia. Older parts have many small perithecia embedded in the surface (Fig. 23.14).

Discomycetidae

Members of the Discomycetidae have fruiting bodies which are flat or cup-shaped apothecia (Fig. 23.15).

Pyronema omphaloides, previously known as *P. confluens*, commonly occurs in forest and savanna after burning, when areas of charred wood and soil several square metres in extent may be seen, covered with bright pink mycelium. Ascus formation follows plasmogamy between an antheridium and an ascogonium, as described earlier, and small apothecia are formed which are less than 1 mm in diameter.

Cookeina tricholoma is common on dead wood. The ascocarp is often 2 cm in diameter and shaped like a shallow cup. It is pale brown in colour, with dark brown hairs around the margin and on the outside.

× 2

black perithecial surface, with numerous tiny openings

outer layer of small perithecia

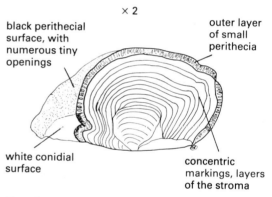

white conidial surface

concentric markings, layers of the stroma

23.14 Fruiting body of *Daldinia* sp. cut across to show layering

smooth inner surface, hymenial layer of asci and paraphyses

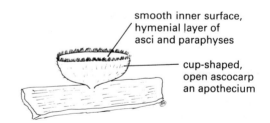

cup-shaped, open ascocarp an apothecium

23.15 Discomycetidae: an apothecium of *Cookeina* sp.

Basidiomycetes

The mycelium of members of the Basidiomycetes is multicellular. As in the Ascomycetes, the main constituent of the cell wall is chitin and the septa between cells are perforated, each by a relatively large pore. Cells may be uninucleate (monokaryotic) or binucleate (dikaryotic). There are no motile stages in the life cycle.

Sexual reproduction results in the production of **basidiospores**, which are uninucleate and haploid. A basidiospore germinates to form a primary mycelium, in which the cells are uninucleate. At some stage, plasmogamy occurs between cells of different mycelia, or different hyphae of the same mycelium, and a dikaryotic secondary mycelium results. As in Ascomycetes, karyogamy is delayed for some time.

The two nuclei of a dikaryotic cell divide at the same time. Mitosis may take place with the two nuclei more or less side-by-side, followed by the formation of a septum between the two pairs of daughter nuclei, or may involve the formation of **clamp connections** (Fig. 23.16).

The formation of clamp connections seems to be a mechanism which ensures that the nuclei of each of the daughter cells are always a pair, one of which was originally derived from one primary mycelium and one from the other. In species which generally lack clamp connections, clamp connections may be formed during one or more of the divisions which immediately precede the formation of basidia (Fig. 23.16).

Basidiomycetes are named after the club-shaped organ (**basidium**) from which spores are produced

(a)

chlamydospores

promycelium of four haploid cells

newly formed clamp connection

end cell inactive

clamp connection results in a dikaryotic cell, which produces a dikaryotic mycelium

basidiospore (haploid) produces a monokaryotic mycelium (see life cycle, Fig. 23.19)

(b)

(i)

(ii) formation of clamp

(iii)

clamp cell contains nucleus

(iv)

(v)

clamp connection remains thus

clamp connection open, nucleus passes through

23.16 Basidiomycetes, clamp connections: (a) in cells of the promycelium of *Ustilago* sp. (Smut); (b) diagram of formation of clamp connections in a dikaryotic mycelium

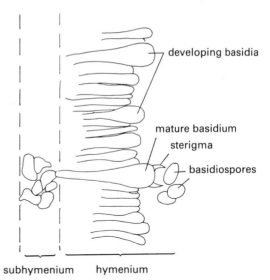

developing basidia

mature basidium

sterigma

basidiospores

subhymenium   hymenium

23.17 Basidiomycetes, hymenial layer: diagram of TS hymenial layer showing spores and a basidium (HP). See also Fig. 23.20

(Fig. 23.17). In most Basidiomycetes, the basidium is formed at the surface of a fruiting body and is, at first, dikaryotic. The two nuclei then fuse (karyogamy) and the diploid nucleus divides by meiosis. The four products of meiosis then migrate into four outgrowths of the surface which form basidiospores. Basidiospores are released by the breakage of the connection between basidiospores and basidium.

In some genera, basidiospores are reduced to two, or increased to larger numbers.

The class Basidiomycetes is divided into three subclasses, Hemibasidiomycetidae, Hymenomycetidae and Gasteromycetidae.

Hemibasidiomycetidae

The Hemibasidiomycetidae differ from other Basidiomycetes in that their basidiospores are formed immediately after the germination of another type of spore formed during the life cycle and not from cells of an organized fertile tissue (hymenium). The subclass includes a number of organisms which are responsible for important diseases of crop plants, including rusts (e.g. *Puccinia* spp.) and smuts (e.g. *Ustilago* spp.) both of which attack cereals as well as other plants. Hyphae are intercellular, with haustoria extending into host cells.

The full life cycle of *Puccinia* spp. is complex and, where fully known, involves two distinct hosts (Fig. 23.18).

Basidiospores are uninucleate and haploid. They germinate to produce hyphae which penetrate and parasitize a host flowering plant. When the mycelium is well established, two types of sexual organs are formed together in pear-shaped hollows in the surface layers of the host tissue. Receptive hyphae

(b)

(c)

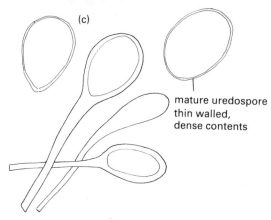

mature uredospore
thin walled,
dense contents

(a)

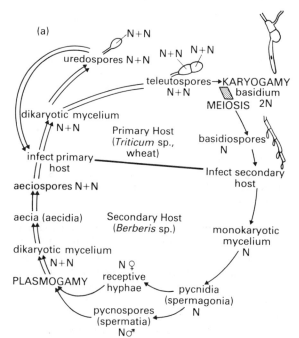

23.18 Hemibasidiomycetidae, rusts, *Puccinia* spp.: (a) life cycle of *Puccinia graminis tritici*; (b) undersurface of leaf of *Euphorbia geniculata* with uredospore stage of *Puccinia* sp.; (c) uredospores scraped from the leaf (LP)

at the edge of each hollow are the female organs. In the centre of the hollow, the male hyphae (**spermatiophores**) produce chains of spermatia (**pycnospores**), which are the male gametes. The hollows producing these two types of hyphae are called **spermagonia** or **pycnidia**. *Puccinia* spp. are heterothallic and spermatia are not compatible with receptive hyphae in the same spermogonium, or originating from the same strain of fungus. A nectar-like liquid is secreted, which oozes from the open end of the spermogonium and it is probable that spermatia are carried from one spermogonium to another by insects.

If a compatible spermatium is brought into contact with a receptive hypha, plasmogamy occurs. The resulting dikaryotic cell gives rise to a dikaryotic secondary mycelium which spreads further through the host tissue.

The dikaryotic mycelium then forms **aecia** (**aecidia**) on another part of the host. An aecium (aecidium) is a cup-like depression, lined with mycelium, which contains many hyphae producing **aeciospores** in long chains, in a similar manner to that in which conidia are produced in many other fungi. The epidermis of the host is broken and the dikaryotic aeciospores are released into the air.

In species for which the full life cycle is known (e.g. *P. graminis tritici*, Wheat Rust), the aeciospores infect the grass host (*Triticum* sp., Wheat), in which a dikaryotic mycelium is produced. Soon after infection, the stage which gives the fungus its common name (Rust) develops. Hyphae which grow up close to the surface, under the epidermis, form spherical spores at their tips. The epidermis of the host breaks and exposes the group (**sorus**) of reddish-brown, unicellular, dikaryotic conidia (**uredospores**). When released, uredospores infect the same species of host and it is at this stage that the disease spreads most rapidly.

Later in the season, the type of spore formed in a sorus changes. **Teleutospores** are formed in the same way as uredospores, but each consists of two dikaryotic cells. Teleutospores do not infect a host directly. Under suitable conditions, the two nuclei in each cell of the spore fuse (karyogamy) and the diploid nuclei produced undergo meiosis. A short hypha grows from a pore in the wall of each of the two cells of the spore. The nucleus divides by meiosis and the hypha divides to form four uninucleate cells. Each cell puts out a small, spherical

23.19 Hemibasidiomycetidae, smuts, *Ustilago* spp.: (a) Maize Smut on a cob of *Zea mays*; (b) *Cynodon dactylon*, (i) healthy plant, (ii) plant infected with smut; (c) smut on spikelets of *Sorghum* sp. (Millet); (d) spores and mycelium of smut from Millet (continued overleaf)

(a)

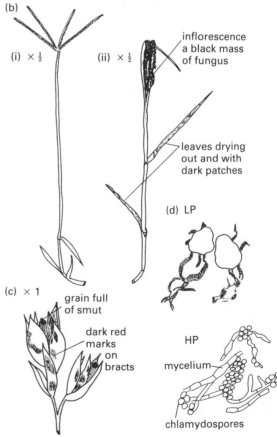

(b)

(i) × ½   (ii) × ½

inflorescence a black mass of fungus

leaves drying out and with dark patches

(d) LP

(c) × 1

grain full of smut

dark red marks on bracts

HP

mycelium

chlamydospores

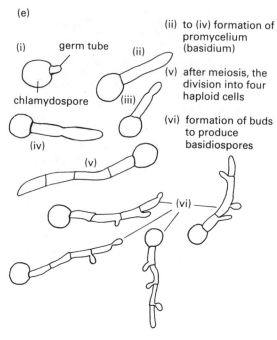

(e)

(i) germ tube (ii)

(ii) to (iv) formation of promycelium (basidium)

chlamydospore (iii)

(v) after meiosis, the division into four haploid cells

(iv)

(vi) formation of buds to produce basidiospores

(v)

(vi)

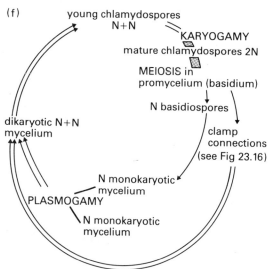

(f)

young chlamydospores N+N

KARYOGAMY

mature chlamydospores 2N

MEIOSIS in promycelium (basidium)

N basidiospores

clamp connections (see Fig 23.16)

dikaryotic N+N mycelium

N monokaryotic mycelium

PLASMOGAMY

N monokaryotic mycelium

23.19 (continued) (e) germination of chlamydospores from Millet; (f) life cycle of *Ustilago* sp.

protrusion, into which its nucleus migrates, as the nucleus of a basidiospore.

In Wheat Rust, the primary host *Triticum* sp. is initially infected by aeciospores, later by uredospores. Basidiospores infect the secondary host, *Berberis vulgaris*, in which plasmogamy occurs and

upon which aeciospores are formed. Under natural conditions, teleutospores are the stage in which the Rust survives the adverse (cold) season.

In the tropics, many grasses suffer from infections of *Puccinia* spp. and species of related genera, but the full cycle, including alternate hosts, is not known. However, examples of plants which are hosts to different stages of the life cycle in different Rust species are commonly seen. *Zea mays* and *Sorghum* spp., and other grasses, are frequently infected with the uredospore-forming stage (Fig. 23.18). Teleutospores of *P. lateritia* are common on the leaves of *Borreria scabra*, a common weed of cultivated ground. Aecia are less common, but occur on some members of the Araceae and on *Canna* sp.

Smuts, *Ustilago* spp. and species of related genera, have a similar life cycle to *Puccinia* spp., but only one host is involved. Plasmogamy occurs between monokaryotic hyphae, soon after infection of a host by basidiospores. The dikaryotic mycelium produces binucleate conidia (cf. uredospores), which are dispersed and spread the disease. In cereal smuts (e.g. on *Zea mays* and many other grasses) mycelium invades the ovaries of the flowers and produces many unicellular, dikaryotic chlamydospores at the time when seed would normally be formed by the grass. Seed heads are covered with the black, powdery mass of spores (Fig. 23.19).

Hymenomycetidae

Members of the Hymenomycetidae are mostly saprophytic but some are parasitic on higher plants. There are very few examples of fungi belonging to this group which form conidia, but most form basidia and produce basidiospores in the manner described at the beginning of this chapter. Basidiospores separate violently from the basidium and are immediately dispersed.

A very few parasitic forms (e.g. *Exobasidium* spp.) have no fruiting body, but form basidia at the surface of the host tissue. Others form fruiting bodies (basidiocarps) with basidia forming a hymenium on surfaces which are exposed at maturity on gills, in pores, or on a smooth surface.

Basidiospores germinate to form monokaryotic hyphae. Plasmogamy occurs by fusion between hyphal cells and almost all of the mycelium is composed of dikaryotic cells.

Lepiota spp. (Fig. 1.6) are **gill fungi**. Each basi-

(a)

(i) young stage

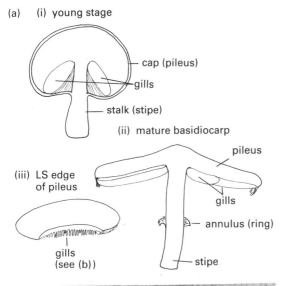

— cap (pileus)

— gills

— stalk (stipe)

(ii) mature basidiocarp

pileus

(iii) LS edge
of pileus

gills

annulus (ring)

gills
(see (b))

stipe

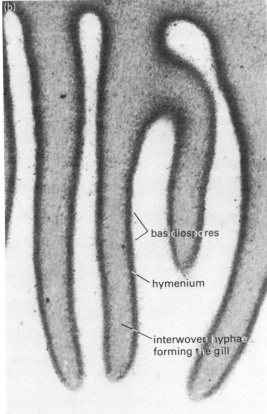

basidiospores

hymenium

interwoven hyphae
forming the gill

23.20 Structure of the basidiocarp of *Lepiota* sp.:
(a) LS basidiocarp; (b) VS gills (LP) (see Fig. 23.17 for HP)

23.21 *Polystictus* sp., a common bracket fungus on dead
wood (× ¼)

diocarp has an upright stalk (**stipe**) and an umbrella-
shaped head (**pileus**). The hymenium covers the
surfaces of radiating gills (Fig. 23.20) underneath the
pileus. Other common gill fungi are *Agaricus* spp.,
Coprinus spp. and *Amanita* spp.

Bracket fungi (e.g. *Polyporus* spp.) are parasites
on trees or saprophytes on dead trees (Fig. 23.21).
A bracket fungus has no stipe but consists of a more
or less semicircular head attached in a horizontal
position to the side of its substrate. In this case, the
hymenium lines the insides of many small pores on
the undersurface of the head.

Stinkhorn fungi (e.g. *Phallus* spp.; *Dictyophora*
spp.) are quite common in moist, forest country.
Here, the basidiocarp is first formed as an egg-
shaped body, 2 to 5 cm high, on the surface of the
ground, usually close to buried, rotting wood. The
'egg' contains a stipe, which extends to about 6 to
15 cm in length over a very short period of time.
There is a swollen receptacle at the top of the stipe,
with an irregular surface covered with dark-coloured
slime containing many basidiospores. Stinkhorns are

most easily located in the wet season, by their very unpleasant smell. In *Dictyophora* spp., there is a net-like, bell-shaped structure which hangs downwards from the top of the stipe. Most of the spores drip away onto the ground, but some must be carried away by flies, which are attracted by the smell.

Clavaria spp. are common on grassland. Its basidiocarps are cylindrical, tapered towards the top, yellow or orange in colour and up to about 5 cm high. Hymenium covers the outside of the upper part.

Gasteromycetidae

Gasteromycetidae differ from Hymenomycetidae in that their spores do not leave the basidia violently and are formed within a closed basidiocarp. Spores are released after the basidia themselves have dried up, by rupture of the wall of the basidiocarp. *Lycoperdon* spp. and related genera (**Puffballs**) are often common on humus-rich soil (Fig. 23.22).

In *Cyathus* spp. (Bird's Nest Fungus), mature basidiocarps are cup-shaped, with several small puff-ball-like **peridioles** inside, resembling a group of eggs in a bird's nest.

23.22 Gasteromycetidae: fruiting bodies of *Lycoperdon* sp., a puffball. In puffballs, the basidia are completely enclosed and basidiospores are released only after the top of the basidiocarp dries out and splits. (× ½) (See also cover photograph of *Geastrum* sp.)

Key words

Structure: mycelium; hypha, septate, coenocytic; chitin; eucaryotic cell. **Cell**: monokaryotic; dikaryotic, dikaryon. **Asexual reproduction**: zoospore, aplanospore, chlamydospore; sporangiophore, sporangium, sporangiospore; conidiophore, conidium; sporangiole. **Sexual reproduction**: isogamy, oogamy; heterothallic, plus and minus strains; homothallic; gametangium; plasmogamy, karyogamy; ascogonium, trichogyne; ascogenous hypha, crozier; ascus, ascospore; ascocarp, apothecium, perithecium, cleistothecium; basidium, basidiospore; basidiocarp. **Rust**: spermagonium (pycnidium), spermatium (pycnospore), spermatiophore; aecium (aecidium), aeciospore; uredospore; teleutospore; basidiospore. **Smut**: chlamydospore; basidiospore.

Summary

Fungi are non-green, heterotrophic, eucaryotic organisms. A few are unicellular but most have a mycelium made up of branched, septate or non-septate hyphae. Asexual spores are motile and non-motile in different groups. In Phycomycetes, spores produced following sexual fusion may be diploid zygospores or oospores, which subsequently produce zoospores or aplanospores. In Ascomycetes, sexual fusion and meiosis are followed by the production of ascospores, formed inside asci. In Basidiomycetes, sexual fusion is followed by the formation of basidiospores, which are formed externally on basidia. In Ascomycetes and Basidiomycetes there is sometimes a long delay between plasmogamy and karyogamy. Hyphae with dikaryotic cells are often present. Some Ascomycetes and Basidiomycetes have large fruiting bodies. Many plant diseases and some animal diseases are caused by fungi.

Questions

1 Fungi are eucaryotic organisms, but are separated from the kingdom Plantae. What structural and reproductive features of the Eumycophyta justify their being separated from plants in this way?

2 How do the Oomycetidae and Zygomycetidae differ in reproductive features? Illustrate your answer with reference to named members of the Peronosporales and Mucorales.

3 What is a dikaryotic mycelium? At what stage does such a mycelium occur in the life cycle of a typical ascomycete? How are asci formed?

4 What is the basic difference between ascospore and basidiospore formation? Describe the different types of ascocarps found in Ascomycetes and one type of basidiocarp.

24

Lichens, mycorrhizae and root nodules

Lichens, mycorrhizae (sing. mycorrhiza) and root nodules are all examples of close and mutually beneficial associations between fungi or bacteria and plants. All are examples of symbiosis.

Symbiosis means 'living together'. In its widest sense, it refers to the relationship between any two (or more) organisms which always live in close contact under natural conditions. If we use this definition, we must recognise three types of symbiosis. In **mutualistic** symbiosis (**mutualism**), both organisms benefit from the association. In **commensal** symbiosis (**commensalism**), neither organism suffers from the association and one may benefit. In **parasitic** symbiosis (**parasitism**), one organism benefits at the expense of the other. The broad definition of symbiosis results mostly from the complex situations which occur in the animal kingdom and many botanists would regard mutualism and symbiosis as synonymous. The examples which follow are all of mutualistic symbiosis.

Lichens

Lichens are composite organisms, in which a fungus lives in symbiotic association with a blue–green or green alga. The fungus cannot exist without the alga, because the alga, through photosynthesis, provides all the food for the lichen. The alga benefits from the protection provided by the fungal part of the lichen and could not survive in the position occupied by the lichen without that protection.

Lichens (Fig. 24.1) are very common indeed in all land habitats, except where industrialization has polluted the atmosphere. Lichens are very sensitive to polluted atmosphere and, indeed, the occurrence of various species of lichens has been used as a method of assessing the degree of atmospheric pollution in areas close to industrial cities.

Lichens are mostly epiphytic or epilithic (growing on rocks), but some species grow on soil where there is little competition from larger plants. For instance, they form a major part of the ground vegetation in tundra (Chapter 28). In tropical countries, they grow

24.1 Various types of lichens on a tree branch: (a) crustose, (b) foliose; (c) fruticose

mostly on exposed rock surfaces, on tree trunks and, in rain forest especially, on the upper surfaces of the leaves of trees and shrubs.

In most lichens, **ascolichens**, the fungal partner is an ascomycete. There are a few examples of **basidiolichens**, where the fungal partner is a basidiomycete. Some basidiolichens grow in West African rain forest, but they are rare. The remainder of our study will be on ascolichens.

The algal partner in a lichen may be unicellular or filamentous and either a green alga (e.g. *Trebouxia* sp.; *Trentepohlia* sp.) or a blue–green alga (e.g. *Gleocapsa* sp.; *Nostoc* sp.). In some lichens the algal cells are distributed evenly throughout the thallus but in most they are confined to a distinct algal layer within the thallus. Ascolichens can be divided into three morphological categories, crustose lichens, foliose lichens and fruticose lichens.

Crustose (encrusting) lichens are usually very thin and are so closely attached to their substrate that they cannot be removed without damage. They are usually pale coloured and look more than anything like circular or irregular paint marks on rocks, bark or leaf surfaces. In vertical section, the thallus has an upper **cortex** of compacted fungal hyphae (pseudo-parenchyma), an **algal layer** composed of algal cells or filaments loosely mixed with fungal hyphae and a **medulla** of loosely interwoven hyphae, some of which are attached to the substrate beneath.

Foliose lichens have dorsiventral thalli which are usually irregularly lobed and white, grey, green, orange or black. The thallus is attached to the substrate by thread-like **rhizines**. If the rhizines are cut or broken, the thallus comes away from the substrate without damage. In vertical section (Fig. 24.2), there is an **upper cortex**, an **algal layer** and a **medulla**, as in crustose forms, but there is also a compact **lower cortex**. Rhizines, which are single hyphae or groups of hyphae, arise from the lower cortex.

In foliose **gelatinous** lichens, the thallus is not layered and the filamentous blue–green algal component is mixed with fungal hyphae throughout. Gelatinous lichens, as the name suggests, have a jelly-like thallus. When moist, they are quite conspicuous and often brown. When dry, they shrink to a much smaller size, lose their shape and become very dark in colour.

Fruticose lichens are attached by rhizines at the base, but their branched thalli are cylindrical and

(a)

(b)

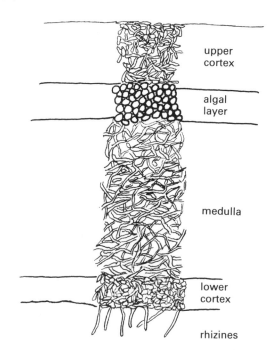

upper
cortex

algal
layer

medulla

lower
cortex

rhizines

24.2 Foliose lichens: (a) growing on a rock surface; (b) diagram of TS thallus (LP)

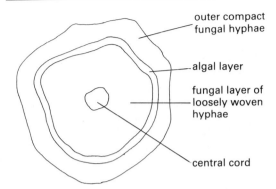

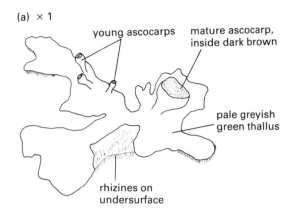

24.3 Diagram of TS thallus of *Usnea* sp., a fruticose lichen (VLP)

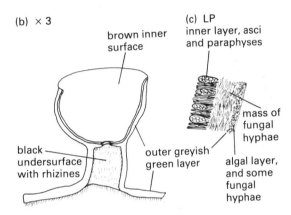

24.4 A mature foliose lichen, *Peltigera* sp.: (a) whole thallus; (b) VS thallus and apothecium; (c) details of apothecium

radially symmetrical. The surface layer is a compact cortex, inside which there is an algal layer, medulla and, in some (e.g. *Usnea* spp., Fig. 24.3), a tough central **cord**.

Asexual reproduction

The fungal element of some simple lichens produces conidia, which propagate the fungus alone, but most lichens produce isidia or soredia, which contain both fungus and alga.

Isidia are short, sometimes branched outgrowths of the surface of the thallus. They are hard to the touch and each has an outer layer of cortex-like tissue, enclosing an inner layer consisting of loosely packed hyphae and algal cells. Isidia are narrowed at the base and quite easily become detached.

Soredia form as eruptions of medulla and algal cells on the surface of the thallus and consist of a powdery mixture of hyphae and algae. Both isidia and soredia are able to produce new lichen thalli if they are transported to a suitable position.

Sexual reproduction

Typical ascomycete reproductive organs are formed within the lichen thallus and typical ascocarps, either perithecia or apothecia, are formed at the surface (Fig. 24.4). Paraphyses and asci line the ascocarp and ascospores are discharged into the air. The fruiting bodies of discomycete lichens are very conspicuous and are often coloured differently from the rest of the lichen.

Like conidia, ascospores propagate only the fungal element of the lichen and depend for their success on arriving at a place where the appropriate free-living alga is present. If the alga is present, the ascospores germinate and the germ tube immediately branches around cells of the alga and forms appresoria. Fungal hyphae grow and branch, algal cells divide and a new lichen thallus is gradually established.

Mycorrhizae

A **mycorrhiza** (pl. **mycorrhizae**) is a symbiotic association between a fungus and the root or rhizome of a higher plant. The concept is also extended to include similar symbiotic associations between fungi and the saprophytic, underground gametophytes of some pteridophytes (e.g. *Psilotum* sp.; *Ophioglossum* sp.) and the green, surface-growing thalli of many liverworts (e.g. *Marchantia* spp.). In most cases it is possible to show that both plant and fungus benefit from the association.

There are two fairly clear categories of mycorrhizae, ectotrophic and endotrophic, and a few intermediate forms.

Ectotrophic mycorrhizae

In **ectotrophic** mycorrhizae, the fungus forms a dense, pseudoparenchymatous **mantle** over the last few millimetres of the tips of some of the roots. The affected root tips stop growing, form no root cap or root hairs and their internal structure is modified to some extent.

The hyphae which form the mantle are part of a system of hyphae spreading through the soil. Some hyphae from the inside of the mantle penetrate a short way between the cells of the surface layers of the root. They do not enter the cells, but acquire organic nutrients (e.g. sugars) from the root. The plant partner also benefits and it has been shown that plant growth is greatly enhanced by the presence of mycorrhizae. Benefit to the plant is due to active absorption of mineral nutrients by the fungus from a very large volume of soil, and the fact that these nutrients are then made available to the plant.

The best known examples of plants with ectotrophic mycorrhizae are conifers (e.g. *Pinus* spp.) and some temperate dicot trees. The fungi concerned include basidiomycetes, some of which form fructifications on the woodland floor (e.g. *Boletus* spp.).

Endotrophic mycorrhizae

The fungal partner of an endotrophic mycorrhiza penetrates well into the cortex of the plant root, where it may form a characteristic band. The presence of the fungus usually causes no morphological changes and the root continues to grow. Much of the mycelium remains between the cells, but hyphae also enter the cells. At this stage, the fungus shows all the signs of being a parasite. At a later stage, the hyphae inside the cortical cells are digested by the plant. Mycorrhizal fungi include phycomycetes and septate fungi.

Under natural conditions, all orchids require an association with a mycorrhizal fungus for the germination of their seeds. At the time of dispersal, the embryo of an orchid seed is at a rudimentary stage of development and there are virtually no food reserves. No germination takes place in the absence of the fungus. When orchids are grown from seed commercially, seed cultures can be innoculated with the appropriate fungus, but it is now more usual to grow them on a sterile medium containing mineral and organic nutrients, including certain vitamins. Under natural conditions, the orchid must receive these organic nutrients from the fungus. As orchid seedlings develop their own photosynthetic apparatus, the need for the fungus becomes less.

For orchids and for many other plants, association with an endotrophic mycorrhizal fungus is obligatory. This certainly applies to totally saprophytic plants, including saprophytic orchids (e.g. some *Eulophia* spp.) and the non-green underground gametophytes of *Psilotum* spp. and *Ophioglossum* spp. In these cases, it is tempting to suggest that the plant 'traps' and subsequently parasitizes the fungus, which obtains nourishment saprophytically in the soil. It is possible, however, that the fungus obtains some vital nutritional requirement, which it cannot itself synthesise, from the plant. It has also been shown that a single mycelium of a mycorrhizal fungus may form mycorrhiza on an autotrophic plant and a saprophytic plant at the same time. Under these circumstances, organic compounds synthesized in the green plant pass through the fungal mycelium in the soil and into the saprophyte.

There is still much to be learned about the exact role of mycorrhizal fungi and, indeed, about the true status of so-called 'saprophytic' plants.

Root nodules

Root nodules are variously shaped, irregular outgrowths on the roots of plants (Fig. 24.5), induced by the presence of a bacterium or a fungus. The bacterium or fungus usually has the ability to fix atmospheric nitrogen in a form which can be assimilated by the plant. The commonest and best known examples are associations between the roots of legumes (Leguminosae, especially Papilionaceae and Mimosaceae) and bacteria belonging to the genus *Rhizobium*.

Rhizobium spp. are present in the soil, where they are relatively inactive and are unable to fix atmospheric nitrogen. In the neighbourhood of roots of an appropriate partner, they produce auxin (IAA), which causes root hairs to grow towards them. Bacteria enter through the tip of a root hair and,

24.5 Root nodules (arrowed) on a young plant of *Mucuna* sp.

once inside, proliferate to form an **infective thread**. The infective thread is a thread of bacteria, several bacteria wide, held together by a gummy secretion. The bacterial thread is always enclosed in an ingrowth of the plasmalemma of the host cell and, at first, has a layer of cellulose wall material laid down around it. The infective thread grows along the root hair and then penetrates from cell to cell into the cortex. Sooner or later, the bacteria at the end of the thread become separated within the protoplast of a cell. The bacteria change their form and become **bacteroids**, which are irregularly branched into various shapes. Each bacteroid is enclosed in a double layer of membrane produced by the plant cell. The infected cell enlarges and starts to divide repeatedly. Neighbouring cells also divide repeat-

edly. Rapid division of cells produces the nodule, which bulges from the side of the root. The **nodule** consists of a mass of large central cells containing bacteroids enclosed in a layer of compact parenchyma in which strands of vascular tissue differentiate, connecting with the vascular tissue of the stele of the root.

The root nodule has a short life. After a time, the bacteroids and infected cells are digested. At the same time some normal bacteria are freed into the soil, where they may infect younger parts of roots.

While nodules are active, it is through the combined enzyme systems of bacterium and legume that atmospheric nitrogen is fixed and amino acids are produced. The fixation process provides all the nitrogen compounds required by the plant and a considerable surplus of fixed nitrogen is also released into the soil.

There are many species and varieties of *Rhizobium* and the correct variety must be present in the soil if a particular species of legume is to benefit. When introducing a new species of legume crop into an area, it is now usual to innoculate the soil with the correct variety of *Rhizobium* at the same time, to ensure immediate nodulation.

Other symbiotic relationships

There are several examples of apparently mutualistic symbiotic associations between plants and nitrogen-fixing blue–green algae. These associations have not all been researched very thoroughly and in some cases the mutualistic nature of the relationship is presumed rather than proved. It is assumed that, as in a lichen, the blue–green alga benefits from the protection it receives. It is also assumed that the plant benefits from the nitrogen compounds produced. In some cases this has been proved.

The blue-green algae involved are *Nostoc* spp. and *Anabaena* spp. Plant partners range from hornworts (e.g. *Anthoceros* spp.), liverworts (e.g. *Blasia* spp.), water ferns (e.g. *Azolla* spp.) and cycads (e.g. *Encephalartos* spp.) to flowering plants (e.g. *Gunnera* spp.). All have morphological adaptations which provide a space in which the blue–green alga can grow. In *Anthoceros* spp. there are mucilage cavities (*Nostoc* cavities) in the thallus. In *Azolla* spp. the upper lobe of each leaf contains a cavity which is always invaded by *Anabaena* sp. In cycads there are

24.6 Symbiosis between a gymnosperm, *Cycas* sp. and the blue-green alga *Anabaena* sp. The much-branched tips of negatively geotropic roots ('coralloid roots') lie very close to the surface of the ground and contain a band of *Anabaena* sp. in the cortex

coralloid roots which form at the surface of the ground and have a zone of *Anabaena* sp. in the cortex (Fig. 24.6).

Key words

Symbiosis: mutualism, commensalism, parasitism. **Mutualism**: lichen, mycorrhiza, root nodule. **Lichen**: fungus, ascomycete, basidiomycete; ascolichen, basidiolichen; green alga, blue–green alga. **Morphology**: crustose, foliose, fruticose. **Structure**: cortex, algal layer, medulla, rhizine; cord. **Asexual reproduction**: conidia, isidia, soredia. **Sexual reproduction**: ascocarp; perithecium, apothecium. **Mycorrhiza**: ectotrophic, endotrophic. **Root nodule**: fungus, bacterium; *Rhizobium* spp., infective thread, bacteroid.

Summary

Mutualistic symbiosis between two organisms benefits both partners in the association. Examples of mutualistic symbiosis include lichens, mycorrhizae, root nodules and certain associations between blue–green algae and higher plants. In lichens, the association is between a fungus and a green or blue-green alga. A mycorrhiza is an association between a fungus and a usually underground part of a eucaryotic plant. Root nodules are formed by an association between a fungus or a bacterium, usually a bacterium, and the roots of a higher plant. In some instances, the basis of mutual benefit for partners in a symbiotic association is clear, in other instances it is not.

Questions

1 Describe the structure of a foliose ascolichen. Use this as an example, to explain what is meant by 'mutualistic symbiosis'.
2 Use different examples from divisions of the plant kingdom to explain what is meant by 'mycorrhiza'.
3 In what way are root nodules beneficial (a) to the plant on which they occur, and (b) to other plants which grow nearby? If you were planting a leguminous crop for the first time, how would you ensure good nodulation?

Further Reading for Section D

Frobisher, M., Hinsdill, R. H., Crabtree, K. T. and Goodheart, C. R. 1974. *Fundamentals of Microbiology*, Saunders: Philadelphia.
Hale, M. E. 1980. *How to Know the Lichens*, 2nd edn, Wm C. Brown: Dubuque, Iowa.
Ingold, C. T. 1984. *The Biology of Fungi*, 5th edn, Hutchinson: London.
Scott, G. D. 1969. *Plant Symbiosis*, Edward Arnold: London.
Webster, J. 1980. *Introduction to Fungi*, 2nd edn, Cambridge U.P.: Cambridge.
Zoberi, M. H. 1972. *Tropical Macrofungi*, Macmillan: London.

Section E

GENETICS AND EVOLUTION

Genetics is the study of inheritance. An understanding of genetical processes depends to a great extent on an understanding of the behaviour of chromosomes. Chromosomes contain the fundamental genetical material, DNA, and pass it on in an orderly manner when cells divide or when organisms reproduce sexually. An understanding of chromosome cytology and genetics is also vital to an understanding of evolutionary processes, especially at the level of the origin of new species.

25

Reproduction and cell division

Growth and asexual reproduction depend on the division of cells. In eucaryotic organisms, this type of cell division follows the type of nuclear division known as mitosis. **Mitosis** ensures that both cells produced by cell division contain exactly the same genetic material as each other, and the same as the parent cell.

Sexual reproduction depends on two nuclear processes. The first of these is the fusion of gametes (**syngamy, fertilization**). When syngamy occurs, the nuclei of two cells fuse and the amount of genetic material contained in the zygote nucleus is twice that contained in each of the gametes. At some stage, the life cycle also includes the type of nuclear division known as meiosis. In **meiosis**, the genetic content of the cell is halved in a very precise way.

Both mitosis and meiosis are almost always followed by **cytokinesis**, which is the division of the protoplast of the cell.

Chromosomes

The genetic material of a eucaryotic cell is organized into a number of threads (**chromosomes**) inside the nuclear membrane. Chromosomes are made up of DNA, the fundamental genetic material, associated in some way with basic proteins (**histones**). When a cell is most active in a metabolic sense and its nucleus is not in the process of dividing (at **interphase**), the chromosome threads are very long and thin. They are so thin that they cannot be seen with

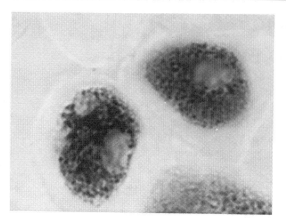

25.1 Mitosis in the root tip of *Chlorophytum* sp. Interphase (HP), at which most of each nucleus is dark- stained, indicating the presence of DNA. The pale patches, one or two per cell, are nucleoli

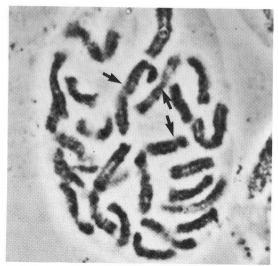

25.2 Chromosome morphology in *Chlorophytum* sp. (HP). This cell has been treated with 1,4-dichlorobenzene solution before fixation, to cause the chromosomes to spread apart. Variation in chromosome length and the positions of some centromeres (arrowed) are easy to see

a light microscope. If we treat an interphase nucleus with a stain that gives a colour reaction only with DNA, the whole of the material within the nuclear envelope, except the nucleolus, takes up the stain (Fig. 25.1). The nucleolus does not stain because it contains only RNA and protein. The colour may be evenly distributed throughout the nucleus, apart from the nucleolus, or there may be stained granules, or bars of stained material (**chromocentres**) within a generally more lightly stained background. These granules and chromocentres are parts of chromosomes which have twisted into a spiral (**spiralized**) until they are thick enough to be visible under the light microscope.

When a nucleus is about to divide, all of the chromosome threads start to spiralize, so that they become visible under the light microscope as a number of continuous threads. If a stain for DNA is used at this stage, all of the staining material is found to be incorporated in the chromosomes.

At the beginning of nuclear division, each chromosome consists of two parallel threads (**chromatids**), joined at a narrow point along the chromosome (**centromere, primary constriction**) where the chromosome is not divided into chromatids. There is usually one chromosome in each set which has a secondary constriction (**nucleolar constriction**). A secondary constriction is a narrow part in each chromatid. Chromatids are not united at a secondary constriction.

If the morphology of a chromosome is sufficiently different from that of others in the same cell, it is possible to recognize corresponding (**homologous**) chromosomes in the same or different cells. The overall length of a chromosome compared with others in the same cell, the relative lengths of parts of the chromosome (chromosome **arms**) at each side of the centromere and the position, presence or absence of a secondary constriction all make up the morphology of a chromosome (Fig. 25.2).

Chromosome number

Every eucaryotic organism has a basic number of chromosomes which is characteristic for the species. This is the **haploid** (**gametic**, '*n*') chromosome number. When two gametes fuse, the zygote nucleus contains two homologous sets of chromosomes. One is a haploid set from the female parent (**maternal** chromosomes) and the other is a haploid set from the male parent (**paternal** chromosomes). Together, they make up the **diploid** chromosome complement ($2n$). Each maternal chromosome has a corresponding chromosome, of identical morphology and sequence of genetic material, in the paternal set.

Chromosome numbers of species of plants vary a great deal, from $n = 3$ in a member of the family Compositae to $n = 501$ in a species of the fern genus, *Ophioglossum*.

Mitosis and cytokinesis

The process of mitosis is a continuous one, running smoothly from start to finish. For convenience, we divide the whole process into five phases, preceded by interphase and ending when the two daughter nuclei enter interphase once more (Figs. 25.3–25.8).

When a nucleus is about to divide, it usually becomes slightly enlarged (Fig. 25.3). At this time, the interphase nucleus contains one, sometimes two, nucleoli within the nuclear envelope. Otherwise, it appears structureless or contains some staining granules or chromocentres. Interphase ends when the chromosomes become visible as double threads.

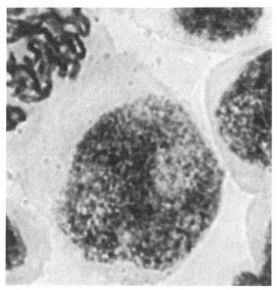

25.3 Mitosis in the root tip of *Chlorophytum* sp. (HP). This interphase nucleus has enlarged and is about to enter prophase

Prophase starts when the chromosomes become visible (Fig. 25.4). The nuclear envelope and nucleoli remain intact as the chromosomes become shorter and thicker. Prophase ends when the nuclear envelope and nucleoli disintegrate.

Prometaphase begins at this point. As the nuclear envelope disappears, protein fibres (**spindle fibres**) appear in the neighbourhood of the chromosomes. Spindle fibres connect each centromere to each of two points in the cytoplasm (**poles**). A line connecting the two poles would be at right angles to the plane of the cell division which is to follow. Each centromere is connected to both poles by spindle fibres (**attached** spindle fibres). There are also spindle fibres which run from one pole to the other, without being connected to chromosomes (**continuous** fibres). At this time, the chromosomes are scattered near the centre of the cell (Fig. 25.5), where they were left when the nuclear envelope broke down. The spindle fibres then start to contract and chromosomes are pulled, by their centromeres, towards a point midway between the poles of the

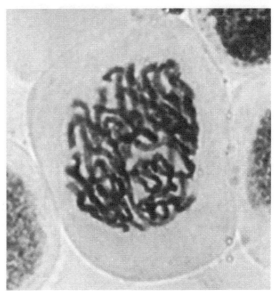

25.4 Mitosis in the root tip of *Chlorophytum* sp. (HP). A nucleus at late prophase

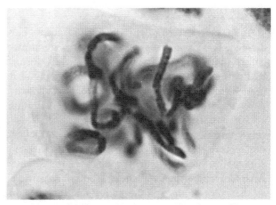

25.5 Mitosis in the root tip of *Chlorophytum* sp. (HP). A nucleus at prometaphase

spindle (the **equator** of the spindle). Prometaphase ends when the centromeres reach the equator of the spindle.

At **metaphase**, the centromeres lie at the equator of the spindle, arranged within an imaginary circle (**metaphase plate**) (Fig. 25.6), but the arms of the chromosomes may lie in any positions, spreading out from their centromeres. Metaphase ends when all the centromeres divide longitudinally, all at the same time.

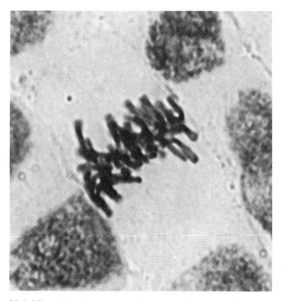

25.6 Mitosis in the root tip of *Chlorophytum* sp. (HP). A side-view of metaphase

Anaphase starts when the centromeres have divided. It is clear that the spindle fibres are already contracting, because the new centromeres, formed by division of the original centromeres, spring apart immediately. Each of the former chromatids now has its own centromere and has become, by definition, a chromosome. Each of these chromosomes is a single thread. Spindle fibres continue to contract and the centromeres, followed by the chromosome arms (Fig. 25.7), are drawn to the poles of the spindle. Anaphase ends when the chromosomes are gathered at the poles of the spindle and a nuclear envelope starts to form around each group of chromosomes. Nucleoli start to form at the same time.

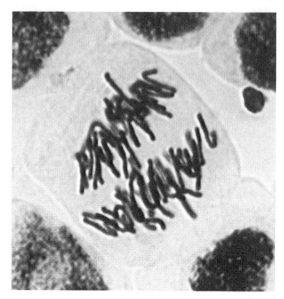

25.7 Mitosis in the root tip of *Chlorophytum* sp. (HP). A side-view of anaphase

During **telophase** (Fig. 25.8), nucleoli continue to grow and the chromosomes despiralize. Nucleoli are formed at the nucleolar constrictions of chromosomes. There is normally one chromosome with a nucleolar constriction in every haploid set of chromo-

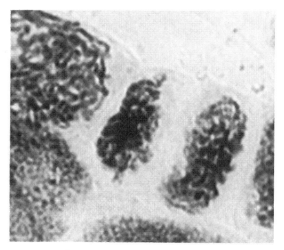

25.8 Mitosis in the root tip of *Chlorophytum* sp. (HP). A side-view of telophase

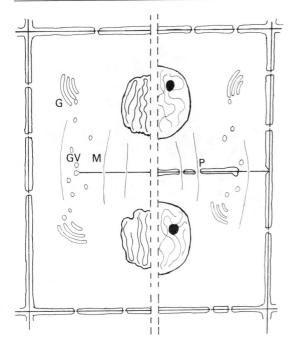

25.9 Cytokinesis following mitosis. A diagram showing the formation of a cell plate (left) and a new cell wall (right) between two nuclei. During telophase (left), the cell plate forms from Golgi vesicles which coalesce along the former equator of the mitotic spindle. Microtubules, of the same structure as spindle fibres, seem to guide the vesicles to the equator. Later, as the nucleus is passing from telophase to interphase (right), the cell plate has passed through the original cell wall at the sides of the cell and united with the middle lamella between the divided cell and its neighbours. Golgi vesicles are depositing cell wall material on both sides of the cell plate. The new cell wall will soon unite with the cell wall material of the parent cell. G = Golgi body; GV = Golgi vesicle; M = microtubule; P = plasmodesma

somes. A haploid nucleus, therefore, forms one nucleolus. A diploid nucleus forms two nucleoli, but these may unite at any stage to form a single, larger, nucleolus. Telophase ends when the chromosomes can no longer be seen under the light microscope. The two new nuclei are then at interphase.

The above are the stages of mitosis and it is clear that they have been defined on the basis of observations made with the light microscope. In fact, electron microscope studies have shown that the prophase spiralization of chromosomes begins before it can be observed with the light microscope and that telophase despiralization ends some time after it

appears to have ended, when observed with the light microscope. Thus, the definitions of some of the stages of mitosis are rather arbitrary, depending as they do on the resolving power of the light microscope.

The prometaphase stage is sometimes omitted from accounts of mitosis. It is then regarded either as the final part of prophase or as a transitional stage between prophase and metaphase.

Chromosomes enter mitosis as double threads and leave mitosis as single threads. Measurements of the amount of DNA in nuclei show that the prophase nucleus contains twice as much DNA as each telophase nucleus. The lower DNA content persists for a time into the interphase following mitosis, after which the DNA content doubles over quite a short period (the **S-phase** of interphase), as DNA replicates.

Cytokinesis in plant cells is by **cell plate formation** (Fig. 25.9). At the beginning of telophase, vesicles containing calcium pectate separate from Golgi bodies and gather at the centre of the equator of the spindle, where they coalesce to form a cell plate. The cell plate gradually widens until it passes through the side walls of the cells, involving enzymatic dissolution of part of the cell wall, and unites at the edges with the middle lamella between the dividing cell and its neighbours. Vesicles of hemicellulose then form from the Golgi bodies and a layer of cellulose is deposited over each surface of the cell plate and becomes united with the side walls of the original cell. Cell division is then complete.

Meiosis

In animals, meiosis takes place during the formation of gametes. The same is true of the few plants which have a diplontic life cycle (e.g. some algae). In nearly all plants, including flowering plants, meiosis takes place during the formation of spores. For instance, in flowering plants, meiosis takes place in anthers, in the formation of pollen grains (microspores) (Fig. 25.10) and in the megasporangium of the ovule, in the formation of the megaspore (Fig. 14.10).

Meiosis comprises two nuclear divisions, accompanied by or followed by two cell divisions. It always results in the formation of four nuclei, but some of the nuclei produced may degenerate, as in the formation of the megaspore in flowering plants.

(a)

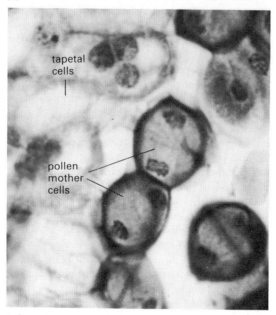

tapetal
cells

pollen
mother
cells

(b)

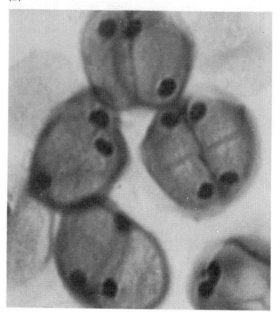

25.10 Meiosis in pollen mother cells (HP): (a) the end of
first division, the two-celled stage; (b) at the end of second
division, the four-celled stage

The first division of meiosis (Fig. 25.10) is more complicated than the second division. The complications take place in the prophase of the first division. Prophase is traditionally divided into several named stages, based on observations with the light microscope.

Prophase of the first division of meiosis starts when the chromosomes become visible due to spiralization. At this stage, **leptotene**, the chromosome threads appear to be single, not divided into chromatids, and are separate from each other.

The second stage, **zygotene**, starts as homologous chromosomes of the diploid complement pair with each other. Pairing (synapsis) starts at certain points along the lengths of homologous chromosomes and continues until they are closely paired along their whole lengths. Spiralization continues and the chromosomes gradually become shorter and thicker. Each pair of homologous chromosomes is now called a **bivalent**.

At the next stage, **pachytene**, homologous chromosomes are closely paired and shortened and each chromosome can now be seen to be composed of two chromatids.

As **diplotene** begins, the chromosomes of each bivalent start to repel each other. They do not, however, separate, because chromatids of homologous chromosomes have exchanged parts at certain points along their lengths and they are prevented from separating completely because of these **chiasmata** (Fig. 25.12). Spiralization continues and homologous chromosomes repel each other more and more strongly.

The last stage of the first prophase is **diakinesis** (Fig. 25.13), when shortening of chromosomes and repulsion between homologous chromosomes of each bivalent reaches its maximum.

Before **prometaphase** of the first division, the nuclear envelope and nucleoli disintegrate. Spindle fibres form and each centromere receives a single spindle fibre. Of the chromosomes of each bivalent, one becomes connected to one pole of the spindle and the other is connected to the other pole of the spindle.

Up to this stage, the chromosomes are held together in bivalents because of the chiasmata and because the chromatids of each chromosome are twined around each other. The bivalents are pulled to the equator of the spindle.

At **metaphase**, the bivalents are at the equator of the spindle.

Centromeres do not divide. **Anaphase** begins when the constituent chromosomes of each bivalent are pulled apart by the spindle fibres. Each chromosome still has two chromatids, connected by a single centromere, as it is drawn to a pole of the spindle.

a diploid cell with four chromosomes

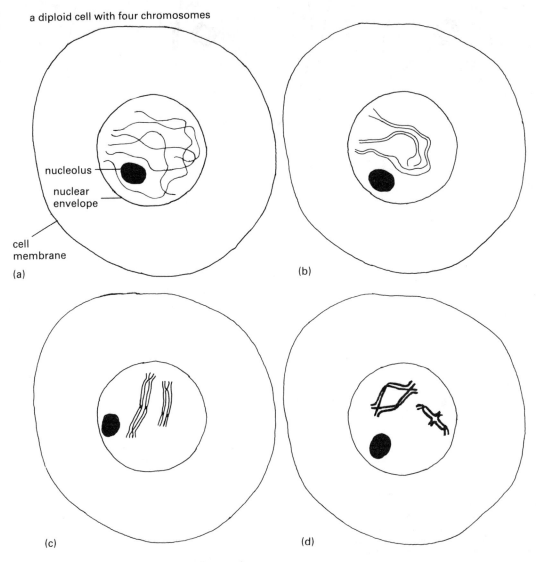

25.11 Diagrams of the stages of the first division of meiosis: (a) leptotene, chromosomes appear as single threads; (b) zygotene, chromosomes coming together in homologous pairs (bivalents); (c) pachytene, chromosomes now visibly double (two chromatids each); (d) diplotene, homologous chromosomes repel each other but are held together at chiasmata (continued overleaf)

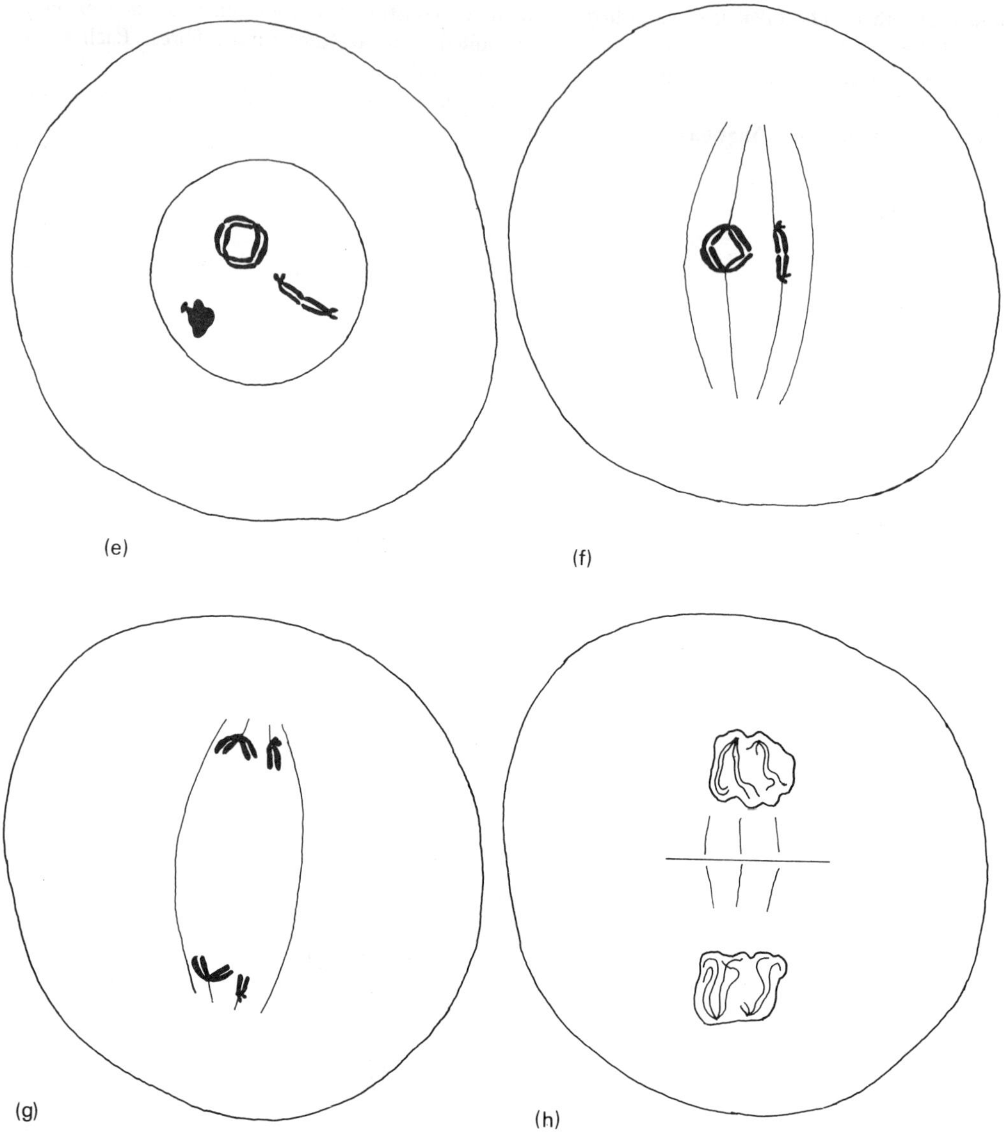

(e)

(f)

(g)

(h)

25.11 (continued) (e) diakinesis, chromosomes
shortened to the maximum degree, chiasmata (usually)
terminalized; (f) metaphase I, spindle formed and
chromosome bivalents at the equator; (g) anaphase I,
chromosomes moving to the poles of the spindle;
(h) telophase I, new nuclei being reconstituted

Telophase follows as nuclear envelopes are
formed. Cytokinesis may or may not take place at
this time, as the nuclei enter interphase.

Interphase between the two divisions of meiosis
is usually brief and does not include an S-phase.

As the second division of meiosis begins, the
chromosomes again become visible due to spializ-
ation. The second division goes through five phases
(prophase, prometaphase, metaphase, anaphase,
telophase) exactly as in mitosis, without further

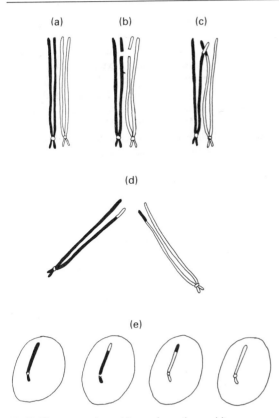

(a)   (b)   (c)

(d)

(e)

25.12 Diagram to show chiasma formation and its consequences. The events shown are very much simplified. (a) At zygotene, homologous chromosomes become closely paired. (b) At pachytene (perhaps earlier), one of each pair of chromatids breaks, both at exactly corresponding points. (c) Immediately, there is a union of the broken ends of non-sister chromatids. (d) At first anaphase of meiosis, homologous chromosomes separate towards opposite poles of the spindle, having exchanged parts. (e) The same chromosomes in separate nuclei at the end of the second division of meiosis. As a result of the exchange of material which has taken place, each of the four nuclei contains a different combination of genetic material. In fact, there would have been several chiasmata between such chromosomes, probably involving both chromatids of each chromosome, so more exchange would have taken place than is shown in this diagram

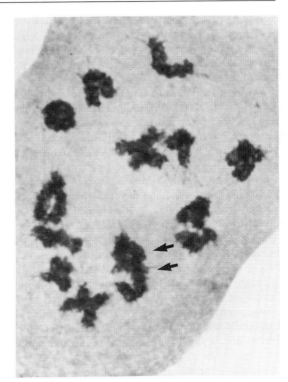

25.13 Meiosis in a pollen mother cell of *Chlorophytum* sp. (HP). Diakinesis of the first division of meiosis, at which fourteen bivalents are seen. Two of the bivalents (arrowed) are overlapping slightly

complication. At the end of the second division of meiosis, there are four haploid nuclei and cytokinesis is completed (Fig. 25.10(b)).

The process described above is as observed with the light microscope. Electron microscope observations have added a few details. In the first place, it is now known that chromosomes are divided into chromatids and that homologous chromosomes have started to pair, at isolated points, before they are visible with the light microscope. In addition to this, before pairing, each chromosome is associated with other material, mostly protein, along its entire length. This material is part of the **synaptonemal complex**. As bivalents are formed, it is these synaptonemal complexes that fuse and hold the homologous chromosomes together. It is probable that chiasmata are formed as the synaptonemal complexes fuse. The remaining information gained from electron microscope observations is that the beginning of diplotene, when homologous chromosomes start to repel each other, coincides with the disppearance of the synaptonemal complex.

The significance of meiosis

Meiosis is complementary to syngamy. Syngamy doubles the number of chromosomes per cell and meiosis halves the number again, in an orderly way. The chromosome number of a species is thus kept constant through the generations. This is, of course, very important. It is not, however, the fundamentally important point.

The most significant feature of meiosis, and probably the very reason for the existence of sexual reproduction, is the formation of new combinations of genetic material which is part of the process.

First, there is the exchange of genetic material between homologous maternal and paternal chromosomes which results from chiasma formation during the first prophase of meiosis. There is almost always some variation in the genetic material possessed by different individuals of a species. Therefore, homologous maternal and paternal chromosomes almost always differ to some extent in the base sequences along their lengths. When chromatids exchange parts at chiasma formation, new combinations of genetic material are formed. Such new combinations are subsequently incorporated in new individuals and contribute to the variability of the species and, therefore, to the ability of the species to adapt to new or changing conditions.

Secondly, there is variability resulting from the random assortment of maternal and paternal chromosomes at first anaphase. A diploid cell contains two homologous sets of chromosomes, one maternal and the other paternal. After these have paired during the first prophase of meiosis, the homologous chromosomes pass randomly to one pole or the other of the spindle. Thus, each pole receives a random mixture of what were the maternal and paternal chromosomes of the diploid cell and, once again, new combinations of genetic material are formed.

Key words

Cell division: mitosis, meiosis, cytokinesis. **Eucaryotic cell**: nucleus, nuclear envelope; nucleolus, RNA; chromosomes, DNA. **Chromosome**: centromere (primary constriction), chromatids; secondary constriction; maternal, paternal, homologous. **Chromosome number**: haploid (*n*), diploid (*2n*). **Mitosis**: growth, asexual reproduction; interphase, prophase, prometaphase, metaphase, anaphase, telophase; spindle fibre. **Sexual reproduction**: meiosis, syngamy (fertilization). **Meiosis**: first division, first prophase, leptotene, zygotene, pachytene, diplotene, diakinesis; synapsis, bivalent, chiasma formation. **EM**: synaptonemal complex. **Cytokinesis**: cell plate formation. **Significance of meiosis**: random assortment, exchange of genetic material.

Summary

Mitosis is a type of nuclear division which, followed by cytokinesis, produces two cells which are identical in the genetic material which they contain. It results in either growth or asexual reproduction. Both haploid and diploid nuclei may divide by mitosis. Sexual reproduction involves syngamy, which results in the formation of a diploid zygote, and meiosis, by which the diploid number is reduced to the haploid number. In organisms with a diplontic life cycle, meiosis leads to gamete formation. In most plants, with a haplontic or haplodiplontic life cycle, meiosis is followed by spore formation. During the process of meiosis, homologous maternal and paternal chromosomes exchange parts by chiasma formation and are also randomly assorted in the haploid cells which are produced.

Questions

1 In what ways do cells produced following mitosis differ from cells produced following meiosis?

2 What are homologous chromosomes? How do they behave in meiosis and why is this important in relation to the survival of a species?

3 List the differences between the first prophase of meiosis and prophase of mitosis.

4 Where, in a flowering plant, would you expect to find (a) meiosis, (b) mitosis in diploid cells and, (c) mitosis in haploid cells.

26

Genetics

It is impossible to treat genetics as purely a study of inheritance in plants. It is strictly a biological subject. Studies of plants, animals, fungi and bacteria have all played a part in the development of our knowledge of the subject. However, genetics is also a very large subject. In introducing the subject here, we shall concentrate our attention on plants as far as possible but introduce animal genetics, especially human genetics, whenever it is necessary or especially desirable.

For centuries, men bred plants and animals without any exact knowledge of the complex mechanisms of inheritance. Until the beginning of this century, most breeding was based on the belief that the best offspring (**progeny**) were obtained by breeding from the best parent stock. In essence, this is true, but much greater knowledge is needed to obtain really predictable results. It was the work of the Austrian monk, Gregor Mendel, that laid the foundations of modern genetics.

In addition to being a monk, Mendel was a mathematician. He was in charge of the monastery garden and used the garden for experiments in genetics. His experiments were carried out over a period of eight years, full records were made and the observations subjected to mathematical treatment. Mendel chose to do his main work on *Pisum sativum* (Garden Pea) for several good reasons. First, he was able to obtain **pure lines** from seed suppliers. These were carefully selected seeds and could be relied upon to produce progeny exactly like the parent stock (to **breed true**). Even then, Mendel grew his chosen varieties for two years, so that he could confirm that they were true breeding, before he started his crossing experiments. Secondly, the varieties he was able to obtain included several pairs of contrasting characteristics for which there were no intermediate forms. Thirdly, if left to themselves, they could be relied upon to pollinate themselves before the flowers opened fully. Lastly, the flowers were large enough to enable him to remove stamens from flower buds, as part of the process of artificial cross-pollination.

Mendel's work was published in 1886, in a respectable Austrian scientific journal. The work went unrecognized at the time. Mendel subsequently became head of his monastery, and died before the world was ready to understand the significance of what he had discovered. It was in 1901, fifteen years after publication, that three prominent geneticists, in different countries, 'discovered' Mendel's work and drew it to the attention of others.

Mendel's peas suited his purpose well. His work on peas has been repeated and added to, but most work has been done on other organisms. The fruit fly, *Drosophila melanogaster*, has proved especially useful and has the advantages that many organisms can be bred in a small space and produce large numbers of progeny in a short period of time. Fungi have also been used, with the same advantages. Much modern work of a very fundamental nature is carried out with bacteria.

Human genetics has also advanced a great deal. Here, one cannot carry out experiments. One must rely instead on the study of pedigrees. A **pedigree** (Fig. 26.1) is a family history in which the occurrence of certain hereditary characteristics (**traits**) is recorded over a period of several generations. They are less reliable than experiments, for various reasons, but have provided a great deal of information.

Variation

Variation between individuals is of two types. First, there is **discontinuous variation**. With discontinuous variation, a certain character has two or more alternative forms, with no intermediates. The traits that Mendel chose to study were of this type. For example, his experimental peas had flowers that were either red or white, with no intermediate forms. The second type is **continuous variation**. Here, a characteristic varies continuously between two extremes. An example is height, or weight, in man. The study of characteristics which show continuous variation provided a great deal of difficulty to early geneticists.

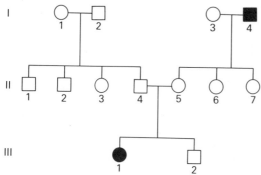

26.1 A diagram of a human pedigree. Circles represent females, squares represent males. A horizontal line between a square and a circle represents a marriage. Children from a marriage are connected to that line, on a horizontal line below in order of age, with the oldest child on the left. Each generation is numbered with a Roman numeral on the left and individuals of a generation are numbered from left to right, for ease of reference. Individual symbols which are shaded indicate the appearance of the trait being studied. Thus, in this pedigree, individuals I–4 and III–1 show the particular trait (e.g. albinism)

Genotype and phenotype

The **genotype** of an individual is what that individual inherits in the genetic material passed on to it by its parent. Sometimes, as in the case of flower colour, the genotype is always expressed in the observable characteristics (**phenotype**) of the individual. In other cases, the genotype interacts with the environment and may or may not appear in the phenotype. For instance, a man may inherit from his parents a tendency to be fat. This is his genotype. However, if he receives too little food, he cannot become fat. His phenotype is thin. In this instance, the environmental factor which modifies the phenotype is food supply.

The earliest (1903) scientific demonstration of the related effects of genotype and environment in producing the phenotype came from the work of the Dutch geneticist, W. Johannsen, on *Phaseolus vulgaris* (Kidney Bean). Like Mendel, he made use of pure lines which could be obtained from seedsmen. Johannsen studied seed size, determined by weighing the seeds. He found that each pure line had a constant average seed size, but that there was quite a large range of continuous variation in seed size within each pure line. By making separate plantings of the largest beans and the smallest beans from

a single pure line, he found that the progeny of both had the same average seed size and the same range of variation as the parent line. Therefore, they had the same genotype. Variation was due to the position which the bean occupied in the pod in which it was formed. The largest beans came from the centre of the pod. The smallest beans came from the tapering ends of the pod.

The special importance of Johannsen's work was that his data were expressed in exact quantities and subjected to statistical analysis.

Genes

The genetic material of all cellular organisms is DNA. DNA is located in chromosomes. A certain sequence of nitrogen bases in the DNA of a chromosome leads to the synthesis of a polypeptide. The polypeptide may, by itself, be an enzyme or structural protein, or it may join with other polypeptides or other substances to form an enzyme or a structural protein. In either case, the sequence of nitrogenous bases in the DNA, which we call the **gene**, is at least partly responsible for producing an observable character of the organism in which it is contained.

Each gene is located at a particular position (**locus**) on a chromosome. Therefore, a haploid organism contains a single locus for each gene. A diploid organism has two loci for the same gene, on separate but homologous chromosomes.

Alleles

If the sequence of nitrogenous bases in a gene is altered, this usually changes the sequence of amino acids in the polypeptide produced. Such a change (**gene mutation**) alters the properties of the gene. The two different forms of the gene, the original form and the mutated form, are **alleles** of the gene. The mutated form may be totally inactive or its presence may result in the formation of a different end product from that of the original form of the gene. If there is a different end product, it will produce a different observable characteristic. In diploid organisms, however, the situation is complicated by the presence of two loci for each gene. An individual may carry the same allele at both loci (may be **homozygous**) or may carry one allele at one locus and another allele at the second locus (may be **heterozygous**).

Dominance

When two different alleles are present on homologous chromosomes in a single diploid individual, the phenotype of the individual depends on the **dominance relationship** between them.

In the case of complete dominance, the phenotype of the heterozygous individual is indistinguishable from that of an individual which is homozygous for one of the two alleles. We say that the allele which is expressed in the phenotype is **dominant** to the other. The allele which is not expressed is **recessive**.

One of the many conventions of genetics is the use of letters of the alphabet to represent genes. The capital letter is used to represent a dominant character and the corresponding small letter is used to represent the recessive allele. The genotype of a diploid organism, with respect to a single pair of allelic characters, is represented by two letters (e.g. AA, Aa or aa). The genotype of a gamete, which is haploid, is represented by a single letter (A or a). We can also use a single letter to represent the phenotype of a diploid individual. The use of letters to represent alleles is shown in Table 26.1.

When an investigation is made of the inheritance of a single pair of alternative characters, we may start by crossing two individuals which are known to be true-breeding for, and therefore homozygous for,

Table 26.1 The use of letters to represent allelic genes when writing genotypes and phenotypes. In this table, the choice of A and a for the dominant and recessive genes respectively is arbitrary. When representing genotypes and phenotypes of specific characters, it is more usual to use initial letters of words which describe the dominant character, e.g. T for tall and t for the recessive allele (short); R for red flowers and r for the recessive allele (white flowers).

Genotype	Representation of genotype	Representation of phenotype
Haploid (gametes) with dominant gene	A	—
with recessive gene	a	—
Diploid homozygous dominant	AA	A
heterozygous	Aa	A
homozygous recessive	aa	a

each of the alternative characters. This, in Mendel's terminology, is a **monohybrid cross**. The two plants with which we start are the **parental (P1)** generation. The progeny obtained from the cross, all of which will have the dominant phenotype, are the **first filial (F1)** generation. If we then self-pollinate flowers of the F1 generation, or make crosses between members of the F1 generation, the progeny are the **second filial (F2)** generation.

Assuming complete dominance, we may predict the phenotypes of the F2 progeny as shown in Table 26.2.

We can now make use of the simple device known as a 'checkerboard' to work out the expected phenotypes in the F2. In this case, we are self-pollinating (**selfing**) plants, both of which have the genotype Aa. Each produces two sorts of gametes (A and a) in equal numbers, so we draw a checkerboard with two spaces across and two spaces down. We write the genotypes of the gametes of one parent at the top and the gametes of the other parent down one side. We then fill in the four squares with letters which tell us the expected genotypes of the F2 generation, as shown in Table 26.3.

Taking the genotypes from the squares, we can now state the ratio of genotypes expected.

$$AA : Aa : aa = 1 : 2 : 1.$$

Since A is completely dominant over a, the ratio of phenotypes is:

$$A : a = 3 : 1.$$

This 3 : 1 phenotype ratio is characteristic of a monohybrid cross in which one character is totally dominant over the alternative character. It was first found by Mendel in his experiments on *Pisum sativum* and has since been found to be true for many other pairs of alternative characteristics in plants, animals and man.

In analysing the result of an experiment such as that described above, there is a problem in finding out the genotypes of individuals showing the dominant character. They may be homozygous or heterozygous. To solve this problem, Mendel invented the **test cross**. To make a test cross, plants of doubtful genotype (e.g. AA and Aa) are crossed with a plant known to be homozygous for the recessive character (aa). The ratio of phenotypes found in the T1 generation is a direct indication of the ratio of genotypes of the gametes of the plant under inves-

Table 26.2 Prediction of genotype and phenotype proportions in the F2 generation.

Genotypes of P1 generation	AA			aa	
Gametes of P1	A			a	Each parent produces only one type of gamete, A or a.
Genotype of F1	Aa			Aa	Following union of gametes, each containing A or a.
Gametes of F1	A	a	A	a	Each parent produces two types of gametes, A and a in the ratio 1:1.
Genotypes of F2	AA	Aa	Aa	aa	Random union of gametes from each parent. AA : Aa : aa = 1 : 2 : 1.
Phenotypes of F2	A	A	A	a	F2 phenotype ratio A : a = 3 : 1

Table 26.3 The use of a checkerboard to predict F2 progeny proportions. Each F1 plant (*Aa*) produces gametes with *A* and *a* in equal numbers. The gamete genotypes are written across the top and down the left-hand side of the checkerboard.

gametes	A	a
A	AA	Aa
a	Aa	aa

Table 26.4 The use of the test cross. When there is complete dominance, the homozygous dominant (*AA*) and heterozygous (*Aa*) progeny have identical phenotypes. If test crosses are made (each individual under test crossed with the homozygous recessive), the proportions of phenotypes in the T1 generation (column (e)) directly correspond to the genotypes of the gametes of the tested plants (column (b)) and, therefore, show the genotypes of the tested plants.

Genotype of plant tested (a)	Genotypes of gametes (b)	Test cross (c)	Genotypes of T1 progeny (d)	Phenotypes of T1 progeny (e)
AA	A (all)	× a	Aa (all)	A (all)
Aa	A and a (1 : 1)	× a	Aa and aa (1 : 1)	A : a (1 : 1)

tigation. The use of the test cross in investigating F2 progeny of a monohybrid cross are shown in Table 26.4. This technique may seem a little elaborate in a simple case such as the monohybrid cross, but we shall see that the test cross is a very useful tool for the investigation of much more complex examples.

It was from his study of the monohybrid cross in *Pisum sativum* that Mendel deduced the existence of **factors** which we call **genes**, responsible for the inheritance of characteristics. This idea was incorporated in Mendel's first law, which may be summarized as: 'In a cross between two parents with differing characteristics, dominant and recessive genes are not lost or diluted, but remain distinct and segregate when gametes are formed'. The dependence of inheritance on the passing on of discrete parts of inherited material (genes) is sometimes described as **particulate inheritance**, to distinguish it from the vaguer ideas of the mechanism which existed before Mendel's work was recognized and accepted.

Having studied complete dominance, we can now look at an example of what is usually called incomplete dominance or partial dominance, or, more logically, **absence of dominance**. In this case, the phenotype of the heterozygous individual is intermediate between the phenotypes of the two alternative homozygous individuals. The classic example is flower colour in the tropical garden plant, *Mirabilis jalapa* (Four O'Clock). Red flowered plants are homozygous and white flowered plants are homozygous. If red-flowered and white-flowered plants are crossed, the F1 progeny all have pink flowers, intermediate between red and white. If we go back to our first checkerboard, let A = red and a = white, we can work out the expected progeny in an F2 generation. Pink (Aa) crossed with pink will give an F2 consisting of red : pink : white in the proportions 1 : 2 : 1. Inheritance is just the same as with complete dominance, except that heterozygous individuals are recognizable by their phenotype.

In the above examples, we have been dealing with cases in which there are two known alleles at each locus. Sometimes, there are more than two alleles at a locus, and we refer to **multiple alleles**. One of the best known examples of multiple alleles relates to the inheritance of the A-B-O series of blood groups in man, in which we also meet **codominance**.

There are three alleles at this locus. Allele A causes the secretion of antigen A in the blood corpuscles. Allele B produces antigen B. The third allele, O, produces no antigens. A is dominant to O. B is dominant to O. When A and B are present together (genotype AB), both alleles are expressed and both antigens are produced. A and B are, therefore, **codominant** to O. The possible combinations of alleles and the blood groups they produce are shown in Table 26.5.

Table 26.5 ABO genotypes and blood groups.

Genotype	Phenotype (blood group)
AA & AO	group A
BB & BO	group B
AB	group AB
OO	group O

We usually identify a gene, or the allele of a gene, by describing its most noticeable effect on the phenotype. Thus, we may talk of 'the gene for red flowers in the garden pea', one of Mendel's examples. However, Mendel noted that the red-flower phenotype was always associated with a brown seed coat and red marks in the axils of leaves. Many genes, like this one, affect more than one character. This is known as the **pleiotropic effect** of a gene.

There are also genes which have alleles which are lethal in the homozygous condition. These **lethal genes** are probably just inactive genes. As long as they are accompanied by the active gene in a heterozygote, the organism does not show signs of harm. In the homozygous condition, however, some vital part of the synthetic machinery of the organism is missing. The classic example here is yellow coat-colour in mice. Yellow mice are always heterozygous. If bred together, they produce mice with normal fur and yellow fur in the ratio 1 : 2. This unusual ratio is really a 1 : 2 : 1 ratio in which the last (homozygous lethal) element is missing. Offspring with the homozygous lethal genotype die long before birth. Yellow coat colour in mice is also an example of the pleiotropic effect of a gene, since the allele affects coat colour as well as viability.

Genes on chromosomes

Since genes are located on chromosomes, we would expect their behaviour to reflect the behaviour of chromosomes. For instance, at anaphase of the first division of meiosis, nonhomologous chromosomes travel to the poles of the spindle independently of each other. Each pole receives a random assortment of maternal and paternal chromosomes. We would therefore expect genes on non-homologous chromosomes to be independently assorted when gametes are formed. We would also expect that genes at different loci on the same chromosome would tend to stay together and to be inherited together, unless they were separated by a chiasma during meiosis.

Independent assortment

Suppose that we are studying the inheritance of two pairs of alleles at the same time. In Mendel's terminology, this is a **dihybrid cross**. Suppose, also, that the two gene loci are situated on two non-homologous chromosomes. The loci are not **linked** (are **unlinked**). If we cross a plant with the genotype *AABB* with a plant with the genotype *aabb*, the F1 progeny will all have the genotype *AaBb*. When gametes are formed, the chromosomes are independently assorted and the gametes have the genotypes *AB*, *Ab*, *aB* and *ab* in the ratio 1 : 1 : 1 : 1.

It is very easy to predict the proportions of phenotypes in a T1 generation, since they must be exactly the same as the genotype frequencies of the gametes,

given above. To predict the genotype and phenotype frequencies in an F2 generation, we use a checkerboard (Table 26.6).

In the checkerboard given, the sixteen squares have been numbered for convenience. We can read off the predicted phenotype frequencies (Table 26.7), which are:

$$AB : Ab : aB : ab = 9 : 3 : 3 : 1.$$

Table 26.7 Predicted phenotype frequencies in the F2 of a dihybrid cross (From Table 26.6).

Phenotype	Square numbers	Proportions
AB	1,2,3,4,5,7,9,10,13	9/16
Ab	6,8,14	3/16
aB	11,12,15	3/16
ab	16	1/16

There is an alternative method of working out the expected proportions of phenotypes, if that is all that is required. This is a mathematical method. We know from the study of the monohybrid cross that each pair of alleles must give a 3 : 1 ratio of phenotypes in the F2. Three-quarters of all offspring will show the dominant phenotypes and one-quarter will show the recessive phenotype. To predict the expected phenotypes in the F2, we list the possible phenotypes and then calculate as shown in Table 26.8.

Table 26.6 Checkerboard to predict F2 genotype and phenotype proportions from a dihybrid cross. The squares are numbered, in brackets, for easy reference (see text).

gametes	AB	Ab	aB	ab
AB	(1) AABB	(2) AABb	(3) AaBB	(4) AaBb
Ab	(5) AABb	(6) AAbb	(7) AaBb	(8) Aabb
aB	(9) AaBB	(10) AaBb	(11) aaBB	(12) aaBb
ab	(13) AaBb	(14) Aabb	(15) aaBb	(16) aabb

Table 26.8 Arithmetical method of predicting F2 phenotype proportions. This method is based on the knowledge that, in respect of each pair of alleles, three-quarters of the progeny will show the dominant character and one-quarter will show the recessive phenotype.

F2 Phenotypes	Calculation	Proportions
AB (dominant, dominant)	3/4 × 3/4	9/16
Ab (dominant, recessive)	3/4 × 1/4	3/16
aB (recessive, dominant)	1/4 × 3/4	3/16
ab (recessive, recessive)	1/4 × 1/4	1/16

This method is especially useful in calculating more complicated examples. The standard method of working out possible gamete genotypes is shown in Table 26.9.

Crosses involving two pairs of alleles (dihybrid crosses), leading to the F2 phenotype proportions shown above, were carried out by Mendel and led to the formulation of his second law, the law of independent segregation. Mendel's second law is: 'In a cross involving several pairs of alleles, each pair of alleles segregates independently of the others.' This was an important discovery at the time, but we can see now that it is not true of pairs of alleles that are close together on the same chromosome.

Linkage

Genes are arranged in linear sequence along the chromosome. Early in the first prophase of meiosis, homologous maternal and paternal chromosomes pair closely. Although chromatids cannot be seen at this time, use of the electron microscope shows that they

exist. The two chromatids of a single chromosome are **sister chromatids**. Those of different chromosomes are **non-sister chromatids**. Chiasma formation results from the breakage of non-sister chromatids at exactly corresponding points followed by the immediate joining together of the broken ends of non-sister chromatids. As a result, non-sister chromatids exchange parts. The visible evidence of this is chiasma formation (Fig.25.12). The genetical consequence, exchange of sequences of gene loci, is known as **crossing over**.

The number of chiasmata formed in a single pair of chromosomes (bivalent) is characteristic of a species. It also depends on the lengths of the chromosome arms involved. Longer chromosome arms have more chiasmata. Chiasmata form in random positions, but the formation of one chiasma interferes with the formation of other chiasmata close by, so chiasmata tend to be spaced out along a chromosome arm.

If we now think of a chromosome as a series of linked genes (**linkage group**), we can see that the frequency of chiasma formation between two known gene loci will be greatest when the loci are far apart and least when they are close together.

If the loci of two pairs of alleles are on the same chromosome and fairly close together, the ratio of gamete genotypes and T1 progeny from an F1 generation will not be 1 : 1 : 1 : 1. The proportions in which the different T1 phenotypes appear depend on the distance between the two loci on the chromosome. Let us suppose that we start with a plant which is known to have the genotype AaBb and make a test cross. Suppose also that the phenotypes of T1 are in the proportions

Table 26.9 Working out gamete genotypes. Where there are more than two pairs of alleles involved, it is convenient to work out gamete genotypes as below. The method can be extended to cover any number of heterozygous loci.

Genotype of parent			AaBbCc				
Diagram							
Gamete genotypes	*ABC* *ABc*	*AbC* *Abc*	*aBC* *aBc*	*abC* *abc*			

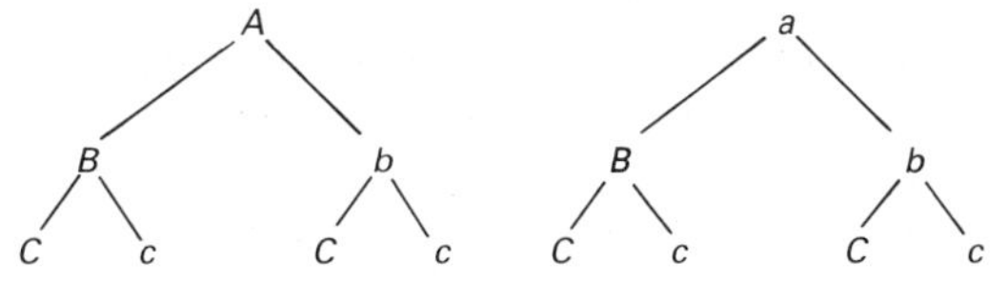

$AB : Ab : aB : ab = 9 : 1 : 1 : 9$. This departure from the $1 : 1 : 1 : 1$ ratio, which is characteristic of independent segregation, tells us two things. First, it tells us that the loci of A and B are linked. Secondly, since there is an excess of T1 phenotypes AB and ab, these must be the combinations of alleles on homologous chromosomes in the plants tested (**parental combinations**). The genotypes Ab and aB must be the genotypes resulting from crossing over, the genetical expression of exchange of material at chiasma formation. These new combinations of alleles are **recombinations**. At this stage we can use the standard way of writing the genotype of the plants we started with. Instead of writing $AaBb$, we write AB/ab. This shows that the two dominants are linked together and the two recessives are linked together. This is **linkage in coupling**. If the genotype had been Ab/aB, we would call this **linkage in repulsion**.

We can learn still more from the T1 phenotypes. To obtain a measure of the distance between the two loci on the chromosome, we first write out the T1 phenotypes and convert the proportions ($9 : 1 : 1 : 9$ above) into decimal fractions that add up to one.

Parental combinations $\quad AB = 0.45 \left.\right\} = 0.90$
$\qquad\qquad\qquad\qquad\quad ab = 0.45$

Recombinations $\qquad\quad Ab = 0.05 \left.\right\} = 0.10$
$\qquad\qquad\qquad\qquad\quad aB = 0.05$

Since there are 10 per cent recombinations, we can say that the loci of A and B are 10 **map units** apart. By definition, one map unit = 1 per cent recombination.

Suppose there is another locus, C, in the same linkage group. If experiments show that the loci of B and C are 20 map units apart and if A, B and C are in alphabetical order on the chromosome, it will also be found that A and C are 30 map units apart (Fig. 26.2). If sufficient linked loci on one chromosome are known, this process can be continued. However, 50 map units is the maximum map distance that can be measured directly, since 50 per cent recombination amounts to independent segregation. Working with successive loci that are less than 50 map units apart, it is possible to make a map of a chromosome. A **chromosome map** is a diagram of a linkage group (chromosome) showing the relative positions of different gene loci.

If we are told the map distance between two loci, it is possible to predict the phenotypes which will be

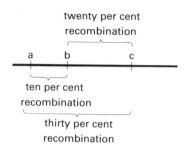

26.2 A simple linkage map. From the data given, we are able to place the loci of *a, b* and *c* in the correct order and with the correct proportional distances between them. Given further information, we would be able to place the loci correctly in relation to the end of the chromosome arm and the centromere

seen in T1 and F2 generations. Let us take the example of loci B and C above, which are 20 map units apart. Let us also assume that the parental genotype shows linkage in repulsion (Bc/bC). First, since we know that one map unit = 1 per cent recombination, we can write out the possible genotypes of the gametes as follows.

$Bc \quad 0.40 \left.\right\} = 0.80 = 80\%$ parental combinations.
$bC \quad 0.40$

$BC \quad 0.10 \left.\right\} = 0.20 = 20\%$ recombinations.
$bc \quad 0.10$

Since T1 phenotypes are the same as gamete genotypes, we already have the predicted proportions ($BC : Bc : bC : bc = 0.10 : 0.40 : 0.40 : 0.10$ or $1 : 4 : 4 : 1$).

To predict the genotypes and phenotypes of the F1, we make a checkerboard, as before, but include the proportions of each gamete genotype (Table 26.10). The proportions of gamete genotypes are multiplied together at the same time as we enter the progeny genotypes.

We now list and sum the figures (Table 26.11). The expected proportions of phenotypes in the F1 are $BC : Bc : bC : bc = 51 : 24 : 24 : 1$, very far from the $9 : 3 : 3 : 1$ ratio which would be expected if the loci of B and C were not linked.

Table 26.10 Checkerboard to predict F2 proportions when loci are linked and 20 map units apart.

Gametes	BC 0.10	Bc 0.40	bC 0.40	bc 0.10
BC 0.10	(1) BC/BC 0.01	(2) BC/Bc 0.04	(3) BC/bC 0.04	(4) BC/bc 0.01
Bc 0.40	(5) BC/Bc 0.04	(6) Bc/Bc 0.16	(7) Bc/bC 0.16	(8) Bc/bc 0.04
bC 0.40	(9) BC/bC 0.04	(10) Bc/bC 0.16	(11) bC/bC 0.16	(12) bC/bc 0.04
bc 0.10	(13) BC/bc 0.01	(14) Bc/bc 0.04	(15) bC/bc 0.04	(16) bc/bc 0.01

Table 26.11 Analysis of checkerboard in Table 26.10 to determine proportions of phenotype frequencies.

Phenotypes	Square numbers for each phenotype	Sums of phenotype proportions in squares
BC	1,2,3,4,5,7,9,10,13	0.51
Bc	6,8,14	0.24
bC	11,12,15	0.24
bc	16	0.01
	Total	1.00

Gene interaction

So far, we have been dealing with alternative characters that are apparently controlled by two or more alleles at a single locus. However, the biochemical processes that take place in cells are so complex that it is unlikely that any phenotypic character could possibly be controlled by one gene alone.

Gene interaction is the involvement of two or more genes, at different loci, in controlling a single phenotypic character.

Epistasis

The synthesis of many biochemical substances requires a series of chemical reactions, each governed by a different enzyme. Each enzyme is formed only if the corresponding gene or genes are present in an active form (Fig. 26.3). If any one of the genes responsible for the formation of a series of enzymes is present only as an inactive allele, the end product will not appear. In this way, the presence of the inactive, recessive form of a gene at one locus can prevent the appearance of a phenotype expected from the action of a gene at a different locus. The type of interaction in which the expression of a gene at one locus is prevented by the presence of an allele of a gene at a different locus is known as **epistasis**. One gene is **epistatic** to the other.

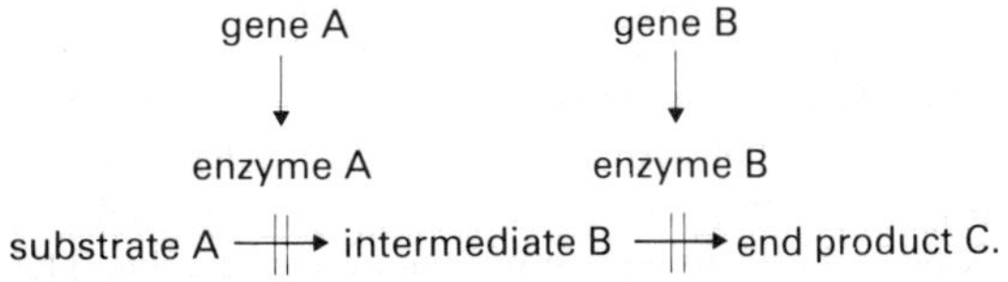

gene A   gene B
  ↓   ↓
enzyme A   enzyme B

substrate A ─╫→ intermediate B ─╫→ end product C.

26.3 Epistasis. This diagram shows a metabolic pathway from substrate A, through an intermediate substance B to the end product C. Each enzyme is produced by the action of a different gene. If gene *A* is faulty, the intermediate substance B is not produced and there is nothing for enzyme B to act upon. In such a case, a faulty (recessive) form of gene *A* would be epistatic to gene *B*

Complementary interaction

In the type of gene interaction known as complementary interaction, genes at two different loci are epistatic to each other when in the recessive form. The classic example in plants comes from the work of W. Bateson and R. C. Punnett on *Lathyrus odoratus* (Sweet Pea). These geneticists crossed two true-breeding white-flowered varieties and produced an F1 generation, all of which had purple flowers. When they selfed the F1, the resulting F2 progeny included purple-flowered and white-flowered plants in the ratio 9 : 7.

The explanation arrived at was that production of flower pigment requires the presence of two dominant alleles, at two loci on different chromosomes. The parental plants had the genotypes *AAbb* and *aaBB* respectively and therefore had white flowers. F1 plants had the genotype *AaBb* and therefore had purple flowers. The F2 genotypes and phenotypes are predicted, using the checkerboard in Table 26.6, as shown in Table 26.12.

Table 26.12 Complementary interaction. Analysis of phenotypes in Table 26.6

Phenotypes	Square nos. (Table 26.6)	Proportions
Purple flowers: genotype includes dominant at both loci	1,2,3,4,5,7,9,10,13	9/16
White flowers: dominant at only one or neither locus	6,8,11,12,14,15,16	7/16

Additive interaction

Additive interaction occurs when a single phenotype character is affected by genes at two or more loci and the degree to which the character is expressed depends on the **number** of dominant alleles present.

Red grain in maize is determined in this way. A plant with deep red grain has the genotype *AABB*. A plant with white grain has the genotype *aabb*. The loci for *A* and *B* are not linked. If we make an F1 (*AaBb*), the grain colour is pale red, intermediate between the grain-colours of the parents. In the F2, there are five grades of grain colour, ranging from white to deep red. We can predict the phenotype proportions if we use the checkerboard in Table

Table 26.13 Additive interaction of genes at two loci. Analysis of phenotypes in Table 26.6.

Phenotypes	Square nos. (Tables 26.6)	Proportion of progeny
Genotypes with four dominants	1	1/16
Genotypes with three dominants	2,3,5,9	4/16
Genotypes with two dominants	4,6,7,10,11,13	6/16
Genotypes with one dominant	8,12,14,15	4/16
Genotypes with no dominants	16	1/16

26.6, and list the numbers of squares containing different numbers of dominant alleles of *A* and *B* (Table 26.13).

The five categories of grain colour and the proportions in which they occur agree with experimental observations. When these figures are put on a histogram, we find that the shape of the histogram is similar to that of a graph showing normal distribution, such as that we might expect to see in a graph of a character showing continuous variation. If we postulate a character inherited in the same way, but involving the additive interaction of a larger number of genes, the shape of the resulting histogram looks even more like a normal distribution and would be even more so if the categories were made less distinct by environmental effects.

It is probable that characters which show continuous variation are governed by several or many genes, with additive interaction and affected by environmental factors. Such genes, all affecting the same continuously variable characteristic, are **polygenes**.

Sex chromosomes and sex determination

Sex chromosomes are chromosomes which determine the sex of the organism in which they are found. The first sex chromosomes to be recognized were those of the fruit fly, *Drosophila melanogaster*. In this insect, the female has three pairs of homologous chromosomes which are not directly involved in sex determination (**autosomes**) and one pair of identical

sex chromosomes (X chromosomes). The male has the same autosomes, plus one X chromosome and a smaller Y chromosome. The X and Y chromosomes are the basis of a sex chromosome mechanism.

The female fly produces ova, all of which contain three autosomes and an X chromosome. Because all of the ova contain the same chromosomes, the female is said to be the **homogametic** sex. The male produces two sorts of sperms, half of which contain an X chromosome and the other half a Y chromosome. The male is the **heterogametic** sex.

An XY sex chromosome mechanism exists also in man and many other animals. In man, as in *D. melanogaster*, the male is the heterogametic sex. In birds, there is a similar mechanism but the female is the heterogametic sex.

The mechanism by which sex is determined by the sex chromosomes differs in different organisms. In *D. melanogaster*, the Y chromosome is inert and the sex of a fly is determined by the ratio between the number of X chromosomes present and the number of sets of autosomes. In man, it appears that the presence of a Y chromosome causes maleness and its absence results in femaleness.

In *D. melanogaster* and in man, the X chromosome is longer than the Y chromosome. Few, if any, gene loci have been identified as belonging to the Y chromosome. There are many gene loci on the X chromosome which do not have corresponding loci on the Y chromosome. This results in a pattern of inheritance different from that of autosomal genes, known as **sex linkage**. An example of the inheritance of a sex linked gene in man is shown in Figure 26.4.

When we look at sex inheritance in plants, we must make a distinction between lower plants, in which sex is expressed in the gametophyte, and higher plants, in which sex is expressed in the sporophyte.

In many algae and all bryophytes, haploid gametophytes may have both types of sex organ on one plant (monoecious gametophytes), or may have separate male and female gametophytes (dioecious gametophytes). The diploid phase of the life cycle is asexual.

In seed plants, sex is expressed, if at all, in the sporophyte plant. Although there are separate male and female gametophytes in all heterosporous plants, the determination of gametophyte sex is not genetic. Instead, it is developmental and whether a spore mother cell produces microspores or megaspores

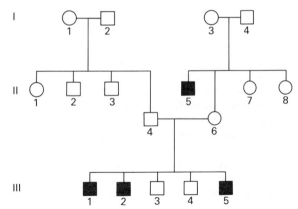

26.4 Sex-linked inheritance in man. Red–green colour-blindness is a recessive trait, the gene for which is located on the X chromosome. Since a man has only one X chromosome and there is no corresponding locus on the Y chromosome, he will be colour-blind if his X chromosome carries the gene. A woman, because she has two X chromosomes, must have the trait in the homozygous condition for her to be colour-blind. In this pedigree, the mother I–3 must be heterozygous for the trait, since her son (II–5) is colour-blind. Similarly, the daughter (II–6) must be heterozygous because her sons (III–1, III–2 and III–5) are colour-blind

depends only on its position in the parent plant, in a microsporangium or in a megasporangium.

Sex in gametophytes

In some isogamous algae, any pair of gametes may fuse and produce a zygote, even if they come from individuals of the same clone. These algae are **homothallic**. Other algae are **heterothallic**. In heterothallic algae, gametes will fuse only with other gametes which belong to a genetically different strain of the species. If these algae are isogamous, we cannot talk of male and female, but heterothallism is a primitive form of sexual differentiation and achieves the same end. Heterothallism and sexual differentiation favour outbreeding rather than self-fertilization.

The first plant shown to have sex chromosomes was the liverwort, *Sphaerocarpos donnellii*. The female gametophyte has seven autosomes and one, large, X chromosome. The male gametophyte has the same autosomes and one, smaller, Y chromosome. The sporophyte has seven pairs of autosomes and X and Y chromosomes. The X and Y chromosomes pair at the first prophase of meiosis and separate at first anaphase. In each tetrad of spores, two grow into female gametophytes and two into male gametophytes.

In *S. donnellii*, there is evidence that the presence of an X chromosome in a gametophyte results in the formation of archegonia and that the presence of a Y chromosome results in the formation of antheridia. Abnormal gametophytes which have both X and Y chromosomes are monoecious. They form archegonia and antheridia and, also, sometimes organs of an intermediate nature.

Similar XY chromosome mechanisms are found in many liverworts, including *Marchantia planiloba*, *Riccardia* spp. and *Symphyogyna* spp. There is a slightly more complex sex chromosome mechanism in the common epiphytic leafy liverwort, *Frullania squarrosa*. In this plant, the female gametophyte has seven autosomes and two X chromosomes. The male has seven autosomes and one, large, Y chromosome. When meiosis takes place before spore formation, the two X chromosomes pair with the Y chromosome. The two X chromosomes then travel to one pole of the spindle and the Y chromosome travels to the other pole. This is an XXY sex chromosome mechanism.

Some dioecious mosses (e.g. *Macromitrium* spp.) have an XY sex chromosome mechanism similar to that of many liverworts.

Sex in sporophytes

It is strictly accurate to describe dioecious sporophytes as megasporangiate and microsporangiate, but more convenient to use the terms female and male.

Sexual differentiation of sporophytes is unknown in pteridophytes. All cycads and some conifers are dioecious. Two species of *Cycas* are reported to have XY sex chromosome mechanisms in which the male is the heterogametic sex, but there seem to be no sex chromosomes in *Encephalartos* spp., as far as the records go.

Nearly all flowering plants are monoecious. Among dioecious genera, several have XY sex chromosome mechanisms in which the male is the heterogametic sex, or more complex mechanisms. Most records are for temperate plants, but *Trichosanthes* sp. (Snake Gourd) is reported to have an XY mechanism.

There is little genetical evidence about the way in which the sex chromosomes of cycads and flowering plants determine the sex of the plants, and no information about sex linkage in plants.

Genetic change

The process of DNA replication is usually accurate, so that the gene content of an organism and of successive generations of organisms is stable. However, mistakes in DNA replication sometimes occur and lead to **gene mutations**. The behaviour of chromosomes throughout the mitotic and meiotic cycles is usually quite regular. Occasional faults may lead to changes in chromosome morphology, gene content and chromosome number. Such changes are **chromosome mutations**.

Gene mutations

Gene mutations happen naturally, without apparent external cause. These **spontaneous** gene mutations are probably due to changes which occur irregularly but continuously in the molecular structure of the nitrogenous bases which make up the most important part of the DNA molecule. If a change takes place to a different (**tautomeric**) form of the base at the moment of replication of that part of the DNA, the wrong base may be included in the new DNA chain which is being formed. For instance, the normal form of adenine (Fig. 5.11) has two bonds and normally pairs with thymine, which also forms two bonds. There is a tautomer of adenine which forms three bonds. If adenine is in the tautomeric form at the moment of replication, it will pair with cytosine or guanine. This changes the sequence of nitrogenous bases in the new DNA chain and, therefore, the properties of the polypeptide that is synthesized by the gene concerned. Similar tautomeric forms of the other nitrogenous bases exist, and may act in the same fashion.

Different genes have different rates of spontaneous mutation. Mutation rates of known genes in maize range from 0.05 per cent, which is very high, down to zero.

There are certain external factors which influence mutation rates. Ionizing radiation (e.g. X-rays, gamma-rays) is one of these. Ionizing radiation is present everywhere at all times and may be responsible for many 'spontaneous' mutations. We can increase mutation rates artificially by using ionizing radiation. Radiation in the ultraviolet waveband also increases mutation rates.

There are also chemical substances that increase mutation rates. Nitrous acid is one of these chemical **mutagens**.

The action of radiation and chemical mutagens is probably due to an increase in the incidence of tautomeric and other changes in nitrogenous bases, due to an increase in the amount of energy present in the molecule (radiation) or chemical stresses (chemicals).

Gene mutations which occur in reproductive tissues may be passed on to the next generation in the gametes. In animals, where reproductive tissues are differentiated very early in the life of an organism, it is only mutations that occur in reproductive tissues that can be passed on. Mutations that occur in other tissues (**somatic** mutations) die with the individual. The situation is different in plants, where reproductive tissues are not separated at an early stage.

A somatic mutation in a plant shoot apex, or in cells that later become meristematic and take part in the formation of leafy shoots, may be passed on to a later generation.

Chromosome mutations

There are two kinds of chromosome mutation. In the first, parts of chromosomes are transferred to other chromosomes, or their position in the original chromosome is changed. These are **translocations, inversions, duplications** and **deletions**. In the second kind, whole chromosomes are added or lost from the chromosome complement, or whole sets of chromosomes are added. These are examples of **polysomy** and **polyploidy** respectively.

Translocations, inversions, duplications and deletions occur naturally and their frequency can be increased greatly by radiation and other agents.

Changes in chromosome structure

If a chromosome thread is broken, for example by radiation, the broken ends normally reunite and no damage is done. If several breaks occur at the same time, however, reunion may take place between the broken ends of different chromosomes or of different parts of the same chromosome. This results in changes in gene content, gene sequence and chromosome morphology (Figs. 26.5, 26.6). All of these types of chromosome mutation are likely to reduce the fertility of the plant concerned, due to their effect on pairing at meiosis (Fig. 26.7). If they survive, they may play a part in the formation of new species (Chapter 27).

Changes in chromosome number

Whole chromosomes may be added to or lost from the normal chromosome complement by a failure of the members of a bivalent to separate at first metaphase of meiosis (**nondisjunction**). This usually leads to inviability and death of the cells in which it occurs, due to imbalance in the gene content of cells. There are cases in which such changes persist without damage and a new (**aneuploid**) chromosome race of a plant is formed.

Where an external factor causes the failure of spindle formation or spindle action at meiosis or mitosis, chromatids separate without division of the nucleus. This results in a doubling of the chromosome number of the cell. If the cell is diploid ($2n$), it becomes **tetraploid** ($4n$). If the cell is haploid (n) it becomes diploid ($2n$) and, if it is later involved in the formation of gametes and crossing with normal haploid gametes, zygotes will be **triploid** ($3n$). Cells or organisms with more than two sets of homologous chromosomes are **polyploid**.

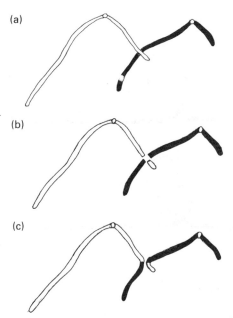

(a)

(b)

(c)

26.5 The origin of a translocation. A greatly simplified diagram showing: (a) two non-homologous chromosomes lying in close contact; (b) a brief (hypothetical) intermediate stage, when both chromosomes have broken where they are in contact; (c) reunion of the 'wrong' broken ends resulting in a reciprocal translocation. Translocation can also take place in a similar manner between homologous chromosomes and between different arms of a single chromosome

(a)

(b)

(c)

(d)

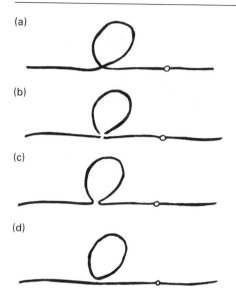

26.6 The origin of an inversion. A greatly simplified diagram showing: (a) a chromosome lying in such a way that a loop is formed; (b) a hypothetical stage, which must be very brief, when the chromosome thread is broken in two places, where different parts of the thread are in close contact; (c) reunion of the 'wrong' broken ends, so that the looped part of the chromosome is reinserted in the chromosome, but inverted; (d) another possible result of similar breakage and reunion, in which part of the chromosome is eliminated, resulting in a deletion

(a)

(b)

(c)

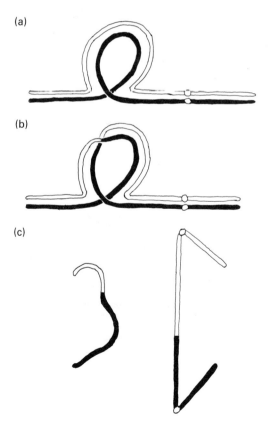

26.7 Chiasma formation in an inversion. A simplified diagram, in which each chromosome is represented as a single chromatid, of one possible consequence of the formation of a chiasma in an organism heterozygous for an inversion. (a) at first prophase of meiosis, pairing of loci is precise and results in the formation of a loop; (b) a chiasma is shown within the loop; (c) separation at first anaphase of meiosis results in the formation of a dicentric chromosome and an acentric fragment. The dicentric chromosome breaks at an arbitrary point. The acentric fragment is soon lost. The products of meiosis resulting from these events are inviable. A different pattern of separation occurs if the inverted segment includes the centromere, but the products are also inviable and the fertility of the organism is reduced

Polyploidy is rare in animals and polyploid animals rarely survive. It is, however, very common in plants (Chapter 27).

Failure of spindle formation resulting in the production of polyploid cells may be caused by low temperature or the sudden onset of other unfavourable conditions. It may also be induced experimentally by the use of chemicals.

Genes in populations

So far, our study of genetics has been centred on organisms and the transfer of genes from one generation to the next. Population genetics is concerned with the behaviour of genes in populations of organisms.

A population, in this sense, is defined as a group of potentially interbreeding individuals. We regard all the genes in a population as belonging to the **gene pool** of the population.

The Hardy-Weinberg equilibrium

The gene pool of a population includes all the alleles at a very large number of loci. In analysing the gene pool, it is necessary to start with a study of alleles at one locus.

Let us suppose that there are (only) two alleles at a particular locus and that one is totally dominant to the other. All the members of the population must have the genotype *AA*, *Aa* or *aa*. The **gene frequencies** of *A* and *a* in the population, expressed as decimal fractions, must add up to one. According to custom, the gene frequency of the dominant allele is represented by '*p*' and that of the recessive allele is represented by '*q*'. The sum of the gene frequencies at the locus is expressed as $p + q = 1.0$.

Under certain conditions, it is possible to predict the frequencies of genotypes (**genotype frequencies**) in a population, if we know the gene frequencies. These conditions are:

- The population must be large.
- All individuals in the population must have an equal chance of breeding (there must be random mating).
- There must be no difference in viability between the genotypes.

Let us suppose that we establish a population by putting together equal numbers of organisms with the genotypes *AA* and *aa* respectively and allowing them to breed for two or more generations. In this case, we know that the gene frequencies are $p = 0.50$ and $q = 0.50$. The gametes of any generation are $(p+q) = 1$. Genotype frequencies in any generation will be

$$(p+q)(p+q) = 1,$$
or $p^2 + 2pq + q^2 = 1.$

In this equation, p^2 is the genotype frequency of *AA*, $2pq$ is the genotype frequency of *Aa*, and q^2 is the genotype frequency of *aa*. Since we know the gene frequencies, we can substitute figures:

$$p^2 = 0.50^2 = 0.25.$$
$$2pq = 2 \times 0.50 \times 0.50 = 0.50.$$
$$q^2 = 0.50^2 = 0.25.$$
$$p^2 : 2pq : q^2 = 0.25 : 0.50 : 0.25.$$

Genotype frequencies, like gene frequencies, must add up to one. The predicted genotype frequencies are in the ratio 1 : 2 : 1 and, with complete dominance, give predicted phenotype frequencies of 3 : 1, just as in the F2 of a monohybrid cross.

As long as the three listed conditions are satisfied, there is no reason why gene frequencies should change from generation to generation.

This is a statement of the Hardy-Weinberg equilibrium, put forward by an English mathematician and a German doctor early this century. At the time, there was disagreement among geneticists about the probable fate of a recessive gene in a population. Some thought that a recessive gene should gradually disappear from a population. Hardy and Weinberg's piece of simple mathematics settled the matter.

In those cases in which it has been possible to show that conditions satisfy the requirements for the application of the Hardy-Weinberg equilibrium, it has been found to apply quite accurately. The formula can also be modified to predict gene and genotype frequencies where there are more than two alleles at one locus.

If there is complete dominance, the only genotype which can be recognized from its phenotype is that of the homozygous recessive. If we know the size of the population and the number of individuals of the recessive phenotype, we can predict the proportions of genotypes in the population.

Let us suppose that homozygous recessive organisms (*aa*) make up one per cent of a population. The genotype frequency, q^2, is thus 1% (= 0.01).

Given: $q^2 = 0.01$,

$q = {}^2\sqrt{0.01} = 0.10$,

thus $p = 1.00 - 0.10 = 0.90$,

and $AA : Aa : aa = p^2 : 2pq : q^2$

$= 0.81 : 0.18 : 0.01$.

We can apply the formula to solve problems. Suppose, for instance, that a plant breeder has a supply of seeds, one per cent of which produce a plant with an undesirable recessive phenotype. If the dominant phenotype were undesirable, it could be eliminated simply by growing the seeds and breeding only from plants with the recessive phenotype. It is more difficult to eliminate a recessive phenotype, because recessive alleles will be retained in heterozygous plants that cannot be distinguished from the homozygous dominant. All the plant breeder can do is to eliminate homozygous recessive plants before they breed, in succeeding generations. How many generations will it take to reduce the incidence of homozygous recessive plants from the original one per cent to a much lower value?

We start by setting out the predicted genotype frequencies, as above.

$p^2 = 0.81$,

$2pq = 0.18$,

$q^2 = 0.01$.

We now remove the homozygous recessive individuals (aa) from the population and calculate the new gene frequencies in the breeding population.

The figure for p^2 represents the proportion of homozygous dominant (AA) individuals. Each of these has two dominant genes. Heterozygous individuals (Aa) have one dominant and one recessive gene each. From this, we can calculate the proportions of dominant and recessive genes in the breeding population:

Proportion of A = $0.81 + 0.81 + 0.18$

                                        $= 1.80$

Proportion of a $= 0.18$

                                        ————

                        Total $= 1.98$

                                        ————

These are the proportions of gene A and gene a respectively in the breeding population, after the elimination of all homozygous recessive individuals. To convert the proportions into gene frequencies (in the breeding population and, therefore, in the next generation), we must convert the proportions into figures which add up to 1.00.

$p = \dfrac{1.80}{1.98} = 0.91$;

$q = \dfrac{0.18}{1.98} = 0.09$;

————

Total $= 1.00$.

————

We can now predict the genotype frequencies in the next generation:

$p^2 = 0.91^2$ $= 0.828$

$2pq = 2 \times 0.91 \times 0.09 = 0.164$

$q^2 = 0.09^2$ $= 0.008$

————

Total $= 1.000$

————

Therefore, in one generation, the proportion of aa phenotypes has been reduced from 1 per cent to 0.8 per cent. The above calculation is worked from first principles. There is, however, a quicker way of reaching the same result. If q is the frequency of the recessive gene in the original generation and individuals with the recessive phenotype are eliminated before they can breed, then the frequency of the double recessive phenotype (new q^2) in the next generation is given by the formula $\left(\dfrac{q}{1+q}\right)^2$. In our example above, the frequency of the double recessive phenotype in the next generation, therefore, equals $\left(\dfrac{0.1}{1+0.1}\right)^2 = 0.008$.

We can also, simply, predict the frequency of double recessive genotypes after several generations by a modification of the same formula. If q is the gene frequency in the original generation, the frequency of the double recessive phenotype after n generations will be $\left(\dfrac{q}{1+nq}\right)^2$. Using this formula, we can show that it would take twenty-one generations to reduce the incidence of the double recessive phenotype from 0.01 in the original generation to a figure of 0.001. The reduction of the incidence of the double recessive phenotype becomes slower and slower with each generation.

Departures from the Hardy-Weinberg equilibrium

There are several possible reasons for gene frequencies in a population to depart from the proportions predicted by the Hardy-Weinberg formula.

It is probable that most alternative phenotypes are not equally viable.

Selection will unbalance and change the equilibrium.

Migration of organisms into or out of the population may alter the balance between gene frequencies, especially if the numbers coming or going are large compared to the size of the population.

Mutation, to or from the dominant form of the gene, or adding a third allele at the same locus, can alter the pattern. It will not have a large effect unless the mutation has a strong positive effect on selection.

All of the above factors have their greatest effect on a small population. If a population is small enough, simple accident can have a striking effect. The gene frequencies of a small population can be drastically changed if, for instance, a random sample forming a significant part of the population is lost as a result of accident or migration. Such changes in gene frequencies, unrelated to selection or any other directional factor, are known as **genetic drift**.

Extra-nuclear inheritance

All of the patterns of inheritance that we have dealt with so far have been related to the behaviour of nuclear DNA (chromosomal DNA). There are some characters that are inherited through the cytoplasm, rather than the nucleus, of the cell.

Female gametes are always large and include much cytoplasm and many cytoplasmic organelles. Most male gametes are small. They contain very little cytoplasm and have a minimum of cytoplasmic organelles. The zygote, therefore, contains many maternal cytoplasmic organelles but few or no paternal cytoplasmic organelles.

Chloroplasts and mitochondria

Both chloroplasts and mitochondria contain their own DNA, in a similar form to that found in a procaryotic cell. As a cell grows, chloroplasts and mitochondria increase in numbers by division. When a cell divides, each daughter cell receives a proportion of the organelles which were present in the parent cell. Chloroplasts and mitochondria are independent entitities to the extent that, if completely lost from a cell, they cannot be replaced by the cell.

Chloroplast DNA contains genes concerned with the formation of chloroplast pigments, but pigment formation is affected by nuclear genes also.

There is a variegated variety of *Mirabilis jalapa* (Four O'Clock), in which mesophyll cells from colourless regions of the leaf contain leucoplasts instead of chloroplasts. These variegated plants have three sorts of branches. There are green branches, in which all the cells contain normal chloroplasts. There are branches with variegated leaves, in which green cells contain normal chloroplasts and colourless cells contain leucoplasts. There are also colourless branches, in which all of the cells contain only leucoplasts. The colourless branches survive because they are supplied with food by the rest of the plant. All three types of branches bear flowers and produce seeds. The types of plants which grow from seeds depend entirely on the female parent and are not affected by the source of the pollen involved. Seeds from green branches grow into green plants. Seeds from variegated branches grow into plants which are green, variegated or totally colourless. Seeds from colourless branches grow into colourless plants. The character of the progeny depends on the type of cell (with chloroplasts only, with leucoplasts only or with both chloroplasts and leucoplasts) which was involved in megaspore formation. This **maternal influence** is characteristic of extra-nuclear inheritance. Extra-nuclear inheritance can be recognized by making reciprocal crosses, using both phenotypes as male and female parents respectively. If there is a consistent adherence to the maternal phenotype, extra-nuclear inheritance may be assumed.

Other examples of extra-nuclear inheritance exist, in higher plants, lower green plants and fungi, but the *M. jalapa* example given above is the most clear-cut.

Key words

Inheritance: parents, progeny; P1 generation, F1 generation, T1 generation. **Genetics**: traits, contrasting characters; continuous variation, discontinuous variation; genotype, phenotype; chromosomes, genes. **Dominance**: dominant, recessive; homozygous, heterozygous; alleles, monohybrid cross, dihybrid cross, independent assortment; pleiotropic effect; absence of dominance, codominance. **Linkage**: linkage group, chiasma, crossing over; parental combination, recombination; linkage in coupling, linkage in repulsion; gene locus, map unit. **Gene interaction**: epistasis, complementary interaction, additive interaction, polygenes. **Sex determination**: sex chromosomes, autosomes; homogametic sex, XX; heterogametic sex, XY; sex linkage. **Gametophytes**: monoecious, dioecious; homothallic, heterothallic. **Sporophytes**: megasporangiate, female; microsporangiate, male. **Gene mutation**: genetic change, spontaneous, radiation, chemical mutagens; somatic mutation. **Chromosome mutation**: translocation, inversion, duplication, deletion; polysomy, polyploidy, haploid, diploid, triploid, etc.; nondisjunction. **Population genetics**: gene pool, Hardy-Weinberg equilibrium, genetic drift. **Extra-nuclear inheritance**: maternal influence.

Summary

The phenotype of an organism depends on the interaction of its genotype with the environment. If the organism is heterozygous, the phenotype also depends on the dominance relationship between gene alleles. In a monohybrid cross, we can predict the proportions of phenotypes in the progeny from a knowledge of the genotypes of the parents and the dominance relationship between the alleles involved. In a dihybrid cross, we can predict the proportions of phenotypes in the progeny from the same information, if the gene loci are unlinked. If the loci are linked, we must also know the map distance between the loci.

If there is gene interaction, genes at two or more loci affect the pattern of inheritance of a single trait.

In many organisms, the sex of an individual depends on the combination of specialized sex chromosomes inherited from its parents.

Genetic change (mutation) is sometimes spontaneous, but may also be caused by radiation or chemical mutagens. It may involve a change in the chemical structure of the DNA of a chromosome, or a change in the structure or number of chromosomes.

Under certain circumstances, we can predict the proportions of genotypes in a population from a knowledge of the gene frequency in the population. We may also calculate the gene frequency from observations of genotype frequencies.

Certain characteristics, especially those involving chloroplasts and mitochondria, are inherited solely from the female parent, regardless of the phenotype or genotype of the male parent. These are examples of extra-nuclear inheritance.

Questions

1 (a) List the genotypes of gametes formed by plants with the genotypes (i) *Aa Bb Cc*, and (ii) *Aa Bb Cc Dd*.

(b) Assuming that the loci are not linked, what would be the phenotypes of testcross progeny from each of the above plants?

(c) Using the mathematical method (Table 26.8), predict the proportions of phenotypes in the progeny of each of the following crosses: (i) *Aa Bb Cc* × *Aa Bb Cc*, (ii) *Aa Bb Cc Dd* × *Aa Bb Cc Dd*.

2 Suppose that the loci of genes *B* and *C* are linked and 20 map units apart. A cross is made between plants with the genotypes *BC/bc* and *BC/bc*. What would be the expected proportions of phenotypes in the progeny? Compare your working and results with Table 26.10. Why are the resulting proportions different from those in Table 26.10?

3 In a certain population, gene *A* is completely dominant to its allele, gene *a*. In a large population, it is found that sixteen per cent of the population is double recessive (*aa*). Assuming that the conditions necessary for a Hardy-Weinberg equilibrium apply, predict (a) the gene frequencies of *A* and *a* in the population, and (b) the genotype frequencies of *AA*, *Aa* and *aa* in the population.

4 A somatic mutation in a animal is never passed on to the next generation. It is possible for a somatic mutation in a plant to be passed on to the next generation. What is the reason for the difference?

27

Evolution

The theory of evolution states that organisms which exist today originated, through change, from organisms which existed previously.

The basic idea of evolution can be traced back a long way, even to writings of some of the ancient Greek philosophers. However, the theory was first formally stated in 1858, in a paper published jointly by Charles Darwin and Alfred Russel Wallace. Darwin's book, *On the Origin of Species by Means of Natural Selection*, was published in 1859.

The title of Darwin's book suggests that he was concerned mainly with the origin of individual species. This was not so. He applied the theory to the whole of the animal and plant kingdoms and postulated that complex organisms had evolved, through time, from the simplest organisms.

This brings us to make a distinction between two aspects of the study of evolution. At one level, we must consider the possible origin of and interrelationships between major groups such as divisions of plants and phyla of animals (**macroevolution**). At the other extreme, we must also study the origin of individual species and variation within species (**microevolution**).

The study of macroevolution uses evidence from many branches of biology, much of which was not available in Darwin's time. Evidence is drawn from the study of fossils (**paleontology**) and from comparative studies of morphology, anatomy, ultrastructure and physiology, including biochemistry. Since we cannot go back in time, all of the evidence is circumstantial. Whether we regard evolution at this level as a fact or as a theory depends on how precise we wish to be in the use of words. Strictly, it remains a theory, but the evidence is so strong that no informed biologist can doubt its truth.

At the other extreme, some examples of microevolution are facts. The origin of some species has been closely observed and, sometimes, repeated under experimental conditions. Further evidence comes from chromosome cytology, genetics, biochemistry and the geographical distribution of species and genera.

The study of evolution and of all the evidence for evolution is a very large subject. Here, we must confine ourselves to looking at the time-scale of evolution, at the probable course of evolution in the plant kingdom and at some of the mechanisms involved in the origin of species.

The time-scale of evolution

Our knowledge of the time-scale of evolution depends on the study of geology, including geochemistry, and the study of fossils.

The origin of life

The earth attained its present form about 4500 million years ago. Studies of earth chemistry show that the first atmosphere was a reducing atmosphere, in the chemical sense. Evidence for this is based on the fact that rocks formed during this period and subsequently protected from exposure to the atmosphere, by burial under later formed rocks, contain inorganic compounds in a reduced form (e.g. iron II compounds), rather than in the oxidized form (e.g. iron III compounds) found in later rocks. The original atmosphere was probably composed mostly of methane, ammonia, hydrogen, carbon dioxide and water vapour. There was no free oxygen.

When such a mixture of gases is prepared under experimental conditions and exposed to electrical discharges or ultraviolet radiation, compounds characteristic of living organisms are produced. These include sugars, amino acids, simple fatty acids and other organic compounds. It is believed that these compounds were formed on earth, under the influence of lightning and unshielded solar radiation. Where these substances became concentrated, further reactions occurred and more complex molecules were formed. The beginnings of life required the formation of DNA, which became enclosed in a membrane and formed the first cell. It is now fairly certain that the first living organisms were cells and not simpler structures (e.g. viruses).

Some of this is difficult to explain and is speculative, but some of the probable stages, including the

Table 27. The sequence of plants in the fossil record. The figures in the table are all expressed in millions of years (Ma) and are based on the estimated ages of known fossils belonging to each group. Note that some groups became extinct, while others are still extant. Also bear in mind that not all plants become fossilized and that not all fossils have yet been found and classified.

Plant group	Age of oldest fossil record	Age of latest fossil, etc.	Geological period of greatest abundance
procaryotic organisms	1600 Ma or more	species still abundant	?
eucaryotic algae	550 Ma or more	species still abundant	?
psilophytes	400 Ma	a very few species still extant	Devonian Period (395–345 Ma ago)
club mosses	395 Ma	a few species still extant	Carboniferous Period (345–280 Ma ago)
horsetails	365 Ma	a very few species still extant	Carboniferous Period (345–280 Ma ago)
bryophytes	350 Ma	species still abundant	?
ferns and fern-like plants	350 Ma	many species extant	Triassic Period (225–190 Ma ago)
seed ferns	300 Ma	100 Ma	Carboniferous Period (345–280 Ma ago)
cycads	260 Ma	a few species still extant	Jurassic Period (225–190 Ma ago)
conifers	260 Ma	still abundant in temperate zones	Jurassic Period (225–190 Ma ago)
angiosperms	125 Ma	dominant in most world vegetation now	now

formation of unit membrane-like structures, have been shown to take place naturally, if the correct proteins and other substances are first brought together in a suitable medium.

The first living organisms must have been heterotrophic, absorbing free organic compounds from their environment.

The fossil record

Fossils are the preserved remains and other traces of organisms which are found in sedimentary rocks and other deposits. They are of several types, which vary in the amount of information they can yield. Material preserved by **silicification** or **calcification** (embedding in silica or calcium carbonate respectively) often shows a lot of cell detail when specimens are sectioned. **Compressions** of plant parts may also show cell detail once they have been partially restored to their original shape and sectioned. Fossils of the types known as **casts** and **impressions** show surface features only.

The sequence of fossils

The oldest traces of living organisms are in calcareous rocks which are believed to have formed at least 2500 million years ago. The type of organism that formed the rock has not been identified but it is thought that it must have been heterotrophic and lived at a time when there was no oxygen in the atmosphere. The oldest recognizable fossils are of cells of blue–green algae which lived about 1600 million years ago. By this time, photosynthesis had evolved but the amount of oxygen in the atmosphere was still very small. The first oxygen produced by photosynthesis would have been used up in oxidizing the components of the original atmosphere and exposed rock surfaces.

The sequence in which major plant groups appear in the fossil record is shown in Table 27.1, from which it can be seen that most of the groups listed still have living representatives. The only group to have disappeared entirely is the seed ferns, which became extinct some 100 million years ago.

The origin and evolution of major groups of plants

The fossil record gives us a time scale for the evolution of plants and also shows the sequence in which most of the divisions and subdivisions evolved. Comparative studies of living plants provide evidence on which we can base speculations about evolutionary trends within the plant kingdom and suggestions of possible links between divisions and subdivisions. We must, however, avoid any suggestion that any type of plant 'evolved from' any other specific type of plant, whether living or fossil. There are many 'missing links' in the plant kingdom, just as there are in the animal kingdom.

Take, for example, the origin of the eucaryotic cell. Eucaryotic cells must have evolved from procaryotic cells. In trying to understand how this may have happened, we have the bacteria and blue–green algae on the one side and eucaryotic algae on the other. We must not assume that any division of eucaryotic algae evolved from blue–green algae. The procaryotic ancestors of other algae were certainly not blue–green algae, but would probably be classified in quite a different division if they still existed. Similar arguments apply when considering relationships between most other divisions.

From procaryotic to eucaryotic

The origin of eucaryotic cells from procaryotic cells involved the development of various membrane systems, including endoplasmic reticulum (ER), nuclear envelope and Golgi bodies. Procaryotic cells often have small, localised invaginations of the plasma membrane (**mesosomes**) and it is easy to imagine that these might increase, with increasing cell size, and give rise to ER. The nuclear membrane is specialized ER and it is possible that Golgi bodies are derived from ER. This leaves mitochondria and chloroplasts to be considered.

Both mitochondria and chloroplasts contain their own DNA. They also have their own characteristic ribosomes, smaller than ribosomes of the cytoplasmic matrix. Although they are influenced by nuclear DNA, both types of organelles are to some extent independent entities within the cell.

A mitochondrion is very similar in structure to the type of organism known as a mycoplasma (Chapter 22). Mycoplasmas are usually classified with bacteria, but have no cell wall and vary in size and shape. The cell is limited by a plasma membrane and DNA is present.

Likewise, a chloroplast resembles a photosynthetic procaryotic organism, similar to a blue–agreen alga but without a cell wall.

It has been suggested that chloroplasts and mitochondria first appeared as free-living procaryotic organisms which became endophytic in larger cells, in the same way as species of the unicellular green alga *Chlorella* are endophytic in the cells of some protozoa (e.g. *Paramecium* spp.) and some coelenterates (e.g. *Hydra viridis*). In each case, the endophytic organism would obtain its outer envelope from the ER of the cell, while retaining its own membrane. There are several recorded examples of continuity between ER and the outer membranes of chloroplasts and mitochondria.

The origin of the eucaryotic cell also involved the organisation of DNA into chromosomes and the evolution of mitosis and meiosis. There are some differences in the details of these processes in different groups of algae but their essential features are the same in all eucaryotic organisms. Meiosis is unknown in euglenoids.

Eucaryotic algae

As has been stated, the divisions of algae differ from each other in a number of structural and biochemical features. Their flagella are different, chloroplast structure is different and their cell wall materials, chloroplast pigments and storage products are different. They must have a common ancestry at some level of evolution, but this may be at the procaryotic level.

Within most of the divisions, we can distinguish what appear to be one or more **evolutionary trends**, as follows:

- from unicellular to multicellular structure;
- from isogamy to anisogamy to oogamy;
- from haplontic to haplodiplontic, sometimes to diplontic life cycles;
- in multicellular forms, from unspecialized cells to increasing specialization of cells.

The advantage of multicellular structure is probably one of size and is achieved in some groups by coenocytic rather than multicellular structure. Increase in size makes a plant less likely to be overshadowed by other organisms and may also make organisms less vulnerable to attacks by predators. Possible evolutionary lines, as seen in the green algae, are shown in Figure 27.1. Other algal divisions show some of the same evolutionary tendencies but not as fully as the green algae.

Oogamy has evolved independently in several algal

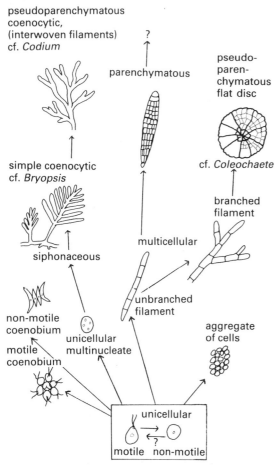

pseudoparenchymatous coenocytic, (interwoven filaments) cf. *Codium*

?

parenchymatous

pseudo-paren-chymatous flat disc

simple coenocytic cf. *Bryopsis*

cf. *Coleochaete*

branched filament

siphonaceous

multicellular

non-motile coenobium

unbranched filament

motile coenobium

unicellular multinucleate

aggregate of cells

unicellular

motile   non-motile

27.1 Evolutionary lines in the Chlorophyta. Starting from unicellular forms, it is possible to recognize several lines of evolution. All these lines lead to an increase in size of the plant body, which is advantageous. Cell aggregates, coenobial and siphonaceous lines have had varying degrees of success in aquatic habitats but not on land. Land plants (Bryophyta, etc.) are all based on parenchymatous structure. Several parenchymatous genera of green algae exist, including *Schizomeris* (Ulvales), in which the plant body is a cylindrical rod of parenchyma, but we cannot assume that the true ancestor of land plants still lives

divisions. It is best thought of as an example of cell specialization, the specialization of gametes in this instance. It is vitally important that compatible gametes should meet, in order that they may fuse, and this is more likely to happen if many gametes are produced. It is also very important that the zygote should contain enough stored food to give the next generation a good start. Male gametes are produced in large numbers and are, necessarily, small. They are almost always motile. Female gametes contain a large food store and are, therefore, large. They are not motile, but the function of finding a gamete of the opposite sex belongs to the male gamete.

The origin of the haplodiplontic life cycle is again an aspect of specialization. Fertilization is always to some extent a matter of chance. With a haplontic life cycle, one fertilization leads to the production of only four haploid spores. By the addition of the diploid generation, with many mitotic divisions between syngamy and meiosis, many more spores are produced after one act of fertilization.

There is also a genetical aspect to consider, especially when we come to a life cycle in which the haploid generation is reduced, or totally eliminated in a diplontic life cycle. In a haploid organism, all genes must be expressed and be subject to selection. A disadvantageous recessive gene will be eliminated. In a diploid organism, recessive genes will be retained in the gene pool in the heterozygous condition. A gene that is disadvantageous under one set of conditions may be advantageous in a different environment or at a different time. If the diploid phase or generation forms a major part of the life cycle of an organism, genes which add to that variability of the species and to its evolutionary potential are more likely to be retained in the gene pool. The overall advantage of being diploid is reflected in the increasing dominance of the diploid generation throughout higher members of the plant kingdom and by the exclusively diplontic life cycles of multicellular animals.

Specialization of cells and organs in relation to different functions adds to the efficiency of a multicellular organism. If a few cells or organs of a plant take over the function of reproduction, other cells or organs are free to become more specialized in relation to photosynthesis and other functions.

Algae are essentially water plants. Sub-aerial species survive on land because of their ability to survive desiccation. While they are desiccated, they are unable to grow or reproduce. The majority of algae survive dry conditions only as thick-walled spores, often zygospores or oospores.

It is the Chlorophyta that provided the ancestors of land plants. On this, there is no disagreement. However, there is no agreement about exactly which type of green alga was the ancestor. They were certainly from a filamentous line (Fig. 27.1 above) and probably from a line which evolved a parenchymatous, thalloid structure. There is certainly no living plant which fulfills all the requirements.

Bryophytes

There is no evidence that bryophytes were the first land plants, but they are certainly the simplest land plants. Like algae, they have little power to resist desiccation, though many are able to survive desiccation without damage. They show several features which may be seen as adaptations to life on land.

Bryophytes are attached to the substrate by rhizoids. Although the plants can absorb water over almost all of their exposed surfaces, the rhizoids of some bryophytes have been shown to absorb water from the substrate.

The most delicate parts, apices and male and female reproductive organs, are sometimes protected from desiccation by mucilage which is secreted by hair-like organs.

The male and female reproductive organs, antheridia and archegonia, have a jacket of sterile cells which protects the fertile cells within.

Some parts of the plant have a cuticle, which slows down the loss of water by evaporation. The sporophytes of hornworts and mosses have stomata in the epidermis, where there is chlorenchyma within.

Different bryophytes vary in the amount of specialization of organs and tissues. Some liverworts (e.g. *Symphyogyna* spp.) have conducting strands composed of cells which greatly resemble tracheids, with thick walls and transversely elongated pits. The conducting strands of a few mosses (e.g. *Polytrichum* spp.) contain tissues which resemble xylem and phloem. The central strands of most mosses, however, probably have no more than a mechanical function and, even where a conducting function has been demonstrated, conduction of water within must be insignificant compared with conduction and absorption along outer surfaces.

Lignin is unknown in bryophytes.

The division Bryophyta is not a natural group. The three classes, Anthocerotae, Hepaticae and Musci, represent three evolutionary lines with the same life cycle but, otherwise, with fundamental differences in structure. The Anthocerotae and Musci each show great uniformity. The Hepaticae include several evolutionary lines (e.g. simple thalli, thalli with complex internal structure and leafy plants), but are sufficiently uniform to be regarded as a natural group, with a common ancestry which would lie within the class.

The bryophytes are a dead end, in an evolutionary sense. There is no evidence that they gave rise to any higher group. Their connection with pteridophytes probably lies in an algal ancestor.

The failure of the bryophytes to compete with higher plant groups probably rests in their failure to acquire efficient conducting tissues, which would have allowed them to become less dependent on surface water and thus to grow larger. Their overall failure may also be related to their life cycle, with its dominant haploid generation.

Pteridophytes

All we know of the ancestry of pteridophytes is that they, like the bryophytes, must have had ancestors in the Chlorophyta.

The earliest of the fossil pteridophytes had no roots, but had xylem and phloem. The xylem consisted entirely of tracheids. The phloem was rudimentary and consisted of elongated, thin-walled cells with oblique end walls. Xylem and phloem formed a protostele (Fig. 19.1). These plants have been classified in the order Psilophytales of the subdivision Psilopsida, division Tracheophyta.

The simplest of these plants (e.g. *Rhynia* spp., Fig. 27.2), consisted of a rhizome, attached only by rhizoids, with dichotomous upright stems. There were no leaves and some of the branches ended in large sporangia.

Fossil relatives of *Rhynia* spp. had stems bearing many overlapping microphylls. The microphylls contained no vascular strand, but leaf traces passed through the cortex and ended just below each leaf base.

The living members of the Psilopsida, e.g. *Psilotum* sp. (Fig. 19.3), are superficially similar except for the position of their sporangia and are classified in the order Psilotales.

Subsequent evolution within the pteridophytes

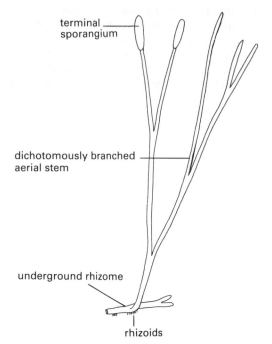

terminal
sporangium

dichotomously branched
aerial stem

underground rhizome

rhizoids

27.2 A reconstruction of *Rhynia* sp. The rootless rhizome of the sporophyte had a protostelic vascular system and formed green, upright, dichotomous branches. The aboveground branches had no leaves or other appendages and some were terminated by large sporangia. The gametophyte is unknown

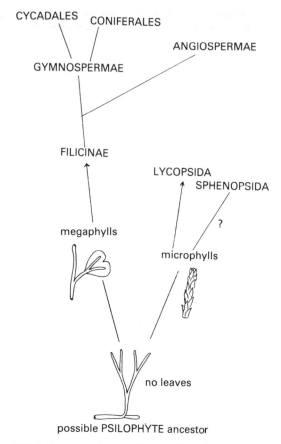

CYCADALES   CONIFERALES

ANGIOSPERMAE

GYMNOSPERMAE

FILICINAE

LYCOPSIDA
SPHENOPSIDA

megaphylls

microphylls

?

no leaves

possible PSILOPHYTE ancestor

27.3 A diagram to show the hypothetical origin of microphylls and megaphylls

went in two directions (Fig. 27.3). One line, or several lines, led into the microphyllous subdivisions (Lycopsida and Sphenopsida) and another line led to the Pteropsida, which are megaphyllous.

Microphylls are regarded as outgrowths of the surface of the stem. According to the telome theory, megaphylls are branches of limited growth, specialized in relation to photosynthesis. A megaphyll has a greater vascular supply than a microphyll, probably because of its evolutionary origin as a specialized branch of a stem.

All pteridophytes have true vascular tissues in the sporophyte generation. Xylem is composed of tracheids, or tracheids and parenchyma. Tracheids have lignified secondary cell walls with the same patterns of thickening as those of flowering plants (annular thickening, to thickening with bordered pits). Phloem consists of sieve cells and parenchyma.

Primitive pteridophytes have protosteles. Larger plants belonging to more advanced genera have solenosteles or dictyosteles, or steles of more complex form.

Vessels occur in some species of *Selaginella* and *Equisetum* (Sphenopsida) and in some ferns (e.g. *Pteridium* sp., Bracken), and must have evolved separately in each group.

Many fossil members of the Lycopsida and Sphenopsida were trees and had secondary vascular tissues. Amongst living pteridophytes, only *Isoetes* spp. have secondary vascular tissues, and they have only a small amount.

Most pteridophytes, both living and fossil, are homosporous. The **origin of heterospory** is a very important subject indeed, since heterospory is the basis of the seed habit. Heterospory occurs in all ligulate members of the Lycopsida, a few fossil members of the Sphenopsida and some Filicales, including living water ferns.

Heterospory, leading to the production of dioecious gametophytes, must promote outbreeding. It also has advantages over homospory, similar to the advantages of oogamy over isogamy. The dispersal of very many small spores, each producing a reduced male gametophyte, helps to ensure that the archegonia of female gametophytes will be fertilized. The large female gametophyte, with much stored food, provides nourishment for the embryo until it is well established.

The history of the Psilophytales was fairly short (Table 27.1), but members of the Lycopsida, Sphenopsida and to a lesser extent the Filicales and related spore-dispersed megaphyllous plants dominated the vegetation of the land for about 100 million years. Some were large trees. Modern representatives of all pteridophyte groups form a very minor part of present day vegetation, even though ferns are quite numerous.

It is thought to be ferns that provided the ancestors of seed plants.

Gymnosperms

Some fossils that were at first considered to be ferns were later found to be heterosporous and to have ovules very similar to those of modern cycads. These were the seed ferns (Cycadofilicales).

Once heterospory evolved, the final step in the development of the **seed habit** was the retention of the megaspore and the female gametophyte within the megasporangium, until after the development of the embryo of the next sporophyte generation. The origin of the seed habit also involved the protection of the megasporangium by one or more integuments, forming the **ovule**. In living and fossil gymnosperms, the female gametophyte is quite large and has two or more archegonia at the end nearest the micropyle of the ovule. The archegonia are similar to those of pteridophytes but with the neck, in living forms, reduced to two cells at the surface of the gametophyte or to a mere channel through the surface tissue.

The seed habit also involves further reduction of the male gametophyte. It also involves the development of a mechanism by which sperms can be passed to the archegonia of the female gametophyte through the overlying tissue of the megasporangium (nucellus).

The evolution of the **pollen tube** is of great importance. In cycads, the pollen tube can be interpreted as an attaching and absorbing organ, like the rhizoid that it resembles. It does not carry the sperms to the archegonia, but causes the breakdown of the tissues of the nucellus, exposing the fertile end of the female gametophyte to the sperms, when they are released. The sperms are motile and swim to the archegonia.

In conifers, the pollen tube has reached its final form and delivers the non-motile sperms into the neck of an archegonium.

In cycads and conifers, the (haploid) female gametophyte remains as the endosperm of the developing seed.

Cycads and conifers represent two lines of evolution within the gymnosperms, both of which have survived to the present day. Both have secondary vascular tissues, with xylem composed of tracheids and phloem composed of sieve cells and parenchyma.

In addition to the Cycadales and Coniferales, there are other orders of gymnosperms, including the Gnetales. Members of this order are notable for the presence of vessels in their xylem. A species of *Gnetum* grows as a liane in rain forest, in West Africa.

Angiosperms

The evolutionary origin of the flowering plants is a matter of controversy, but certainly lies among plants which we would classify as gymnosperms.

The origin of the group involved several important changes. These include the origin of the **flower**, the enclosing of the ovules within the folded megasporophyll and further refinements to the structure of the ovule, female gametophyte and seed.

The primitive flower is regarded as a stem-tip (receptacle) bearing floral parts in spiral sequence. Sepals and petals are similar to each other. Stamens are petaloid, with broad filaments and connectives. The gynaecium consists of many separate carpels, each with several ovules. Flowers showing one or more of these characters are found in the order Ranales, which includes families such as the Annonaceae, Nymphaeaceae (Figs. 14.2, 15.1) and Ranunculaceae. In one family, the Degeneriaceae, the carpels are folded but not fused along the margin, showing exactly the hypothetical primitive form. The Degeneriaceae are represented by a single species, *Degeneria vitiensis*, which grows in Fiji. *Magnolia* spp., belonging to the Magnoliaceae, have primitive flowers and also have xylem which lacks vessels, another primitive character.

The origin of the flowering plants involved a reduction of the female gametophyte to the extent that there are usually only three mitotic divisions between meiosis and the formation of the ovum. Of the eight nuclear embryo sac nuclei so formed, the three antipodal cells probably represent vegetative tissue. The group of ovum and synergids is sometimes interpreted as a reduced archegonium, with the synergids representing the neck. There is no evidence to support this, except that the synergids certainly play a part in attracting the pollen tube. The polar nuclei, on the grounds of their subsequent behaviour, form a new feature which cannot be compared with any part of more primitive gametophytes.

Sperms are delivered by the pollen tube, all the way into the embryo sac, through one of the synergids. Double fertilization and the formation of triploid ($3n$) endosperm are features found only in the angiosperms.

The male gametophyte (pollen grain, pollen tube, tube nucleus and sperms) shows further reduction on the gymnosperm pattern, since it has no prothallial cells. There are only two mitotic divisions between meiosis and the formation of sperms.

Evolution within the angiosperms is another controversial subject, though it is certain that monocots evolved from dicots. The monocotyledonous condition arose by suppression of one of the two cotyledons in a dicotyledonous embryo. There are several families of monocots which regularly produce embryos with two cotyledons (e.g. Araceae, Dioscoreaceae).

Microevolution

We may define a species as a group of morphologically similar organisms which are potentially capable of interbreeding under natural conditions. The origin of a new species therefore involves change, and involves the formation of a barrier to interbreeding between the new species and its nearest relatives. The formation of such a barrier leads to **reproductive isolation**.

Variation
Gene mutation is the primary source of variation.

Mutant genes are usually recessive and, if so, are not expressed in the phenotype of an organism until they appear in the homozygous condition. If a mutant gene is lethal or greatly disadvantageous in the haploid stage of the life cycle, it will be quickly eliminated (**gametic selection**). Otherwise, it will eventually appear in the phenotype of the diploid generation and will be subject to selection at that level (**zygotic selection**).

Selection is based on the fitness of an organism to survive and breed in a given environment. In the case of a plant, the environment includes all the physical, chemical and biological (**biotic**) factors which affect the plant. Biotic factors include (a) competition from members of the same and other species for space, light, water and mineral nutrients, (b) relationships with predators, and (c) in many flowering plants, relationships with animals which pollinate the flowers.

As long as a mutant gene is not disadvantageous at the haploid level or in the heterozygous, diploid condition, it will remain in the population indefinitely. If it is advantageous, the gene frequency will increase and it will spread through the population.

It is combinations of genes, rather than single genes, which are the subject of selection. When a mutant gene first appears, it is part of the total gene content (**genome**) of an individual. Chiasma formation at meiosis leads to recombination and each mutant gene is then subjected to selection as part of a variety of genomes.

An organism that is more fit than another is favoured by selection and is likely to produce more offspring. One that is less fit will be selected against and will produce fewer offspring. An organism that is fit in one environment may be less fit in another environment.

Selection and variation within a species
Mutation rates for different genes vary but it is probable that new mutant genes are present in the genomes of all individual organisms. There are also many recessive genes which are not new mutations. These factors, along with recombination which takes place in each generation, allow for much variation.

A population of a species which has a wide distribution will be subjected to different selection pressures in different parts of its range. It will adapt to its immediate environment in each part of its range and may thus become diverse in structural and physiological features. In some cases, plants from one extreme of the range (e.g. high altitude) may differ considerably from plants at the other extreme (e.g. low altitude). Plants from extreme parts of the range are capable of interbreeding, if brought together, but do not interbreed in nature. Plants that are close together can interbreed naturally and, because the population is continuous, genes can pass from one plant to another through the whole of the distribution range. They belong to the same **gene pool**. A population of this sort is a **cline**. Members of a cline which differ from each other and are adapted to different environments within the distribution range of the cline are **ecotypes** of the species.

Reproductive isolation

The formation of a new species requires that parts of a population of one species become separated from each other by some barrier to interbreeding. There are several ways in which reproductive isolation can be brought about.

Geographical isolation may result from unusually long-distance dispersal of seeds or spores. Plants dispersed far beyond the normal range immediately form a new population, with its own separate gene pool. Selection in the new environment may result in adaptation along quite different lines from that in the parent population.

Geographical isolation may also result from climatic changes or from human action. For instance, if a species has a wide distribution throughout a forested area, including lowland and mountainous parts, climatic change may result in the destruction of the lowland forest. This would leave the mountain communities undisturbed, but isolated from each other. A similar situation could arise if the lowland areas were deforested by man.

Reproductive isolation can develop similarly in several other ways. **Ecological isolation** may result, in flowering plants, from the development of different flower sizes, so that different parts of the original populations require different pollinating insects. **Allochronic** (seasonal) **isolation** results from the development of different breeding seasons. The development of self-pollination in part of a population can also result in reproductive isolation.

All of these mechanisms may result in the formation of communities which are reproductively isolated from each other and therefore have separate gene pools. This separation will tend to accelerate

further changes, as the isolated populations adapt more closely to their immediate environments. If the communities become sufficiently different in morphology, they may be considered to be different species. However, most species are also separated by genetical barriers which either make interbreeding impossible, or lead to the production of hybrids with very much reduced fertility.

Genetical isolation sometimes results from mutations which make members of different species incompatible with each other. For instance, the pollen of one species may fail to germinate on the stigma of a member of another species. Alternatively, it may result from chromosomal rearrangements or from changes in chromosome number.

The principal chromosomal rearrangements which cause genetical isolation are inversions and translocations (Chapter 26). Where a length of a chromosome becomes inverted relative to the normal order of genes, this results in adverse effects on meiosis, in organisms heterozygous for the inversion. The end result is much reduced fertility of the heterozygous individual which acts as a reproductive barrier between organisms with and without the inversion respectively. Translocations have similar effects on meiosis and on the fertility of hybrid organisms.

Many closely related species of plants and animals are genetically isolated because of inversions and translocations.

Polyploidy is very important in the origin of new species of plants. A polyploid organism is an organism which has three or more times the haploid (n) chromosome number in what is normally the diploid stage of the life cycle. It is estimated that polyploidy was involved in the origin of about 50 per cent of living species of flowering plants. It is much less common in animals. There are two types of polyploidy, autopolyploidy and allopolyploidy.

An **autopolyploid** is a polyploid in which all the chromosomes come from a single species. It results from the·failure of spindle action at mitosis or at the first or second division of meiosis. If the spindle fails to form, or fails to operate normally, the chromosome bivalents which should have separated into two daughter nuclei become enclosed in a single nuclear envelope. A nucleus which is formed in this way, containing twice the normal number of chromosomes, is known as a **restitution nucleus**.

If a restitution nucleus is formed at mitosis in a diploid ($2n$) cell, the result is a tetraploid ($4n$) cell.

If this cell subsequently continues to divide normally, it may give rise to a tetraploid branch and even, through vegetative reproduction, to a tetraploid (**autotetraploid**) plant.

If a restitution nucleus is formed during the first or second division of meiosis, or during mitosis in a haploid cell of a gametophyte, diploid gametes may result. If a diploid gamete fuses with a normal, haploid gamete, a triploid ($3n$) zygote is formed. If a diploid gamete fuses with another diploid gamete, the zygote is tetraploid. In either case, the zygote produces an autopolyploid (autotriploid, autotetraploid) plant.

Autopolyploid plants are almost always sterile. This is because the first division of meiosis is complicated by the presence of more than two homologous sets of chromosomes. Pairing of homologous chromosomes is irregular. Some normal bivalents may be formed, but there are also groups of three or, in a tetraploid, four chromosomes (**trivalents** and **quadrivalents**) and some chromosomes which fail to pair (Fig. 27.4). Separation of chromosomes at first anaphase is irregular and few of the cells produced are viable.

Allopolyploidy involves the hybridization (crossing) of two species before a doubling of the chromosomes. Hybrids between species (**species hybrids**) are usually sterile because their chromosomes do not pair properly at meiosis, due to the presence of inversions, translocations and even different chromosome numbers in the haploid chromosomes sets from those of the two parent species of the hybrid. Some parts of chromosomes are sufficiently similar to pair but others are not. Therefore, meiosis fails.

If such a hybrid becomes tetraploid, the position is immediately altered. There are now two haploid sets of chromosomes from each parent and these can pair normally. Some irregularities may arise, but an allopolyploid is usually fertile. A fertile allopolyploid immediately qualifies as a new species, since it differs from both of its parents and will not interbreed with either.

Macroevolution

Many of the factors involved in the origin of variation within a species, and in the origin of new species, are now understood. We also know a great deal about some of the major changes which have

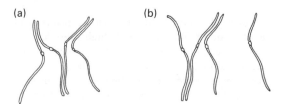

27.4 Synapsis in an autotetraploid: (a) synapsis in which four homologous chromosomes form a quadrivalent: (b) synapsis with one trivalent and one (unpaired) univalent. Normal bivalents may also be formed. The formation of quadrivalents, trivalents and univalents leads to very irregular separation of chromosomes at anaphase of the first division of meiosis and results in greatly reduced fertility

taken place during the evolution of the plant and animal kingdoms. However, there are many gaps in our knowledge of evolutionary connections between major groups of organisms.

Some of the gaps in our knowledge result from the fact that the fossil record is very incomplete. Very special conditions are required if an organism is to become a fossil and the chances of a particular plant becoming fossilized are very small. The chances of a fossil being found intact and coming to the notice of a competent expert in the study of plant fossils (paleobotanist) are also small. It is not surprising that there are gaps in the record. The absence of plants of a particular type from the fossil record must not be taken as always indicating that they did not exist.

There are two schools of thought about how major steps in the evolution of plants and animals came about. Some biologists believe that the type of gradual change that we see in the origin of individual species is not enough to explain the major changes required for the origin of new divisions. They consider that some additional mechanism is needed, which would produce major (**cataclysmic**) changes in one step. Perhaps major changes in genomes could be caused by sudden increases in radiation. Such changes might occasionally give rise to totally new, viable genomes which would produce organisms which were quite different from their parents and form the starting points for new plant divisions.

The other school of thought is that the small changes that we can observe today are enough to explain the whole process, taking into account the enormous periods of time involved. There have been periods in the history of the earth when the plant and animal populations have been very stable. There have been other periods during which evolutionary change has been rapid. These periods correspond, respectively, to periods of stability and instability in the climate of the earth. Evolution is most rapid under a changing environment, when organisms are continually adapting, through selection, to new conditions.

Key words

Evolution: macroevolution, microevolution. **Evidence for evolution**: fossils, paleontology, paleobotany; comparitive studies, morphology, anatomy, ultrastructure, biochemistry. **Algae**: procaryotic, eucaryotic; unicellular, filamentous, thalloid; sexual reproduction, isogamy, anisogamy (heterogamy), oogamy; life cycles, haplontic, haplodiplontic, diplontic; specialization. **Bryophytes**: cuticle, rhizoids, archegonia, antheridia. **Pteridophytes**: vascular tissues; root, stem, leaf; microphyll, megaphyll; homosporous, heterosporous. **Seed plants**: ovule, retention of the megaspore, reduction of the male gametophyte, pollen tube, endosperm, seed. **Microevolution**: species, variation, mutation, selection, fitness, cline, ecotype; mutant gene, genome; reproductive isolation, geographical isolation, ecological isolation, allochronic isolation, genetical isolation. **Macroevolution**: cataclysmic evolution, gradual evolution, climatic changes.

Summary

Complex plants evolved from simple plants over a great period of time. There is much evidence to show that land plants evolved from green algae. Within the algal divisions, evolution involved increases in size and cell number, changes in the details of sexual reproduction and of life cycles and some specialization of cells. In land plants, evolution led to further increase in size, greater specialization of cells, tissues and organs and, through heterospory, the origin of the seed habit. Microevolution, the origin of variation within species and the origin of new species, can sometimes be studied directly by observation and experiment. Macroevolution, the origin of major groups of organisms, can be studied only indirectly through fossils and through comparitive studies of structure and biochemistry.

Questions

1 Why is it that microevolution can be studied directly, while macroevolution cannot?
2 What is a cline? In what ways may parts of a cline become reproductively isolated from each other?
3 Periods of rapid evolution, in the distant past, have been associated with periods of climatic change. Explain why this might be expected.

Further reading for section E

Solbrig, O. T. and Solbrig, D. J. 1979. *Introduction to Population Biology and Evolution*, Addison-Wesley: Reading, Massachusetts.
Stewart, W. N. 1983. *Paleobotany and the Evolution of Plants*, Cambridge U.P.: Cambridge.
Strickberger, M. W. 1976. *Genetics*, 2nd edn, Collier Macmillan: New York.
Suzuki, D. T., Griffiths, A. J. F. and Lewontin, R. C. 1981. *An Introduction to Genetic Analysis*, 2nd edn, Freeman: San Francisco.

Section F
PLANT COMMUNITIES

Up to this point our study of botany has been centred on individual organisms and species of organisms. However, organisms do not live in isolation, but in communities which are usually quite complex. We are now to look at plant communities, especially those found in tropical Africa. Before we come to tropical plant communities we shall first look briefly at types of vegetation which occur in different climates and at the climatic conditions which govern the distribution of different types of vegetation in Africa.

28
The geography of vegetation

Before we study tropical vegetation, it is necessary to understand something of the types of vegetation in all parts of the world, and the factors which determine the type of vegetation which occurs in each part of the world.

Vegetation types

Table 28.1 is a summary of the principal vegetation types found throughout the world. As we shall see when we come to tropical vegetation, the categories in this table are very broad and most of the categories can be subdivided further.

Forest is characteristic of areas with high rainfall fairly evenly distributed throughout the year so that there are no prolonged dry periods, and a climate which is hot or warm for at least part of the year.

Grassland is characteristic of areas in which the amount and distribution of rainfall throughout the year, or other climate factors, are not suitable for the development of forest. It also replaces forest where soil conditions, fire or the activities of animals or man prevent forest from developing.

Semi-desert vegetation occurs where dryness or coldness with a very short growing season prevent the development of denser vegetation.

Desert conditions prevail in places which are hot and dry or very cold. Even cold deserts are dry, in the sense that any water present is frozen, and therefore not available to plants, for most of the year.

Table 28.1 The major types of world vegetation.

Vegetation types	Descriptions and examples
A. Forest	Vegetation dominated by trees which mostly meet overhead (closed canopy).
1. Tropical forest	Includes mangrove forest, fresh-water swamp forest, rain forest and montane forest.
2. Temperate forest	Vegetation dominated by dicot trees. Sometimes referred to rather as woodland.
3. Coniferous forest	Vegetation dominated by coniferous trees.
B. Grassland	Vegetation dominated by grasses. Trees may be present but do not form a closed canopy.
1. Savanna	Tropical grassland. Several categories are recognized according to the frequency of trees and shrubs (Table 29.1).
2. Temperate grassland	Often to some extent cultivated or grazed, e.g. North American prairies, South American pampas.
C. Semi-desert	Open shrubland or sparse herbaceous vegetation.
1. Steppe	Open shrubland with annual herbs, or dry grassland, temperate.
2. Tundra	Grasses, sedges and lichens under stunted trees, polar.
D. Desert	Very sparse vegetation.
1. Hot, dry desert	Tropical and sub-tropical, vegetation mostly succulent, sometimes with grass tussocks.
2. Cold, dry desert	Temperate or polar, sparse vegetation of herbaceous plants and lichens.

Climate

The type of vegetation which develops in any area depends, in the first place, on climate. Climate itself is governed by a number of factors.

Solar radiation

As the earth travels round the sun, the angle at which the sun's rays reach each part of the surface of the earth changes (Fig. 28.1). As seen from the earth, the sun is directly above the Tropic of Cancer (23.5° North) on 21st or 22nd June each year. After that, the sun appears to move south, is directly over the Equator on 23rd September, and over the Tropic of Capricorn (23.5° South) on 21st or 22nd December. The sun again passes over the Equator on 21st March, and so on from year to year.

The latitude at which the sun's rays reach the earth at right angles (**Thermal Equator**) receives most light and heat. Those parts of the earth on which the sun shines most obliquely are the coldest.

Winds

The basic pattern of wind systems over the surface of the earth results from differences in the amounts of solar radiation received in different places. Hot air expands and rises, creating a zone of **low pressure** over the Thermal Equator. As the hot air rises, cooler air is drawn towards the Thermal Equator from both north and south. At the surface of the earth, winds blow from the north and from the south towards the Thermal Equator.

The air which rises from the Thermal Equator divides, at high altitude, and blows back towards the pole in each hemisphere. The two airstreams cool as they travel towards the poles and the air again descends to the surface at about 20 to 30° North and 20 to 30° South. **High pressure** zones are therefore created in sub-tropical latitudes and surface winds blow from there, towards the Equator and towards the poles, in each hemisphere.

The directions of prevailing winds are also affected by the rotation of the earth, which deflects them towards the west (Fig. 28.2). North–easterly and south–easterly winds meet more or less over the Thermal Equator. The latitude at which they meet changes as the Thermal Equator moves north and south.

Another factor affecting wind direction is the **Coriolis effect**. When a stream of air, or a current

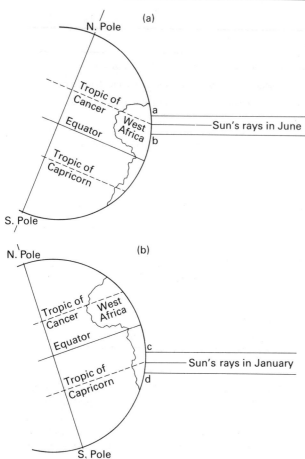

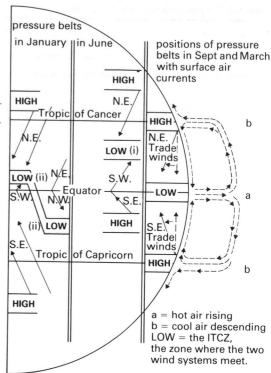

28.1 Solar radiation. A simplified diagram showing the 'movements' of the sun to give changing positions of the Thermal Equator in Africa. The June position of the sun gives the Thermal Equator in the area a–b. Compare this with the hottest region, over 32 °C, on Map 28.1 (a). Similarly, compare the position of the Thermal Equator in January, area c–d, with Map 28.1(b)

28.2 Movements of the pressure and wind belts in tropical African regions. As the sun 'moves' North (April to June) the pressure and wind belts follow the sun, so the low pressure (LOW (i)) does not reach all the way to the Tropic of Cancer before the sun starts to move South again. The positions of the pressure belts are also influenced by the relative positions of land and sea. The large land mass of Africa modifies the January position of the low pressure (LOW (ii))

of water, from the Northern Hemisphere crosses the Equator (0° latitude), it tends to swing to the left. A stream from the Southern Hemisphere swings to the right. Wind direction close to a continent is also modified by the fact that the land takes up solar heat more rapidly than the sea. The air over land therefore becomes heated more than that over the sea. The heated air rises and draws air from over the sea. For this reason, a prevailing wind which is passing over the sea tends to be diverted towards large land masses. For example, all of these factors combine off the coast of West Africa to change the prevailing south-easterly wind, from south of the Equator, into a south-westerly wind, north of the Equator, when the Thermal Equator is far north (June/July) (Map 28.1).

Rainfall

When air passes over the sea, it becomes heavily laden with water vapour. Over land, water condenses as rain (**precipitation**). It condenses most strongly when it is cooled by passing over high ground or when it meets a cool stream of air.

The effects of wind systems in determining climate are very clearly seen in West Africa, where the coastline is under the influence of two wind systems. The

(a) June: West African wet season,
Southern African dry season

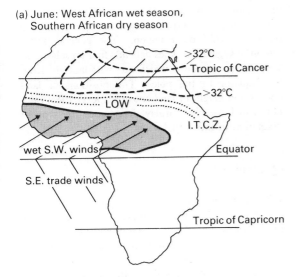

(b) January: West African dry season,
Southern African wet season

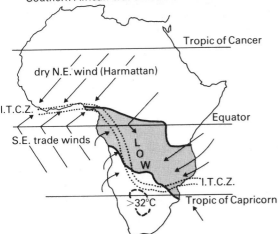

Map 28.1 Different positions of wind belts, Thermal Equator and low pressure areas in Africa. Shaded areas represent the rain belts in June (wet season between May and October) and in January (wet season between November and April). ITCZ is the inter-tropical convergence zone, the low pressure belt where wind systems meet. The areas of highest temperature (> 32 °C) mark the Thermal Equator

cool, dry north-easterly harmattan wind meets the warm, wet, southerly wind somewhere between the Equator and the Tropic of Cancer. The north-easterly wind brings the dry season to West Africa when the Thermal Equator is towards the south, in December and January. At this time the rain-belt is in the south, from the Equator to the Tropic of Capricorn. As the Thermal Equator moves north (June and July) the rain belts spread north and bring the wet season to West Africa. Here the coastline and the land closest to the coastline receive most rain, over the longest period. Further north, and inland, the rains are lighter. The rains also start later, and end earlier as the Thermal Equator again moves south.

Similar rainfall patterns exist in other parts of Africa and on other continents.

Altitude

The temperature at the surface of the earth falls by 0.6°C for every 100 m altitude above sea level. In addition to this temperature effect, land at high altitudes (e.g. 800–2500 m in the tropics) receives higher rainfall than the land below. Humidity is also higher, especially where cloud-belts hang around the sides of mountains. At still higher altitudes, rainfall and humidity are much lower and the vegetation has to tolerate dryness, as well as low temperatures.

At altitudes above 2000 m in the tropics, the vegetation is very different from that at sea level and is similar to that of very wet areas in temperate regions. Genera and even species that are characteristic of temperate vegetation occur naturally. In Africa, these include species of flowering plant genera such as *Senecio*, *Rubus* and *Ranunculus* and pteridophytes such as *Pteridium aquilinum* (Bracken Fern) and *Equisetum* sp. (Horsetail). Also, due to the temperature effect, the usual tropical crop plants are not successful and temperate crop plants such as the Irish Potato produce a high yield. Irish Potatoes planted at low altitudes in the tropics grow into large plants but yield only very small tubers, of no commercial value.

Vegetation and climate

The various climatic factors can thus be seen to affect the vegetation of any region in the following ways.

Solar radiation determines the temperature of an area. This affects the air pressure and the movement of air. Winds, moving with the air pressure, take hot or cool air, which may be wet or dry, over land surfaces. These winds affect the amount of rainfall. Rainfall is the most important factor affecting the vegetation type. The slope of the land and its altitude also affect the amount of rainfall. With the change from lowland to highland there is a change in vegetation type, from sea-level to the tops of mountains. Similarly, there is a change from the hot, wet equatorial regions through the cooler, drier subtropics and onwards, giving vegetation belts through tropical, subtropical, temperate and polar regions.

Key words

Vegetation: forest, grassland, semi-desert, desert. **Climate**: latitude, solar radiation, Thermal Equator; winds, air pressure; rainfall (precipitation); seasonal changes, wet season, dry season; altitude; temperature.

Summary

The natural vegetation of the earth's land masses can be divided into three types, forest, grassland and desert. These grade from one to another, according to climate, as we pass from the hot, wet or dry regions near the Equator and through temperate regions to cold polar regions. The general distribution of vegetation types depends first on solar radiation and temperature, which vary with latitude, from the Equator to the poles. The positions of land masses and height of land above sea level affect rainfall, which is what largely determines the type of vegetation in individual areas.

Questions

1 What is solar radiation? How does it affect the distribution of different types of vegetation by producing wind belts and precipitation?
2 What causes the change from wet season to dry season along the coast of West Africa?
3 Vegetation on tropical mountains may include some genera and even species which otherwise are found only in temperate vegetation. Why is it that these essentially temperate plants can grow at high altitudes in the tropics?

29

Tropical vegetation

As already stated, climate (the **climatic** factor) is of primary importance in determining the type of vegetation which may develop in any particular area. However, other factors also play a part in modifying the development of vegetation and in determining the vegetation present at any particular place and time.

Characteristics of the soil (the **edaphic** factor) are, of course, important. The nature of the rocks from which a soil is formed affects the species present in vegetation rather than the type of vegetation, but characteristics such as depth and drainage can have much greater effects.

The activities of animals, including man, make up the **biotic** factor. Animals other than man are important in some cases. For instance, forest may be replaced by savanna if grazing animals prevent the regeneration of forest after many trees have died. The activities of man may be treated as part of the biotic factor, or separately, as the anthropogenic factor.

The **anthropogenic** factor is very important indeed, and has been so for a very long time. For instance, in some of the finest forests in Africa, one has only to dig a hole down to a depth of one or two metres to find pottery fragments. These fragments show that the area has been populated in the past, even if it is far from existing settlements. The present influence of man on vegetation is more obvious, including the clearing of land for farming, road-building and urban and industrial expansion.

Partly associated with man, **fire** is a very important factor. Fires occur naturally, following lightning strikes or volcanic eruptions, but are also started by man as part of farming and hunting procedures. Much of the savanna in West Africa probably grows where forest would be growing, if repeated burning by man did not prevent forest from developing.

In describing tropical vegetation, it is convenient to start with coastal vegetation, under conditions of high rainfall, and to deal in turn with the types of vegetation which occur in successively drier areas. The sequence is similar in most respects to what one would observe in making a journey, for example, from the coast of Nigeria, travelling inland and north as far as the Sahara Desert.

Coastal and riverine vegetation

Large tropical rivers carry a great deal of sand and silt down to the sea. This suspended solid matter tends to settle to the bottom of the water when the river current meets the sea.

In the case of the largest rivers, the sand and silt may settle in the river mouth, to form a **delta**. The delta fills the river but is traversed by many channels (**distributaries**), through which the river water passes to the sea. Such a delta continues to increase in size with the passage of time and the land surface extends seawards.

If there are strong ocean currents running along the coastline, sediment is carried from the mouth of a river and deposited along the coast, where it forms sandy beaches (**strand**). These beaches also gradually extend seawards and add to the coastal plain. Under some conditions, some of the sediment is also deposited as a **sand bar**, parallel to the coast and usually about 400 m from the coast. Such a sand bar encloses a **lagoon**, the water of which is a mixture of river water and sea water (**brackish** water). The salt content of lagoon water varies throughout the year. It is highest in the dry season, when flow from the rivers is lowest and water enters the lagoon from the sea. River water predominates in the wet season.

Beaches and sand bars are not formed where there are headlands and the land projects outwards into the sea. Here, the land is eroded by the sea and the coastline is rocky. It is on rocky parts of the coastline that the greatest variety of red and brown algae are found.

Strand vegetation

The sand close to the sea's edge is unstable and saturated with sea water. Unless it is colonised by mangroves (see below), it has no vegetation. As sand is built up by wind action a few metres inland, its

29.1 *Ipomoea pes-caprae* spreading over a sandy beach. This is a common sand-binding plant in West Africa

29.2 *Calotropis procera*, a woody shrub of sandy coastal areas, also found on sandy areas in savanna. It has a·very upright habit and its grey-green leaves and inflated fruits are very conspicuous

surface becomes raised above the seawater table and salt is leached away by rain. Sand-binding plants grow at this level, on both beaches and sand bars.

In tropical Africa, the commonest sand binders are two creeping plants, *Ipomoea pes-caprae* (Convolvulaceae) (Fig. 29.1) and *Canavalia rosea* (Papilionaceae). Both have tough, creeping stems and tough, xeromorphic leaves. These are often associated with other low-growing plants with xeromorphic characters, including *Cyperus maritimus* (Cyperaceae), *Paspalum vaginatum* (Gramineae) and *Alternanthera maritima* (Amaranthaceae). All of these plants are able to grow up to the surface after being buried by sand. They also trap wind-blown sand and thus help to build up the level of the beach.

Coastal scrub

A short distance inland, the sand is sufficiently compacted to support the growth of shrubs and small trees, which form dense vegetation at the top of the beach. In West Africa, the shrubs include *Calotropis procera* (Asclepiadaceae) (Fig. 29.2) and *Dodonaea viscosa* (Sapindaceae), and the trees include *Baphia nitida* (Papilionaceae) and *Phoenix reclinata* (Palmae).

It is at the level of coastal scrub that Coconut Palms (*Cocos nucifera*) are often planted.

As in strand vegetation, all the plants have xero-
morphic characters. Although the rainfall is high, sand is extremely porous and the plants also have to withstand strong winds.

Mangrove forest

At any point on the coast, the level of the sea rises to a maximum (high tide) and sinks back to a minimum (low tide), twice in every twenty-four hours. The level to which the tide rises varies throughout the year and is also affected by weather conditions. A strong onshore wind drives the sea higher than would otherwise be expected.

At low tide, the water in the lower part of a river is fresh water, down to the sea, and the sea water is diluted outside the estuary (river mouth). Sea water contains about 3.5 per cent by weight of dissolved salts, mostly sodium chloride. River water contains a negligible amount of dissolved minerals.

As the tide rises, sea water enters the river mouth and brackish water extends for some distance up the lowest part of the river. This is the **tidal region** of the estuary. As previously stated, sediment is deposited where river water reaches the sea, so silt and sand are deposited along the banks of the river, in the delta if there is one, and along the shoreline close to the river mouth. It is on this sediment, regularly or occasionally flooded with brackish water, that mangrove forest grows. The plants which make up mangrove forest must therefore tolerate great variations in salinity and a soft soil which is waterlogged and contains little oxygen.

Mangrove forest is dominated by trees, with very little ground flora. *Rhizophora* spp. (Red Mangroves, Rhizophoraceae) (Fig. 4.4) predominate in West and East Africa and along the tropical coast of South America. *Avicennia nitida* (White Mangrove, Verbenaceae) is common on firmer, sandier soil. Species of both these genera occur in mangrove forests on most tropical coasts.

In West Africa, the most prominent species is *R. racemosa*, which may reach a height of forty-five metres but is usually no more than about ten metres. *R. mangle* and *R. harrisonii* grow to about ten metres. All species are characterized by having prominent prop roots. Prop roots arise from the trunk and from the undersides of branches and grow downwards at an angle towards the surface below. They usually branch before they reach the surface, but branch most profusely after they have penetrated the mud. The prop roots of neighbouring trees interlock, and their underground branches form a raft of interlace roots on which the trees stand. Aerial parts of prop roots have many, large lenticels, by way of which the underground roots are aerated.

Rhizophora spp. have salt glands on their leaves, through which excess salt is exuded in solution. Another adaptation to the mangrove environment is vivipary (germination of seeds while still attached to the parent tree, Chapter 2). The density of mangrove growth is probably explained by the fact that the seedlings, when they drop, may be immediately embedded in the muddy soil, where they may root before they are disturbed by being flooded.

Avicennia nitida has no prop roots, but has a raft-like root system close to the surface. The root system has many vertical branches which grow up out of the soil to a height a little above the highest level of the water. These 'breathing roots' have many lenticels and serve to ventilate the whole root system. Like those of *Rhizophora* spp., the leaves of *Avicennia* sp. have salt glands.

The soil level of mangrove forest may build up gradually, until the surface is above high tide level, after which the mangroves are slowly replaced by land vegetation.

Similar mangroves, including *Rhizophora* spp. and other members of the Rhizophoraceae and other families (e.g. Combretaceae) occur in many parts of the world. *Nipa fruticans* (Sugar Palm) grows naturally in mangrove forest in India and more easterly countries. This palm has a prostrate stem and is of economic importance as a source of sugar. The sugar is obtained from sap exuded from cut inflorescence stalks. This plant was introduced into Nigeria in 1906, where it has since become naturalized.

Mangrove forest extends up the banks of rivers as far as the tidal zone extends. Towards the upper part of its range, where the water is only slightly brackish, other plants (e.g. *Pandanus candelabrum*, Screw Pine) appear, growing amongst *Rhizophora* spp.

Fresh-water swamp forest
Where a large river passes towards the sea across a fairly flat coastal plain, the lower parts of the plain, close to the river, are flooded with fresh water during the wet season. Pools of water, indicating a high water table, may remain even during the dry season. Plants growing in such regions must cope with waterlogged and rather unstable soil. The vegetation is fresh-water swamp forest.

Many of the trees in swamp forest have **buttress roots** (Fig. 29.3) or **stilt roots**. Each of these types of root system gives a tree a very wide base and greater stability than that given by an ordinary root system. Some trees are tall and grow up to as much as 45 m in height. Most are smaller, up to about 10 m. The trees do not form a dense canopy and there are places where direct sunlight reaches the ground. Here, very dense growths of lianes, shrubs and large herbs occur.

In West Africa, the tall trees include *Mitragyna ciliata* (Abura, Rubiaceae), a tree of economic importance which has buttress roots and pneumatophores. Smaller trees include *Uapaca* spp.

29.3 Buttress roots of a swamp-forest tree

29.4 Large aroids (*Cyrtosperma* sp.) and thick ground vegetation in fresh-water swamp forest, near Benin, Nigeria

(Euphorbiaceae), which have stilt roots. *Raphia vinifera* (Raphia Palm) and *Pandanus candelabrum* (Screw Pine) are both characteristic of river margins and other places where there are large breaks in the tree canopy. *Raphia* sp. has numerous small pneumatophores which grow to a height of a few centimetres above the soil. *Pandanus* sp. has prop roots.

Lianes include climbing palms (e.g. *Calamus deeratus*) and the herbs include large members of the Araceae (e.g. *Cyrtosperma senegalense*, Fig. 29.4).

Like mangrove forest, fresh-water swamp forest is a region of land building. River sediments are deposited each time the land is flooded and the level gradually builds up until the surface is above the water table. In the higher parts, species characteristic of rain forest gradually appear (e.g. *Lophira alata*, Ochnaceae). At its inland margin, swamp-forest species become fewer, except along river banks, and the vegetation merges into rain forest.

Rain forest

The description 'rain forest' is used here in its widest sense, to include all tropical forest developed under fairly high rainfall, including lowland forest and montane forest. In tropical Africa, lowland forest may extend to an altitude of up to about 1900 m. Where lowland forest is continuous with montane forest, the intermediate zone is occupied by an intermediate type of vegetation. However, such continuity is rare in Africa and regions of montane vegetation

are usually separated from lowland forest by belts of savanna.

Both types of rain forest exist under conditions of high rainfall, on ground that is not continuously waterlogged. Lowland forest develops where the average temperature is relatively high, montane forest at lower temperatures, due to altitude.

Lowland rain forest

Lowland rain forest (Fig. 29.5) is characterized by the **stratification** of different elements of the vegetation. The tallest stratum (Stratum A) is of tall trees, reaching a height of 40 m or more. These **emergent** trees are generally scattered, so that their crowns rarely meet. The trunks of the trees are straight and branch only at the top, to form a crown which is wide and shallow. Below this, Stratum B

29.5 Profile of rain forest, as seen at the cut edge of forest near Benin, Nigeria. The long, straight trunks and continuous canopy of stratum B trees are clearly seen. Stratum A trees have been removed, but stratum C trees are still present. The Cassava farm in the foreground shows a ground layer of tree seedlings and weeds

is formed of trees which grow to a height of about 30 m. The crowns of these trees may form a continuous canopy, but there are sometimes gaps. The trees have straight trunks which branch near the top to form a crown which is vertically elongated. Stratum C is of trees which grow to a height of about 10 m and have vertically tapered crowns. The trees commonly form a closed canopy. Stratum D is a layer of shrubs and young trees, with an average height of about 1 m. It may also contain tall herbs and ferns. Stratum E, the ground flora, contains tree seedlings and small herbs, including small ferns and *Selaginella* spp. Unlike the ground flora in montane and temperate forest, bryophytes are almost entirely absent from the ground flora, though they appear as epiphytes and on parts of the soil which have been disturbed by the actions of animals, for example

termites, or by erosion, for example at the edges of paths and beside the roots of trees.

The strata described above are not always clearly defined and there are always trees and other plants of intermediate sizes. However, an aerial view of the forest shows a clear main canopy (Stratum B), with emergent trees (Stratum A) standing well above. A view from the forest floor shows the canopy formed by Stratum C and the lower layers and the trunks of trees in Stratum A and Stratum B. The full stratification can be seen clearly only where part of a forest has been felled and the forest is seen, as it were, in section (Fig. 29.5).

The trees

Rainforest trees, especially those of Stratum A and Stratum B, have certain characteristics which are seldom seen in other types of vegetation.

Many of the trees have stilt roots or buttress roots. Buttress roots are especially common. African examples of trees with buttress roots are *Triplochiton scleroxylon* (Obeche, Sterculiaceae) and *Entandrophragma* spp. (Sapele, Meliaceae). Stilt roots occur on *Uapaca guineensis* (Euphorbiaceae). It seems obvious that either type of root gives the best possible support for a tall tree, but it is not clear why they are so common in tropical rain forest and freshwater swamp forest and are so rarely seen in tall trees in temperate vegetation.

The bark of most forest trees is very thin and often merely a powdery layer over green underlying tissue. Fire does not normally occur in rain forest. When it does occur, the trees are easily killed, since there is little protection for the underlying phloem and cambium. Several species of forest trees also have 'thorns' (prickles), which start on the young tree as epidermal emergences and are added to later, at the base, by the production of localised areas of hard phellem (Chapter 10). *Ceiba pentandra* (Silk Cotton Tree, Bombacaceae) is an example of a forest tree with prickles (Fig. 4.8).

Climbing plants

Climbing plants are very common in the rain forest, but are always best developed where there is a break in the canopy, where a tree has fallen, or along a river bank. Small, closely attached climbers include several members of the Araceae (e.g. *Culcasia scandens*) and ferns (e.g. *Lygodium* spp.).

Large, woody climbers and scramblers (**lianes**) are

29.6 Rain forest epiphytes. *Rhipsalis* sp., the only member of the Cactaceae (Cactus family) indigenous in Africa. The succulent stems have only scale leaves and the small white berries are quite conspicuous

29.7 A strangling fig, *Ficus* sp., encircling the host with clasping aerial roots

also common. Like most other climbers, these are rooted in the ground and grow upwards through the shrubs and Stratum C trees, as far as the canopies of Stratum B trees, where they often form dense tangles. There are lianes belonging to several families of plants. West African examples include several species of *Combretum* (Combretaceae) and *Ancyclobotrys* (Apocynaceae) and several climbing palms.

Epiphytes

There are numerous epiphytic flowering plants. The most conspicuous are epiphytic orchids (e.g. *Bulbophyllum* spp.) (Fig. 4.1). The only member of the Cactaceae which is indigenous in Africa, *Rhipsalis cassutha* (Fig. 29.6), grows as an epiphyte in the crowns of the tallest trees.

A number of figs (*Ficus* spp.) are forest epiphytes and, once established, develop aerial roots which reach the ground. In some of these, **strangling figs** (Fig. 29.7), the aerial roots closely surround the trunk of the host tree and gradually fuse with each other to form an almost continuous cylinder of woody root tissue around the trunk of the host tree. The host eventually dies and leaves the fig standing in its place. Some species of *Ficus* may start life either as epiphytes or growing on the ground and, in either case, grow into large trees 20–30 m high.

Ferns are very common as epiphytes. Filmy ferns (e.g. *Hymenophyllum* spp.; *Trichomanes* spp.) grow in the wettest places, low down on the trunks of trees, as **shade epiphytes**. Other, larger, ferns are **sun epiphytes** and grow higher up, where sunlight penetrates (e.g. *Asplenium africanum; Microsorium punctatum*).

Bryophytes are very common indeed. The bases of trees are usually colonised by creeping mosses.

29.8 Rain forest epiphytes, *Blechnum* sp., a large fern, growing as an epiphyte on the trunk of a tree fern, *Cyathea* sp., beside a river. Smaller epiphytes are growing on the tree-branch above

Different species (Fig. 18.14) form a close covering on the bark of stems and branches (e.g. *Ectropothecium regulare*), hang down from branches in festoons (*Pilotrichella sordido-viridis*), or form usually small, compact cushions of upright stems (e.g. *Calymperes lecomtei; Octobelepharum albidum*). Leafy liverworts and some small, thalloid liverworts are also common as epiphytes. Species belonging to several genera of the leafy liverwort family Lejeuneaceae are especially common and usually form closely adhering, usually circular colonies on the bark of trees. Other members of the Lejeuneaceae and other families form festoons, hanging from the trunks and branches of trees (e.g. *Plagiochila* spp., Plagiochilaceae). Small thalloid species belonging to the genus *Metzgeria* are also common, growing on bark.

In the wettest places, there are epiphyllous liverworts, which grow on the upper surfaces of the leaves of small trees, shrubs and even herbs. Several of these belong to the Lejeuneaceae (e.g. *Cololejeunea* spp.) but, in West Africa, the commonest is *Radula flaccida* (Radulaceae), a leafy liverwort with two rows of lateral leaves but no underleaves.

Several green algae also grow as epiphytes. *Cephaleuros* sp., a branched filamentous form, forms small, greyish–green, wool-like tufts on tree trunks and the margins of leaves. *Trentepohlia* sp., a relative of the last, forms a bright orange coating of short, branched filaments on tree trunks. *Coleochaete* spp. form small (1 mm) discs of pseudoparenchymatous tissue, closely applied to leaf surfaces.

Lichens are common as epiphytes, including crustose, foliose and fruticose forms. Crustose lichens grow on leaves and bark, foliose and fruticose lichens grow on bark. Pale, greyish–green festoons of *Usnea* spp. hang from tree-trunks and branches, especially where sunlight penetrates the canopy (Fig. 1.7).

Parasites and saprophytes

Parasitic and saprophytic flowering plants (Chapter 16) are commoner in the rain forest than in any other type of vegetation.

Lowland forest and climate

Although the main characteristics of lowland forest remain much the same as we travel away from the coast of West Africa, species composition changes as rainfall becomes less. In the wettest part, where there is only a short dry season, the trees are evergreen and the forest is **evergreen** lowland forest. Further inland, where the dry season is more pronounced, many of the trees lose their leaves during the dry season. This is **semi-deciduous** or **mixed-deciduous**, lowland forest. Semi-deciduous forest contains fewer epiphytes than evergreen forest.

The change from evergreen forest to semi-deciduous forest is gradual and more deciduous species appear as one travels further from the coast.

Lowland savanna

Savanna is a type of vegetation in which grasses dominate the ground flora. In Africa, it occurs naturally where the average annual rainfall is low (e.g. around 1000 mm) and where there is a long dry season. It may also develop in areas with higher rainfall under certain soil conditions, for instance where

Table 29.1 Types of lowland savanna in Africa.

Type of savanna	Description
savanna woodland	Trees up to 15 m high form an almost closed but not very dense canopy. There may be occasional taller trees (e.g. *Adansonia digitata*, Baobab). Under the trees there are scattered shrubs and the ground flora is dominated by tall grasses.
tree savanna	Trees up to about 10 m high are scattered, their branches rarely in contact and never forming a complete canopy. There are also occasional shrubs and a ground flora dominated by tall grasses.
shrub savanna	Trees are absent. Shrubs are present, scattered amongst a ground flora dominated by tall grasses.
grass savanna	Trees and shrubs absent. The vegetation is entirely dominated by tall grasses.

the soil is very permeable and does not hold water for long. Prolonged drought stress is the deciding factor.

Types of savanna are broadly classified according to the types of plants present. In **savanna woodland**, there are many trees and shrubs and the trees grow close together, though without forming a continuous canopy. In **tree savanna**, trees and shrubs are present but the trees are scattered. In **shrub savanna**, shrubs are the only woody plants and there are no trees. **Grass savanna** has grasses and other herbaceous plants, but there are no woody plants. Further subdivision of savanna types varies from region to region. A classification of savanna types is given in Table 29.1.

In West Africa, savanna lies mostly inland of a forest belt, but reaches the coast where the coastline runs more or less parallel to the direction of the moist winds which bring the rain in the wet season, for example, in the so-called 'Dahomey Gap' in the Republic of Benin.

It is probable that only a small part of the savanna in Africa is natural savanna, in which the presence of savanna is the result of the action of climatic and edaphic factors alone. Most of the savanna is **secondary savanna** and grows in areas where the natural vegetation would be forest. Here, the savanna nature of the vegetation is maintained by regular burning, usually in the middle of the dry season.

Smaller, isolated, patches of savanna lie within the forest belt. These also are secondary savanna and result from clearing of the forest for farming, following by regular burning. Burning prevents the regeneration of forest species. Under these conditions, grass savanna is established in the first place. If the area is too far from established savanna wood-

land or tree savanna for the dispersal of seeds of savanna trees, or if burning is severe and takes place each year, it remains as grass savanna. If the seeds of savanna trees are available and if burning is not at all severe or regular, it develops into tree savanna or savanna woodland.

Secondary savanna also occurs along the inland margin of the forest belt. In Nigeria, a belt of secondary savanna with an average width of about 150 km runs parallel to the inland margin of the forest. Although this savanna is continuous with similar vegetation further inland, there are several pieces of evidence which show that it is secondary and not natural savanna. Experiments have shown that, if protected from fire, areas of savanna close to the forest margin are rapidly invaded by forest species. It would probably take a century or more for the vegetation to become typical forest, but the tree and other species appear rapidly. Also, forest extends along river banks (**fringing forest**) for many kilometres out into the savanna and patches of forest (**forest outliers**) occur in places where some natural feature has protected them from the spread of fire. An implication of this evidence is that forest would spread into these areas if they were protected from fire. A further implication, based especially on the presence of forest outliers, is that the whole of these areas was covered with forest at some time in the past and that forest outliers are a remnant (**relict**) of what was once continuous forest cover.

Savanna extends to quite high altitudes. At about 1500 m altitude, it either borders submontane or montane forest or undergoes gradual transition into montane grass savanna.

Savanna plants

Savanna woodland contains many trees, up to about

(a)

(b)

20 m tall, with occasional taller trees (e.g. *Adansonia digitata*, Baobab, Fig. 29.9). Almost all of the trees are deciduous, with tough, xeromorphic leaves. The leaves of trees in the wetter parts of the savanna (e.g. Guinea savanna of West Africa) are usually broad, or compound with broad leaflets, and dark green. In drier regions, leaves and the leaflets of compound leaves are very small, and often greyish–green. Savanna trees have thick, furrowed bark containing much rhytidome. Bark of this type gives good insulation and protects the delicate inner tissues when fire passes fiercely, but usually briefly, through the vegetation.

The species of savanna trees vary from place to place. Species of *Lophira*, *Butyrospermum* and *Daniellia* are common in the moister parts of the savanna in West Africa. Further north, in drier parts, species of *Acacia*, *Isoberlinia* and *Combretum* are amongst the commonest trees. Species of *Brachystegia* occur in West Africa, but form a larger part of savanna vegetation in East and Central Africa.

Savanna shrubs grow to a height of up to 4 m. Like the trees, they have xeromorphic leaves. Many have spines or thorns (e.g. *Acacia* spp., Fig. 4.10).

Savanna grasses grow to a height of 3 m in the wet season, or slightly shorter in drier parts. They have flat (not rolled) leaves. Common African genera include *Andropogon*, *Hyparrhenia*, *Pennisetum* and *Schizachyrium*. All the most important savanna grasses grow in clumps, with the bases of many stems pressed closely together and with bare ground between the clumps. In the dry season, when the upper parts of the leafy stems have dried, the clumped stem bases protect the living apices, which are at ground level, from fire.

A common feature of the herb flora of savanna is the presence of many monocots with underground storage organs (bulbs, corms, rhizomes). These plants die back to ground level at the end of each growing season and the underground parts are protected from fire. When the rains come again, they produce inflorescences and leaves, often in that order, before there are leaves on the trees and before the grasses have grown very high. Common examples include species of *Albuca* (Liliaceae), *Haemanthus* (Amaryllidaceae), *Tacca* (Taccaceae) and *Eulophia* (Orchidaceae).

Epiphytes are uncommon in the savanna, especially in drier areas. Epiphytic lichens and bryophytes occur, but mostly on the trunks and

29.9 Savanna plants: (a) *Adansonia digitata (Baobab)*; (b) *Cochlospermum* sp., the bright yellow flowers appear at ground level after burning. (See also *Haemanthus* sp., Fig. 15.31.)

29.10 Savanna plants of rocky outcrops: the succulent, spiny stems of *Euphorbia* sp. and a plant of *Aloe* sp. growing with grasses. (See also Figs. 15.30, *Urginea* sp., and 19.6, *Selaginella njam-njamensis*.)

branches of trees and shrubs which are close to more or less permanent streams.

Rocky outcrops in the savanna provide a specialized habitat and have a flora of their own. Such outcrops carry patches of shallow soil, in depressions or cracks in the rock. The soil is completely dry for most of the dry season, but is waterlogged during the wet season. The vegetation on this soil is **seasonal flush** vegetation. It includes small grasses and sedges and, often, a surface covering of thalloid liverworts (*Riccia* spp.). Species of the pteridophyte genera *Ophioglossum*, *Isoetes* and *Selaginella* (*S. njam-njamensis*) are often found and the carnivorous genera *Drosera* (Droseraceae) and *Utricularia* (Lentibulariaceae) occur in the wettest parts. Apart from the seasonal flushes, rocky outcrops carry occasional savanna trees, where the soil is deepest, and there are more climbing plants (e.g. *Smilax* spp.) than are found in the main part of the savanna. Succulent

plants (e.g. *Euphorbia* spp.; *Aloe* spp.) are also common on and around rocky outcrops (Fig. 29.10).

Montane vegetation

The names given to different types of montane vegetation vary from author to author. Here, it will be sufficient to distinguish between **montane forest** and **montane grassland** (montane savanna). It is impossible to give truly meaningful altitudinal limits for the different vegetation types because there is a great deal of variation. The limits vary between different mountainous areas, due to differences in rainfall and latitude. They also vary on each mountain, due to differences in aspect, slope and soil. The figures mentioned below are examples only.

The factors which determine the presence of montane vegetation are temperature, mist and rainfall. Temperature drops with altitude. Cloud banks form persistent mist at certain levels, even in the dry season. Rainfall is usually high.

Lowland vegetation, forest or savanna, extends to an altitude of about 1900 m. Above this level, there is a transitional zone which may reach as high as 2100 m. Truly montane vegetation occurs above this level. Where there is forest, montane forest extends up to the tree line, at about 2600 m. Between 2100 m and 2600 m, the natural vegetation is montane forest. However, fire is often a regular feature of some areas and much of the forest may be replaced by montane grassland, with forest restricted to gulleys containing streams and small rivers (fringing forest, gallery forest). Above the tree-line is a sparser type of montane grassland.

Montane forest

Montane forest differs from lowland forest in several ways. Emergent trees are absent and there are two tree strata only. Trees of the upper stratum reach a height of about 35 m. Those of the lower stratum are up to 20 m in height. Neither stratum forms a continuous canopy. All the trees are evergreen. Few or none of the trees have stilt roots or buttress roots.

Below the lower stratum of trees there is a shrub layer, up to a height of about 4 m. The ground flora is sparse. On the forest floor, bryophytes are rare except around the bases of trees. Lianes are rare or altogether absent.

Montane forest is noted for its epiphytes

29.11 Montane forest epiphytes: *Arthropteris* sp. (large fronds) and *Asplenium* sp. (small fronds)

29.12 Montane grassland at an altitude of about 2000 m. The distant small river valleys and the edges of the dissected plateau still bear montane forest (gallery forest)

(Fig. 29.11). Their abundance is accounted for by the fact that, at this altitude, clouds envelop the vegetation in mist for long periods, even in the dry season. Epiphytes include flowering plants, ferns, lycopods, bryophytes and lichens. The trunks and lower branches of trees, also the branches of shrubs, are festooned with mosses and liverworts. The highest branches of trees are festooned with the lichen, *Usnea* sp.

Towards the upper limit of the forest, above the mist and in drier conditions, the species present and the character of the vegetation are quite different. The trees are smaller, with twisted and branched trunks, and are more widely separated. This sparser vegetation may be referred to as montane woodland, rather than forest, because the trees are smaller and more widely separated. The branches of the trees are covered with epiphytes, but these are mostly lichens (e.g. *Usnea* sp.). The replacement of bryophytes with lichens reflects the drier environment.

Montane grassland

Montane grassland (Montane grass savanna) occurs in two distinct zones.

The lower zone occurs at levels below the upper limit of the trees and is probably secondary vegetation, determined by fire. Like those of lowland savanna, the grasses grow in clumps and have flat leaves. They grow to a maximum height of about 0.8 m. There are other herbs and occasional stunted and charred trees. The grassland (Fig. 29.12) is dissected by streams and small rivers, which are lined with fringing forest. Tree ferns (*Cyathea* spp.) and *Phoenix reclinata* (Wild Date) are prominent in narrow patches of fringing forest.

The second type of montane grassland grows above the level of forest and is probably a natural vegetation, determined by the dry environment and low temperature. The grasses are shorter than in the last type of grassland and have rolled leaves. At higher altitdes (e.g. over 3000 m), the grass is shorter still and the vegetation is best described as steppe, though other terms have been used.

Marine vegetation

Land and fresh water environments are dominated by plants of the Chlorophyta–Tracheophyta evolutionary line. Fresh-water environments also contain representatives of other divisions but plants belonging to the Phaeophyta and Rhodophyta, although few and rare in fresh water, are common in the sea.

Marine phytoplankton

Marine phytoplankton includes representatives of all of the divisions which occur in fresh water phytoplankton. There are rather few Chlorophyta (e.g. *Chlamydomonas* spp.) and several of the orders of this division are totally absent (e.g. Zygnematales, Oedogoniales). The commonest constituents of marine phytoplankton are diatoms and dinoflagellates.

Marine benthic vegetation

As in fresh water, benthic vegetation occurs in relatively shallow waters, especially along the coasts of land masses. Deeper parts of the sea are without vegetation. Benthic marine algae are known to occur down to a depth of at least 40 m off the coast of West Africa.

The plants we are dealing with here grow attached to rock surfaces between the highest point that sea water reaches by splashing, above the highest point reached by the sea as a whole, to a depth of many metres. Sandy seashores and ocean floors are more or less devoid of algae, except where there are protruding rocks or similar solid surfaces such as the shells of large molluscs. Sand itself does not provide a sufficiently firm surface, because it shifts with movements of the water.

The sea is subject to **tidal movements**. The level of the sea reaches a maximum (**high tide** level) twice in every 24 hours and, alternately, descends to a minimum twice each day (**low tide** level). The magnitude of the rise and fall of the tide is affected by many factors. The basic difference between high tide level and low tide level along the coast of West Africa varies with the season from about 0.4 m to 1.8 m at different times of the year, but the rise and fall tends to be greater where there are bays and inlets and in wide river mouths. Tide levels are also affected by strong winds.

The effect of the rise and fall of the tide is to produce a belt of land, parallel to the coast, which is alternately covered and uncovered by the sea,

twice each day. This is the **intertidal zone**. The extent of the intertidal zone varies. It is widest where the slope of the shore is slight and narrowest where the ground slopes steeply into the sea. Its width also varies with the seasons, according to the difference between high tide level and low tide level.

Plants which grow in the intertidal zone are exposed to alternate periods of desiccation and hydration, twice each day. The relative lengths of the periods of exposure and immersion depend on the position in which a plant is growing in relation to high tide level and low tide level. This results in zonation of the algae within and close to the intertidal zone. In general, the zones or belts run parallel to the coast.

Starting from the land, the first zone or belt is the **littoral fringe**. This is just above high tide level for most of the year but is covered by the highest tides and is subject to splashing with sea water at other times, especially when the sea is rough. This zone is fairly barren, except in crevices and well-sheltered corners. The algae which occur are all small, including stunted red algae and traces of filamentous green and blue–green algae. The zone is characterized more by the species of snails which it supports than by the algae. The littoral fringe is widest on exposed coasts, where there is much wave action.

The **eulittoral zone** is the part of the shore which is more or less regularly covered and uncovered by tidal movement. It is not a uniform habitat. In its upper part, it is subjected to long periods of desiccation alternating with short periods of hydration. The balance between desiccation and hydration changes continuously down the shore towards low tide level. In the upper part, the eulittoral zone is sparsely colonized by small clumps of green algae (e.g. *Ulva* spp., *Enteromorpha* spp.), filamentous brown algae (e.g *Ectocarpus* spp.) and small, stunted red algae. Some of the brown algae are of the encrusting type, in which most of the branched filaments are prostrate and intertwined so that the plants form circular or irregularly shaped cushions on the rocky surface. Lower in the eulittoral zone, where the duration of desiccation is less, the algal cover becomes progressively richer. Encrusting brown and red algae are present, but they are accompanied by taller plants. Green algae include tufts of the rather coarse, unbranched filaments of *Chaetomorpha* sp., the more delicate branched filaments of *Cladophora* spp. and plants of the siphonaceous green algae *Bryopsis* spp. and *Caulerpa* spp.

Brown algae are not conspicuous and the remainder of the almost continuous algal cover is made up of small red algae.

The **sublittoral fringe** occurs at about mean low tide level and is uncovered only at times of extreme low tides, and then only for a short time. It is here that brown algae are most conspicuous. The zone is characterized by the presence of *Sargassum* spp. and *Dictyopteris delicatula*. However, the rather large plants of *Sargassum* spp. (Fig. 1.9) do not tolerate much wave action. *Dictyopteris* sp., which is a smaller plant with narrow, dichotomous thalli similar to those of *Dictyota* spp. (Fig. 17.23), is more tolerant of wave action and predominates where the coast is exposed.

Sargassum spp. and *Dictyopteris* sp. also form a large part of the flora of rock pools (tidal pools). **Rock pools** are formed in hollows in the surface of the eulittoral zone. They are always full of water and offer a similar environment to that of the sublittoral fringe.

Further from the shore, the **sublittoral zone** is dominated by a great variety of red algae. Some green algae (e.g. *Caulerpa* spp.; *Codium* spp.) are also present, as are brown algae similar to those of the sublittoral fringe, but these form a relatively insignificant part of the flora.

The zonation described above is most easily observed on a gradually sloping, rocky shore. However, the same zonation is present, although compressed, even on a vertical surface. Also, where there are large rocks which stand with their tops above high tide level, in the eulittoral or sublittoral zone, each face of each rock shows the same, compressed zonation.

Key words

Factors affecting the development of vegetation: climatic, edaphic, biotic, anthropogenic, fire. **Tropical vegetation**: coastal, riverine; forest, savanna. **Coastal**: strand, scrub, mangrove, sand, lagoon, brackish water, delta, tidal region; prop roots, pneumatophores. **Forest**: fresh-water swamp; rain forest, evergreen; semi-deciduous, mixed deciduous; lowland forest, montane forest. **Trees**: stratification, emergent; stilt roots, buttress roots, thorns; epiphytes, lianes. **Savanna**: natural, secondary; woodland, tree, shrub, grass. **Savanna plants**: deciduous trees, thick bark; perennating organs. **Aquatic vegetation**: fresh water; sea water, marine, intertidal. **Inter-tidal region**: littoral fringe, eulittoral zone, sublittoral fringe, sublittoral zone. **Intertidal habitat factors**: exposure, desiccation; immersion, wave action; rocky shore, rock pools; sandy shore, sand movement; tidal levels, slope of shore.

Summary

The coastal climate of West Africa is very wet, with rainfall fairly evenly distributed throughout the year. As we travel inland, the climate becomes progressively drier, with longer dry seasons. These differences in climate are reflected in the type of vegetation which develops. On the coast and in the tidal regions of rivers, we have mangrove swamp, which develops where the water is salty. Behind the mangrove swamp, where the land is flooded with fresh water, we have fresh-water swamp forest. We next arrive at a belt of rain forest, in which, as in mangrove and fresh-water swamp forest, all the trees are evergreen. As we travel inland through the rain forest into regions of lower rainfall and longer dry seasons, the forest changes as more and more deciduous species of trees appear. Beyond the driest forest lies savanna, with a ground flora dominated by tall grasses. Some of the savanna is natural, its presence determined by climate, but much grows where forest would grow if it were not for frequent burning. If we move from forest onto higher ground, where rainfall and humidity are higher and temperatures are lower, we see a transition into montane forest and, at still higher altitudes, into montane grassland. At every point, plants are adapted to their environment. Plants which inhabit aquatic habitats are similarly adapted. In marine habitats we see a gradual change in the type and species-composition of vegetation as we move from the water on to drier and drier land.

Questions

1 What are the factors affecting the distribution of mangrove forest, fresh-water swamp forest and rain forest in West Africa?

2 How do trees of savanna woodland differ from trees in rain forest? Compare the ground flora of lowland rain forest with that of savanna woodland.

30

Changes in vegetation

Until fairly recently, any type of vegetation that had apparently reached its greatest degree of development under prevailing environmental conditions was regarded as a type of **climax** vegetation. In parts of the tropics with high and evenly distributed rainfall, rain forest was regarded as climatic climax vegetation (the **climatic climax**). Much secondary savanna, where regular burning was the factor which prevented the development of forest, was described as a **fire climax**. **Edaphic** and **biotic climaxes** were also recognized. Many ecologists now regard the concept of climax vegetation as unacceptable. This is because all vegetation is in dynamic equilibrium not just with one environmental factor, but with the sum of all environmental factors. It is also subject to small changes, for instance in the proportions of the species of which it is composed, as some plants die and are replaced by others of different species. Vegetation which has developed to a fairly steady state in equilibrium with the total environment is now usually termed **mature vegetation**. However, reference to climatic climax, fire climax, edaphic climax and biotic climax vegetation is still to be found in books.

Another concept which is not acceptable in its literal sense is that of primary vegetation. Primary vegetation, strictly, is vegetation which has never been disturbed by man. In this sense, there is probably little vegetation anywhere in the world which is strictly primary and, almost certainly, there is none in Africa. In spite of this, mature rain forest is very often referred to as **primary forest**. Vegetation which develops (**regenerates**) after the destruction of mature or other vegetation is referred to as secondary vegetation (e.g. **secondary forest, secondary savanna**).

All vegetation is in dynamic equilibrium with its environment, but sometimes the environment changes.

Changes in the environment to which plants are exposed may be slow, as, for example, when changes in climate occur. It is thought that this may be happening in the Sahel region of Africa at the present time. More rapid changes result from disasters, such as fires or tornadoes, or from the actions of man. When vegetation is partly or completely destroyed, the environment at ground level is greatly changed. Again, as new vegetation develops, this development itself again slowly modifies the environment in which the plants are growing. This brings us to our next topic, that of **plant succession**.

The term succession describes progressive changes in the species composition of a plant community as bare ground, or almost bare ground, is first colonized and the vegetation develops, over a period of time, into mature vegetation. The first plants to colonize the ground (**pioneer** species) are gradually replaced by later arrivals and the species composition continuously changes until a more or less steady state is reached. Two types of succession are recognized, primary succession and secondary succession.

Primary succession

When ground which has not previously been colonized by land plants is colonized and the vegetation then changes, with time, until mature vegetation is established, this is **primary succession**. Like the concept of climax vegetation, the concept of primary succession has been questioned, but rather on the basis that the description is often applied when there is no real evidence of change in vegetation with time and not because examples of primary succession do not exist. This will be referred to again, when specific examples are given.

We recognize several different types of succession (**seres**), according to the type of ground which is first colonized by plants. A **hydrosere** begins on bare ground which is saturated with fresh water. **A xerosere** begins on bare ground which is not saturated with water, or on bare rock. A sere which starts on bare rock may also be called a **lithosere**. A **halosere** begins on soil saturated with salty water (sea water or brackish water).

30.1 Hydrosere: *Cyperus* sp. on mud close to the Warri river, Nigeria

Hydroseres

The starting point of a hydrosere is in shallow water, where silt is being deposited. This may be at the margin of a lake, lagoon or river, where land-building is taking place. Rooted and floating water plants also play a part in trapping sediment and in building up the level of the sediment until it reaches the surface of the water. At this point, true water plants are excluded and the pioneer plants that colonize the soil are hydrophytes only to the extent that they grow on waterlogged soil. These plants include grasses (e.g. *Anadelphia* sp.), sedges (e.g. *Cyperus* spp, Fig. 30.1) and ferns (e.g. *Dryopteris* spp.). As the land surface rises and becomes consolidated with further deposits of organic and inorganic material, palms (e.g. *Raphia hookeri*, Raphia Palm; *Phoenix reclinata*, Wild Date Palm) and *Pandanus candelabrum* (Screw Pine) appear. If land building

continues, as the plants trap blown sand and dust, the vegetation continues to develop. The final stage of a hydrosere is usually the same type of vegetation which exists in the neighbouring region, according to climate and other factors.

If we take a line (**transect**) starting at the water's edge and leading away from the water as far as nearby mature vegetation, we pass through a number of zones of intergrading but distinctly different vegetation of increasingly complex types. However, very similar zonation of vegetation can be found near the margins of bodies of fresh water where land building is not taking place and the various plant communities are perfectly stable. This type of zonation can be taken as evidence of the presence of a hydrosere only if we can be certain that progressive changes in the composition of the plant communities are taking place with time, at each point along the transect. To prove the existence of a sere, it is necessary to map the borders of the zones over a period of years, perhaps many years, or to bore or dig down into the soil and find evidence of previous plant communities under those established on the surface. The same caution is necessary when examining other types of supposed seres.

Xeroseres

A xerosere begins on dry ground, for instance where a major landslide has completely buried previous vegetation or where land-building is taking place on the seashore or sand bar, or along the shore of a lake. In each case, we have sand or soil which is completely free from vegetation and is not waterlogged.

On a seashore or sand bar, the first stage of the sere is strand vegetation, consisting of sandbinding plants. As the level of the land builds up, from wind-blown sand and the accumulation of organic debris, strand vegetation is succeeded by coastal scrub. After this, plant succession usually leads to the gradual establishment of mature vegetation similar to that in the surrounding region. Along most of the coast of Nigeria, this is rain forest. At some points on the coast of West Africa it is savanna. Development from strand vegetation to mature vegetation must take a long time, centuries in the case of rain forest.

Very similar xeroseres can also develop on the sandy shores of inland lakes, where silt and sand from inflowing rivers and wave action build up sandy beaches. The sequence of vegetation and even some

30.2 Xerosere: blue–green algae (dark) and lichens on rocks (lithosere)

of the species which occur (e.g. *Ipomoea pes-caprae*; *Calotropis procera*) are the same as those on the coast.

Lithoseres

A lithosere is a xerosere which starts on bare rock. Most rock surfaces which are available for colonising are either steeply sloping or closely surrounded by ground which has a steep down-ward slope and lithoseres which develop in such circumstances are very subject to erosion and do not develop very far.

Rocks which are very dry for part of the year (Fig. 30.2) are usually first colonized by crustose or foliose lichens (e.g. *Parmelia* spp.). On moist rock surfaces, blue-green algae (e.g. *Scytonema* spp.) are usually the pioneers. Crevices in the rock trap dust and small particles of organic matter and are generally first colonized by small mosses (e.g. *Bryum argenteum*). Moist mats of blue-green algae, which may be several millimetres in thickness, are also colonized by mosses. Lichen mats provide anchorage for small xerophytic flowering plants. In many places, *Selaginella njam-njamensis* appears early, rooted in crevices and spreading out over the bare

rock, where debris accumulates under the protection of its branches. A soil with a high organic content builds up, centred especially on rock crevices, and bulbous plants (e.g. *Albuca* spp.), grasses and sedges form a sparse vegetation. If there are large crevices, shrubs and stunted savanna trees may become established. However, the soil which starts to form is usually eroded away by heavy rains and little more than seasonal flush vegetation often results.

What has been described are the early stages of the development of a lithosere. If regularly eroded, the sparse vegetation persists almost unchanged, except for seasonal fluctuations. It can then be considered to be mature vegetation since it is a relatively stable community in dynamic equilibrium with its harsh environment. Further development of a true lithosere can only take place on a more or less level surface such as that left in an abandoned quarry or on the surface of a solidified lava flow.

Haloseres

With the same stipulation that the existence of a sere can only be assumed if there is evidence of a real succession of plant and vegetation types with time, mangrove forest may be the first stage in the development of a halosere. Again, land-building must be involved. Development of vegetation from mangrove forest, through fresh-water swamp forest, possibly as far as rain forest, has probably taken place in the estuaries and deltas of large rivers such as the River Niger. However, there is little real evidence for this.

Secondary vegetation

Secondary vegetation may develop over a small area, for instance where a single forest tree has fallen and has left a gap in the canopy. Alternatively, it may develop where a large area of land has been cleared for farming and has later been abandoned.

Regeneration of rain forest

The seedlings of trees which make up the main part of rain forest require shade for their early development and are slow growing. There are, however, species of trees that normally grow where the main tree canopy has been broken by the death of a forest tree and the environment has been changed by the admission of more light. Two of the commonest examples of such trees are *Musanga cecropioides*

30.4 Regeneration (secondary forest) at margin of farmland, showing new growth of *Musanga* sp.

30.3 *Musanga cecropioides* (Umbrella Tree), close-up showing leaf shape

(Fig. 30.3) and *Fagara macrophylla*. These trees have efficient mechanisms of seed dispersal and are fast-growing. Also, they are light-demanding and will not grow in deep shade, though *Musanga* sp. requires the shade of herb cover for its early development.

Once a cover of herbaceous plants and small shrubs has developed on more or less bare ground in a forest area, these and other species of fast-growing trees become established and dominate the vegetation for several decades (Fig. 30.4). After fifteen or twenty years, they form secondary forest with a canopy at a height of about 20 m. Before they reach this height, they produce a suitable environment for the development of young plants of the tree species of true rain forest, but not for seedlings of their own species. It is thought that true ('primary') rainforest trees emerge above the secondary forest species after about sixty to one hundred years, after

which the secondary forest species, being light-demanding, die out. No figures are available, but complete regeneration of typical rainforest structure, with its typical stratification, must take a very long time indeed.

Regeneration of savanna

After savanna vegetation has been cleared, farmed and abandoned, or otherwise destroyed, regeneration as far as savanna grassland is very rapid. Seeds of grasses are always present in abundance and their growth quickly shades out any species of farmland weeds which may be present. Further regeneration depends largely on the frequency and intensity of fire.

Fires which start early in the dry season are much less destructive than fires which start late in the season. When early burning takes place, many of the grasses still hold moisture and burning is fairly superficial. With late burning, the dead grass stems and leaves are very dry and fire is more intense. There can also be differences between fires which burn at night or early in the morning, when there may be much moisture condensed on the grass, and fires which burn later in the day.

Re-establishment of savanna trees is from several sources, depending on the degree of destruction of the original savanna. If clearing was complete, including root systems, seeds are the only source. If parts of trunks and root systems of trees are left intact, replacement may be comparatively rapid. Most savanna trees will produce adventitious shoots from the cut or broken bases of their trunks and some (e.g. *Brachystegia* spp.) produce adventitious shoots from their root systems. Regrowth of trees from adventitious shoots is more rapid than that from seedlings, since adventitious shoots have underground stores of food on which they can draw.

The seedlings and young shoots of savanna trees are as sensitive to fire as those of forest species. Some protection is required if they are to survive to an age at which they become more or less fire-resistant. For example, a seedling that is growing in the middle of a group of rocks, or next to or upon a termitarium, is likely to be to some extent protected from fire.

The same factors operate, of course, in well-developed savanna woodland or tree savanna, where old trees are replaced by new trees from time to time. Indeed, where there is annual burning in middle or late dry season, savanna woodland is changed to tree savanna over a relatively small number of years.

If there are no underground parts of trees which can produce root or shoot suckers, the regeneration of any type of vegetation depends on the availability of seeds of appropriate species. This is illustrated when an area of rain forest which lies far from savanna is cleared and afterwards subjected to occasional burning. The vegetation which develops is grass savanna. The regeneration of forest is prevented by burning and the growth of more fire-resistant species is not possible, because their seeds are not available.

Key words

Mature vegetation: climatic climax, edaphic climax, biotic climax, fire climax. **Primary vegetation**: primary forest. **Secondary vegetation**: secondary forest, secondary savanna. **Plant succession**: primary

succession, pioneer species, hydrosere, xerosere, lithosere, halosere, transect; secondary succession, regeneration.

Summary

All vegetation is subject to change. When newly available land surfaces are colonized by plants, the first, simple vegetation is generally succeeded by more and more complex types of vegetation. When established vegetation is destroyed, it is replaced by secondary vegetation, which may at first be quite different from that which was destroyed.

Questions

1 What are the most important differences between a hydrosere and a xerosere? How do you think these differences are reflected in the types of plants which first colonize the different types of habitat?

2 Suppose you make a transect from the margin of a lake as far as complex vegetation nearby and find that the line passes through a series of zones of vegetation similar to those expected in a hydrosere. What steps would you take to find out whether the zonation of vegetation represented a true hydrosere or whether it was merely the result of varying soil conditions (e.g. soil moisture) at different distances from the lake?

3 What factors govern the regeneration of (a) rain forest, and (b) savanna? Why is it that the regeneration of true rain forest takes so much longer than regeneration of savanna?

Further reading for Section F

Good, R. 1974. *The Geography of Flowering Plants*, 4th edn, Longman: London.

Hopkins, B. 1979. *Forest and Savanna*, new edn Heinemann: London.

Longman, K. A. and Jenik, J. 1974. *Tropical Forest and its Environment*, Longman: London.

Pritchard, J. M. 1979. *Africa: A Study Geography for Advanced Students*, 2nd edn, Longman: London.

Richards, P. W. 1952. *The Tropical Rain Forest*, Cambridge U.P.: Cambridge.

Walter, H. 1971. *Ecology of Tropical and Subtropical Vegetation*, Oliver & Boyd: Edinburgh.

Index

This index is designed to serve also as a glossary. For this reason, many adjectives and descriptive words appear in the main alphabetical sequence. Page numbers printed in **bold type** refer to pages on which scientific terms are defined or explained, and often illustrated. Page numbers printed in *italics* refer to pages on which there is an illustration of a structure, plant, etc. The letter 'f' following a page number means 'and the following page' and 'ff' means 'and following pages'. Names of plants are given to genus level. Names of species are not included.